WELL DESIGN

Drilling and Production

WELL DESIGN

Drilling and Production

B. C. Craft & W. R. Holden

Petroleum Engineering Department
Louisiana State University

& E. D. Graves, Jr.

The Dow Chemical Company

Prentice-Hall, Inc.

Englewood Cliffs, New Jersey 1962

Current printing (last digit):
15 14 13

 Printed in the United States of America. Library of Congress Catalog Card Number 62-9949.
95002 — C

Dedicated to
Bertha, Barbara, and Jessie

Preface

The purpose of this book is to present, in as clear and concise a form as possible, the fundamental principles underlying the analysis, design, and engineering of petroleum production systems. One factor that has been kept in mind in writing this book is the close association that exists between petroleum reservoir engineering and production engineering. Developments of the past few years have so broadened the latter field that an exhaustive treatment of all its aspects cannot be contained in a single volume. For example, several recently published works deal exclusively with fluid flow, cementing, pumping, and gas lift, respectively.

Although this book has been developed from notes used for several years in a course given to senior petroleum engineers, the topics are introduced at a level that will make them accessible to those specializing in other branches of engineering. Theoretical equations are derived and illustrated by numerical examples as an aid to developing an understanding of the problems encountered in modern production practice. Derivations of the less-important equations have been omitted from the exposition but are included as exercises among the problems at the end of each chapter. References are given at the end of each chapter for readers who want more information on the topics presented. Insofar as possible, the symbols adopted as standard by the Society of Petroleum Engineers of the American

Institute of Mining, Metallurgical, and Petroleum Engineers are used throughout the text.

We wish to pay tribute to the many engineers and scientists who have made so many fine contributions to the field of production engineering, and to express our gratitude to the Society of Petroleum Engineers of the AIME and to the Division of Production of the API in whose transactions so many of the contributions have been published.

We wish to thank the following for their corrections, contributions, and criticisms of specific chapters: Phil C. Montgomery, Halliburton Oil Well Cementing Company; W. R. Cox, Shell Oil Company; Roy M. Willis, The Atlantic Refining Company; Harold W. Winkler, Camco, Incorporated; A. P. Batchelor, Consulting Engineer; Denton R. Wieland, University of Tulsa; Bob C. Crittendon, Roden Oil Company; Raymond G. Loper, The Western Company; and H. L. Mosley, Trinity Portland Cement Division, General Portland Cement Company.

We are particularly indebted to the following companies for their encouragement and financial assistance; Dowell Division of the Dow Chemical Company, Gulf Oil Corporation, Socony Mobil Oil Company, Inc., The California Company, Standard Oil Company of Texas, Shell Oil Company, and The Atlantic Refining Company.

Our thanks are due also to Donald W. Hecker for checking numerous derivations and problems, to J. R. Rollo for drafting the figures, and to Professor Murray F. Hawkins for invaluable constructive criticism and timely suggestions.

B. C. C.

W. R. H.

E. D. G., Jr.

Contents

CHAPTER ONE **Fluid Flow in Pipes** 1

1.1 Introduction 1; 1.2 Dimensional deceptions, 2; 1.3 Basic laws of fluid flow, 9; 1.4 Classification of fluids and flow regimes, 16; 1.5 Pipe flow of Newtonian liquids, 18; 1.6 Annular flow of Newtonian liquids, 26; 1.7 Rheological classification of fluids, 31; 1.8 Pipe flow of Bingham plastics, 34; 1.9 Annular flow of Bingham plastics, 41; 1.10 Rotary drilling hydraulics, 48; 1.11 Generalized equations of flow for non-Newtonian fluids, 55; 1.12 Two-phase flow, 79.

CHAPTER TWO **Design of Casing Strings** 101

2.1 Introduction, 101; 2.2 Grades of casing, 102; 2.3 Casing dimensions, 103; 2.4 Casing threads and couplings, 106; 2.5 Weights of casing, 115; 2.6 Liners, 124; 2.7 Effects of external pressure, 124; 2.8 Effects of internal pressure, 130; 2.9 Effects of axial loading, 133; 2.10 Design factors, 139; 2.11 Selection of casing sizes, 143; 2.12 Design of a combination string, 146.

CHAPTER THREE ***Cementing*** **158**

3.1 Introduction, 158; 3.2 Composition of Portland cement, 159; 3.3 Standardization of oil-well cements, 163; 3.4 Strength of cement, 175; 3.5 Additives for oil-well cements, 183; 3.6 Surface and subsurface cementing equipment, 206; 3.7 Casing-cementing program, 215; 3.8 Squeeze cementing, 229; 3.9 Liner cementing, 236; 3.10 Open-hole plugback cementing, 239; 3.11 Cementing hydraulics, 243.

CHAPTER FOUR ***Casing Landing*** **256**

4.1 General landing practices, 256; 4.2 Buckling criteria, 261; 4.3 Wellhead loads, 270; 4.4 Application, 272.

CHAPTER FIVE ***Sucker Rod Pumping*** **280**

5.1 Introduction, 280; 5.2 The subsurface pump, 281; 5.3 The sucker rod string, 283; 5.4 The surface pumping equipment, 284; 5.5 The prime mover, 286; 5.6 Theoretical analysis of rod motion, 286; 5.7 The effective plunger stroke, 288; 5.8 Calculation of polished rod loads, 293; 5.9 Design of the sucker rod string, 295; 5.10 Pump displacement and production rate, 299; 5.11 Counterbalance design, 300; 5.12 Calculation of torque, 302; 5.13 Speed reduction from prime mover to crankshaft, 304; 5.14 Power requirements of the prime mover, 304; 5.15 Summary of design equations, 306; 5.16 Steps in the design of a pumping installation, 312; 5.17 Factors in sucker rod and tubing failure, 320; 5.18 Use of the recording dynamometer in well analysis, 333; 5.19 Interpretation of a dynamometer card, 336; 5.20 Applications of dynamometer cards, 343; 5.21 Problem well testing, 349; 5.22 Determination of pumping efficiencies, 352.

CHAPTER SIX ***Gas Lift*** **368**

6.1 Introduction, 368; 6.2 Application, 372; 6.3 Valve mechanics, 377; 6.4 Design of a continuous-flow gas-lift system, 400; 6.5 Intermittent flow, 429; 6.6 Accumulation chambers, 434; 6.7 Free piston system of artificial lift, 439.

CHAPTER SEVEN ***Fluid Separators*** **453**

7.1 Oil and gas separators, 453; 7.2 Heater treaters, 470; 7.3 Oil skimmers, 475.

CHAPTER EIGHT ***Hydraulic Fracturing*** **483**

8.1 Introduction, 483; 8.2 Inducing fractures, 485; 8.3 Productivity ratio, 491; 8.4 Fracture area, 495; 8.5 Fracturing fluid coefficients, 501; 8.6 Fracture efficiency, 514; 8.7 Fracturing hydraulics, 516; 8.8 Fracture design calculation, 519.

CHAPTER NINE ***Acidizing*** **536**

9.1 Introduction, 536; 9.2 Types of treatments, 537; 9.3 Acid-fracturing design, 542.

APPENDICES

A Cementing tables, 547
B Constants and conversion factors, 560

INDEX, 563

WELL DESIGN
Drilling and Production

CHAPTER ONE

Fluid Flow in Pipes

1.1. Introduction

Almost every phase of petroleum drilling and production operations involves the movement of fluids such as gases, liquids, and dispersed solids through circular pipe and fittings. As a consequence, the petroleum engineer must be able to apply the principles of fluid mechanics to a wide range of flow problems. These problems will include the flow of drilling muds, cement slurries, and other *non-Newtonian* fluids whose resistance to shearing deformation is anomalous in comparison with the behavior of familiar *Newtonian* fluids, e.g., oil, gas, and water. This complication requires some familiarity with the related science of *rheology*. In addition, two-phase gathering systems for the simultaneous transport of both oil and gas are currently being used to link remote producing wells (e.g., in offshore, marsh, or desert locations) with centralized metering, separation, and storage facilities. The peculiarities of these two-phase pipelines are at present only partially understood, but research efforts sponsored by the gas and oil industry promise continuing improvement in multiphase design methods.

A review of the basic principles of fluid flow is given in this chapter together with presentations of practical techniques for handling non-Newtonian systems and of the current state of the art of designing multiphase pipelines.

1.2. Dimensional Deceptions

An inherent source of error and confusion within the allied fields of mechanics and fluid flow is the indiscriminate use of inconsistent unit systems. By a process of what might be called "dimensional deception," innumerable forms and simplifications of the same basic relationship can be made to appear as a host of independent equations.

Admittedly, an engineer must use both metric and English units because of his close liaison with both science and technology. Routine conversions between these units are only mildly troublesome, but the intermixing of gravitational and absolute unit systems can easily complicate matters. For example, when pounds force and pounds mass appear in the same equation, confusion will frequently arise; when pounds weight are included, confusion is usually assured. To avoid these complications, the differences between these quantities — force, mass, and weight — should be clearly understood, and their units and dimensions watched closely in all formulas and calculations.

Units and dimensions. Measurement of a physical quantity involves two elements: (1) a *dimension*, which identifies the nature of the quantity and also the choice of the *unit* used as a basis for comparison, and (2) a number which states how many of these units are contained in the quantity being measured. Consider, for example, the three equal quantities 1.0 mile, 1760 yards, and 1609 meters. All have the same dimensions, namely length, but all differ in numerical value since different units were used in their measurement. As another example, 2 acres and 10 square miles, though not equal, do have the same dimensions, namely area or length squared.

The majority of physical quantities have dimensions which can be derived from, or expressed in terms of, the dimensions of a limited number of fundamental or *primary* quantities. The time rate of change of position, for example, has the dimensions of velocity by definition. Velocity, however, can be expressed as a length divided by a time interval. Then the unit of velocity used in subsequent measurements of like quantities would be fixed by the choice of length unit and time unit used in its definition. A sufficient choice of primary quantities for most engineering work is mass, length, time, and temperature. All other quantities, hereinafter called *secondary* or *derived* quantities, can be expressed in terms of these four.

Units of length and mass. The fundamental standards of length and mass designed for the entire world are the distance between two lines inscribed on a particular platinum-iridium bar and the mass of a certain block of platinum. Known as the international meter bar and kilogram mass, these standards are maintained at the International Bureau of Weights and Measures in Sèvres, France. At present, the commonly used unit of time is the second, defined as 1/86,400 part of a mean solar day.* Customary English units of length and mass for the United States are defined by reference to the standard yard and avoirdupois pound. But these in turn have been standardized by the U. S. Congress in terms of the international meter and kilogram mass. By a law of 1866 and subsequent executive order of 1893, the following equivalents were established:

$$1 \text{ yard} = 3600/3937 \text{ meter}$$

$$1 \text{ pound mass} = 0.4535924277 \text{ kilogram mass}$$

Thus far, the interchange of English and metric units should not give rise to arguments unless one has some quarrel with the law. Conversion factors for a variety of multiple and submultiple units derived from these fundamental standards are listed in Appendix B.

Example 1.1. A car is traveling at a speed of 90 miles per hour. Express its speed in (a) feet per second and (b) meters per minute.

SOLUTION:

(a)
$$90\,\frac{\text{mi}}{\text{hr}} = 90\,\frac{\text{mi}}{\text{hr}} \times \frac{\text{hr}}{60\text{ min}} \times \frac{\text{min}}{60\text{ sec}} \times \frac{5280\text{ ft}}{\text{mi}}$$

$$90\,\frac{\text{mi}}{\text{hr}} = \frac{90 \times 5280}{60 \times 60} = 132\text{ ft/sec}$$

(b)
$$90\,\frac{\text{mi}}{\text{hr}} = 90\,\frac{\text{mi}}{\text{hr}} \times \frac{\text{hr}}{60\text{ min}} \times \frac{1760\text{ yd}}{\text{mi}} \times \frac{3600\text{ m}}{3937\text{ yd}}$$

$$90\,\frac{\text{mi}}{\text{hr}} = \frac{90 \times 1760 \times 3600}{60 \times 3937} = 2414\text{ m/min}$$

When this procedure is used, the probability of error is greatly reduced. Each conversion involves the multiplication of the original quantity by a series of well-chosen, dimensionless factors each having a numerical value of unity. To verify this last statement, notice that the numerator of each factor is equal to its denominator.

Units of force. It would be difficult, if it were possible at all, to maintain a fixed quantity of force as an unvarying physical standard. For by its nature, the very existence of a force can only be *inferred* from the changes it produces in the state of motion of a material body. By Newton's law of motion, the acceleration a produced in a body of constant mass m is attributed to the action of a force F whose magnitude is proportional to the mass and acceleration of the body, or

$$F \propto ma$$

*In 1960, a General Conference on Weights and Measures redefined the standard meter as 1,650,763.73 wavelengths of the red-orange line in the emission spectra of krypton 86. The unit of time, the second, was also defined more specifically as 1/31,556,925.9747th of the year 1900.

Upon introducing a constant of proportionality, g_c, the previous expression can be written as an equality,

$$F = \frac{1}{g_c} ma \tag{1.1}$$

This equation now becomes a defining expression for any of the four quantities of force, mass, length, or time in terms of the remaining three. If mass, length, and time are recognized as primary quantities, Eq. (1.1) defines the dimensions of the derived quantity, force, as a mass times a length divided by the square of a time interval. If, however, these fundamental dimensions of force were carried over in the definition of secondary quantities, such as pressure and energy, needless complexity would result. Instead, force is used as a convenient, but not essential, secondary dimension. The size of a unit force is then fixed by an arbitrary choice of acceleration effects it must produce in a given mass. Once these standard effects have been specified, the proportionality constant g_c is adjusted to preserve the numerical and dimensional equality of Eq. (1.1).

Example 1.2. A particular unit of force, called a *dyne,* is defined as that force required to accelerate a gram of mass at the rate of one centimeter per second squared. Calculate the value of the constant g_c to be used with this set of units.

SOLUTION: When the above definition is introduced into Eq. (1.1), we have

$$F = \frac{1}{g_c} \times 1 \text{ gm} \times 1 \frac{\text{cm}}{\text{sec}^2} = 1 \text{ dyne} \quad \text{(by definition)}$$

To preserve dimensional as well as numerical equality, the constant must have the value

$$g_c = 1 \frac{\text{gm cm/sec}^2}{\text{dyne}}$$

Thus we see that the constant serves the twofold purpose of (1) defining the standard effects expected of a certain unit force, and (2) identifying this unit by name.

Since mass and force are not independent quantities, the conversion factor g_c will appear in all problems that involve dynamics, for it allows the simultaneous use of both quantities in the same equation.

Example 1.3. Develop an expression for the kinetic energy E_k of a body of mass m moving at a constant velocity v.

SOLUTION: Consider the body to be originally at rest. Upon application of a steady unbalanced force F acting through a sufficient distance s, the body will acquire the required velocity v.

In accordance with Eq. (1.1), this force gives the body a constant acceleration a,

$$F = \frac{ma}{g_c}$$

and the distance traversed by the body while acquiring the velocity v is

$$s = \frac{v^2}{2a}$$

The energy imparted to the body is then

$$E_k = Fs = \frac{ma}{g_c} \times \frac{v^2}{2a}$$

or,

$$E_k = \frac{mv^2}{2g_c}$$

This last equation, as given in elementary physics, probably did not contain g_c. But the dimensions of the result, mL^2/t^2, or gm cm²/sec², are not associated with energy until one makes the mental conversion 1 dyne = gm cm/sec². Then the result can be rearranged as (gm cm/sec²) cm or dyne cm, and finally recognized as an energy term with the dimensions of force times distance. Unreliable mental conversions are eliminated by the presence of g_c.

Weight. The force of attraction which the earth exerts on a material body is called the weight of the body. Being a force quantity, the weight W of a body of mass m may be calculated using Newton's law in the form

$$W = \frac{1}{g_c} mg \tag{1.2}$$

where g is the acceleration which gravity would impart to the body in a free fall. Unfortunately, the term "weight" is sometimes used improperly as a synonym for mass. To make matters worse, in two of the unit systems to be discussed, the unit of weight has the same name as the unit of mass, and the necessary distinction between these quantities becomes even more elusive. Therefore, it cannot be overemphasized that *weight has the dimensions of force.* Equation (1.2) provides the only sure conversion between the quantities weight and mass.

Unit systems. Each time an arbitrary choice of basic units is made for mass, length, and time, a new *system* of units is established together with an opportunity for defining a new unit of force. In Table 1.1 are listed five systems commonly used in scientific and engineering calculations. While each system is self-consistent, the various systems are not compatible with each other.

The derived units of force (and of mass in system 5, Table 1.1) are defined by the following statements:

1. A force of one *dyne* will accelerate a one gram mass at the rate of 1 cm/sec².
2. A force of one *poundal* will accelerate a one pound mass at the rate of 1 ft/sec².

Table 1.1. Common unit systems

Quantity	Symbol	Absolute		Gravitational		
		(1) Metric	(2) English	(3) Engineering metric	(4) Engineering English	(5) Technical English
Mass	m	gram, gm	pound, lbm	gm	lbm	slug
Length	L	centimeter, cm	foot, ft	cm	ft	ft
Time	t	second, sec	sec	sec	sec	sec
Force	F	dyne	poundal, pdl	gram, gf	pound, lbf	lbf
Conversion constant	$g_c = \frac{mL}{Ft^2}$	$1 \frac{\text{gm cm}}{\text{dyne sec}^2}$	$1 \frac{\text{lbm ft}}{\text{pdl sec}^2}$	$980.7 \frac{\text{gm cm}}{\text{gf sec}^2}$	$32.17 \frac{\text{lbm ft}}{\text{lbf sec}^2}$	$1 \frac{\text{slug ft}}{\text{lbf sec}^2}$
Density	$\rho = \frac{m}{L^3}$	gm/cu cm	lbm/cu ft	gm/cu cm	lbm/cu ft	slug/cu ft
Specific weight	$\rho' = \frac{F}{L^3}$			gf/cu cm	lbf/cu ft	lbf/cu ft
Specific volume	$V = \frac{L^3}{m}$	cu cm/gm	cu ft/lbm	cu cm/gm	cu ft/lbm	cu ft/slug
	$V' = \frac{L^3}{F}$			cu cm/gf	cu ft/lbf	
Absolute viscosity	$\mu = \frac{m}{Lt}$	$\frac{\text{gm}}{\text{cm sec}}$ (poise)	$\frac{\text{lbm}}{\text{ft sec}}$	$\frac{\text{gm}}{\text{cm sec}}$ (poise)	$\frac{\text{lbm}}{\text{ft sec}}$	$\frac{\text{slug}}{\text{ft sec}}$
Energy	$E = FL$	dyne cm (erg)	ft pdl	gf cm	ft lbf	ft lbf
Power	$P = \frac{FL}{t}$	erg/sec	ft pdl/sec	gf cm/sec	horsepower = 550 ft lbf/sec	horsepower = 550 ft lbf/sec

3. A force of one *gram force* will accelerate a one gram mass at the rate of 980.665 cm/sec^2.
4. A force of one *pound force* will accelerate a one pound mass at the rate of 32.1740 ft/sec^2.
5. A force of one pound force will accelerate a mass of one *slug* at the rate of 1 ft/sec^2.

The units and numerical value of g_c are obtained by substitution of these defining statements into the general form of Newton's equation, Eq. (1.1), as previously illustrated in Example 1.2. Two of the values given in Table 1.1, however, have been abbreviated to be consistent with the accuracy required for ordinary calculations.

It must be remembered that the constant g_c is not an acceleration term, but a conversion factor between the unit of force and its assigned primary units.

Example 1.4. Calculate a factor for converting pounds of force to equivalent grams of force.

SOLUTION:

$$1 \text{ lbf} = 1 \text{ lbf} \times \frac{32.17 \text{ lbm ft}}{\text{lbf sec}^2} \times \frac{453.6 \text{ gm}}{\text{lbm}} \times \frac{30.48 \text{ cm}}{\text{ft}} \times \frac{\text{gf sec}^2}{980.7 \text{ gm cm}}$$

$$1 \text{ lbf} = \frac{32.17 \times 453.6 \times 30.48}{980.7} = 453.6 \text{ gf}$$

Notice that it was necessary to convert pounds force into *primary* English units, by use of g_c, before the legal conversion from pounds mass to grams mass could be applied. Similarly, an intermediate answer having been obtained in primary metric units, g_c was again used to obtain the final result in grams force.

In both engineering English and engineering metric systems, the unit force and unit mass have the same name. To eliminate confusion, the pound mass is often abbreviated as lbm, the pound force as lbf, the gram mass as gm, and the gram force as gf. Synonymous terms for force, the pound weight and gram weight, are denoted by lbw and gw. The particular notation adopted for density, specific weight, and specific volume is an attempt to dissociate those troublesome units which employ a unit weight in their definition from those which use a unit mass.

Consistent units and equations. Mechanical units not listed in Table 1.1 are derived from the units of mass, length, time, and force specified for each system. Thus, instead of an arbitrary choice of, say, square yards as the unit of area for the engineering English system, area is recognized as the square of a length, and since the foot has already been selected as the unit of length, the square foot is adopted as the consistent unit of area.

In this same system, unit pressure is defined, in like manner, as pounds force per square foot (psf); volumetric flow rate as cubic feet per second; and work or energy as foot pounds force.

In the derivation of a physical equation, exclusive use of a consistent system of units, including the ubiquitous g_c, has a distinct advantage. The resulting expression can be used with any other system of consistent units without further modifications.

Example 1.5. The pressure drop required to maintain a fluid in laminar flow through a horizontal pipe is given by the equation

$$\Delta p = \frac{32\mu L v}{g_c d^2}$$

where Δp = pressure drop
μ = absolute viscosity
L = length of pipe
v = linear flow velocity
g_c = conversion factor
d = inside diameter of pipe

Let us investigate the dimensional properties of this equation by substituting consistent units from the engineering English system for all quantities:

$$\frac{\text{lbf}}{\text{ft}^2} = \frac{\text{lbm}}{\text{ft sec}} \times \text{ft} \times \frac{\text{ft}}{\text{sec}} \times \frac{\text{lbf sec}^2}{\text{lbm ft}} \times \frac{1}{\text{ft}^2}$$

The dimensions of the pressure term on the left are seen to be the same as the combined dimensions of all terms on the right. The equation is therefore consistent or dimensionally homogeneous. As a consequence, consistent units from any other system may be used in this equation as given, without need for recourse to any other conversion factors or additional qualifying remarks.

Empirical equations developed from correlations of experimental data, as well as physical equations which have been modified to accept certain quantities expressed in more "practical" units, are usually inconsistent. These equations cannot stand alone, but must be accompanied by a detailed listing of particular units to be used with each different quantity.

Example 1.6. Modify the equation in Example 1.5 to accept the following field units:

$\Delta p'$ = pounds force per square inch, psi
μ' = centipoises, cp
L' = feet
v' = feet per second
d' = inches

Since the equation as given will accept any consistent set of units, the field units will be converted to consistent engineering English units.

$$\Delta p' \frac{\text{lbf}}{\text{in.}^2} \times \frac{144 \text{ in.}^2}{\text{ft}^2} = \Delta p \text{ lbf/ft}^2$$

$$\mu' \text{ cp} \times \frac{6.72 \times 10^{-4} \text{ lbm/ft sec}}{\text{cp}} = \mu \text{ lbm/ft sec}$$

$$L' \text{ ft} = L \text{ ft}$$

$$v' \text{ ft/sec} = v \text{ ft/sec}$$

$$d' \text{ in.} \times \frac{\text{ft}}{12 \text{ in.}} = d \text{ ft}$$

These consistent units may then be substituted into the original equation.

$$\Delta p'(144) = \frac{32\mu'(6.72 \times 10^{-4})L'v'}{32.17(d'/12)^2}$$

or

$$\Delta p' = \frac{0.000668\, \mu' L' v'}{d'^2}$$

Our final equation is no longer consistent, for all quantities must now be expressed in the field units specified, or else the equation will give a wrong answer.

1.3. Basic Laws of Fluid Flow

By unrestricted choice of symbolic notation and mathematical expression, a bewildering assortment of "basic" formulas have been developed to describe the flow of fluids. Before we add to this assortment, it should be emphasized that only three physical laws apply:

1. Conservation of mass or material
2. Conservation of energy
3. Conservation of momentum or Newton's second law of motion

These three laws are basic to the solutions of all pipe flow problems. The third law is particularly useful in estimating reaction forces on pipe structures and analyzing momentum transfer devices such as turbines, fans, and centrifugal pumps.

Material balance. Pipe flow in general is a steady-state process wherein the flow velocity, temperature, and pressure at any point along the flow do not change with time. Under these conditions, the mass rate of flow past any given cross section of pipe is constant. In the absence of an accumulator or leaks, the law of conservation of mass requires that the mass of fluid entering one end of a pipe in unit time be balanced by an equal mass of fluid leaving the other end of the pipe. In short, what goes in must come out.

Flow rates are usually given in one of three equivalent forms: the mass rate of flow w, the volume rate of flow q, or an average velocity v defined as q/A where A is the cross-sectional area of the pipe. With subscripts 1 and 2 used to denote conditions at an upstream cross section A_1 and a downstream cross section A_2, respectively, a continuity equation or material balance can be written for consistent units as

$$\begin{aligned} w_1 &= w_2 \\ \rho_1 q_1 &= \rho_2 q_2 \\ \rho_1 A_1 v_1 &= \rho_2 A_2 v_2 \end{aligned} \tag{1.3}$$

where ρ is the density of the fluid at the prevailing temperature and pressure of the flow at any particular cross section. By use of the transform $(g/g_c)m = W$ or $(g/g_c)\rho = \rho'$ from Eq. (1.2), the previous equations can be rewritten in terms of a constant weight rate of flow. If we introduce the inverse properties V and V' for ρ and ρ', it is possible to express the first conservation law in ten different mathematical forms.

Example 1.7. Consider the isothermal steady flow of a liquid through a pipe. The density ρ of the liquid (assumed incompressible) would be constant along the flow, as would the volume rate of flow q past any cross section. For a pipe of uniform size, the average velocity v of the liquid would also be constant along the flow.

Gas flow under similar conditions would be different. If the upstream pressure were greater than the downstream pressure (the usual case), the following conditions would prevail along the direction of flow: a decreasing density, an increasing volume rate of flow, and, for a uniform pipe, an increasing average flow velocity.

Total energy balance. For steady-state pipe flow, the second conservation law requires that the energy carried in by the fluid at an upstream section A_1, plus any additional energy imparted to the fluid while in transit to a downstream section A_2, must be carried away by the fluid as it leaves section A_2. In other words, with no provisions for the accumulation or depletion of either mass or energy along a pipe system, the total energy input per unit time must equal the total energy output.

An energy balance equation can be written for the general fluid flow system shown in Fig. 1.1 after a complete accounting has been made of all the energy transferred to or from the system. In the time interval t required to introduce a unit mass of fluid at section A_1, the following energy is carried into the system by the fluid:

u_1 = internal or intrinsic energy, a function of the thermodynamic state of the fluid while at section A_1

$\left(\frac{g}{g_c}\right)Z_1$ = energy of position or potential energy, where Z_1 is the elevation of the flow above some arbitrary datum level and g is the local acceleration of gravity

$\frac{v_1^2}{2g_c}$ = energy of motion or kinetic energy of a unit mass moving with a flow velocity v_1

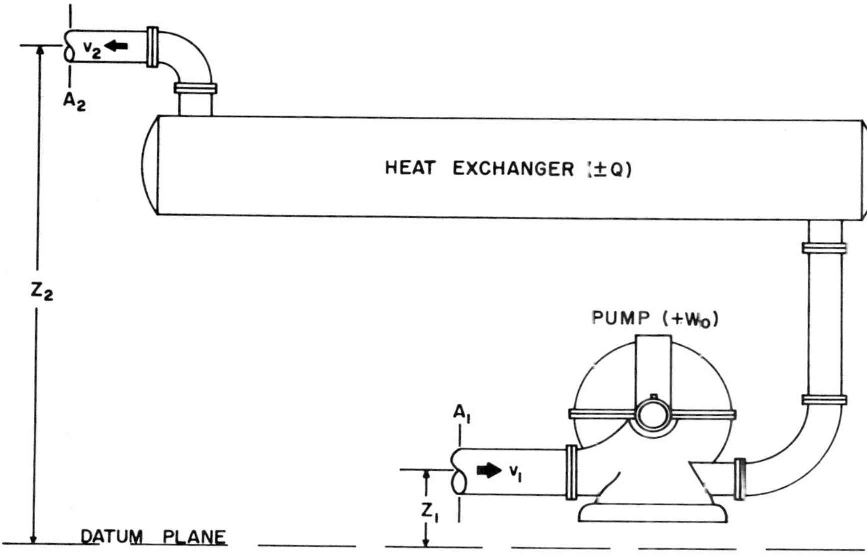

Fig. 1.1 General fluid flow system.

In this same time interval t, a unit mass of fluid leaves the system at A_2 and carries away energy in the same forms but perhaps in different amounts from those entering at A_1.

Sources external to the system are responsible for the additional energy transfer:

p_1V_1 = work or mechanical energy expended by adjacent fluid in forcing a unit mass into the system at A_1 against the prevailing system pressure p_1

p_2V_2 = work or mechanical energy expended by the system in voiding a unit mass at A_2 against the pressure p_2

Q = net heat energy absorbed by the system from external sources during the time interval t; for a net heat loss, Q is a negative quantity

W_o = net work or energy imparted to the system from outside mechanical sources (such as pumps or compressors) in the time interval t

An energy balance over the time interval t is given by the equation

Energy Input = Energy Output

$$u_1 + \frac{g}{g_c} Z_1 + \frac{v_1^2}{2g_c} + p_1V_1 + Q + W_o = u_2 + \frac{g}{g_c} Z_2 + \frac{v_2^2}{2g_c} + p_2V_2 \tag{1.4}$$

or in terms of finite differences, where Δ refers to a quantity at an exit section minus the same quantity at an entrance section,

$$\Delta u + \frac{g}{g_c} \Delta Z + \frac{1}{2g_c} \Delta v^2 + \Delta(pV) = Q + W_o \tag{1.5}$$

All terms when expressed in consistent units have the common dimensions of energy per unit mass. The usual English units are ft lbf/lbm and mean exactly that. It will be noticed that no friction term appears in Eq. (1.5), for friction is responsible only for the conversion of some mechanical energy into heat energy and does not change the overall energy content of a flow system.

When the enthalpy function $h = u + pV$ is substituted into Eq. (1.5), the resulting expression

$$\Delta h + \frac{g}{g_c} \Delta Z + \frac{1}{2g_c} \Delta v^2 = Q + W_o \tag{1.6}$$

will be recognized as the first law of thermodynamics applied to a steady-flow process. This last equation is particularly suited for the analysis of flow systems which involve either heat transfer or adiabatic ($Q = 0$) processes and fluids whose thermodynamic properties have been previously tabulated.

Example 1.8. Room air (assumed dry) is cooled from 74°F to 58°F while flowing past the evaporator coils of a room air conditioner which is handling 465 cu ft of room air per minute. Atmospheric air behaves as a perfect gas and has a mean heat capacity of 0.24 Btu/lbm at constant pressure. Calculate the cooling capacity of the unit in Btu/hr.

SOLUTION: While the air is flowing past the coils no external work is done on it, and its potential and kinetic energy remain essentially constant. As a consequence, these three terms may be dropped from Eq. (1.6). The change in enthalpy is

$$(h_2 - h_1) = c_p(t_2 - t_1) = 0.24(58 - 74) = -3.84 \text{ Btu/lbm}$$

The transferred heat Q is a negative quantity, indicating that heat energy flows from the air to the cold coils as expected. Applying the ideal gas law, the density of the room air is

$$\rho = \frac{pM}{RT} = \frac{14.7 \times 29}{10.73 \times (460 + 74)} = 0.074 \text{ lbm/ft}^3$$

so that the unit is handling a mass of

$$0.074 \times 465 = 34.4 \text{ lbm air/min}$$

The cooling capacity is then

$$34.4 \frac{\text{lbm}}{\text{min}} \times 60 \frac{\text{min}}{\text{hr}} \times 3.84 \frac{\text{Btu}}{\text{lbm}} = 7926 \text{ Btu/hr}$$

Mechanical energy balance. Only within the last 150 years has the law of conservation of energy been extended to include changes in internal energy and the flow of heat. The recognition of heat flow as a form of energy transfer was implied in a posthumous publication of Carnot (d. 1832) and finally (1840–45) placed on a firm experimental basis by Joule. Prior to this time, however, engineers had already developed a valid fluid-flow equation which involved a balance of *mechanical* energies only.

When a unit mass of fluid enters the general system of Fig. 1.1, the *net* mechanical energy E_m imparted to the system from external sources is

$$E_m = p_1V_1 + W_o - p_2V_2$$

Conversely, the mechanical energy E'_m used to increase the potential and/or kinetic energy of the fluid or possibly compress the fluid while in transit through the system is given by

$$E'_m = \frac{g}{g_c}(Z_2 - Z_1) + \frac{1}{2g_c}(v_2^2 - v_1^2) - \int_1^2 p\,\mathrm{d}V$$

where the integral expression represents the mechanical work required to compress the volume of a unit mass from some initial value V_1 to a smaller value V_2.

For mechanical energy to be conserved, we might expect the quantities E_m and E'_m to be identical. But because of frictional effects within the flow, a part of the mechanical energy input will always be dissipated into heat energy. As a consequence, E_m and E'_m are never equal. However, a mechanical energy balance for a system can still be written as

$$p_1V_1 + W_o - p_2V_2 = \frac{g}{g_c}\Delta Z + \frac{1}{2g_c}\Delta v^2 - \int_1^2 p\,\mathrm{d}V + F \tag{1.7}$$

provided we let the term F represent the "lost work" or mechanical energy wasted by irreversible processes within the flowing fluid.

The energy term $(p_2V_2 - p_1V_1)$ or ΔpV is a complete differential equivalent to

$$\int_1^2 \mathrm{d}(pV) = \int_1^2 p\,\mathrm{d}V + \int_1^2 V\,\mathrm{d}p$$

which upon substitution into Eq. (1.7) yields a more compact expression,

$$\int_1^2 V\,\mathrm{d}p + \frac{g}{g_c}\Delta Z + \frac{1}{2g_c}\Delta v^2 = W_o - F \tag{1.8}$$

In this final form, our mechanical energy balance will be recognized as Bernoulli's equation expanded to include friction losses. Equation (1.8) is a completely general expression containing no limiting assumptions other than the exclusion of phase boundary, magnetic, electric, and chemical effects, and may be applied to any fluid-flow system whether or not heat is transferred. Although Eq. (1.8) appears to be independent of Q, comparison with Eq. (1.5), a total energy balance, will show that transferred heat definitely influences the friction loss term as

$$F = \Delta u + \int_1^2 p\,\mathrm{d}V - Q$$

which shows that the *increase* in internal energy, Δu, *not* contributed by the reversible work of compression,

$$-\left[-\int_1^2 p\,\mathrm{d}V\right]$$

and *not* contributed by the absorption of heat energy from an external source, $-Q$, must have come from a portion of the input mechanical energy, F. Choice of either type of energy balance equation, therefore, is mainly a matter of convenience. Dimensions of the individual energy terms in Eqs. (1.7) and (1.8) are energy per unit mass of flowing fluid. With consistent English units they become ft lbf/lbm.

A mechanical energy balance is especially useful when the flowing fluid can be considered as almost incompressible, as are most liquids. For approximate isothermal flow of liquids, the specific volume V is essentially constant and Eq. (1.8) reduces to

$$\frac{\Delta p}{\rho} + \frac{g}{g_c}\Delta Z + \frac{\Delta v^2}{2g_c} = W_o - F \tag{1.9}$$

For compressible fluids such as gases, evaluation of the integral

$$\int_1^2 V\,\mathrm{d}p$$

may be difficult unless the exact path of the compression or expansion is known.

Pressure balance. For the flow of incompressible fluids, Eq. (1.9) can be further simplified to yield a pressure balance equation. Upon multiplying all terms by ρ, which is assumed constant, one obtains

$$\Delta p + \frac{g}{g_c}\rho\Delta Z + \frac{\rho\Delta v^2}{2g_c} = \rho W_o - \rho F$$

where all terms now have the dimensions of energy per unit volume or FL/L^3 which, when expressed in consistent English units, becomes lbf/ft^2 or units of pressure. By expanding and rearranging terms, this last expression can be written as

$$p_1 + \frac{g}{g_c}\rho(Z_1 - Z_2) + \frac{\rho}{2g_c}(v_1^2 - v_2^2) + \rho W_o - \rho F = p_2 \tag{1.10}$$

which better identifies the sources of pressure change along the flow. If we can trust our intuition for the choice of proper algebraic signs, Eq. (1.10) can be stated in a more flexible form as

$$p_1 + \Delta p_s + \Delta p_v + \Delta p_P - \Delta p_f = p_2 \tag{1.11}$$

where $\Delta p_s = (g/g_c)\rho(Z_1 - Z_2)$, the increase in hydrostatic pressure which accompanies a loss in elevation

$\Delta p_v = (\rho/2g_c)(v_1^2 - v_2^2)$, the increase in pressure which accompanies a loss in velocity

$\Delta p_P = \rho W_o$, the increase in pressure across a pump in the direction of flow

$\Delta p_f = \rho F$, pressure loss caused by all frictional or irreversible processes along the flow, exclusive of those within the pump

Example 1.9. A 36° API crude oil with a viscosity of 7.0 cp flows by gravity through a 3-in. ID line. Inlet pressure is 22 psia. The line discharges to atmospheric storage which is 480 ft lower than the inlet. Calculate the mechanical energy loss along the line in ft lbf/lbm of oil.

Solution: In the absence of a pump, $W_o = 0$, and for a uniform pipe carrying an incompressible fluid $v_1^2 = v_2^2$. Equation (1.9) reduces therefore to

$$\frac{p_2 - p_1}{\rho} + \frac{g}{g_c}(Z_2 - Z_1) = -F$$

$$\text{Density of oil} = \frac{141.5}{131.5 + 36} \times 62.4 = 52.7 \text{ lbm/ft}^3$$

$$\frac{p_2 - p_1}{\rho} = \frac{(14.7 - 22)144}{52.7} = -20 \text{ ft lbf/lbm}$$

Assuming $g = 32.17$ ft/sec²,

$$\frac{g}{g_c}(Z_2 - Z_1) = (1)\,(0 - 480) = -480 \text{ ft lbf/lbm}$$

By substitution into Eq. (1.9), $F = 500$ ft lbf/lbm.

Alternate Solution: Using Eq. (1.11),

$$p_1 + \Delta p_s + \Delta p_v + \Delta p_P - \Delta p_f = p_2$$

$$\Delta p_s = \underbrace{\frac{141.5}{131.5 + 36}}_{\substack{\text{specific}\\ \text{gravity}\\ \text{of oil}}} \times \underbrace{0.433}_{\substack{\text{gradient}\\ \text{of water,}\\ \text{psi/ft}}} \times \underbrace{480}_{\substack{\text{elevation,}\\ \text{feet}}} = \underbrace{176 \text{ psi}}_{\substack{\text{positive,}\\ \text{loss in}\\ \text{elevation}}}$$

$$\Delta p_v = \Delta p_P = 0$$

$$22 \text{ psia} + 176 \text{ psi} + 0 + 0 - \Delta p_f = 14.7 \text{ psia}$$

$$\Delta p_f = 183.3 \text{ psi pressure loss caused by friction}$$

$$F = \frac{\Delta p_f}{\rho} = \frac{183.3 \times 144}{52.7} = 500 \text{ ft lbf/lbm}$$

Head loss. Equations (1.7) through (1.9) may be transformed to a unit weight basis if all terms are multiplied by (g_c/g). Equation (1.9) for incompressible fluid flow then becomes

$$\frac{\Delta p}{\rho'} + \Delta Z + \frac{\Delta v^2}{2g} = W_o' - F' \tag{1.12}$$

In this form all terms have the dimensions of energy per unit weight of flowing fluid, which in consistent English units is ft lbf/lbf or ft. For this reason the energy terms of Eq. (1.12) are frequently stated in units of "head" or feet of flowing fluid. Using this terminology, $(\Delta p/\rho')$ is a pressure head, (ΔZ) a static head, $(\Delta v^2/2g)$ a velocity head, and F' the head loss caused by friction. Head units should be used with caution as they are frequently and erroneously applied to the terms of Eq. (1.9).

1.4. Classification of Fluids and Flow Regimes

Real fluids tend to resist continuous deformation or flow, to a greater or lesser extent, depending upon their viscous nature. As a fluid flows along a stationary boundary, it clings to the surface and flow proceeds by the sliding action of adjacent fluid layers past the one in contact with the boundary. Induced slippage between fluid layers is accompanied by a frictional drag or shear stress, and a continuous supply of mechanical energy is required to maintain the flow. This energy is dissipated by viscous action and ultimately reappears as heat energy.

Viscosity. That fluid property responsible for the frictional drag or shear resistance which develops when one layer of a fluid is caused to slide over another is called *viscosity*. The viscosity at any point in a fluid is quantitatively defined as the ratio of the shearing stress to the resulting shearing

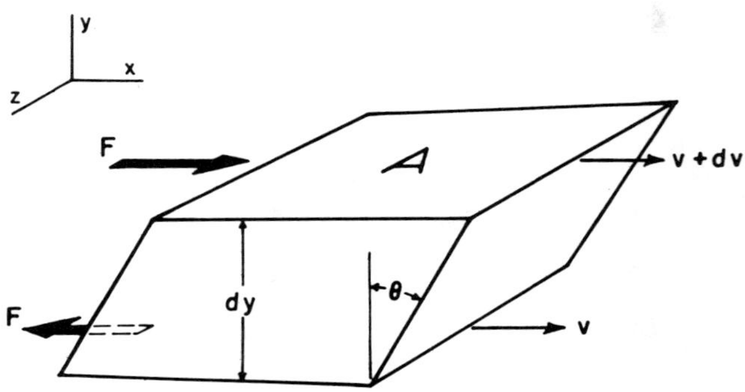

Fig. 1.2 Simple planar shearing of a fluid.

rate. An example of simple planar shear between two parallel layers of fluid is shown in Fig. 1.2. The shearing force F acting over area A produces a shearing stress $\tau = F/A$. The shearing rate θ caused by the stress τ is equivalent to the velocity gradient dv/dy. The absolute viscosity μ is given by the relation

$$\mu = g_c \frac{F/A}{\theta} = g_c \frac{\tau}{dv/dy} \tag{1.13}$$

where the conversion factor g_c is necessary since absolute viscosities have the dimensions of m/Lt as shown in Table 1.1.

Newtonian and non-Newtonian fluids. If the viscosity of a fluid is influenced only by pressure and temperature, the fluid is classified as *Newtonian,* as exemplified by water, gases, and thin oils. At a given temperature and pressure, one experimental determination is sufficient to define completely the viscous properties of a Newtonian fluid, for the ratio of shear stress to shear rate is a constant, namely μ. All fluids which do not exhibit a direct proportionality between shear stress and shear rate at constant temperature and pressure are classified as *non-Newtonian.* Suspensions of solids in liquids are generally non-Newtonian, particularly if the solid tends to swell, solvate, or otherwise associate with the liquid phase. Drilling muds and related clay suspensions as well as cement slurries and various gelled fracturing fluids are examples of non-Newtonian systems of interest to the petroleum engineer. The viscosity of these fluids will vary with the magnitude of applied shear stress and, in some instances, with the duration of shear.

Laminar and turbulent flow. Fluid will move through a conduit according to one or the other of two flow regimes — laminar or turbulent. When all fluid particles move in straight lines parallel to the conduit axis, and adjacent layers of fluid slip past each other with no mixing or interchange of fluid from one layer to the next, the flow pattern is called *laminar* (also *streamline* or *viscous*). Steady laminar flow in circular pipes of constant diameter can be visualized as a series of thin concentric cylinders each sliding past its neighbors like the tubes of a telescope. The cylinder in contact with the pipe wall remains stationary (condition of no slip) while the inner cylinders move progressively faster as their diameters become smaller.

At higher average flow velocities, when the fluid particles move downstream in a tumbling, chaotic motion so that vorticies and eddies are formed in the fluid, the flow is called *turbulent.* In turbulent pipe flow, there is no orderly shear between fluid layers but a random shearing and impact of the fluid masses caught up in the swirls and eddies of the flow.

1.5. Pipe Flow of Newtonian Liquids

Both theoretical analysis and experimental evidence have established that within a pipe carrying a Newtonian fluid in laminar flow, the actual point velocity v_r of all fluid particles situated at a common radius r is given by the expression

$$v_r = v_{\max}\left[1 - \frac{r^2}{r_w^2}\right] \tag{1.14}$$

The velocity is a maximum, $v_{\max}$, at the center of the pipe and decreases to zero at the pipe wall where $r = r_w$. As a consequence of this parabolic distribution law, Eq. (1.14), the *average* flow velocity v defined as q/A is exactly half the value of the maximum particle velocity, as shown in Fig. 1.3.

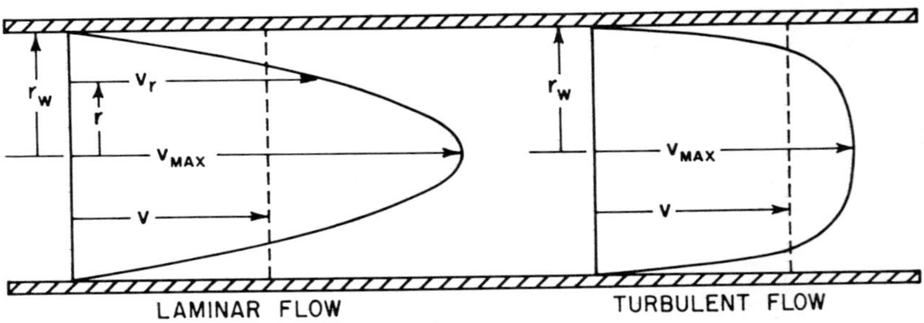

Fig. 1.3 Velocity profiles for pipe flow.

In turbulent flow, the velocity profile is flatter and the velocity gradient near the pipe wall much larger than the laminar flow profile for a corresponding value of average velocity. In addition, the turbulent velocity profile of Fig. 1.3 is indicative of the actual point velocity of only those fluid particles in the immediate vicinity of the pipe wall where a thin film of fluid remains in laminar flow. In the central turbulent core of the flow, individual particles exhibit erratic fluctuations in both speed and direction which are superimposed on a time-averaged or specific drift velocity in the direction of flow.

The kinetic energy terms introduced in Sec. 1.3 strictly apply only when the velocity distribution across any section of pipe is uniform. Since the velocity profile in turbulent flow is nearly flat or uniform, use of the average velocity v in the expression $v^2/2g_c$ is approximately valid, being about 3 to 8 per cent too low. With the parabolic distribution of laminar flow, however, the proper kinetic energy term is v^2/g_c.

Reynolds criterion. The classic experiments of Osborne Reynolds (1883) established a criterion whose magnitude indicates whether the pipe flow of a

Newtonian fluid is in the laminar or turbulent region. This dimensionless flow parameter, known as the Reynolds number, is defined by

$$N_R = \frac{dv\rho}{\mu} \tag{1.15}$$

where N_R = Reynolds number of the flow

d = inside diameter of pipe

v = average flow velocity

ρ = density of fluid

μ = absolute viscosity (Newtonian fluid)

A consistent system of units must be used with Eq. (1.15) to preserve its dimensionless character; however, it is frequently presented in an alternate form to accept the following practical units:

d = diameter, inches

v = velocity, ft/sec

ρ = density, lbm/gal

μ = viscosity, centipoises

and Eq. (1.15) takes the dimensional form

$$N_R = 928\,\frac{dv\rho}{\mu} \tag{1.15a}$$

When the Reynolds number exceeds a certain critical value of approximately 2000 for flow in straight round pipes, turbulent flow ensues, while for lower values of N_R the flow is laminar. Aside from the annular flow of rotary drilling fluids, almost all flow problems are concerned with turbulent flow, for prohibitively large diameters are required to realize laminar flow in pipes at reasonable volumetric rates.

Example 1.10. A service station pump can deliver gasoline at 20 gal/min. Assuming gasoline has a density of 6.15 lbm/gal and a viscosity of 0.60 cp, calculate the minimum size delivery hose that will maintain laminar flow.

Solution: For laminar flow in the hose, the Reynolds number should not exceed 2000, so that

$$2000 = 928\,\frac{dv\rho}{\mu}$$

where the flow velocity v must be calculated in ft/sec.

$$v = \frac{q}{A} = \frac{20\,\dfrac{\text{gal}}{\text{min}} \times \dfrac{\text{min}}{60\text{ sec}} \times \dfrac{\text{ft}^3}{7.48\text{ gal}}}{\dfrac{\pi d^2}{4}\text{ in.}^2 \times \dfrac{\text{ft}^2}{144\text{ in.}^2}} = \frac{8.17}{d^2}\text{ ft/sec } (d = \text{in.})$$

$$2000 = 928\,\frac{d \times 8.17 \times 6.15}{d^2 \times 0.60}$$

$$d = 38.9\text{ in. ID}$$

With a conventional $1\frac{1}{2}$-in. ID hose, the Reynolds number would be 51,800 at a reasonable flow velocity of 3.6 ft/sec.

Alternate equations. To enjoy their inherent ease of derivation and subsequent simplification, we shall usually base the equations to be developed in the sections that follow on a consistent system of units. On the other hand, consistent equations are seldom directly applicable to the solution of practical problems without the tedious conversion of working or "field" data to consistent units. As a compromise, primary equations will be restated in an alternate dimensional form following the procedure already adopted for Eq.(1.15) and Eq.(1.15a). Alternate equations will be tailored to accommodate directly the suggested system of engineering units given in Table 1.2 (p. 44). This scheme will also eliminate the need for a detailed listing of practical units following every alternate equation.

Laminar flow. Frictional losses in laminar Newtonian flows are due entirely to the shear stresses produced by slippage between adjacent layers of fluid. For isothermal flow in a straight run of circular pipe, the pressure drop caused by friction can be calculated from the Hagen-Poiseuille law

$$q = \frac{g_c \pi d^4\, \Delta p_f}{128 \mu L} \tag{1.16}$$

where q = volumetric flow rate

g_c = unit system conversion constant

d = inside diameter of pipe

Δp_f = pressure loss

μ = absolute viscosity (Newtonian fluid)

L = length of pipe

For circular pipes, q may be replaced by the quantity $v(\pi d^2/4)$ and Eq. (1.16) rearranged to yield the pressure loss as

$$\Delta p_f = \frac{32 \mu L v}{g_c d^2} \tag{1.17}$$

When this last expression is modified to accept the engineering units of Table 1.2, we have the alternate equation

$$\Delta p_f = \frac{\mu L v}{1500 d^2} \tag{1.17a}$$

This law was discovered by experiment and verified by many investigators long before it was derived theoretically. In the same spirit, we will likewise delay its formal derivation until Sec. 1.8. Notice that the pressure loss increases linearly with increases in average flow velocity or volumetric flow rate as long as the flow remains laminar. At Reynolds numbers greater than 2000, however, the flow becomes turbulent and Eqs.(1.16) and (1.17) are no longer valid. As we shall see, the pressure loss in turbulent flow varies approximately with the square of the average flow velocity.

Turbulent flow. Outside the orderly realm of laminar flow, the theoretical analysis of mechanical energy losses becomes extremely complicated and, as yet, it is far from being completed. However, the complexity of turbulent flow in conduits of simple geometry has been partially resolved by employing the techniques of dimensional analysis to generalize the results of experimental investigations.

For steady flow through long straight pipes of circular section, experience has shown that frictional losses are affected by the physical properties of the flowing fluid and the geometry of the pipe system. Fluid properties of particular importance are the average flow velocity v, the density ρ, and the absolute viscosity μ. Pipe geometry is specified by the inside diameter d, length L, and the condition of the inner surface of the pipe. By applying the method of dimensional analysis, the friction loss F is found to be related to these variables by the expression

$$F = \frac{\Delta p_f}{\rho} = \frac{2fLv^2}{g_c d} \tag{1.18}$$

which is known as the *Fanning equation*. Upon multiplying all terms in Eq. (1.18) by ρ and introducing practical engineering units, we obtain an alternate form of the Fanning equation

$$\Delta p_f = \frac{fL\rho v^2}{25.80 d} \tag{1.18a}$$

The dimensionless parameter f, called the Fanning friction factor, depends on both the Reynolds number N_R and the surface conditions of the pipe wall characterized by a relative roughness parameter ϵ/d, or

$$f = \phi(N_R, \epsilon/d)$$

In the definition of relative roughness, the quantity ϵ represents the average depth of pipe-wall irregularities and is defined as *absolute* roughness. Given two pipes having the same absolute roughness but different diameters, the larger pipe would have the smaller value of relative roughness.

Friction factor curves. The nature of the function $\phi(N_R, \epsilon/d)$ has been established in part from theoretical analysis and in part from correlations of experimental data. The lower family of curves in Fig. 1.4 is a log-log plot of this relationship for straight runs of smooth tubing (e.g., glass, lead, copper, and brass) and several sizes of clean steel pipe. In the laminar region where $N_R \leq 2000$, the friction factor is not affected by pipe roughness. As a consequence, the friction factor–Reynolds number correlation plots as a single line. When the equation for this straight line relationship, $f = 16/N_R$, is substituted into the Fanning equation, it is reassuring to find that the resulting expression simplifies to the Hagen-Poiseuille law of Eq. (1.17).

For Reynolds numbers in the range of 2000 to 4000, the curves (shown as broken lines) are seen to overlap. In this *critical region*, flow is unstable and can fluctuate between laminar and turbulent conditions depending upon local disturbances in the pipe system, such as entrance effects and the proximity of fittings, bends, valves, etc. However, the usual practice is to assume that turbulent flow exists at all Reynolds numbers greater than 2000.

In the region of stable turbulence at Reynolds numbers greater than 4000, relative roughness of the pipe begins to have a noticeable effect on the friction factor, especially at the higher Reynolds numbers. The curves of Fig. 1.4 apply to pipe diameters as indicated, provided the absolute roughness of the pipe surface $\epsilon = 0.00065$ in. This value agrees with most of the published data obtained on clean steel pipes in gas well and pipeline service as reviewed by Cullender and Smith.*[1] Friction factor curves are available that incorporate a greater range of relative roughness values,[2] but those presented in Fig. 1.4 are adequate for most engineering calculations involving oil-well tubular goods and field gathering systems.

Example 1.11. One mile of 6-in. ID pipeline connects a central tank battery with a barge loading terminal. Calculate the frictional pressure loss in the line when it is handling 4000 bbl/hr of a 40° API crude oil which has a viscosity of 5.0 cp.

SOLUTION: The flow will approximate isothermal, incompressible conditions, and engineering units will be used in all calculations.

$$\text{Oil density} = \frac{141.5}{131.5 + 40} \times 8.34 = 6.88 \text{ ppg}$$

$$\text{Volume rate} = 4000 \frac{\text{bbl}}{\text{hr}} \times \frac{\text{hr}}{3600 \text{ sec}} \times \frac{5.615 \text{ ft}^3}{\text{bbl}} = 6.24 \text{ ft}^3/\text{sec}$$

$$\text{Cross-sectional area} = \frac{\pi \times 6^2}{4 \times 144} = 0.196 \text{ ft}^2$$

$$\text{Average velocity} = \frac{6.24}{0.196} = 31.8 \text{ ft/sec}$$

*References throughout the text are given at the end of each chapter.

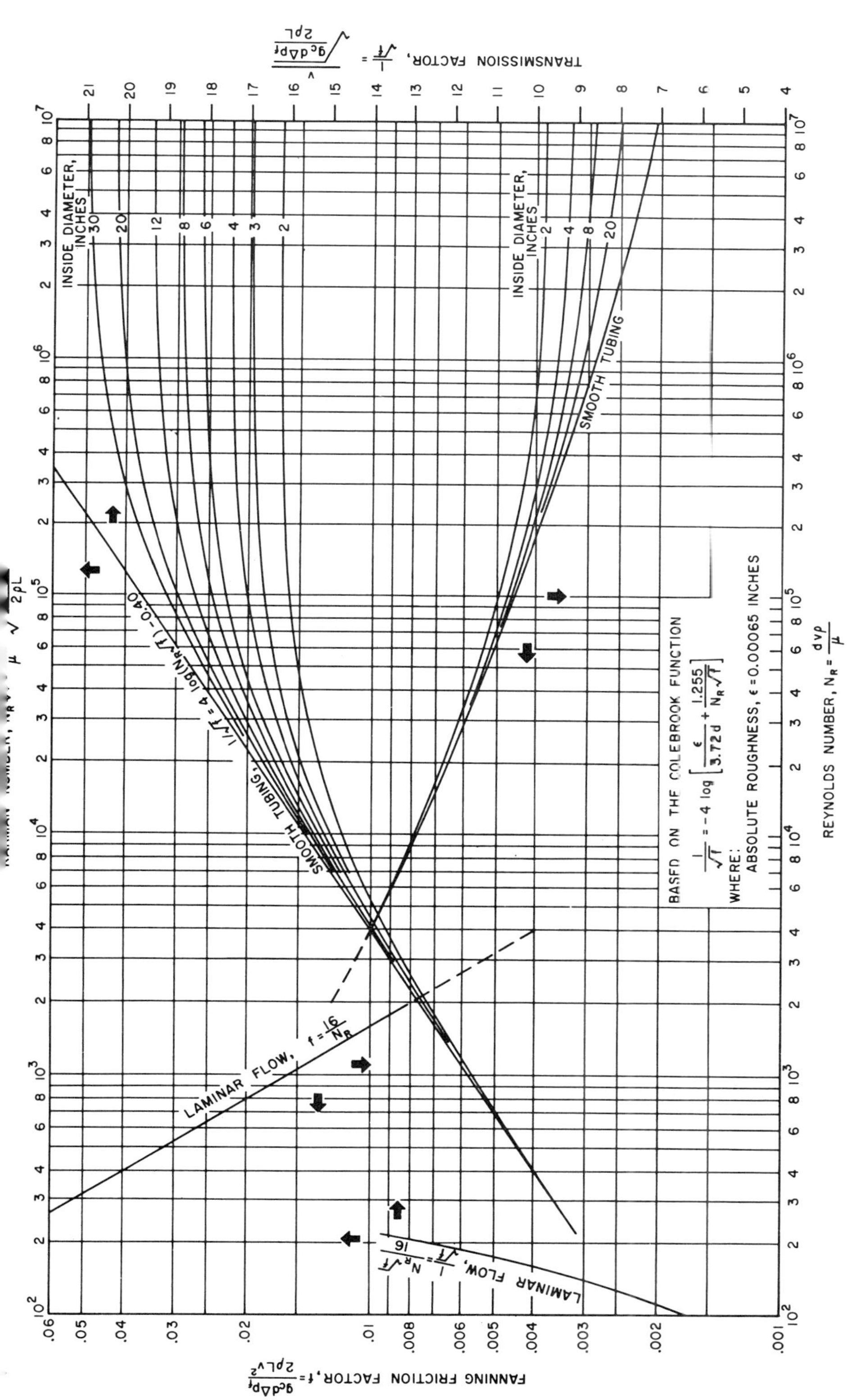

Fig. 1.4 Friction factors for clean steel pipe.

From Eq. (1.15a),

$$N_R = 928\,\frac{6 \times 31.8 \times 6.88}{5.0} = 2.44 \times 10^5$$

From Fig. 1.4,

$$f = 0.0040$$

From Eq. (1.18a),

$$\Delta p_f = \frac{0.0040 \times 5280 \times 6.88 \times (31.8)^2}{25.80 \times 6} = 949 \text{ psi}$$

Using a plot of f versus N_R, one calculates the average flow velocity within a pipe, for a given pressure drop, by trial and error, for the unknown velocity v appears in the equation for f as well as N_R. However, a direct solution is possible through use of the upper family of curves in Fig. 1.4 which incorporate the functional dependence of f on N_R in a semilog plot of two dimensionless parameters, the transmission factor $1/\sqrt{f}$ and the Karman number $N_R\sqrt{f}$. Using Eqs. (1.15a) and (1.18a) in practical engineering units, we obtain the following expressions for these two parameters:

$$\frac{1}{\sqrt{f}} = \frac{v}{\sqrt{\dfrac{25.80d\ \Delta p_f}{\rho L}}} \tag{1.19a}$$

$$N_R\sqrt{f} = 928\,\frac{d\rho}{\mu}\sqrt{\frac{25.80d\ \Delta p_f}{\rho L}} \tag{1.20a}$$

Since Eq. (1.20a) does not contain v, it may be solved for the Karman number and subsequently used with the curves of Fig. 1.4 to evaluate the transmission factor. The unknown velocity v is then calculated by use of Eq. (1.19a).

Example 1.12. Suppose the frictional pressure loss in the pipeline of Example 1.11 must be limited to 475 psi. Calculate the crude oil capacity of the line in barrels per hour.

SOLUTION: From Eq. (1.20a),

$$N_R\sqrt{f} = 928\,\frac{6 \times 6.88}{5.0} \times \left[\sqrt{\frac{25.80 \times 6 \times 475}{6.88 \times 5280}}\right]$$

$$= 7662 \times [1.423] = 1.09 \times 10^4$$

From Fig. 1.4,

$$1/\sqrt{f} = 15.3$$

From Eq. (1.19a),

$$v = 15.3 \times 1.423 = 21.8 \text{ ft/sec}$$

$$q = vA = 21.8 \times 0.196 = 4.27 \text{ ft}^3\text{/sec}$$

$$\text{Capacity} = \frac{4.27 \times 3600}{5.615} = 2738 \text{ bbl/hr}$$

Comparing these results with those of Example 1.11 shows that a 50 per cent reduction in frictional pressure loss is obtained by a 31.5 per cent reduction in flow rate. In this example, pressure loss varies with the 1.8 power of flow rate.

The line size for a given turbulent flow capacity and pressure drop is also calculated by trial and error. Computation can be shortened, however, by an approximate, graphical solution. The required capacity is held constant while the pressure loss Δp_f is calculated for two assumed diameters. These results are then plotted (linearly) on log-log paper and extrapolated to the diameter which yields the required pressure loss.

Friction factor curves similar to Fig. 1.4 are found in the technical literature in which the symbol f represents a factor which is four times larger than the Fanning friction factor defined by Eq. (1.18). To avoid confusion, plots of the Fanning friction factor are readily identified by a two-point check:

$$f = 0.016 \text{ at } N_R = 1000$$

$$f \cong 0.012 \text{ at } N_R = 3000$$

Pipeline efficiency. We can hardly expect all pipe, regardless of size, grade, and manufacture, to exhibit a common value of absolute roughness. Even new steel pipe direct from a mill, and certainly pipe that has aged in a yard, will show random variations in ϵ values.[3] And as yet we have not considered the cumulative effects of corrosion, erosion, and scaling which can take place while a line is in service. It should, therefore, be obvious why flow rates predicted from generalized correlations seldom agree precisely with values measured in an actual line.

To offset this disagreement, an efficiency factor E (sometimes called an experience factor) is usually introduced in all calculations. This factor is defined as the fraction (or percentage) of the total calculated flow rate actually handled by a given line. In the absence of test data, proper choice of E is largely a matter of experience; however, a value of 0.90 is a safe estimate for design calculations based on the correlations of Fig. 1.4.

To calculate the actual capacity of a line, its theoretical capacity is simply multiplied by the E factor. To correct a pressure loss calculation, the theoretical pressure loss (as calculated in Example 1.11) is divided by the square of the E factor. This procedure has the same effect as changing the friction factor f, as read from Fig. 1.4, to a new value f/E^2.

Friction losses in fittings. Elbows, valves, vessel connections, and other types of fittings increase the friction losses in a pipe system. And at best, these additional losses can be predicted with only a fair degree of accuracy. They may be included in the analysis of a system by considering each fitting as being replaced by an equivalent length of straight pipe L_e which will produce the same friction loss as the actual fitting. The equivalent length of each fitting is then added to the axial length of the pipe system, L, before calculating the total friction loss of the system. By use of the nomograph prepared by the Crane Company,[4] as reproduced in Fig. 1.5, approximate values of L_e for common valves and fittings are readily determined for conditions of turbulent flow. For example, a 2-in. swing check valve (open) is equivalent to 13 ft of straight pipe.

1.6 Annular Flow of Newtonian Liquids

In the analysis of wellbore circulatory systems, we must be able to predict frictional pressure losses for flow through annular spaces in addition to flow through pipes. As in the previous section, we will restrict our attention to isothermal, steady-state flows of Newtonian liquids and investigate their laminar and turbulent flow behavior within an annulus formed by two concentric, stationary, circular pipes.

Average annular flow velocity v is again defined as q/A where q is the volumetric rate of flow and A the cross-sectional area of the annulus. Letting d_o represent the ID of the outer pipe and d_i the OD of the inner pipe,

$$v = \frac{4q}{\pi(d_o^2 - d_i^2)}, \quad \text{or} \quad v = \frac{4q}{\pi d_o^2(1 - \alpha^2)} \tag{1.21}$$

where α is a dimensionless parameter of the annular geometry, defined by $\alpha = d_i/d_o$. In practical engineering units (Table 1.2), Eq. (1.21) can be expressed in several alternate forms to accept directly the more common expressions for volumetric flow rate q,

$$v = \frac{17.16(q, \text{bbl/min})}{(d_o^2 - d_i^2)} \qquad v = \frac{3.056(q, \text{ft}^3/\text{min})}{(d_o^2 - d_i^2)}$$

$$v = \frac{(q, \text{gpm})}{2.448(d_o^2 - d_i^2)} \tag{1.21a}$$

These alternate expressions also apply to pipe flow systems when d_i is set equal to zero.

Laminar flow. An exact expression relating frictional pressure drop and volumetric flow rate for laminar annular flow was determined analytically by Lamb:[5]

$$q = \frac{g_c \pi d_o^4 \Delta p_f}{128 \mu L} \left\{ (1 - \alpha^2)\left[(1 + \alpha^2) - \frac{(1 - \alpha^2)}{\ln 1/\alpha} \right] \right\} \tag{1.22}$$

where the quantity in braces is a function of the annular geometry only. Representing this quantity, which we shall call the *Lamb coefficient*, by C_L, we can write Eq. (1.22) in a more compact form,

$$q = \frac{g_c \pi d_o^4 \Delta p_f}{128 \mu L} C_L \tag{1.23}$$

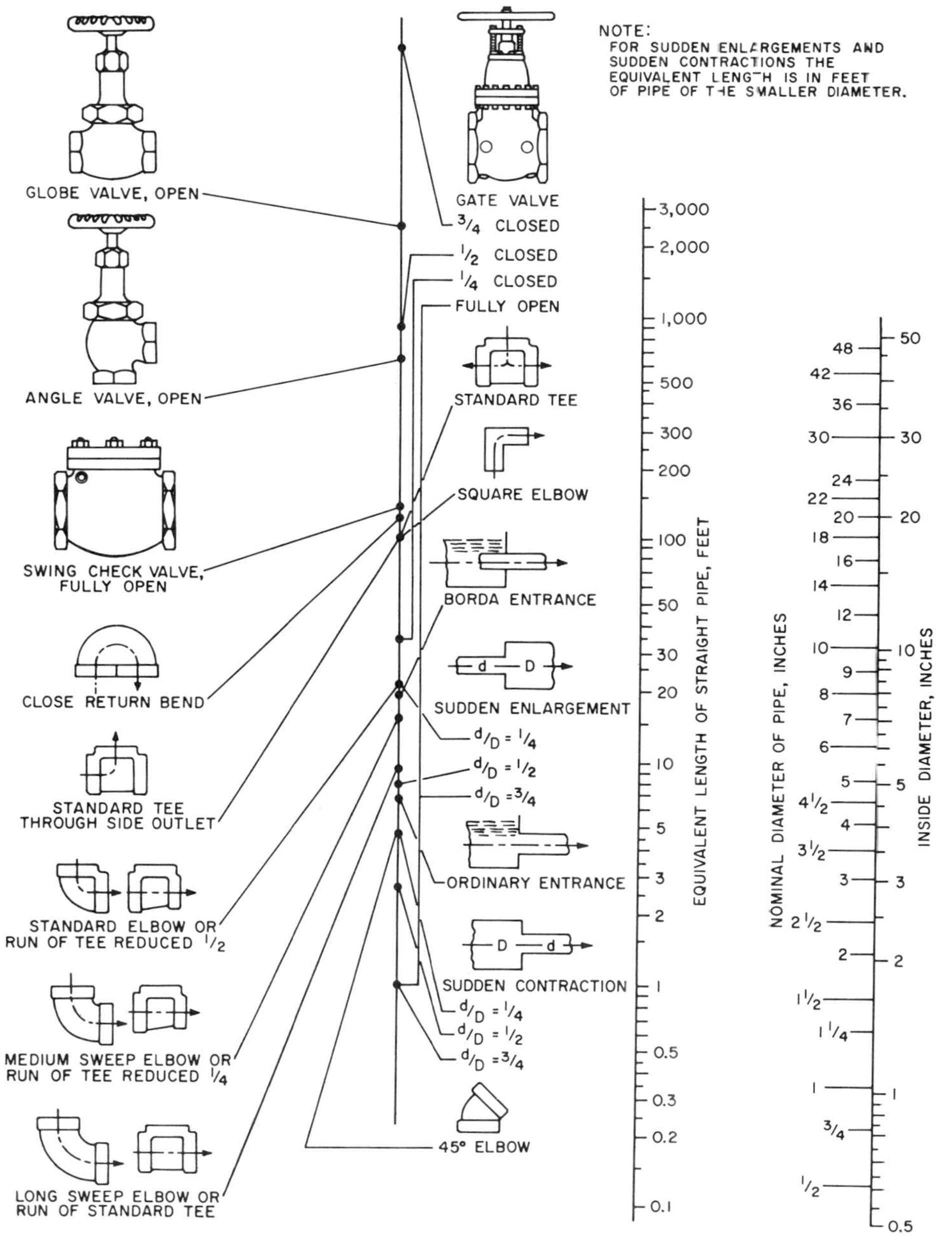

Fig. 1.5 Turbulent flow resistance of valves and various fittings. (Courtesy of Crane Company.[4])

A plot of the coefficient C_L as a function of α is given in Fig. 1.6. When $\alpha = 0$, i.e., for flow in a pipe, $C_L = 1$ and Eq.(1.23) reduces to the Hagen-Poiseuille law of Eq. (1.16). From the steep, negative slope of the curve at

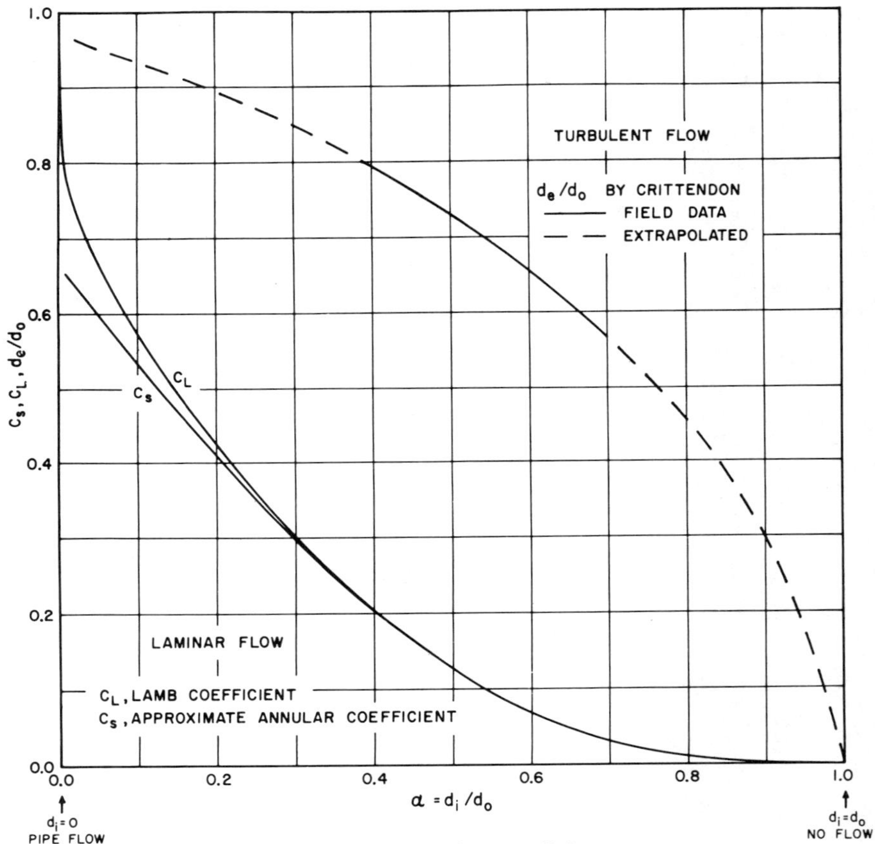

Fig. 1.6 Annular flow coefficients.

small values of α, we see that the laminar flow capacity of even a very wide annulus ($d_o \gg d_i$) is considerably less than the capacity of an unrestricted pipe of diameter d_o. For example, if a $\frac{1}{8}$-in. diameter rod were centered along the axis of a 2-in. ID pipe, the capacity of the annular section ($\alpha = 0.06$) would be only 65 per cent of the flow capacity of the pipe alone.

By use of the method of analysis developed in Sec. 1.9, an equation for laminar flow through a narrow slot of width W, cross-sectional area A, and length L can be obtained as

$$q = \frac{g_c A W^2 \Delta p_f}{12 \mu L} \tag{1.24}$$

While this conduit geometry is never encountered in wellbore or pipeline systems, it may be used to approximate an annular system with remarkable accuracy.

If we consider an annular space as being representative of a narrow slot, the area and width of the slot would be given by

$$A = \frac{\pi}{4}(d_o^2 - d_i^2) = \frac{\pi d_o^2}{4}(1 - \alpha^2)$$

$$W = \left(\frac{d_o}{2} - \frac{d_i}{2}\right) = \frac{d_o}{2}(1 - \alpha)$$

When these expressions are substituted into the slot flow equation, Eq. (1.24), we obtain

$$q = \frac{g_c \Delta p_f}{12\mu L} \times \frac{\pi d_o^2}{4}(1 - \alpha^2) \times \frac{d_o^2}{4}(1 - \alpha)^2$$

which can be simplified to

$$q = \frac{g_c \pi d_o^4 \Delta p_f}{128\mu L}\left\{\frac{(1 + \alpha)(1 - \alpha)^3}{1.5}\right\} \tag{1.25}$$

If we represent the quantity in braces by C_s, which we shall call the approximate annular coefficient, we see that Eq. (1.25) bears a striking resemblance to the exact expression for annular flow given by Eq. (1.23).

When the coefficient C_s is plotted as a function of α, as shown in Fig. 1.6, and compared with a plot of the exact coefficient C_L, the two curves are seen to match very well, except at small values of α. For values of α greater than 0.4, C_s approximates C_L with an error of less than 2 per cent. Our approximate solution may be used, therefore, for all annular flow situations which are normally encountered in wellbore circulatory systems.

Replacing the volumetric flow rate q in Eq. (1.25) by the quantity vA and solving for v, we obtain

$$v = \frac{g_c d_o^2 \Delta p_f}{32\mu L} \times \frac{(1 - \alpha)^2}{1.5}$$

which may be rearranged to yield an expression for frictional pressure drop in consistent units,

$$\Delta p_f = \frac{48\mu L v}{g_c(d_o - d_i)^2} \tag{1.26}$$

or an alternate form using practical engineering units,

$$\Delta p_f = \frac{\mu L v}{1000(d_o - d_i)^2} \tag{1.26a}$$

Turbulent flow. The friction factor correlations developed for flow through round pipes will sometimes yield satisfactory results when applied to conduits of noncircular cross section. When a value of pipe diameter is required in either the Fanning equation or the expression for Reynolds number, an equivalent diameter d_e is used for a noncircular pipe. However, the average flow velocity v, which also appears in these same two expressions, must be calculated from the equation $v = q/A$, where A is the *actual* cross-sectional area of the noncircular pipe.

The equivalent diameter is defined as four times the hydraulic radius of the conduit which is denoted by R, where

$$R = \frac{\text{Area of flow cross section}}{\text{Wetted perimeter}}$$

A rectangular passage, with sides a and b, has an equivalent diameter of

$$d_e = 4R = 4\left[\frac{ab}{2(a+b)}\right] = \frac{2ab}{(a+b)}$$

For a circular concentric annulus running full,

$$d_e = 4\left[\frac{\pi(d_o^2 - d_i^2)/4}{\pi(d_o + d_i)}\right] = (d_o - d_i)$$

and for the degenerate case of a circular pipe,

$$d_e = 4\left[\frac{\pi d^2/4}{\pi d}\right] = d$$

This simple adaptation of our previous pipe flow equations will give fairly reliable estimates of pressure loss and line capacity provided its use is restricted to *turbulent* flows in conduits of simple cross-sectional shape, e.g., triangular, rectangular, and elliptical sections.

For turbulent annular flows, however, a very promising correlation has been developed by Crittendon[6] from a study of more than 100 hydraulic fracture treatments of producing wells in which lease crude was used as a fracturing fluid. The following semiempirical relationship was developed to define the diameter, d_e, of a circular pipe whose flow rate–pressure loss characteristics will duplicate those of an annular system:

$$\frac{d_e}{d_o} = \frac{1}{2}[C_L^{0.25} + (1 - \alpha^2)^{0.5}] \tag{1.27}$$

where C_L is the Lamb coefficient which was defined in Eq. (1.22). In using the friction factor curves of Fig. 1.4, which differ slightly from those employed by Crittendon, it is suggested that an E value of 0.90 be taken for cased annuli. The onset of turbulence in annular flow takes place at a Reynolds number of about 2000 as calculated for the flow through the equivalent pipe. As a convenience, a plot of Eq. (1.27) is given in Fig. 1.6.

Example 1.13. A workover rig is using gun-barrel salt water as a circulating fluid. For a circulation rate of 200 gpm, calculate the frictional pressure loss per 1000 ft of annular flow using the following well data:

Water density = 8.8 ppg	Casing ID = 4.892 in.
Water viscosity = 0.80 cp	Tubing OD = 2.375 in.

SOLUTION:

$$\alpha = \frac{2.375}{4.892} = 0.485$$

From Fig. 1.6, $d_e/d_o = 0.74$, so that the diameter of the equivalent pipe is

$$d_e = 0.74 \times 4.892 = 3.62 \text{ in.}$$

From Eq. (1.21a), the average flow velocity in the equivalent pipe is

$$v = \frac{200}{2.448(3.62)^2} = 6.23 \text{ ft/sec}$$

Notice, this last figure is not the actual velocity in the annular section, but rather the velocity that would be realized in the equivalent pipe (ID = 3.62 in.) whose flow rate – pressure loss characteristics will duplicate those of the annular section.

$$N_R = 928\,\frac{3.62 \times 6.23 \times 8.8}{0.80} = 2.30 \times 10^5 \text{ (Turbulent)}$$

From Fig. 1.4, $f = 0.0042$ (with $d_e = 3.62$ in.), and so

$$\Delta p_f = \frac{0.0042 \times 1000 \times 8.8 \times (6.23)^2}{25.80(3.62)} = 15.36 \text{ psi/1000 ft}$$

Line efficiency correction:

$$\Delta p_f = \frac{15.36}{(0.9)^2} = 19.0 \text{ psi/1000 ft}$$

1.7 Rheological Classification of Fluids

With pressure and temperature held constant, a Newtonian fluid will exhibit a direct proportionality between applied shear stress and resulting rate of shear. In equation form, this defining relationship becomes

$$\tau = \frac{\mu}{g_c}\left(-\frac{dv_r}{dr}\right) \tag{1.28}$$

where the proportionality constant, μ, is the absolute viscosity of the fluid and $(-\,dv_r/dr)$ is an expression of shear rate for laminar flow within a circular pipe. A plot of Eq. (1.28) on Cartesian coordinates produces a *flow curve* characteristic of all Newtonian fluids. As shown in Fig. 1.7, the flow curve is always a straight line passing through the origin with a slope governed by μ/g_c. In other words, the flow curve is uniquely established

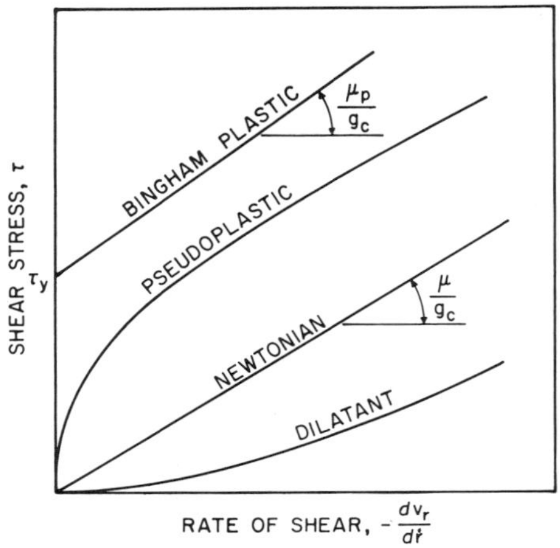

Fig. 1.7 Flow curves of Newtonian and non-Newtonian fluids.

by a single physical property, viz., the absolute viscosity μ. Accordingly, a single experimental determination of shear rate at only one value of applied shear stress is sufficient to define completely the laminar flow properties of the fluid; additional measurements at other values of shear stress will merely confirm the accuracy of the first determination. All materials which do not conform to a direct proportionality between shear stress and shear rate are classified as non-Newtonian. Study of the flow and deformation of matter, particularly the plastic flow of solids and the flow of non-Newtonian liquids, constitutes that branch of science called *rheology*, which takes its name from the words "rheo" (flow) and "logos" (theory).

As yet no general relationship has been proposed which will describe exactly the shear stress–shear rate behavior of all non-Newtonian systems over all ranges of shear rate. Experimental rheologists, however, have been fairly successful in correlating the behavior of many non-Newtonian liquids and suspensions with the ideal behavior of several model liquids of empirical origin.

Ideal or Bingham plastics. The model proposed by E. C. Bingham is best described with reference to its flow curve as shown in Fig. 1.7. For a curve of this type, the defining equation is

$$(\tau - \tau_y) = \frac{\mu_p}{g_c}\left(-\frac{dv_r}{dr}\right), \qquad [\tau > \tau_y] \tag{1.29}$$

where τ_y is the positive intercept on the shear stress axis and μ_p is proportional to the constant slope of the curve. Unlike a Newtonian fluid, a Bingham plastic will not deform continuously (or flow) until the applied shear stress, τ, exceeds a certain minimum value τ_y, known as the *yield point*. After the yield point value has been exceeded, equal increments of additional shear stress will produce equal increments of shear rate in proportion to μ_p, called the *plastic viscosity*. These two rheological parameters, plastic viscosity and yield point, specify completely the flow properties of a Bingham plastic. Conversely, at least two experimental determinations of shear rate at different values of applied shear stress are necessary to establish the flow curve.

As with all non-Newtonian fluids, the *apparent viscosity* of a Bingham plastic, defined as the ratio of shear stress to shear rate, is not a constant but varies with applied shear stress. By use of a geometric construction, the apparent viscosity, μ_a, can be obtained from the slope of a straight line drawn from the origin and terminating at some point on the flow curve whose position coincides with a given value of shear stress. From the slope of this line, which is equivalent to μ_a/g_c, it can be seen that μ_a decreases with increasing values of shear stress (in excess of τ_y).

Pseudoplastic and Dilatant fluids. Aside from Bingham plastics, all other rheological models may be assigned the class designation of "generalized non-Newtonian." Fluids of this type do not have yield points but their apparent viscosity is some nonlinear function of shear stress and possibly duration of shear. Within this general class are two major subgroups referred to as *pseudoplastic* and *dilatant* materials. The apparent viscosity of a pseudoplastic fluid will decrease with increasing values of shear stress. As indicated in Fig. 1.7, the slope of the flow curve characteristic of these fluids continually decreases, and quite often will approach a constant value at high rates of shear. A dilatant fluid displays rheological behavior opposite to that of a pseudoplastic, in that its apparent viscosity increases with increasing shear stress.

Time-Dependent behavior. Non-Newtonian fluids whose apparent viscosity at a fixed value of shear rate and temperature do not remain constant, but change with the duration of shear, are further classified as "time-dependent." While subjected to a constant rate of shear, a *thixotropic* fluid exhibits a decrease in shear stress with time of shear. The converse behavior, a shear stress increase with time of shear at a constant shearing rate, is the distinguishing property of a *rheopectic* fluid.

When a sequence of increasing shear rates is imposed, a thixotropic material will display a breakdown in consistency or "thickness" which is a function of time as well as of the maximum imposed shear rate. If the same material is then subjected to a sequence of decreasing shear rates, no further breakdown will usually take place. This phenomenon produces a double or looped flow curve, an "up-curve" being generated while shear rates are increasing followed by a "down-curve" produced while shear rates are decreasing. Following exposure to high rates of shear, a thixotropic fluid will regain its original consistency after a suitable quiescent period.

While thixotropy is of considerable importance in certain commercial products (e.g., paints and so called nondrip plastisols), it is of little consequence in a study of steady-state flow devices wherein only two extremes of shear duration are normally encountered, viz., zero and infinite times of shear associated with the start-up and subsequent maintenance of steady flow.

Power-Law or power-model fluids. An empirical, two-constant power function of the form

$$\tau = K\left(-\frac{dv_r}{dr}\right)^n \tag{1.30}$$

has been found experimentally to represent the relationship between shear stress and shear rate for a great many non-Newtonian fluids over wide ranges of shear rate. An ideal "power-model" fluid would obey this power-law relationship at all rates of shear. Values of n between zero and unity characterize pseudoplastic fluids, whereas their opposites, the dilatant fluids, have n values greater than unity. This power-law model also includes Newtonian behavior as a special case with n equal to unity and K equal to μ/g_c. In summary, the exponent n is a quantitative index which may be used to evaluate the degree of non-Newtonian behavior. The greater the difference between n and unity, in either direction, the more pronounced are the non-Newtonian characteristics of a given fluid. The coefficient K is related more to the consistency of a fluid, the larger the K value the "thicker" or "more viscous" the fluid.

In general, the true rheological properties of a given fluid will seldom duplicate exactly those characteristics ascribed to any one of the model fluids presented, except perhaps over a select range of shear rates. For this reason, a preconceived notion about a fluid's rheological properties in terms of a particular model fluid can easily result in the erroneous interpretation of experimental viscometric data. A general method of analysis will be developed in Sec. 1.11 for establishing the absolute flow properties of all time-independent non-Newtonian fluids without recourse to any arbitrary concepts of model-fluid classification.

1.8 Pipe Flow of Bingham Plastics

As a prelude to a development of the aforementioned general method of analysis, we will investigate the flow mechanics of a Bingham plastic material. Although we limit our study to the simplest of all non-Newtonian models, we can introduce the analytical techniques required for a more general analysis at the outset with a minimum of formidable complexity. In addition, the results of this preliminary investigation will find direct application in our study of rotary drilling hydraulics which follows in Sec. 1.10.

Laminar flow. For isothermal, steady-state flow through a straight run of circular pipe, the pressure loss caused by friction can be determined analytically, subject to the following assumptions:

1. There is no slippage at the wall, i.e., the velocity of the fluid contacting the pipe wall is zero.
2. All particles situated in a cylindrical shell of radius r and infinitesimal thickness dr travel as a group at a constant velocity v_r parallel to the axis of the pipe.
3. The shearing rate or velocity gradient produced at any point in the flow, as one shell of fluid slides past another, is a function only of the shear stress at that point.

Consider the steady flow of fluid through a section of horizontal pipe of radius r_w. Upon isolating a concentric cylinder of radius r and length L, as illustrated in Fig. 1.8, we see that this cylinder is being acted upon by two

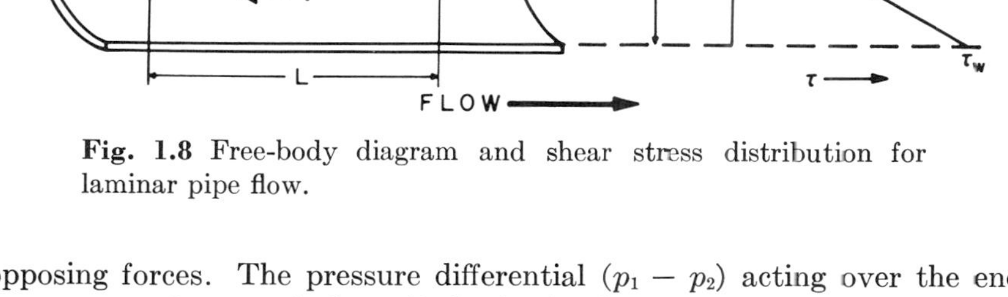

Fig. 1.8 Free-body diagram and shear stress distribution for laminar pipe flow.

opposing forces. The pressure differential $(p_1 - p_2)$ acting over the end areas πr^2 tends to push the cylinder in the direction of flow. At the same time, the fluid particles residing just within the curved surface $2\pi rL$ are being sheared past those particles immediately outside this surface and, as a result, they experience a shear stress τ which opposes the forward motion of the cylinder. If fluid within this cylinder is not accelerating (the condition for steady flow), these two forces must cancel each other, or

$$(p_1 - p_2)\pi r^2 = \Delta p_f \pi r^2 = \tau(2\pi r L)$$

where the pressure loss $(p_1 - p_2)$ is caused entirely by friction, Δp_f, since the flow is horizontal. The shear stress produced along a curved surface of radius r is therefore

$$\tau = \frac{r}{2}\left(\frac{\Delta p_f}{L}\right) \tag{1.31}$$

and, in particular, the shear stress developed at the pipe wall is given by

$$\tau_w = \frac{r_w}{2}\left(\frac{\Delta p_f}{L}\right) = \frac{d\Delta p_f}{4L} \tag{1.32}$$

a quantity which is *directly* measurable.

From Eqs. (1.31) and (1.32), an expression for the distribution of shear stress within the fluid is obtained in terms of the known shear stress at the wall and the radial position of a fluid element under consideration,

$$\tau = \frac{2\tau_w}{d}\, r$$

or

$$d\tau = \frac{2\tau_w}{d}\, dr \tag{1.33}$$

Thus we see that the shear stress is a linear function of radius, as shown in Fig. 1.8. Along the axis of the flow, the stress is zero and it increases with radius to a maximum value of τ_w, as given by Eq. (1.32), at the inside wall.

Since shear rate dv_r/dr was assumed to be a function of shear stress only,

$$-\frac{dv_r}{dr} = k\phi(\tau)$$

or

$$dv_r = -k\phi(\tau)\, dr \tag{1.34}$$

where k is a constant and $\phi(\tau)$ represents some unspecified function of τ. The minus sign is introduced here because dv_r/dr is a negative quantity, i.e., v_r decreases as r increases.

Using Eq. (1.33), we can express the differential velocity distribution of Eq. (1.34) in terms of τ in place of the position coordinate r, since these two quantities are linearly related:

$$dv_r = -k\phi(\tau) \times \frac{d}{2\tau_w}\, d\tau$$

$$dv_r = -\frac{kd}{2\tau_w}\,\phi(\tau)\, d\tau \tag{1.35}$$

Integrating this last expression either (a) from some point at radius r to the wall where $r = r_w$ and $v_r = 0$, or (b) from a point where the shear stress is τ

to the wall where $\tau = \tau_w$ and $v_\tau = 0$, we obtain the particle velocity distribution in the pipe,

$$\int_r^{\tau_w} dv_r = \int_\tau^{\tau_w} dv_\tau = -\frac{kd}{2\tau_w}\int_\tau^{\tau_w} \phi(\tau)\, d\tau$$

$$v_\tau = \frac{kd}{2\tau_w}\int_\tau^{\tau_w} \phi(\tau)\, d\tau \tag{1.36}$$

The volumetric rate of fluid transfer within a cylindrical shell of radius r and thickness dr which is moving at a velocity v_r, is given by

$$dq = v_r\, dA = v_r\,(2\pi r\, dr) \tag{1.37}$$

By replacing v_r with the equivalent expression of Eq. (1.36) and recognizing that

$$2\pi r\, dr = \frac{\pi d^2}{2\tau_w^2}\,\tau\, d\tau$$

which can be verified by Eq. (1.33), the differential flow rate previously stated as Eq. (1.37) can be expressed in terms of τ exclusively, or

$$dq = \frac{\pi k d^3}{4\tau_w^3}\,\tau\left[\int_\tau^{\tau_w} \phi(\tau)\, d\tau\right] d\tau \tag{1.38}$$

Integration of this equation over the entire cross section of the pipe from $\tau = 0$ (at the axis) to $\tau = \tau_w$ (at the wall) will yield the total volumetric rate of flow as

$$q = \frac{\pi k d^3}{4\tau_w^3}\int_0^{\tau_w} \tau\left[\int_\tau^{\tau_w} \phi(\tau)\, d\tau\right] d\tau \tag{1.39}$$

If the average velocity of the flow, v, be defined as

$$v = \frac{q}{A} = \frac{4q}{\pi d^2}$$

then Eq. (1.39) can be rearranged to give

$$\frac{v}{d} = \frac{k}{\tau_w^3}\int_0^{\tau_w} \tau\left[\int_\tau^{\tau_w} \phi(\tau)\, d\tau\right] d\tau \tag{1.40}$$

The definite (double) integral in Eq. (1.40) can be simplified, upon integration by parts, to yield the general expression for laminar pipe flow:

$$\frac{v}{d} = \frac{k}{2\tau_w^3}\int_0^{\tau_w} \tau^2\, \phi(\tau)\, d\tau \tag{1.41}$$

This final equation indicates that a plot of v/d versus τ_w (or $d\Delta p_f/4L$) should produce a unique curve for a given type of fluid, a curve which is solely dependent upon the rheological properties of the fluid or the functional dependence of shear rate upon shear stress as expressed by $k\phi(\tau)$.

For a Bingham plastic material, this functional dependence is known to be

$$\phi(\tau) = (\tau - \tau_y) \quad \text{for } \tau > \tau_y$$

and

$$\phi(\tau) = 0 \quad \text{for } \tau \leq \tau_y$$

where

$$k = \frac{g_c}{\mu_p}$$

Substituting these relations into Eq. (1.41) and performing the integration, we obtain

$$\frac{v}{d} = \frac{g_c}{2\mu_p \tau_w^3}\left[\int_0^{\tau_y} \tau^2 \,(0)\, \mathrm{d}\tau + \int_{\tau_y}^{\tau_w} \tau^2\,(\tau - \tau_y)\,\mathrm{d}\tau\right]$$

or

$$\frac{8v}{g_c d} = \frac{\tau_w}{\mu_p}\left[1 - \frac{4}{3}\left(\frac{\tau_y}{\tau_w}\right) + \frac{1}{3}\left(\frac{\tau_y}{\tau_w}\right)^4\right] \tag{1.42}$$

When $\tau_w = \tau_y$, the term in brackets becomes zero which implies that the shear stress imposed on the fluid in contact with the wall, $\tau_w = d\Delta p_f/4L$, must always exceed the yield point of the fluid in order to maintain flow. From the shear stress distribution of Eq. (1.33) and Fig. 1.8, however, we see that even when the wall stress is sufficient to maintain steady flow ($\tau_w > \tau_y$), there will always remain a central core of diameter $d(\tau_y/\tau_w)$ within which the stress is everywhere less than τ_y. The fluid within this central core will move as a unit or "plug" and only the fluid outside this core will slip or flow in laminar fashion.

Since the quantity $(\tau_y/\tau_w)^4$ is usually relatively small (except when small shear stresses or pressure gradients are imposed upon the fluid), the term $(\frac{1}{3})(\tau_y/\tau_w)^4$ can be dropped from Eq. (1.42) to obtain the approximate expression

$$\frac{8v}{g_c d} = \frac{1}{\mu_p}\,(\tau_w - \tfrac{4}{3}\tau_y) \tag{1.43}$$

which is linear in τ_w and has as its intercept at $v = 0$ the value $\tau_w = \frac{4}{3}\tau_y$. The solid curve of Fig. 1.9 corresponds to a plot of Eq. (1.42) while the asymptote (dashed line) is a plot of the approximate expression Eq. (1.43). Whenever (τ_y/τ_w) is less than 0.4, or when τ_w exceeds $2.5\tau_y$, the approximate equation may be used in place of Eq. (1.42), with introduction of less than 2 per cent error. To insure the validity of Eq. (1.43), the average velocity of the flow must always *exceed* a certain lower critical value v_{lc} given by

$$v_{lc} = \frac{g_c d}{8\mu_p}\,[2.50\tau_y - 1.33\tau_y]$$

or

$$v_{lc} = 0.146\,\frac{g_c d \tau_y}{\mu_p} \tag{1.44}$$

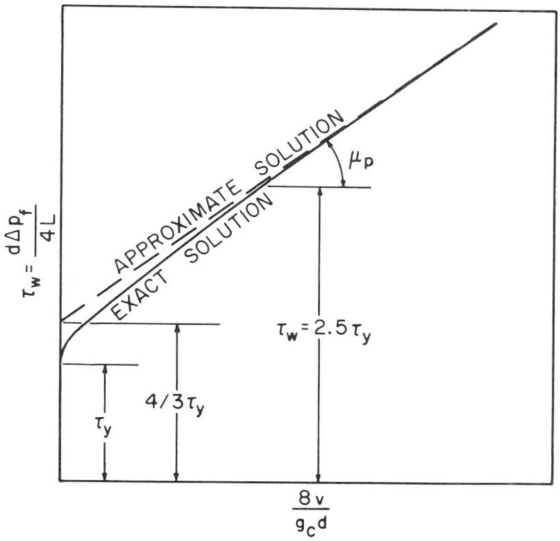

Fig. 1.9

When modified to accept practical engineering units (Table 1.2), this equation of condition reduces to

$$v_{lc} = 5.82 \frac{\tau_y d}{\mu_p} \tag{1.44a}$$

Solving Eq. (1.43) for τ_w and introducing its equivalent in terms of Δp_f, we obtain an expression for frictional pressure loss,

$$\Delta p_f = \frac{32\mu_p Lv}{g_c d^2} + \frac{16\tau_y L}{3d} \tag{1.45}$$

or an alternate form to be used with practical units,

$$\Delta p_f = \frac{\mu_p Lv}{1500d^2} + \frac{\tau_y L}{225d} \tag{1.45a}$$

For pipe flow of a Newtonian fluid, which has no yield point by definition, Eq. (1.45) will reduce to a form of the Hagen-Poiseuille law previously given in Eq. (1.17) and repeated here as a convenience,

$$\Delta p_f = \frac{32\mu Lv}{g_c d^2} \tag{1.17}$$

where μ_p has been replaced by the absolute viscosity of the Newtonian fluid.

Effective viscosity. The similarity between Eqs. (1.45) and (1.17) suggests the possibility of assigning an effective viscosity μ_e to a Bingham plastic material which when used with Eq. (1.17) would yield the same results as Eq. (1.45). Proper choice of μ_e would have to satisfy the relation

$$\frac{32\mu_e Lv}{g_c d^2} = \frac{32\mu_p Lv}{g_c d^2} + \frac{16\tau_y L}{3d}$$

which when solved for μ_e gives

$$\mu_e = \mu_p + \frac{g_c \tau_y d}{6v} \tag{1.46}$$

or an alternate result suitable for practical units,

$$\mu_e = \mu_p + 6.649\tau_y \frac{d}{v} \tag{1.46a}$$

Notice that no fixed value of μ_e can be assigned a given Bingham material, for Eq. (1.46) admits that μ_e is dependent upon the average velocity of the flow and also upon the diameter of the pipe. From a plot of Eq. (1.43), the dashed line in Fig. 1.9, it can be shown that μ_e is also the slope of a straight line which joins the origin with some point on the (approximate) solution curve that corresponds to a particular value of $8v/g_c d$ in question. This graphical determination again points out that μ_e is a function of not only μ_p and τ_y (the rheological properties of the plastic) but also the ratio v/d. Our present definition of effective viscosity must not be confused with apparent viscosity μ_a, defined by $g_c\tau/(-dv_r/dr)$. Although $d\Delta p_f/4L$ always represents the shear stress at the pipe wall, the quantity $8v/g_c d$ is the wall shear rate *only* for a simple Newtonian fluid.

For predicting frictional pressure losses, it would appear that Eq. (1.45) suggests a more direct solution than the calculation of μ_e and subsequent use of Eq. (1.17). However, the validity of either equation is subject to the condition that the flow be laminar, and, as we shall see, the onset of turbulence is governed by μ_e.

Turbulent flow. From pipe flow experiments using Bingham-type suspensions,[7,8] it has been determined that turbulent pipe flow usually occurs whenever the *equivalent* Reynolds number of the flow, N_{Re}, exceeds a value of 2000, where

$$N_{Re} = \frac{dv\rho}{\mu_e} \tag{1.47}$$

or in practical units,

$$N_{Re} = 928\frac{dv\rho}{\mu_e} \tag{1.47a}$$

and μ_e, the equivalent viscosity of the pipe-fluid system, is defined by Eqs. (1.46) and (1.46a), respectively. This same criterion is also compatible with the theoretical analysis to be developed in Sec. 1.11. For a given pipe size and Bingham suspension, the onset of turbulence may also be predicted by comparing the average velocity of the flow with a parameter called the critical velocity v_c defined by the expression

$$v_c = \frac{1.078\mu_p + 1.078\sqrt{\mu_p^2 + 12.34d^2\tau_y\rho}}{\rho d} \tag{1.48a}$$

which is already adapted for practical units. When the average velocity of the flow exceeds this critical velocity, turbulence is assured. Equation (1.48a) assumes that turbulence occurs at $N_{Re} = 2000$, and may be obtained by substituting Eq. (1.46a) for μ_e in Eq. (1.47a) and solving the resulting quadratic expression for v. Equation (1.48a) is also suited for Newtonian fluids with τ_y set equal to zero.

It is generally agreed that the Fanning equation together with the Newtonian friction factor curves of Fig. 1.4 can also be used to calculate friction losses for Bingham plastics in turbulent flow. A debatable question, however, is what value of turbulent viscosity should be used in the expression for N_R in order to obtain a correct value of f.[9] We will adopt the method proposed by Hedstrom,[10] as suggested in API RP 39,[11] wherein the plastic viscosity μ_p is used in place of μ in Eqs. (1.15) and (1.15a) to calculate N_R. Criticism of this premise should not be too serious since f in the turbulent flow region is not a very strong function of μ.

Calculation of frictional pressure losses may be summarized as follows:

1. Calculate N_{Re} or v_c from Eq. (1.47a) or (1.48a).
2. If $N_{Re} < 2000$ or $v < v_c$, flow is laminar and Eq. (1.45a) or Eq. (1.46a) in conjunction with Eq. (1.17a) applies.
3. If $N_{Re} > 2000$ or $v > v_c$, flow is turbulent, requiring:
   (a) Calculation of N_R using μ_p in place of μ in Eq. (1.15a);
   (b) Determination of f using the curves in Fig. 1.4;
   (c) Calculation of pressure loss using Eq. (1.18a), the Fanning equation.

1.9. Annular Flow of Bingham Plastics

Laminar flow. Exact analytical expressions for the isothermal, laminar flow of a Bingham model fluid within a concentric annulus of circular section have been developed by several authors.[12,13] Melrose and coworkers[14] derived more generalized solutions, using suitable dimensionless coordinates, and were able to show that the flow equations for an annular system could be approximated with good precision by those equations which apply to laminar flow through a narrow slot. By use of analytical techniques similar to those developed in the previous section, a general equation

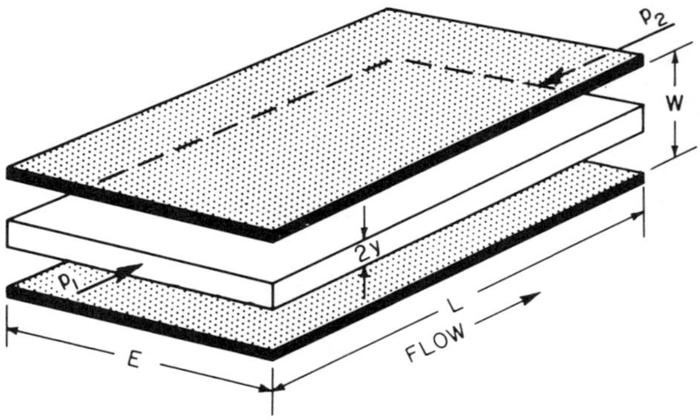

Fig. 1.10 Plane rectilinear flow between fixed parallel plates.

can be derived for laminar, slot flow. For the system shown in Fig. 1.10, where it is assumed that the lateral extent of the slot, E, is far greater than its width, W, the general flow equation is

$$q = \frac{kEW^2}{2\tau_w^2}\int_0^{\tau_w}\left[\int_\tau^{\tau_w}\phi(\tau)\,\mathrm{d}\tau\right]\mathrm{d}\tau \tag{1.49}$$

Upon integration by parts and substitution of $v(EW)$ for the volumetric flow rate q, the general equation simplifies to

$$\frac{v}{W} = \frac{k}{2\tau_w^2}\int_0^{\tau_w}\tau\phi(\tau)\,\mathrm{d}\tau \tag{1.50}$$

where τ_w, the shear stress at either (major) wall of the slot, is given by

$$\tau_w = \frac{W\Delta p_f}{2L} \tag{1.51}$$

Equation (1.50) is applicable to the slot flow of any type of fluid provided the functional dependence of shear rate upon shear stress, $k\phi(\tau)$, is known. For Newtonian fluids, Eq. (1.50) in combination with Eq. (1.51) will reduce to Eq. (1.24) which was introduced in Sec. 1.6. Substituting the Bingham plastic relation

$$k\phi(\tau) = \frac{g_c}{\mu_p}(\tau - \tau_y)$$

into Eq. (1.50) and performing the integration, we obtain

$$\frac{6v}{g_cW} = \frac{\tau_w}{\mu_p}\left[1 - \frac{3}{2}\left(\frac{\tau_y}{\tau_w}\right) + \frac{1}{2}\left(\frac{\tau_y}{\tau_w}\right)^3\right] \tag{1.52}$$

which is the counterpart of Eq. (1.42) for a circular pipe system.

When τ_w is larger than $2.5\tau_y$ (the usual case), the last term in Eq. (1.52) can be neglected without introducing an error greater than 7 per cent. Allowing this simplification, Eq. (1.52) together with Eq. (1.51) can be reduced to a simple expression for frictional pressure loss in a slot,

$$\Delta p_f = \frac{12\mu_p L v}{g_c W^2} + \frac{3\tau_y L}{W} \tag{1.53}$$

Approximating an annular section by a slot whose width $W = \frac{1}{2}(d_o - d_i)$, we obtain an expression in consistent units,

$$\Delta p_f = \frac{48\mu_p L v}{g_c(d_o - d_i)^2} + \frac{6\tau_y L}{(d_o - d_i)} \tag{1.54}$$

or an alternate form compatible with practical engineering units,

$$\Delta p_f = \frac{\mu_p L v}{1000(d_o - d_i)^2} + \frac{\tau_y L}{200(d_o - d_i)} \tag{1.54a}$$

These last two equations when applied to Newtonian fluids ($\tau_y = 0$) reduce to Eqs. (1.26) and (1.26a).

Turbulent flow. With little fundamental information available on annular turbulent flow, we can proceed only by making several assumptions tempered where possible with sound theoretical reasoning.

Using the hydraulic radius concept, a Reynolds number for the slot flow of a Newtonian fluid may be defined as

$$N_{Rs} = \frac{(2W)v\rho}{\mu} \tag{1.55}$$

where the equivalent diameter for a slot is given by

$$d_e = 4R = 4\left[\frac{EW}{2(E + W)}\right] = 2W \qquad \text{for } E \gg W$$

Substitution of $d_e = 2W$ in the Fanning equation for frictional pressure loss, Eq. (1.18), provides us with an expression

$$\Delta p_f = \frac{fL\rho v^2}{g_c W} \tag{1.56}$$

which can be used in place of Eq. (1.24) for Newtonian slot flow provided the friction factor–Reynolds number relationship in the laminar region is defined by

$$f = 24/N_{Rs} \tag{1.57}$$

In addition, it is generally agreed that turbulent slot flow can be expected whenever the Reynolds number as defined in Eq. (1.55) exceeds 2000.

By an extension of the analytical techniques to be introduced in Sec. 1.11, it can be shown that Eqs. (1.55) through (1.57) also apply to the

Table 1.2. SYMBOLS, DEFINITIONS, AND PRACTICAL ENGINEERING UNITS USED IN ALTERNATE EQUATIONS

Symbol	*Quantity*	*Units*
A	Cross-sectional area of bit nozzle	sq in.
C_d	Coefficient of discharge for bit nozzle	dimensionless
d	Inside diameter of pipe or tube	in.
d_i	Outside diameter of the inner pipe, annulus	in.
d_o	Inside diameter of the outer pipe, annulus	in.
f	Fanning friction factor	dimensionless
H_h	Hydraulic power	hp
K_a'	Equivalent consistency index, annular flow	lbf $sec^{n'}/ft^2$
K_p'	Consistency index, pipe flow	lbf $sec^{n'}/ft^2$
K_v'	Equivalent consistency index, rotational viscometer	lbf $sec^{n'}/ft^2$
L	Length of pipe or tube	ft
n'	Flow behavior index	dimensionless
N	Range extension factor for V-G meter dial readings	dimensionless
N_R	Reynolds number, Newtonian fluid	dimensionless
N_{Ra}	Equivalent Reynolds number, laminar annular flow of Bingham plastic	dimensionless
N_{Re}	Equivalent Reynolds number, laminar pipe flow of Bingham plastic	dimensionless
N_R'	Generalized Reynolds number, pipe flow	dimensionless
N_{Ra}'	Generalized Reynolds number, annular flow	dimensionless
Δp_f	Frictional pressure loss	psi
Δp_j	Pressure drop across bit nozzle	psi
Δp_P	Increase in pressure across a pump	psi
q	Volumetric flow rate	gpm
v	Average or bulk velocity	ft/sec
v_c	Critical velocity for onset of turbulence, pipe flow of Bingham plastic	ft/sec
v_{ca}	Critical velocity for onset of turbulence, annular flow of Bingham plastic	ft/sec
v_{lc}	Lower critical velocity for Bingham plastic flow	ft/sec
θ	V-G meter dial reading	degrees
μ	Absolute viscosity, Newtonian fluid	cp
μ_e	Effective viscosity of Bingham plastic, laminar pipe flow	cp
μ_{ea}	Effective viscosity of Bingham plastic, laminar annular flow	cp
μ_p	Plastic viscosity of Bingham plastic	cp
ρ	Density of fluid	lbm/gal, ppg
τ_b	Shear stress at surface of viscometer bob	lbf/ft^2
τ_w	Shear stress at wall(s) of pipe or annulus	lbf/ft^2
τ_y	Yield point of Bingham plastic	$lbf/100\ ft^2$
τ_{300}	Equals θ_{300}, approximate shear stress at bob of V-G meter	$lbf/100\ ft^2$
τ_{600}	Equals θ_{600}, approximate shear stress at bob of V-G meter	$lbf/100\ ft^2$

Table 1.3. FLOW EQUATIONS FOR BINGHAM PLASTIC AND NEWTONIAN FLUIDS USING PRACTICAL ENGINEERING UNITS OF TABLE 1.2

Fluid	Pipe flow	Annular flow
	Average velocity v	
All fluids	Use annulus equations with $d_i = 0$	$v = \dfrac{17.16(q, \text{bbl/min})}{(d_o^2 - d_i^2)}$ $v = \dfrac{3.056(q, \text{ft}^3/\text{min})}{(d_o^2 - d_i^2)}$ $v = \dfrac{(q, \text{gpm})}{2.448(d_o^2 - d_i^2)}$
	Critical velocity v_c	
Bingham plastic	$v_c = \dfrac{1.078\mu_p + 1.078\sqrt{\mu_p^2 + 12.34d^2\tau_y\rho}}{\rho d}$	$v_{ca} = \dfrac{1.078\mu_p + 1.078\sqrt{\mu_p^2 + 9.256(d_o - d_i)^2\tau_y\rho}}{\rho(d_o - d_i)}$
Newtonian	Set $\tau_y = 0$ and $\mu_p = \mu$	Set $\tau_y = 0$ and $\mu_p = \mu$
	Laminar flow, $v < v_c$	
Bingham plastic	$\Delta p_f = \dfrac{\mu_p L v}{1500d^2} + \dfrac{\tau_y L}{225d}$	$\Delta p_f = \dfrac{\mu_p L v}{1000(d_o - d_i)^2} + \dfrac{\tau_y L}{200(d_o - d_i)}$
Newtonian	Set $\tau_y = 0$ and $\mu_p = \mu$	Set $\tau_y = 0$ and $\mu_p = \mu$
	Turbulent flow, $v > v_c$	
Bingham plastic	$\Delta p_f = \dfrac{fL\rho v^2}{25.80d}$ $N_R = 928\dfrac{dv\rho}{\mu_p}$ $f = \phi(N_R, \epsilon/d)$, see Fig. 1.4	$\Delta p_f = \dfrac{fL\rho v^2}{25.80(d_o - d_i)}$ $N_{Ra} = 928\dfrac{(d_o - d_i)v\rho}{\mu_p}$ $f = \phi(N_{Ra}, \epsilon/d)$, see Fig. 1.4
Newtonian	Set $\mu_p = \mu$	Set $\mu_p = \mu$ and $(d_o - d_i) = d_e$ and calculate a new value of v using d_e (obtain d_e from Fig. 1.6) Use an E factor of 0.90

laminar slot flow of a Bingham material with the stipulation that the Newtonian viscosity in Eq. (1.55) be replaced by an effective viscosity for slot flow, μ_{es}, given by

$$\mu_{es} = \mu_p + \frac{g_c \tau_y W}{4v} \tag{1.58}$$

Proceeding with the premise that a narrow annulus can be represented by the geometry of a thin slot, we adapt Eqs. (1.55) through (1.58) for annular flow of a Bingham material using the equivalence $W = \frac{1}{2}(d_o - d_i)$. In other words, the Fanning equation in the form

$$\Delta p_f = \frac{2fL\rho v^2}{g_c(d_o - d_i)} \tag{1.59}$$

or

$$\Delta p_f = \frac{fL\rho v^2}{25.80(d_o - d_i)} \tag{1.59a}$$

can be used in place of our previous expressions for laminar annular flow, Eqs. (1.54) and (1.54a), provided we define our friction factor for laminar flow by the relation $f = 24/N_{Ra}$ where

$$N_{Ra} = \frac{(d_o - d_i)v\rho}{\mu_{ea}} \tag{1.60}$$

or

$$N_{Ra} = 928\,\frac{(d_o - d_i)v\rho}{\mu_{ea}} \tag{1.60a}$$

and the effective viscosity of annular flow is obtained from

$$\mu_{ea} = \mu_p + \frac{g_c \tau_y (d_o - d_i)}{8v} \tag{1.61}$$

or

$$\mu_{ea} = \mu_p + 4.987\,\frac{\tau_y (d_o - d_i)}{v} \tag{1.61a}$$

The onset of turbulence is expected when N_{Ra} exceeds 2000, and, as a consequence, a limiting or critical velocity for laminar annular flow is given by

$$v_{ca} = \frac{1.078\mu_p + 1.078\sqrt{\mu_p^2 + 9.256(d_o - d_i)^2 \tau_y \rho}}{\rho(d_o - d_i)} \tag{1.62a}$$

which is the counterpart of Eq. (1.48a) for laminar pipe flow and is likewise presented in practical engineering units.

In the turbulent region, for want of reliable information, we must further assume that the friction factor curves of Fig. 1.4 together with Eqs. (1.59) and (1.59a) will provide satisfactory approximate values for annular friction losses. Furthermore, in fully developed turbulent flow, the Reynolds number N_{Ra} for a Bingham material is calculated using the plastic viscosity μ_p in place of μ_{ea} in Eqs. (1.60) and (1.60a), again following the suggestion of Hedstrom.[10]

Coupled with the uncertainties involved in our present treatment of annular flows, there are several other factors which complicate the calculation of frictional losses in the annulus of a wellbore. The probable effects of (1) drill pipe rotation, (2) eccentricity (the off-axis location of the drill

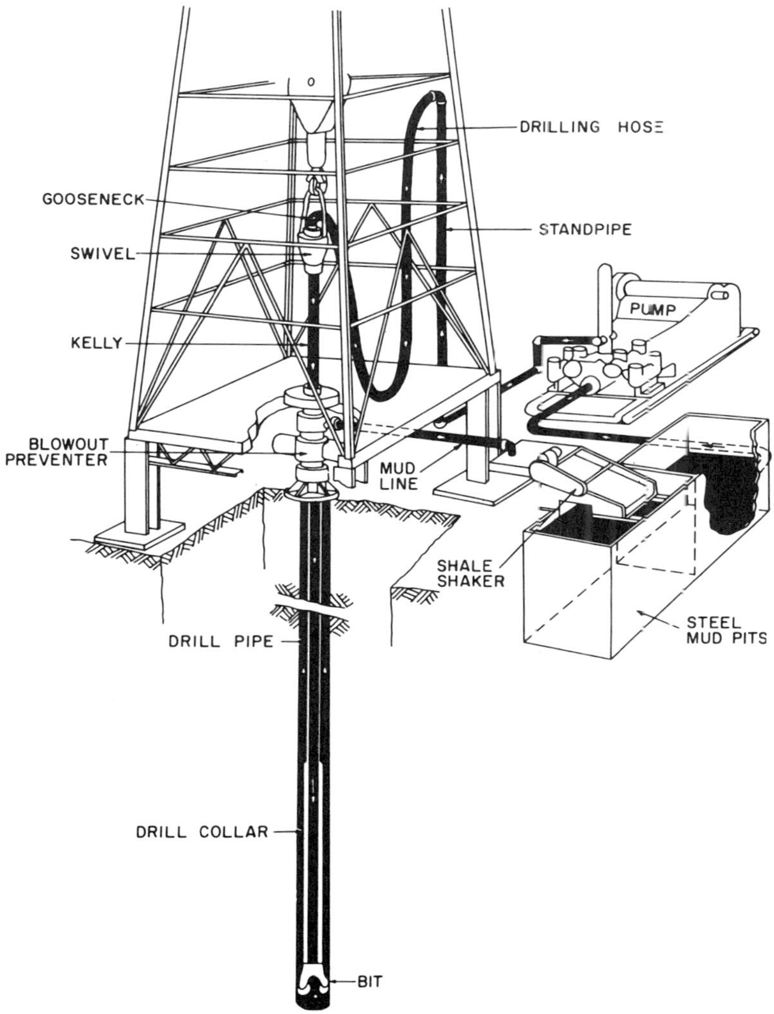

Fig. 1.11 Circulatory system for a rotary drilling rig. (After Bobo, Hoch, Boudreaux, and Angel,[15] courtesy of Gulf Publishing Company.)

pipe with respect to the borehole), and (3) downhole changes in rheological properties of the drilling fluid will tend to reduce the reliability of our calculations to the extent that they will provide only reasonable estimates of annular friction losses.

As a convenience, the equations developed for pipe and annular flow are summarized in Table 1.3 in a form which emphasizes their remarkable similarities.

1.10. Rotary Drilling Hydraulics

In some areas, compressed air or natural gas and also plain water are being used as rotary drilling fluids, but water-base muds containing collodial clay solids enjoy a more universal popularity. Since these clay-water muds approximate the behavior of a Bingham plastic material,[7,8] analysis of the circulatory system of a rotary drilling rig (Fig. 1.11) will afford us ample opportunity to employ the techniques developed in Sections 1.8 and 1.9.

Principal sources of pressure (or energy) loss within a circulatory system are as follows:

1. Frictional losses inside the drill pipe and drill collars.
2. Pressure drop across the nozzles or watercourses of the drill bit.
3. Frictional losses in the annular clearance between the borehole and the drill collars or the drill pipe.
4. Frictional losses in surface connections which include flow lines, standpipe, hose, swivel, and kelly.

Summation of these losses will determine the discharge pressure of the mud pump and, coupled with the volumetric flow rate required of the system, they will also govern the necessary power input to the pump.

Frictional losses within the drill string and annulus. To illustrate the rather involved calculations associated with a complete analysis of a circulatory system, let us consider the following representative system:

Well Data

Drilling depth	10,000 ft
Diameter of hole	$9\frac{7}{8}$ in. (Bit size)
Drill pipe	5 in. OD, 19.50 lbw/ft, 4.276 in. ID
Tool joints	$6\frac{1}{4}$ in. OD by $3\frac{3}{4}$ in. ID, X-hole
Drill collars	400 ft, $7\frac{3}{4}$ in. OD by 3 in. ID
Annular velocity	135 ft/min
Mud weight	12.0 ppg
Plastic viscosity	43 cp
Yield point	20 lbf/100 sq ft

1. *Circulation rate.* In the annular space between the drill pipe and borehole, the return velocity of the mud is maintained at a rate sufficient

to entrain the bit cuttings and carry them to the surface. Depending upon drilling rate, return velocities in the range of 100 to 200 ft/min are usually adequate. For an annular velocity of 135 ft/min, the required circulation rate for our present system is calculated from Eq. (1.21a).

$$\frac{135}{60} = \frac{q \text{ (gpm)}}{2.448[(9\tfrac{7}{8})^2 - (5)^2]}$$

$$q = 400 \text{ gpm}$$

2. *Critical velocities.* While the return flow through the annulus is usually laminar, flow through the drill string is ordinarily turbulent. Until we become completely familiar with a given well system, however, the critical velocities in all sections should be investigated. The critical flow velocity inside the drill pipe is calculated using Eq. (1.48a):

$$v_c = \frac{1.078(43) + 1.078\sqrt{(43)^2 + 12.34(4.276)^2 20(12)}}{12(4.276)}$$

$$v_c \text{ (pipe)} = 5.87 \text{ ft/sec}$$

Similarly, the critical velocity through the bore of the drill collars is

$$v_c \text{ (collars)} = 6.34 \text{ ft/sec}$$

In the annular space around the drill collars, critical velocity is calculated from Eq. (1.62a):

$$v_{ca} = \frac{1.078(43) + 1.078\sqrt{(43)^2 + 9.256(9\tfrac{7}{8} - 7\tfrac{3}{4})^2 20(12)}}{12(9\tfrac{7}{8} - 7\tfrac{3}{4})}$$

$$v_{ca}\begin{pmatrix}\text{around}\\ \text{collars}\end{pmatrix} = 6.42 \text{ ft/sec}$$

For the larger clearance around the drill pipe,

$$v_{ca}\begin{pmatrix}\text{around}\\ \text{pipe}\end{pmatrix} = 5.10 \text{ ft/sec}$$

3. *Actual velocities.* With a circulation rate of 400 gpm, average flow velocities through the drill string and annulus are obtained using Eq. (1.21a). The average flow velocity down the drill pipe is then

$$v \text{ (pipe)} = \frac{400}{2.448(4.276)^2} = 8.94 \text{ ft/sec}$$

Since 8.94 > 5.87 (the critical velocity), the flow is turbulent. In like manner it can be shown that

$$v \text{ (collars)} = 18.15 \text{ ft/sec} > 6.34, \text{ Turbulent}$$

$$v\begin{pmatrix}\text{around}\\ \text{collars}\end{pmatrix} = 4.36 \text{ ft/sec} < 6.42, \text{ Laminar}$$

$$v\begin{pmatrix}\text{around}\\ \text{pipe}\end{pmatrix} = 2.25 \text{ ft/sec} < 5.10, \text{ Laminar}$$

4. *Frictional pressure losses.* For the turbulent flow through the drill pipe, Eq. (1.15a) applies with μ_p substituted for μ.

$$N_R = 928 \frac{4.276(8.94)12}{43} = 9900$$

From Fig. 1.4, $f = 0.0077$

Pressure loss through 9600 ft of drill pipe (10,000 ft − 400 ft of drill collars) is calculated using Eq. (1.18a):

$$\Delta p_f \text{ (pipe)} = \frac{0.0077(9600)12(8.94)^2}{25.80(4.276)} = 642.6 \text{ psi}$$

Considering the flow through 400 ft of drill collars,

$$N_R = 928 \frac{3(18.15)12}{43} = 14{,}100$$

which may also be obtained by an alternate procedure once the Reynolds number for the pipe flow has been calculated:

$$N_R = 9900 \times \frac{4.276}{3} = 14{,}100$$

Again from Fig. 1.4, $f = 0.0072$

$$\Delta p_f \text{ (collars)} = \frac{0.0072(400)12(18.15)^2}{25.80(3)} = 147.1 \text{ psi}$$

Annular losses are calculated using Eq. (1.54a):

$$\Delta p_f \begin{pmatrix}\text{around}\\ \text{collars}\end{pmatrix} = \frac{43(400)4.36}{1000(9.875 - 7.750)^2} + \frac{20(400)}{200(9.875 - 7.750)}$$

$$= 35.4 \text{ psi}$$

$$\Delta p_f \begin{pmatrix}\text{around}\\ \text{pipe}\end{pmatrix} = \frac{43(9600)2.25}{1000(9.875 - 5)^2} + \frac{20(9600)}{200(9.875 - 5)}$$

$$= 236.0 \text{ psi}$$

The total pressure loss in the system exclusive of the surface connections and the bit nozzles is then

$$\Delta p_f = 642.6 + 147.1 + 35.4 + 236.0 = 1061 \text{ psi}$$

Frictional losses in surface connections. This factor is a variable since individual rigs are seldom outfitted with identical surface equipment and piping. While seldom attempted, measurements of surface losses could be made directly by running the pump with only the kelly in the hole and observing the mud gauge readings at various circulation rates. Instead, the usual practice is to estimate these losses considering only the standpipe, drilling hose, swivel washpipe, and kelly and making due allowance for the bend at the top of the standpipe as well as the swivel gooseneck. Four typical combinations of surface connections are given in Table 1.4 together with their equivalent lengths in terms of feet of drill pipe. Using values from this table, surface losses can be estimated with an accuracy sufficient for most field applications.

Table 1.4. TURBULENT FLOW RESISTANCE OF SURFACE CONNECTIONS

Components of surface connections		Typical combinations							
		No. 1		No. 2		No. 3		No. 4	
		ID, in.	L, ft	ID, in.	L, ft	ID, in.	L, ft	ID, in.	L, ft
Standpipe		3	40	$3\frac{1}{2}$	40	4	45	4	45
Drilling hose		2	45	$2\frac{1}{2}$	55	3	55	3	55
Swivel washpipe and gooseneck		2	4	$2\frac{1}{2}$	5	$2\frac{1}{2}$	5	3	6
Kelly		$2\frac{1}{4}$	40	$3\frac{1}{4}$	40	$3\frac{1}{4}$	40	4	40
Drill pipe		Equivalent length of surface connections in feet of drill pipe							
OD, in.	Weight, lbw/ft								
$3\frac{1}{2}$	13.3	437		161					
$4\frac{1}{2}$	16.6			761		479		340	
5	19.5					816		579	

Assuming combination No. 4 in the table is representative of our present example, surface connection losses could have been included in our previous calculation of drill pipe loss by adding an additional 579 ft to the pipe length of 9600 ft. Otherwise,

$$\Delta p_f \text{ (surface)} = \frac{0.0077(579)12(8.94)^2}{25.8(4.276)} = 38.8 \text{ psi}$$

or

$$\frac{579}{9600} \times 642.6 = 38.8 \text{ psi}$$

The total pressure loss in the system is now 1100 psi exclusive of the pressure drop across the bit.

Pressure drop across bit nozzles. For the ideal or frictionless flow of an incompressible fluid through a short constriction, such as a nozzle or orifice, the pressure balance relation of Eq. (1.10) reduces to

$$p_1 + \frac{\rho}{2g_c}(v_1^2 - v_2^2) = p_2$$

where subscript 1 refers to steady-state conditions upstream from the constriction and subscript 2 to those conditions which prevail at the exit.

When the exit area A_2 is much smaller than the entrance area A_1, $v_2^2 \gg v_1^2$; therefore,

$$v_2 \cong \sqrt{\frac{2g_c}{\rho}(p_1 - p_2)}$$

However, for an observed pressure drop of $(p_1 - p_2)$, the exit velocity predicted by this last expression is never realized since we have conveniently overlooked an unavoidable loss in mechanical energy caused by friction, the term (ρF) in Eq. (1.10). To offset this difference, a common practice in calibrating an orifice nozzle or plate is to introduce a correction factor C_d, known as a coefficient of discharge, so that the modified expression

$$v_2 = C_d\sqrt{\frac{2g_c}{\rho}(p_1 - p_2)} \tag{1.63}$$

will result in a correct value for exit velocity. With the exit velocity v_2 known, Eq. (1.63) may be rearranged to predict the drop in pressure,

$$(p_1 - p_2) = \Delta p_j = \frac{\rho v_2^2}{2g_c C_d^2}$$

where Δp_j now represents the drop in pressure across the constriction. Replacing v_2 by the quantity q/A_2, we obtain

$$\Delta p_j = \frac{\rho q^2}{2g_c C_d^2 A^2} \tag{1.64}$$

or an alternate form to be used with our practical engineering units,

$$\Delta p_j = \frac{\rho q^2}{12032 C_d^2 A^2} \tag{1.64a}$$

where it is understood that A now represents the cross-sectional area of the exit.

Eckel and Bielstein[16] have shown that C_d for high-velocity jet bit nozzles may be as high as 0.98, but a value of 0.95 is a more practical limit. Although their tests were conducted using water, a subsequent study by Havenar[9] indicated that the C_d value of a nozzle is the same for mud or water flow, and is only influenced by the geometry of the nozzle. Regular watercourses which do not feature hydraulic streamlining are usually assigned a value of $C_d = 0.80$.[17]

A rock bit usually has the same number of jet nozzles as cones; however, this will not complicate the calculation of pressure drop across the bit, as the sum of the individual nozzle areas may be substituted for A in Eqs. (1.64) and (1.64a). As a convenience, the exit areas for representative sizes of jet bit nozzles are given in Table 1.5.

Table 1.5. Areas of jet nozzles

Diameter, in. fraction	Diameter, in. decimal	Area, sq in.	Diameter, in. fraction	Diameter, in. decimal	Area, sq in.
$\frac{1}{4}$	0.2500	0.0491	$\frac{15}{32}$	0.4687	0.1726
$\frac{9}{32}$	0.2812	0.0621	$\frac{1}{2}$	0.5000	0.1963
$\frac{5}{16}$	0.3125	0.0767	$\frac{9}{16}$	0.5625	0.2485
$\frac{11}{32}$	0.3437	0.0928	$\frac{5}{8}$	0.6250	0.3068
$\frac{3}{8}$	0.3750	0.1104	$\frac{11}{16}$	0.6875	0.3712
$\frac{13}{32}$	0.4062	0.1296	$\frac{3}{4}$	0.7500	0.4418
$\frac{7}{16}$	0.4375	0.1503	1	1.0000	0.7854

As the drilling mud is forced to pass through the nozzles of the bit, nearly all of its loss in pressure energy (per unit mass) is converted into kinetic energy as shown by Eq. (1.63) rewritten in the form

$$C_d^2\left(\frac{p_1 - p_2}{\rho}\right) = \frac{v_2^2}{2g_c}$$

For nozzles of proper design, this process of mechanical energy exchange is better than 90 per cent efficient, since

$$C_d^2 \geq (0.95)^2 = 0.9025$$

In other words, while the mud is in transit through the nozzles, less than 10 per cent of its loss in pressure energy ($\Delta p_j/\rho$) is dissipated in the form of heat energy. Once the mud is ejected from the nozzles, however, its recent increase in kinetic energy is rapidly dissipated as the mud stream impinges on the bottom of the hole and sweeps the cuttings from beneath the bit. As a consequence, all the pressure energy which was lost in transit through the bit nozzles is ultimately converted into heat energy. This permanent loss of mechanical energy from the mud system results in a *permanent* loss in pressure, Δp_j.

Continuing with our previous example, if a 3-cone bit equipped with $\frac{3}{8}$-in. jets were being used,

$$A = 3(0.1104) = 0.3312 \text{ sq in.}$$

and

$$\Delta p_j = \frac{12(400)^2}{12032(0.95)^2(0.3312)^2} = 1612 \text{ psi}$$

As the mud issues from the nozzles, its exit or "jet" velocity can be calculated from Eq. (1.21a), remembering that in this present circumstance only one third of the total circulation is handled by each nozzle:

$$v_j = \frac{(400/3)}{2.448(0.375)^2} = 387 \text{ ft/sec}$$

When the nozzles are of different size, the *fraction* of total circulation handled by a given nozzle of exit area A_i is given by A_i/A_t, where A_t is the total nozzle area of the bit.

Power requirements of the mud pump. Hydrostatic pressures (or changes in elevation) are of no consequence in a calculation of the discharge pressure of the pump. Applying a pressure balance to a complete traverse around the circulatory system, starting for example at the pump discharge and returning to this same station, our balance equation

$$p_1 + \Delta p_s + \Delta p_v + \Delta p_P - \Delta p_f = p_2 \tag{1.11}$$

reduces to

$$\Delta p_P = \Delta p_f$$

because the upsteam section A_1 is coincident with the downsteam section A_2 for a complete traverse. Consequently, the increase in pressure across the pump, Δp_P, is merely the summation of all frictional pressure losses around the system. For our example problem,

$$\Delta p_P = \Sigma \, \Delta p_f = 2712 \text{ psi}$$

This figure will approximate the gauge pressure at the pump discharge since the suction is usually at or very near atmospheric pressure.

To meet the demands of frictional losses, the mud pump must provide the circulating fluid with a continuous supply of mechanical energy in the amount of $W_o = \Delta p_P/\rho$ per unit mass of fluid. For a circulation rate q, the pump handles fluid at a mass rate of $w = \rho q$. Therefore the rate of energy output from the pump, or the power input to the circulating fluid, is given by

$$\text{Power} = wW_o = q\Delta p_P \tag{1.65}$$

or in practical units

$$H_h = \frac{q\Delta p_P}{1714} \text{ (hp)} \tag{1.65a}$$

where H_h, usually expressed in units of horsepower, is referred to as the hydraulic power output of the pump or, conversely, the hydraulic power input to the circulating fluid.

At a circulation rate of 400 gpm and a discharge pressure of 2712 psig, the hydraulic horsepower developed by the pump would be

$$H_h = \frac{400(2712)}{1714} = 633 \text{ hp}$$

Of course, the power input to the pump will necessarily be larger, for some power is lost to mechanical friction in the driving end or gear box. In calculating the required *input* horsepower to a pump, most manufacturers recommend using a mechanical efficiency of 85 per cent. Thus, for our pump to deliver 633 hydraulic horsepower, it would require a power input of 633/0.85 or 745 hp.

Having calculated the hydraulic power input to our circulatory system, it will be of interest to determine where this power is expended. In sweeping the bottom-hole surface free of cuttings, the power consumed by the jetting action of the mud can be calculated from Eq. (1.65a) by replacing Δp_P with Δp_j, the pressure drop across the bit nozzles:

$$\frac{400(1612)}{1714} = 376 \text{ hp}$$

This so-called jet bit horsepower represents a useful expenditure of energy as it contributes to the bit's performance. Power consumed by frictional losses within the drill string, return annulus, and surface connections represents wasted or parasitic energy. Again using Eq. (1.65a) and our previously calculated value of 1100 psi frictional pressure loss, this wasted power accounts for

$$\frac{400(1100)}{1714} = 257 \text{ hp}$$

and when added to the jet bit horsepower totals 633 hp, the hydraulic power output of the pump.

1.11. *Generalized Equations of Flow for Non-Newtonian Fluids*

From an academic point of view, no real fluid can be expected to exhibit precisely the rheological properties assigned to a hypothetical "model" fluid. As a practical consideration, however, there will be many situations where a model fluid can be found to approximate the behavior of an actual fluid with an accuracy commensurate with the reproducibility of measured field data.

Limitations of model-fluid approximations. Consider, for example, the viscometric data presented in Fig. 1.12 which were obtained from measurements on a typical clay-water drilling mud using a multispeed, rotational-type viscometer (Fann V-G Meter, Model 35). For shear rates in the range of approximately 479 to 957 reciprocal seconds (300 to 600 rpm), shear stress increases almost linearly as shown by the dashed line passing through the points B and C. The model fluid whose flow curve satisfies this straight-line approximation would be a Bingham plastic with $\mu_p = 27$ cp and $\tau_y = 44$ lbf/100 sq ft. Similarly, in the range of shear rates between points A and B, a Bingham plastic would again be a satisfactory approximation provided $\mu_p = 42$ cp and $\tau_y = 29$ lbf/100 sq ft. Finally, over the extended range between points A and C, a power-model fluid with

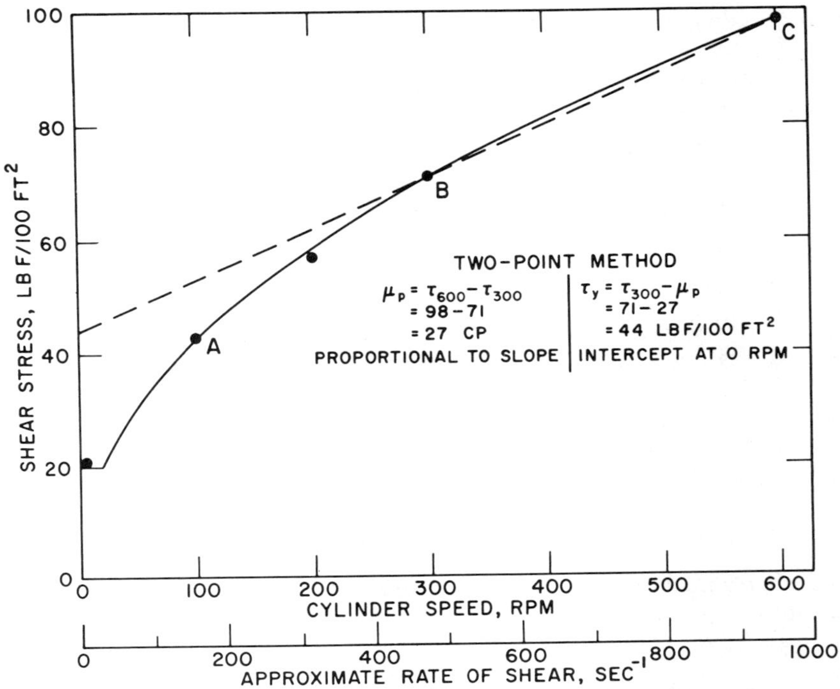

Fig. 1.12 Approximate flow curves for a clay-water drilling mud.

$n = 0.464$ and $K = 0.0404$ would produce the solid curve shown in the figure. Proper choice of model fluid to represent this particular mud would necessarily be governed by the range of shear rates the mud could be expected to experience in a particular flow system as well as by the accuracy required of the approximating analysis.

For the usual problems attending the hydraulics of rotary drilling muds, a Bingham model analysis has proven to be entirely adequate. To complement this method of analysis, the instrument constants of the Fann V-G Meter were purposely adjusted so that only two instrument readings (at 300 and 600 rpm) are required to determine the equivalent plastic flow properties of a mud. This is known as the "two-point" method,[18] by which the plastic viscosity in centipoises is obtained by subtracting the shear stress reading at 300 rpm (τ_{300}) from the shear stress reading at 600 rpm (τ_{600}),

$$\mu_p = (\tau_{600} - \tau_{300}) \text{ cp} \tag{1.66a}$$

Then, subtracting the plastic viscosity from the 300 rpm reading provides the yield point,

$$\tau_y = (\tau_{300} - \mu_p) \text{ lbf/100 ft}^2 \tag{1.67a}$$

In using the plastic flow parameters obtained by this two-point method, it is implied that the shear stress imposed on the mud at viscometer speeds of 300 and 600 rpm (i.e., τ_{300} and τ_{600}) will include the probable range of wall shear stress τ_w the mud will experience while traversing the pipes of the circulatory system. For a fluid flowing in a pipe, the shear stress acting on the fluid varies linearly with radial location from zero along the axis to a maximum value given by $(d\Delta p_f/4L)$ at the pipe wall. But in turbulent pipe flow (Fig. 1.3), average flow velocity is dependent primarily on the shear rates and resulting velocity gradients in very close proximity to the wall, and is essentially independent of the shear rates nearer to the center of the pipe. Consequently, it is the shear stress–shear rate behavior of the fluid in the immediate vicinity of the wall which establishes the bulk or average velocity of the flow. Ideally then, these same conditions of (wall) shear stress are those which should be duplicated within the viscometer.

Returning to the well system studied in Sec. 1.10, the plastic flow properties of the mud, viz., $\mu_p = 43$ cp and $\tau_y = 20$ lbf/100 sq ft, were probably determined by the two-point method of reduction of viscometric data obtained from a Fann V-G Meter. Using Eq. (1.67a) in reverse,

$$\tau_{300} = (\tau_y + \mu_p) = (20 + 43) = 63 \text{ lbf/100 ft}^2$$

and from Eq. (1.66a),

$$\tau_{600} = (\mu_p + \tau_{300}) = (43 + 63) = 106 \text{ lbf/100 ft}^2$$

Utilization of these two-point data for pipe flow calculations should be restricted to situations where the wall shear stress is likewise in the range of 63 to 106 lbf/100 sq ft. Recall that our calculated value of frictional pressure loss through 9600 ft of 4.276-in. ID drill pipe was 642.6 psi. The wall shear stress experienced by the mud is therefore

$$\tau_w = \frac{d\Delta p_f}{4L} = \frac{(4.276/12)(642.6 \times 144)}{4(9600)}$$

$$= 0.859 \text{ lbf/ft}^2 \text{ or } 85.9 \text{ lbf/100 ft}^2$$

a value about midway between the conditions of shear stress produced in the viscometer. Pressure loss in the annular space around the drill pipe was calculated as 236.0 psi, but the wall shear stress as determined by our slot-flow approximation of an annulus would be

$$\tau_w = \frac{(d_o - d_i)}{2}\left(\frac{\Delta p_f}{2L}\right) = \frac{(9.875 - 5)}{2 \times 12}\left(\frac{236 \times 144}{2 \times 9600}\right)$$

$$= 0.359 \text{ lbf/ft}^2 \text{ or } 35.9 \text{ lbf/100 ft}^2$$

Consequently, our calculation of annular pressure loss includes a preconceived notion of mud flow behavior which is unsupported by the viscometric data obtained at instrument speeds of 300 and 600 rpm. A more refined calculation of annular losses is seldom warranted, however, since these losses normally represent no more than 10 per cent of the total frictional loss in a circulatory system. Nevertheless, we should always be aware of the limitations inherent in the classification of a fluid into a certain category of model behavior, for a particular classification and even the numerical values assigned its rheological parameters can be extremely dependent upon the flow conditions under which viscometric measurements are made.

With the increasing use of salt-saturated and "low solids" drilling muds; oil emulsion, inverted emulsion, and oil-base drilling fluids; as well as the blocking agents, aqueous gels, and gelled crude oils currently used in hydraulic fracturing operations; the power-model fluid is gaining recognition as a more versatile and representative approximation for these special drilling and completion fluids. In lieu of developing specific equations for the pipe flow of a power-model fluid, we will investigate a more general and theoretically rigorous method of analysis of the flow properties of all non-Newtonian fluids which includes, as mere special cases, the particular behavior previously associated with Newtonian, Bingham plastic, and power-model fluids.

Absolute, capillary viscometry. Given the flow curve for a non-Newtonian fluid, it is always possible, at least in principle, to predict the laminar flow properties of such a fluid in conduits of simple cross section. Equations (1.41) and (1.50) for laminar pipe and narrow slot flow, respectively, are typical examples, since these equations are applicable to any type of fluid whose shear stress – shear rate relationship is known. Conversely, the flow curve for a fluid can be rigorously and easily derived from pressure drop and flow rate data obtained with a capillary-tube or pipe viscometer (a narrow slot device being not so readily available).

Following the original development of Rabinowitsch[19] and subsequent reiteration by Mooney,[20] our general expression for laminar pipe flow,

$$\left(\frac{2v}{kd}\right)\tau_w^3 = \int_0^{\tau_w} \tau^2\,\phi(\tau)\,d\tau \tag{1.41}$$

when differentiated with respect to τ_w yields

$$\left(\frac{2v}{kd}\right)3\tau_w^2 + \tau_w^3\,\frac{d(2v/kd)}{d\tau_w} = \tau_w^2\,\phi(\tau_w)$$

which can be further simplified by multiplying all terms by (k/τ_w^2) to obtain

$$3\left(\frac{2v}{d}\right) + \tau_w\,\frac{d(2v/d)}{d\tau_w} = k\phi(\tau_w)$$

In the derivation of Eq. (1.41), shear rate, dv_r/dr, was assumed to be some unspecified function of shear stress represented by $k\phi(\tau)$. This same function again appears in the equation above, only now it is evaluated at the particular value of shear stress developed at the pipe wall, τ_w, and consequently represents the rate of shear at the wall, denoted by $(-dv_r/dr)_w$.

Introducing this equivalent expression for $k\phi(\tau_w)$ and replacing τ_w by $(d\Delta p_f/4L)$, the preceding equation can be rearranged as

$$\left(-\frac{dv_r}{dr}\right)_w = \frac{3}{4}\left(\frac{8v}{d}\right) + \frac{1}{4}\left(\frac{d\Delta p_f}{4L}\right)\frac{d(8v/d)}{d(d\Delta p_f/4L)} \tag{1.68}$$

an entirely general expression for the evaluation of rate of shear at the pipe wall caused by an applied shear stress τ_w.

In order to simplify the calculation of the derivative ratio in Eq. (1.68), Metzner and Reed[21] introduced an additional modification. Upon rearrangement of the last term in Eq. (1.68),

$$\left(-\frac{dv_r}{dr}\right)_w = \frac{3}{4}\left(\frac{8v}{d}\right) + \frac{1}{4}\left(\frac{8v}{d}\right)\frac{[d(8v/d)]/(8v/d)}{[d(d\Delta p_f/4L)]/(d\Delta p_f/4L)}$$

which is no great improvement until it is recalled that $d(\ln x) = dx/x$. Recognizing this type of expression in the preceding equation, we may write

$$\left(-\frac{dv_r}{dr}\right)_w = \frac{3}{4}\left(\frac{8v}{d}\right) + \frac{1}{4}\left(\frac{8v}{d}\right)\frac{d\ln(8v/d)}{d\ln(d\Delta p_f/4L)} \tag{1.69}$$

Representing the derivative ratio in Eq. (1.69) by the symbol $1/n'$, Metzner and Reed obtained a final expression in the form

$$\left(-\frac{dv_r}{dr}\right)_w = \frac{3n'+1}{4n'}\left(\frac{8v}{d}\right) \tag{1.70}$$

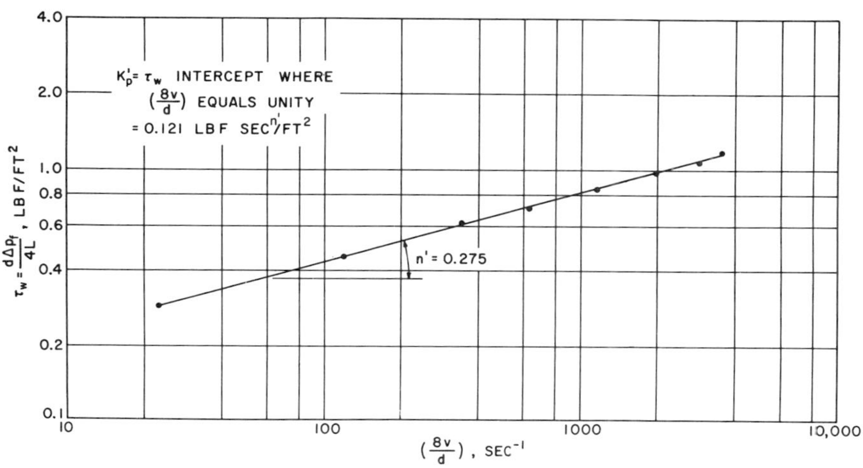

Fig. 1.13 Pipe viscometer data, water suspension containing 25 per cent by weight calcium carbonate, weighing 73.5 lbm/cu ft.

From the slope of a logarithmic plot of viscometric data in the form $(d\Delta p_f/4L)$ versus $(8v/d)$, the derivative ratio, n', is readily evaluated, as shown in Fig. 1.13. While the slope of such a curve has been found to be constant over wide ranges of shear stress for a great variety of non-Newtonian fluids, in the exceptional circumstance where n' is found to be a function of τ_w it is not especially difficult to make a point-by-point evaluation of the slope. This graphical evaluation of the parameter n', together with Eq. (1.70), makes it possible to establish the absolute flow curve for any fluid (excluding thixotropic and rheopectic behavior) from laminar pipe-flow data. In this connection, the use of a small-diameter or capillary-tube device is preferred, since a greater range of applied (wall) shear stress can be investigated before the onset of turbulent flow.

Example 1.14. The laminar pipe-flow data given in Cols. (1) and (2) of the table which follows were reported by Melton and Saunders[22] for a water suspension of calcium carbonate, weighing 73.5 lbm/cu ft. Their pipe viscometer consisted of a 6.92-ft horizontal length of 0.278-in. ID steel pipe, preceded and followed by calming sections of the same material.

Experimental data		Calculated results		
(1) Flow rate $q \times 10^3$, cu ft/min	(2) Pressure drop Δp_f, psi	(3) $\left(\frac{8v}{d}\right)$, sec^{-1}	(4) $\left(\frac{d\Delta p_f}{4L}\right)$, lbf/sq ft	(5) $\left(-\frac{dv_r}{dr}\right)_w$, sec^{-1}
1.674	2.37	22.85	0.2856	37.91
8.789	3.73	120.0	0.4495	199.1
25.11	5.07	342.9	0.6109	568.9
46.04	5.83	628.7	0.7025	1043
84.54	6.94	1154	0.8363	1914
144.8	8.06	1977	0.9712	3280
212.6	8.84	2903	1.0652	4816
260.7	9.74	3560	1.1737	5906

Calculate the rate of shear at the wall produced by the eight values of wall shear stress imposed on the fluid.

SOLUTION: Using consistent units throughout, the average flow velocity is calculated from Eq. (1.21a):

$$v = \frac{3.056(q\ \text{ft}^3/\text{min})}{(d\ \text{in.})^2} = \frac{3.056 \times 10^{-3}}{(0.278)^2}(q \times 10^3)\ \text{ft/sec}$$

$$\frac{8v}{d} = \frac{8}{(0.278/12)}v = \frac{8(12)3.056 \times 10^{-3}}{(0.278)^3}(q \times 10^3)$$

$$= 13.655(q \times 10^3)\ \text{sec}^{-1}$$

Values in Col. (1) multiplied by 13.655 are listed in Col. (3).

$$\frac{d\Delta p_f}{4L} = \frac{(0.278/12)(\Delta p_f,\ \text{psi} \times 144)}{4(6.92)} = 0.1205(\Delta p_f)\ \text{lbf/ft}^2$$

Pressure drop data from Col. (2) multiplied by 0.1205 are given in Col. (4); when plotted against corresponding values of $(8v/d)$ on log-log paper they yield the curve given in Fig. 1.13. From the geometric slope of this curve, which in this instance is constant, $n' = 0.275$ (dimensionless).

From Eq. (1.70), the rate of shear at the wall is given by

$$\left(-\frac{dv_r}{dr}\right)_w = \frac{3(0.275)+1}{4(0.275)}\left(\frac{8v}{d}\right) = 1.659\left(\frac{8v}{d}\right)$$

Multiplying entries in Col. (3) by 1.659 yields the rate of shear at the wall, listed in Col. (5), as produced by the applied (wall) shear stress, given in Col. (4). If required, the data from Cols. (4) and (5) could be used to construct the absolute flow curve for this fluid [i.e., a Cartesian plot of shear stress τ versus the corresponding shear rate $(-dv_r/dr)$], a curve which would be very similar to the pseudoplastic curve illustrated in Fig. 1.7.

On logarithmic coordinates, τ would plot as a linear function of $(-dv_r/dr)$ with the same slope as the τ_w versus $8v/d$ plot shown in Fig. 1.13; for when n' is constant, Eq. (1.70) admits that $(-dv_r/dr)_w$ and $8v/d$ are identical except for the constant multiplying factor $(3n'+1)/(4n')$. In view of these facts, an analytic expression for the shear stress–shear rate behavior of the suspension could be written as

$$\tau = K\left(-\frac{dv_r}{dr}\right)^n$$

a power-law relationship with $n = n'$.

Oddly enough, now that we have developed a method for determining the absolute flow curve or shear stress–shear rate behavior of a fluid, it will be shown that in the analysis of both laminar and turbulent pipe-flow problems the logarithmic plot of τ_w versus $8v/d$ is of greater utility.

Indices of flow behavior and consistency. Upon investigating the physical significance of the flow parameter n', Metzner and Reed[21] concluded that this parameter could be used as a quantitative index for comparing the degrees of non-Newtonian behavior exhibited by all fluids (other than time-dependent ones). More recently, their proposed scale has been recognized as a standard for classifying the rheological properties of hydraulic fracturing fluids as outlined in API RP 39.[11]

Recall that n' was defined as

$$n' = \frac{d\ln(d\Delta p_f/4L)}{d\ln(8v/d)}$$

and is evaluated from the slope of a straight line drawn tangent to the logarithmic curve of τ_w versus $8v/d$ at any particular value of τ_w in question. The equation of this tangent line may be written as

$$\tau_w = \frac{d\Delta p_f}{4L} = K_p'\left(\frac{8v}{d}\right)^{n'} \tag{1.71}$$

where K'_p is the intercept value of the tangent line on the τ_w axis where $8v/d$ equals unity.

As illustrated in Fig. 1.13, the logarithmic plot of τ_w versus $8v/d$ for many fluids will be linear over a wide (10- to 100-fold) range of shear stresses. Within this range both n' and K'_p remain constant (i.e., they are independent of τ_w), so that Eq. (1.71) is the equation of the logarithmic curve as well as of its tangent. For exceptional fluids whose logarithmic curves are nonlinear, n' and K'_p are functions of τ_w, but a continual revaluation of these parameters in step with changing values of τ_w (or $8v/d$) permits one still to use Eq. (1.71) to represent the logarithmic curve. In other words, the flow characteristics of all fluids may be defined completely by the laminar $(d\Delta p_f/4L)$—$(8v/d)$ relationship given by Eq. (1.71) regardless of whether n' and K'_p are constants.

When the expression for $8v/d$ in Eq. (1.70) is substituted into Eq. (1.71), there results

$$\tau_w = K'_p \left(\frac{4n'}{3n' + 1}\right)^{n'} \left(-\frac{dv_r}{dr}\right)_w^{n'} \tag{1.72}$$

We can now demonstrate the physical significance of the flow parameter n'; for if n' is constant with value unity, Eq. (1.72) becomes

$$\tau_w = K'_p \left(-\frac{dv_r}{dr}\right)_w$$

the familiar linear relationship between shear stress and shear rate characteristic of Newtonian fluids, as previously introduced in Eq. (1.28) where $K'_p = \mu/g_c$. For n' constant but less than unity, we obtain

$$\tau_w = K\left(-\frac{dv_r}{dr}\right)_w^{n} \tag{1.30}$$

the power-law relationship which closely approximates the behavior of many pseudoplastic fluids, where

$$K = K'_p \left(\frac{4n'}{3n' + 1}\right)^{n'}, \quad \text{and} \quad n = n' \tag{1.73}$$

Finally, if n' is constant but greater than unity, the fluid is dilatant in character.

This dimensionless parameter n', called a *flow-behavior index,* is recognized as that physical property of a fluid which characterizes the degree of its non-Newtonian behavior. The further removed this index from unity (in either direction), the more non-Newtonian is a fluid's behavior. The companion parameter K'_p, which has the rather awkward dimensions of lbf $\text{sec}^{n'}/\text{ft}^2$ in consistent English units, is likewise a physical property of a fluid, called a *consistency index*. While two fluids might exhibit the same flow-behavior index, the thicker or apparently more viscous of the two would have the larger consistency index. In terms of these two fundamental physical properties, we can describe the flow behavior of all fluids in a completely general and rigorous manner; the only special recognition to be given Newtonian fluids is that their flow-behavior index is always unity.

From the repeated implication that non-Newtonian fluids in general will exhibit constant values of n' and K'_p over wide ranges of shear stress, one may justifiably conclude that these same fluids would also follow the power-law relationship given by Eq. (1.30) with n and K constant. Going a step further and considering fluids whose n' and K'_p parameters are functions of shear stress rather than true constants, these fluids too could be represented by Eq. (1.30) with the understanding that n and K would likewise be functions of shear stress. A question arises, therefore, as to which pair of fluid flow indices, n' and K'_p or n and K, should be regarded as the more fundamental in describing the physical properties of a fluid.

Probably n and K should be recognized as more fundamental, for they are independent of the type of apparatus in which rheological measurements are made, as true physical indices should be. Unfortunately, Eq. (1.30) which utilizes these parameters relates shear stress only to shear rate and not flow rate. To obtain an equation for the average flow velocity within a pipe would require a double integration of Eq. (1.30) over a range of shear stresses from $\tau = 0$ to $\tau = \tau_w$; and unless n were constant over this range, this would entail multiple integrations.

On the other hand, n' and K'_p may be obtained directly from laminar pipe-flow data (Example 1.14 and Fig. 1.13); and Eq. (1.71) clearly is a direct relationship between pressure loss Δp_f and average flow velocity v for laminar pipe flow, which has as its graphical solution the logarithmic plot of τ_w versus $8v/d$. It should be apparent, therefore, that Eq. (1.71) is inherently superior to Eq. (1.30) for the analysis of pipe-flow systems involving power-law and nonpower-law fluids alike.

At any particular shear stress, n' and n may be related by the following analysis originally developed by Metzner.[23] Taking logarithms of Eq. (1.70) and differentiating, we obtain

$$\mathrm{d}\left[\ln\left(-\frac{\mathrm{d}v_r}{\mathrm{d}r}\right)_w\right] = \mathrm{d}\left(\ln\frac{3n'+1}{4n'}\right) + \mathrm{d}\left(\ln\frac{8v}{d}\right)$$

Dividing all terms in this expression by $\mathrm{d}(\ln \tau_w)$ yields

$$\frac{\mathrm{d}\left[\ln\left(-\frac{\mathrm{d}v_r}{\mathrm{d}r}\right)_w\right]}{\mathrm{d}(\ln \tau_w)} = \frac{\mathrm{d}\left(\ln\frac{3n'+1}{4n'}\right)}{\mathrm{d}(\ln \tau_w)} + \frac{\mathrm{d}\left(\ln\frac{8v}{d}\right)}{\mathrm{d}(\ln \tau_w)}$$

The first term in this expression is the inverse slope of a logarithmic plot of τ_w versus $(-\mathrm{d}v_r/\mathrm{d}r)_w$ and is equivalent to $1/n$, while the last term was previously defined as $1/n'$. Making these simplifications, we may write

$$\frac{1}{n} = \frac{\mathrm{d}\left(\ln \dfrac{3n'+1}{4n'}\right)}{\mathrm{d}(\ln \tau_w)} + \frac{1}{n'} \tag{1.74}$$

As n' is understood to refer to a particular value of shear stress regardless of where this stress may have occurred, the subscript w may be dropped and Eq. (1.74) rearranged to obtain

$$n = \frac{n'}{1 - \left(\dfrac{1}{3n'+1} \times \dfrac{\mathrm{d}n'}{\mathrm{d}(\ln \tau)}\right)} \tag{1.75}$$

The derivative ratio $\mathrm{d}n'/\mathrm{d}(\ln \tau)$ represents the slope of a plot of n' versus $\ln \tau$. If n' is a constant independent of shear stress, which requires that the initial logarithmic plot of τ_w versus $8v/d$ be essentially linear, the second term in the denominator of Eq. (1.75) is zero and n becomes identical with n'. In the completely general case, however, considering nonpower-law fluids, this will not occur and n and n' will be numerically distinct.

Generalized equations for pipe flow. Although Eq. (1.71) has already been recognized as a general expression relating pressure loss and flow velocity by means of pipe dimensions and the physical properties of the flowing fluid n' and K_p', the Fanning equation may also be generalized by the introduction of a modified friction factor. However, of far greater importance is the coincident development of a generalized Reynolds number.

Rearranging Eq. (1.18) to read

$$f = \frac{(d\Delta p_f/4L)}{(\rho v^2/2g_c)}$$

and substituting the general expression for $d\Delta p_f/4L$ from Eq. (1.71), we obtain

$$f = \frac{2g_c K_p'}{\rho v^2}\left(\frac{8v}{d}\right)^{n'}$$

which further simplifies to

$$f = \frac{16 g_c K_p'(8^{n'-1})}{\rho v^{2-n'}\, d^{n'}} \tag{1.76}$$

If we require that $f = 16/N_R$ as for Newtonian fluids in laminar flow, Eq.

(1.76) suggests an expression for a generalized Reynolds number N'_R given by the dimensionless grouping

$$N'_R = \frac{d^{n'} v^{2-n'} \rho}{g_c K'_p (8^{n'-1})} \tag{1.77}$$

Since the original generating equation for this entire development, Eq. (1.70) or its forerunner Eq. (1.41), involved only three implicit assumptions concerning the nature of the flowing fluid (viz., the absence of thixotropic or rheopectic behavior and no fluid slip at the pipe wall), we conclude from the foregoing analysis that in laminar pipe flow all fluids must follow a universal relationship $f = 16/N'_R$, where N'_R is the generalized Reynolds number defined by Eq. (1.77). This argument enjoys such wide scope of application and sound theoretical background that it may be used to check the accuracy of all pipe viscometer data. Any deviation from the criterion that $f = 16/N'_R$ in the laminar flow region indicates that either (1) the data or calculations are in error, (2) the fluids show evidence of thixotropic or rheopectic behavior, or (3) there is slippage at the wall of the pipe. The effects of fluid slip, however, can be isolated and evaluated separately by use of a technique outlined by Mooney.[20]

For Newtonian fluids, N'_R will reduce to the simpler expression $dv\rho/\mu$, which shows that this traditional dimensionless group is only a special form of the more general expression given in Eq. (1.77). It may also be shown that the equivalent Reynolds number $N_{Re} = dv\rho/\mu_e$ introduced in Sec. 1.8 is likewise a restricted form of Eq. (1.77) adapted to the rheological parameters μ_p and τ_y of a Bingham-model fluid.

In a survey of published flow data covering sixteen different non-Newtonian fluids, pipe diameters varying from $\frac{1}{8}$ to 12 in., and (generalized) Reynolds numbers over a range of 2×10^9, Metzner and Reed[21] reported excellent agreement of all laminar-flow data with the criterion $f = 16/N'_R$. In addition, they tentatively established the onset of turbulent flow for all fluids at $N'_R = 2100$.

Restricting our attention to power-model fluids (i.e., those fluids for which n' and K'_p are true constants), we may use either the Fanning equation or Eq. (1.71) to calculate frictional pressure losses for laminar pipe flow. Regardless of our choice, however, the generalized Reynolds number must always be evaluated to ascertain that laminar conditions prevail. Consequently, a logical procedure is to calculate f from the equation $f = 16/N'_R$ and then proceed with the Fanning equation. To partially simplify the evaluation of N'_R, Eq. (1.77) presented in consistent units may be reduced to

$$N'_R = 1.86 \frac{(d/12)^{n'} v^{2-n'} \rho}{8^{n'} K'_p} \tag{1.77a}$$

upon introducing the practical engineering units of Table 1.2, but the units of K'_p are still fixed as lbf sec$^{n'}$/ft^2. Still another simplification of Eq. (1.77) is possible, for if the combination $K'_p(8v/d)^{n'}$ is replaced by τ_w [from Eq. (1.71)] our Reynolds number expression reduces to

$$N'_R = \frac{8v^2\rho}{g_c\tau_w} \tag{1.78}$$

using consistent units throughout, or

$$N'_R = 1.86\,\frac{v^2\rho}{\tau_w} \tag{1.78a}$$

in practical units.

Equation (1.78) is a practical convenience only when a plot of τ_w versus $8v/d$ is available from capillary viscometer data as illustrated in Fig. 1.13. Once the quantity $8v/d$ has been evaluated for a particular flow situation, the corresponding value of τ_w to be used in Eq. (1.78) or Eq. (1.78a) is simply read from the logarithmic plot. Then, if N'_R proves to be less than 2100, this same value of τ_w can be used to calculate Δp_f directly, since $\tau_w = d\Delta p_f/4L$. Both solution techniques are illustrated in Example 1.15.

Example 1.15. From the data and results of Example 1.14, compute the frictional pressure loss which would accompany the flow of the calcium carbonate suspension through 100 ft of 0.500-in. ID pipe at a steady rate of 4.5 gpm.

SOLUTION: From Eq. (1.21a),

$$v = \frac{4.5 \text{ gpm}}{2.448(0.5)^2} = 7.35 \text{ ft/sec}$$

$$\rho = 73.5 \text{ lbm/ft}^3 = 9.83 \text{ ppg}$$

From Eq. (1.77a),

$$N'_R = 1.86\,\frac{(0.5/12)^{0.275}(7.35)^{1.725}9.83}{0.121(8)^{0.275}} = 1111$$

The flow is laminar since $N'_R < 2100$; therefore,

$$f = \frac{16}{N'_R} = \frac{16}{1111} = 0.0144$$

From Eq. (1.18a),

$$\Delta p_f = \frac{0.0144(100)9.83(7.35)^2}{25.80(0.5)} = 59.3 \text{ psi}$$

ALTERNATE SOLUTION: Using consistent units,

$$\frac{8v}{d} = \frac{8(7.35)}{(0.5/12)} = 1411.2 \text{ sec}^{-1}$$

From the logarithmic plot of viscometer data given in Fig. 1.13,

$$\tau_w = \frac{d\Delta p_f}{4L} = 0.89 \text{ lbf/ft}^2$$

Recalculating N'_R from Eq. (1.78a),

$$N'_R = 1.86 \frac{(7.35)^2 9.83}{0.89} = 1110 \text{ (Laminar)}$$

$$\Delta p_f = \frac{4L\tau_w}{d} = \frac{4(100)(0.89)}{(0.5/12)} = 8544 \text{ lbf/ft}^2 = 59.3 \text{ psi}$$

Nonpower-law fluids pose no additional problems in laminar flow situations, for the relationship $f = 16/N'_R$ is rigorously valid regardless of whether n' and K'_p are constants. However, the evaluation of N'_R by use of Eq. (1.77) or its alternate would require a trial-and-error procedure since n' and K'_p would be functions of τ_w, and normally τ_w is the quantity being sought. Fortunately, our alternate approach of using Eq. (1.78) in combination with a logarithmic plot of τ_w versus $8v/d$ avoids this difficulty, for in laminar flow the variations of n' and K'_p with changes in τ_w are graphically displayed by this logarithmic plot.

Dodge and Metzner,[24] on the basis of both theoretical and experimental investigations, were able to define more specifically the onset of turbulence in pipe flow and to establish a friction factor – Reynolds number relationship for the fully developed turbulent flow of several representative non-Newtonian fluids, among which are polymeric solutions, solid-liquid suspensions, and power-law and nonpower-law fluids. Their results are presented in Fig. 1.14 as a generalized friction factor plot which is applicable to Newtonian and non-Newtonian systems alike. As correlating parameters they used the generalized Reynolds number and the flow-behavior index n' introduced previously in the analysis of laminar pipe flow. As a consequence, the curves of Fig. 1.14 are ideally suited for the analysis of pipe flow systems handling power-law fluids. Although the transition-region portions of the curves of Fig. 1.14 were located somewhat arbitrarily, the Reynolds number corresponding to the onset of fully developed turbulent flow was found to increase slowly with decreasing values of n'.

Once the Reynolds number has been evaluated and a friction factor read from the appropriate n' curve of Fig. 1.14, calculation of the frictional pressure loss accompanying the turbulent flow of a power-law fluid becomes identical with the usual procedure for simple Newtonian fluids.

Example 1.16. The suspension studied in Example 1.14 is pumped at a rate of 5 bbl/min through 1000 ft of 2-in. ID tubing. Calculate the frictional pressure loss.

SOLUTION: From Eq. (1.21a),

$$v = \frac{17.16(5 \text{ bbl/min})}{(2)^2} = 21.45 \text{ ft/sec}$$

$$\rho = 9.83 \text{ ppg}, \quad n' = 0.275, \quad K'_p = 0.121 \text{ lbf sec}^{n'}/\text{ft}^2$$

From Eq. (1.77a),

$$N'_R = 1.86 \frac{(2/12)^{0.275}(21.45)^{1.725} 9.83}{0.121(8)^{0.275}} = 10{,}320$$

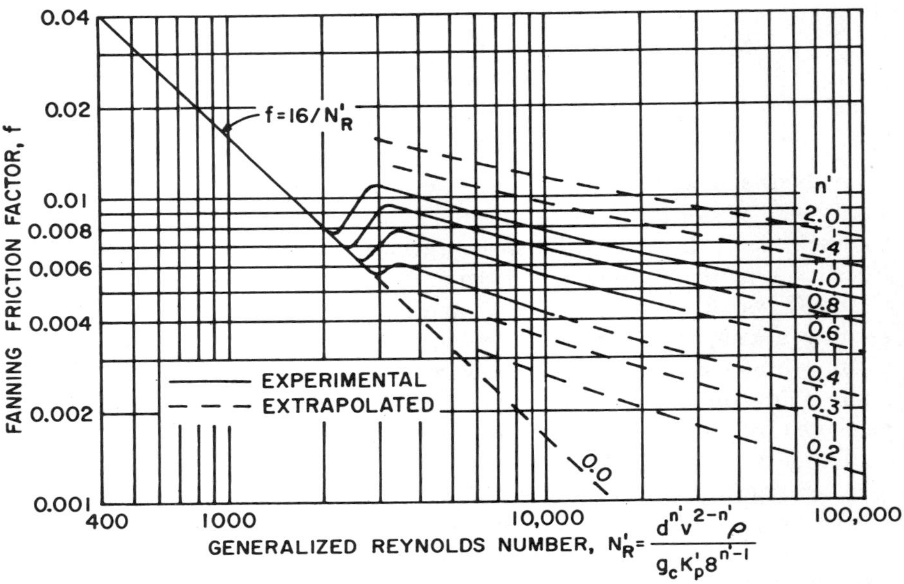

Fig 1.14 Friction factors for general non-Newtonian fluids. (After Dodge and Metzner,[24] A.I.Ch.E. Journal)

Referring to Fig. 1.14, the approximate value of N'_R for the onset of fully developed turbulent flow with $n' = 0.275$ is 4000. Thus the flow is turbulent and $f = 0.0032$, whence

$$\Delta p_f = \frac{0.0032(1000)9.83(21.45)^2}{25.80(2)} = 280 \text{ psi}$$

It is advisable to calculate the wall shear stress under these flow conditions and compare it with the range of viscometer data used to establish the n' and K'_p indices of the suspension. In consistent units,

$$\frac{d\Delta p_f}{4L} = \frac{(2/12)(280 \times 144)}{4(1000)} = 1.68 \text{ lbf/ft}^2$$

This value shows a slight, but not serious, extrapolation of the flow behavior observed in the pipe viscometer, wherein the wall shear stress did not exceed 1.17 lbf/ft².

Alternate Solution: As a convenient procedure for evaluating only the Reynolds number, Eq. (1.78a) together with a plot of viscometer data (Fig. 1.13) may be used:

$$\frac{8v}{d} = \frac{8(21.45)}{(2/12)} = 1030 \text{ sec}^{-1}$$

From Fig. 1.13,

$$\tau_w = 0.815 \text{ lbf/ft}^2$$

Hence

$$N'_R = 1.86 \frac{(21.45)^2 9.83}{0.815} = 10{,}322$$

As the flow is turbulent, Δp_f must not be evaluated from the value of τ_w used in the calculation of N'_R but can only be evaluated by use of the Fanning equation and a friction factor obtained from Fig. 1.14.

Perhaps our alternate procedure for evaluating a Reynolds number in turbulent flow (as used in the preceding example) deserves further consideration, especially when we realize that the value of τ_w utilized in Eq. (1.78a) was read from a logarithmic plot of τ_w versus $8v/d$ which was constructed from *laminar* pipe flow data. In short, we have used laminar flow data in the evaluation of a turbulent flow parameter N'_R. It must be emphasized that this technique is valid only in the case of (quasi) power-law fluids, for we depend upon their n' and K'_p indices to remain essentially constant over a range of shear stresses that will include all values that τ_w might assume in a particular turbulent flow system under study. Recall that, in the same spirit, the viscosity μ of a Newtonian fluid, also evaluated under laminar flow conditions, is assumed to remain constant, and it is subsequently used in the relation $dv\rho/\mu$ to establish the Reynolds number for laminar and turbulent flow situations alike.

The friction factor–Reynolds number curves of Fig. 1.14 are valid also for nonpower-law fluids, provided the variable parameters n' and K'_p are evaluated at the existing wall shear stress τ_w, for it will be recalled that the shear rate response of the fluid in the immediate vicinity of the wall has primary responsibility for the bulk or average velocity of the flow. For turbulent flows, however, a trial-and-error procedure cannot be avoided; for if Δp_f is the quantity being sought, then τ_w is unknown and the corresponding values of n' and K'_p to be used in evaluating N'_R are likewise unknown. Fortunately, the majority of non-Newtonian fluids behave essentially as power-law fluids, and exceptional ones will seldom depart radically from this model behavior. Thus, their n' and K'_p indices change slowly with variations in wall shear stress, which hastens the convergence of a trial-and-error solution.

Generalized equations for annular flow. In reviewing the analytical expressions derived by Fredrickson and Bird[13] to describe the annular flow of power-law fluids, Savins[25] proved that these rather elaborate equations could be approximated by simpler expressions developed for narrow slot flow. For annular geometry factors α ($= d_i/d_o$) in the range of $0.3 \leqslant \alpha \leqslant 0.9$ and flow-behavior indices greater than 0.3, Savins showed that a slot-flow approximation introduces a probable error no greater than 7 per cent.

Reassured by these results, we may proceed with the derivation of a generalized friction factor – Reynolds number criterion for the laminar slot flow of all fluids (excluding time-dependent ones), to be used subsequently for an approximate analysis of annular flow systems.

Following the procedure used in the development of the Rabinowitsch equation, our general expression for laminar slot flow, Eq. (1.50), when differentiated with respect to τ_w and subsequently multiplied by k/τ_w becomes

$$2\left(\frac{2v}{W}\right) + \tau_w \frac{\mathrm{d}(2v/W)}{\mathrm{d}\tau_w} = k\phi(\tau_w) \tag{1.79}$$

As $k\phi(\tau)$ represents the functional dependence of shear rate, $-\mathrm{d}v_y/\mathrm{d}y$, on shear stress, the last term in Eq. (1.79) represents the shear rate of the fluid in contact with either major wall of the slot, which we shall denote by $(-\mathrm{d}v_y/\mathrm{d}y)_w$. Introducing this expression for $k\phi(\tau_w)$ and replacing τ_w by $W\Delta p_f/2L$, we can rearrange Eq. (1.79) to read

$$\left(-\frac{\mathrm{d}v_y}{\mathrm{d}y}\right)_w = \frac{2}{3}\left(\frac{6v}{W}\right) + \frac{1}{3}\left(\frac{6v}{W}\right)\frac{[\mathrm{d}(6v/W)]/(6v/W)}{[\mathrm{d}(W\Delta p_f/2L)]/(W\Delta p_f/2L)}$$

$$= \frac{2}{3}\left(\frac{6v}{W}\right) + \frac{1}{3}\left(\frac{6v}{W}\right)\frac{\mathrm{d}\ln(6v/W)}{\mathrm{d}\ln(W\Delta p_f/2L)} \tag{1.80}$$

a completely general expression for the rate of shear at the slot wall produced by a wall shear stress of $W\Delta p_f/2L$, analogous to Eq. (1.69) which applies to laminar pipe flow.

Representing the derivative ratio in Eq. (1.80) by the symbol $1/n^*$, we obtain a final expression in a more compact form,

$$\left(-\frac{\mathrm{d}v_y}{\mathrm{d}y}\right)_w = \frac{2n^*+1}{3n^*}\left(\frac{6v}{W}\right) \tag{1.81}$$

If measurements of pressure drop and flow rate were available for flow through a narrow-slot viscometer, a logarithmic plot of this data in the form $(W\Delta p_f/2L)$ versus $(6v/W)$ would readily yield n^* as the slope of the curve at any chosen value of τ_w.

As n^* was defined as

$$n^* = \frac{\mathrm{d}\ln(\tau_w)}{\mathrm{d}\ln(6v/W)}$$

over any range of wall shear stresses for which n^* remains constant, perhaps only a differential range if n^* happens to be a function of τ_w, we may integrate this expression and obtain

$$\tau_w = \frac{W\Delta p_f}{2L} = K'_a\left(\frac{6v}{W}\right)^{n*} \tag{1.82}$$

where K'_a is a constant of integration. As Eq. (1.82) is also the equation of a straight line drawn tangent to the logarithmic τ_w versus $(6v/W)$ curve at any particular value of τ_w in question, K'_a is the intercept value of the tangent line on the τ_w axis where $6v/W$ equals unity. Over any range of τ_w in which the logarithmic plot of τ_w versus $6v/W$ is linear, n^* and K'_a remain constant; but if the logarithmic plot is nonlinear, n^* and K'_a will require a point-by-point revaluation in step with changing values of τ_w in order to maintain the validity of Eq. (1.82).

We now have, in Eq. (1.82), a completely general expression relating frictional pressure loss to average flow velocity in laminar slot flow, which involves the two new fluid parameters n^* and K'_a. Introducing the expression for $6v/W$ from Eq. (1.81) into Eq. (1.82), we may write

$$\tau_w = K'_a\left(\frac{3n^*}{2n^*+1}\right)^{n*}\left(-\frac{dv_y}{dy}\right)_w^{n*} \tag{1.83}$$

If n^* is a constant with value unity, Eq. (1.83) degenerates to a linear relationship between shear stress and shear rate, the relationship normally associated with Newtonian fluids where $K'_a = \mu/g_c$. For n^* a constant but less than unity, we obtain

$$\tau_w = K\left(-\frac{dv_y}{dy}\right)_w^{n} \tag{1.30}$$

the familiar power-law relationship between shear stress and shear rate, where the constant coefficients are given by

$$K = K'_a\left(\frac{3n^*}{2n^*+1}\right)^{n*}, \quad \text{and} \quad n = n^* \tag{1.84}$$

We now must relate Δp_f in Eq. (1.82) to the Fanning equation in order to obtain an expression for a generalized Reynolds number which will apply to the slot flow of all fluids. When adapted for the geometry of a narrow slot, the Fanning equation becomes

$$f = \frac{\tau_w}{(\rho v^2/2g_c)} = \left(\frac{W\Delta p_f/2L}{\rho v^2/2g_c}\right) \tag{1.56}$$

as previously introduced in Sec. 1.9. Substituting the general expression for τ_w from Eq. (1.82) and rearranging terms, we obtain

$$f = \frac{2g_c K'_a}{\rho v^2}\left(\frac{6v}{W}\right)^{n*}$$

which may also be written as

$$f = \frac{24 g_c K'_a (12^{n*-1})}{\rho v^{2-n*}(2W)^{n*}} \tag{1.85}$$

By requiring that $f = 24/N_{Rs}$ as given in Eq. (1.57) for Newtonian fluids in laminar slot flow, we may define a generalized Reynolds number N'_{Rs} as the dimensionless grouping

$$N'_{Rs} = \frac{(2W)^{n*} v^{2-n*} \rho}{g_c K'_a (12^{n*-1})} \tag{1.86}$$

From this analysis we conclude that all fluids in laminar slot flow follow a universal law $f = 24/N'_{Rs}$, where N'_{Rs} is a generalized Reynolds number defined by Eq. (1.86). Any deviation from this criterion would indicate either poor data, thixotropic or rheopectic behavior, or fluid slip at the wall.

As most of the drilling and completion fluids which are of interest to petroleum engineers can be represented quite well by a power-law model, we shall restrict our attention, in all subsequent derivations, to this particular class of non-Newtonian fluids. While this will result in some welcome simplifications, it will also place some restraint on the completely general equations developed thus far.

As an example of the simplification which results from dealing exclusively with power-law fluids, Eqs. (1.73) and (1.84) may be combined to obtain an interconnecting relationship between the three pairs of flow indices introduced in Eqs. (1.30), (1.71), and (1.82):

$$n' = n = n^*$$

$$K = K'_p\left(\frac{4n'}{3n'+1}\right)^{n'} = K'_a\left(\frac{3n'}{2n'+1}\right)^{n'}$$

$$\frac{K'_p}{K'_a} = \left(\frac{9n'+3}{8n'+4}\right)^{n'} \tag{1.87}$$

Knowing the flow-behavior and consistency indices of a power-law fluid, as determined from capillary-tube viscometer data, we may use Eq. (1.87) to calculate the K'_a index required for the analysis of narrow-slot or annular flow systems. As a convenience, a graphical solution of Eq. (1.87) is presented in Fig. 1.15.

Continuing with the premise that a thin slot approximates with acceptable accuracy the geometry of a narrow annulus, we can adapt Eqs. (1.79) through (1.86) to the dimensions of an annular system by replacing W with the term $(d_o - d_i)/2$. In particular, Eq. (1.82) becomes

$$\tau_w = \frac{(d_o - d_i)\Delta p_f}{4L} = K'_a \left(\frac{12v}{d_o - d_i}\right)^{n'} \tag{1.88}$$

showing that a logarithmic plot of τ_w versus $12v/(d_o - d_i)$ for the laminar annular flow of a power-law fluid ($n^* = n'$) should be linear with a slope n' and an intercept value of K'_a on the τ_w axis where $12v/(d_o - d_i)$ equals unity.

We may construct a plot of log τ_w versus log $[12v/(d_o - d_i)]$ for a graphical solution of Eq. (1.88) from a similar plot of available pipe viscometer data. For example, consider the plot given in Fig. 1.13. As $K'_p = 0.121$ and $n' = 0.275$, calculation by Eq. (1.87) or its solution curve in Fig. 1.15 yields $K'_a = 0.125$. With this value as an intercept on the τ_w axis where $8v/d$ or $12v/(d_o - d_i)$ equals unity, a straight line drawn parallel to the original pipe-flow plot would represent the annular flow behavior of this suspension in the laminar region. The scale of the abscissa remains the same, representing $8v/d$ for the pipe-flow plot and $12v/(d_o - d_i)$ for the annular flow plot, as both parameters have the common dimensions of reciprocal seconds.

As an approximate expression for a generalized Reynolds number, Eq. (1.86) when adapted to the geometry of an annular system becomes

$$N'_{Ra} = \frac{(d_o - d_i)^{n'} v^{2-n'} \rho}{g_c K'_a (12^{n'-1})} \tag{1.89}$$

and when modified to accept directly our practical engineering units (Table 1.2) it reduces to

$$N'_{Ra} = 2.79 \frac{[(d_o - d_i)/12]^{n'} v^{2-n'} \rho}{12^{n'} K'_a} \tag{1.89a}$$

but K'_a retains its dimensions in consistent units, lbf sec$^{n'}$/ft^2.

For Newtonian fluids, $n' = 1.0$, $K'_a = \mu/g_c$, and Eq. (1.89) reduces to $(d_o - d_i)v\rho/\mu$, our previous expression for an annular-flow Reynolds number introduced in Sec. 1.9. With a little more effort, it can be shown that Eq. (1.89) will also reduce to Eq. (1.60), the Reynolds number expression for the annular flow of a Bingham-model fluid.

Once the logarithmic plot of τ_w versus $12v/(d_o - d_i)$ has been constructed, a semigraphical evaluation of N'_{Ra} is also possible. Replacing the quantity $K'_a[12v/(d_o - d_i)]^{n'}$ in Eq. (1.89) by its equivalent from Eq. (1.88), the generalized Reynolds number expression may be written as

$$N'_{Ra} = \frac{12v^2\rho}{g_c \tau_w} \tag{1.90}$$

in consistent units, or

$$N'_{Ra} = 2.79 \frac{v^2\rho}{\tau_w} \tag{1.90a}$$

in practical engineering units wherein the dimensions of τ_w are retained as lbf/ft^2.

To estimate frictional pressure losses in laminar annular flow, we may use either the Fanning equation [Eq. (1.59) or its alternate] together with the general relation $f = 24/N'_{Ra}$ or the graphical solution of Eq. (1.88) as constructed from a log τ_w versus log $(8v/d)$ plot of pipe viscometer data.

It can only be *assumed* that the transition from laminar to turbulent flow occurs when the generalized Reynolds number, N'_{Ra}, exceeds 2100 and that in turbulent flow, friction factors for both pipe and annular flow correlate on the Dodge and Metzner curves given in Fig. 1.14. These are tentative proposals which still lack experimental confirmation, but continuing research within the petroleum and chemical industries is sure to remove this uncertainty.

Correlation of rotational viscometer and pipe flow behavior. In the absence of thixotropy, rheopexy, and wall slip, the general equation which describes the laminar flow behavior of all fluids within a rotational viscometer of the Fann V-G Meter type is

$$\omega_c = \frac{k}{2}\int_{\tau_b/\beta^2}^{\tau_b} \tau^{-1}\phi(\tau)\, d\tau \tag{1.91}$$

where
r_b = radius of inner (stationary) cylinder or bob
r_c = radius of outer (rotating) cylinder
$\beta = r_c/r_b$
L = effective length of bob
T = torque transmitted to bob
$\tau_b = T/2\pi L r_b^2$, shear stress experienced by fluid in contact with bob
ω_c = angular velocity of outer cylinder
$k\phi(\tau) = dv_r/dr$, shear stress – shear rate behavior of fluid

Although the shear stress at the bob, τ_b, is directly measureable, the corresponding shear rate at the bob, $(dv_r/dr)_b$, becomes rather elusive if one is testing a nonpower-law fluid. For a complete discussion of this type of viscometer, reference is made to the work of Krieger and Maron.[26]

Since a great many non-Newtonian fluids, possibly the majority, have been found to follow closely the power-law relationship given in Eq. (1.30),[24] we shall again restrict our attention to this particular model. The unspecified function $k\phi(\tau)$ in Eq. (1.91) may then be replaced by

$$\left(\frac{dv_r}{dr}\right) = k\phi(\tau) = \left(\frac{1}{K}\right)^{1/n} \tau^{1/n} \tag{1.92}$$

Performing the integration in Eq. (1.91) and solving for the quantity $(1/K)^{1/n}$, we obtain

$$\left(\frac{1}{K}\right)^{1/n} = \frac{2\omega_c\beta^{2/n}}{n(\beta^{2/n} - 1)\tau_b^{1/n}} \tag{1.93}$$

which when multiplied by $(\tau_b)^{1/n}$ will provide an expression for the rate of shear at the bob,

$$\left(\frac{dv_r}{dr}\right)_b = \frac{2\omega_c\beta^{2/n}}{n(\beta^{2/n} - 1)} \tag{1.94}$$

Introducing the parameter γ, defined as

$$\gamma = \left(\frac{\beta^{2/n}}{n\beta^2}\right)\left(\frac{\beta^2 - 1}{\beta^{2/n} - 1}\right) \tag{1.95}$$

Equation (1.94) may be written in a simpler form,

$$\left(\frac{dv_r}{dr}\right)_b = \gamma\left(\frac{2\omega_c\beta^2}{\beta^2 - 1}\right) \tag{1.96}$$

where the term in parentheses on the right-hand side represents the shear rate at the surface of the bob only for simple Newtonian liquids, i.e., when $n = 1 = \gamma$.

Combining Eq. (1.96) with Eq. (1.92) and solving for τ_b, we obtain

$$\tau_b = K\gamma^n\left(\frac{2\omega_c\beta^2}{\beta^2 - 1}\right)^n \tag{1.97}$$

Taking logarithms of Eq. (1.97) yields:

$$\log \tau_b = \log (K\gamma^n) + n \log\left(\frac{2\omega_c\beta^2}{\beta^2 - 1}\right) \tag{1.98}$$

which shows that Eq. (1.97) when plotted on logarithmic paper as τ_b versus $[2\omega_c\,\beta^2/(\beta^2 - 1)]$ will produce a straight line, as the indices K, n, and γ^n are all constants. From the slope of the logarithmic plot n is obtained directly, and the intercept K_v' on the τ_b axis where $[2\omega_c\beta^2/(\beta^2 - 1)]$ equals unity is related to K by

$$K_v' = K\gamma^n \tag{1.99}$$

Since this entire development has been restricted to power-law fluids, n is equivalent to the flow-behavior index n' introduced in our study of pipe flow. Similarly, the intercept value K_v' may also be related to the consistency index K_p'. It follows from Eqs. (1.73) and (1.99) that

$$\frac{K_v'}{K_p'} = \left(\frac{4n'\gamma}{3n' + 1}\right)^{n'} \tag{1.100}$$

where γ is a function of n' as well as of the radii ratio β as stated in Eq. (1.95). A graphical solution of Eq. (1.100) is given in Fig. 1.15 for the range

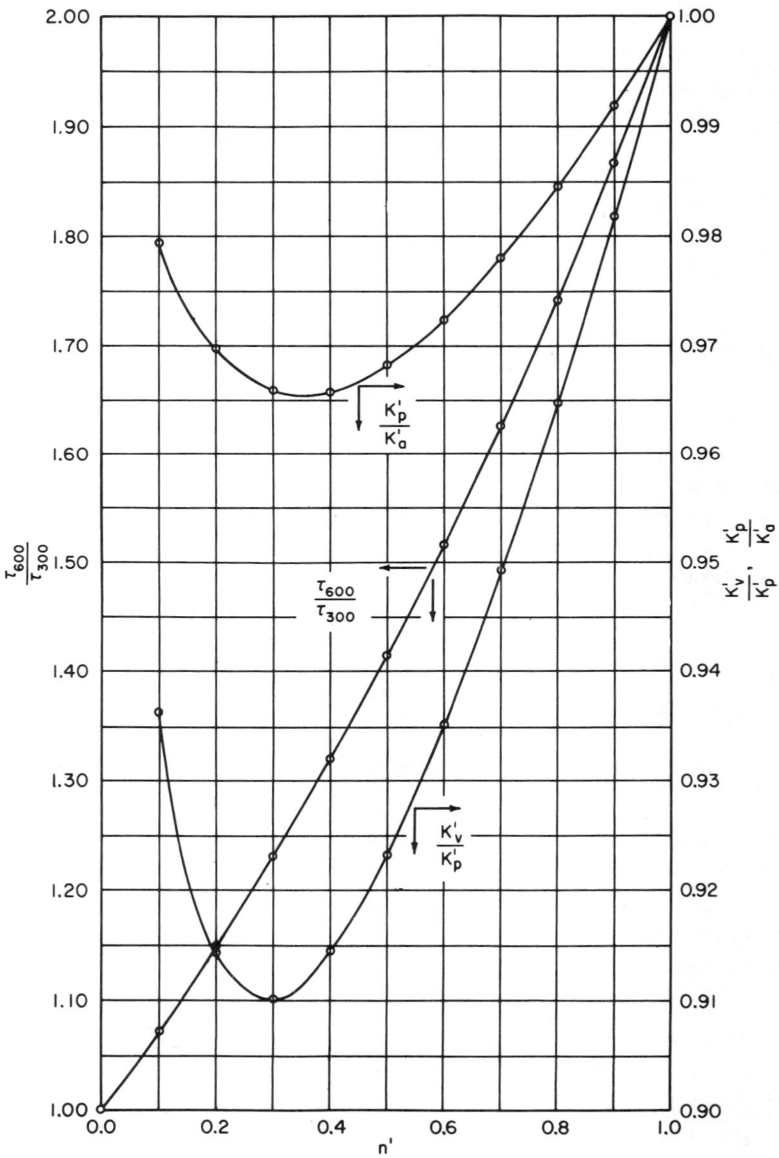

Fig. 1.15 Correlation diagram for pipe, annular section, and rotational viscometer data. (After Savins,[25] *Trans.* AIME.)

0.1–1.0 in n' based on $\beta = 1.0678$, the radii ratio used in the Model 35 Fann V-G Meter.

With regard to this same instrument, there are six fixed values of $[2\omega_c\beta^2/(\beta^2 - 1)]$ corresponding to the six rotational speeds available, as shown in Table 1.6. While this instrument essentially reads τ_b directly in

Table 1.6. Newtonian shear rate at the inner bob as a function of RPM setting[a]

Cylinder speed ω_c, rpm	$\left(\frac{2\omega_c\beta^2}{\beta^2 - 1}\right)$, sec^{-1}
600	1022
300	511
200	340
100	170
6	10.2
3	5.1

[a]Standard rotor-bob combination and speed settings for Model 35 Fann V-G Meter.

lbf/100 ft² when equipped with a standard torsion spring, a more exact relationship between τ_b in lbf/ft² (consistent units) and any dial reading θ is

$$\tau_b = N(0.01066)\theta \tag{1.101a}$$

where N is the range extension factor (e.g., $\frac{1}{5}$, 1, 3, 5) corresponding to optional torque springs available from the manufacturer.

In lieu of preparing a complete logarithmic plot of viscometer data, or when using the Model 34 instrument which has but two operating speeds, viz., 300 and 600 rpm, n' may be obtained from the ratio of the 600-rpm dial reading, θ_{600}, to the 300-rpm dial reading, θ_{300}, by using the relationship

$$n' = 3.32 \log (\theta_{600}/\theta_{300}) \tag{1.102}$$

or the graphical solution of this expression given in Fig. 1.15. Under these conditions K'_v, the intercept value in lbf sec$^{n'}$/ft², is calculated from

$$K'_v = \frac{N(0.01066)\theta_{300}}{511^{n'}} \tag{1.103a}$$

where $N = 1$ when the instrument is equipped with a standard torsion spring.

Example 1.17. The data given in Cols. (1) and (2) of the table which follows were recorded for a fracturing fluid in a standard Model 35 V-G Meter ($N = 1.0$).

Experimental data		Plotting parameters	
(1) Cylinder speed ω_c, rpm	(2) Dial reading, θ	(3) $\left(\dfrac{2\omega_c\,\beta^2}{\beta^2 - 1}\right)$, sec^{-1}	(4) Shear stress τ_b, lbf/sq ft
600	134	1022	1.428
300	95	511	1.013
200	77	340	0.821
100	54	170	0.576
6	13	10.2	0.138
3	9.4	5.1	0.100

Calculate n' and K'_p from a plot of the data, and then construct a companion plot to be used in the analysis of pipe flow systems.

Solution: Entries in Col. (3) are obtained directly from Table 1.6. Values of shear stress τ_b in lbf/ft², listed in Col. (4), are obtained by multiplying each entry in Col. (2) by the factor 0.01066. These τ_b values are plotted in Fig. 1.16, and a straight line fitted to the data points (curve V). From the intercept, $K'_v = 0.0437$ and from the slope (26.9°), $n' = 0.507$.

From Fig. 1.15, $K'_v/K'_p = 0.9237$, and therefore $K'_p = 0.0473$. Constructing a second straight line with slope $n' = 0.507$, or parallel to curve V, and K'_p in-

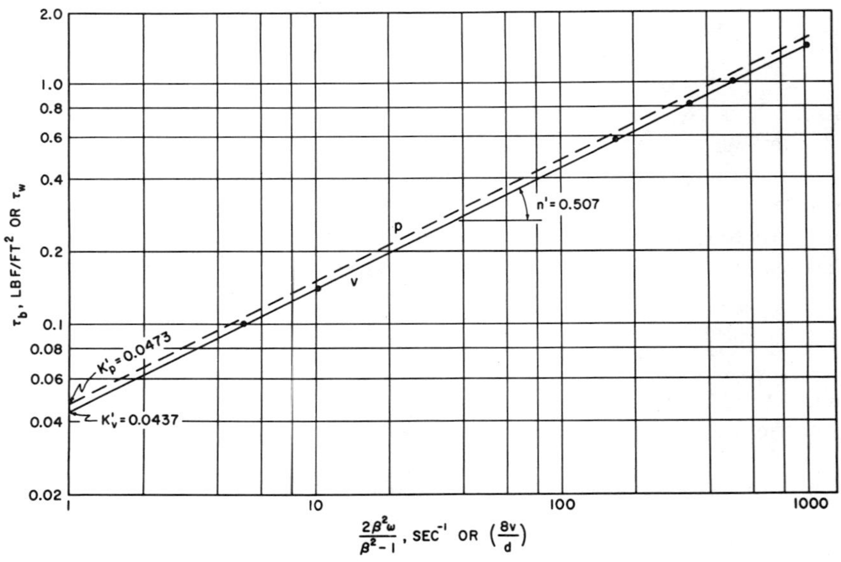

Fig. 1.16 Example plot of V-G meter data (V) and resulting pipe flow curve (P).

tercept at 0.0473, we have curve P of Fig. 1.16 (shown dashed). Using curve P we may proceed with the analysis of laminar and turbulent pipe-flow behavior of this gel fluid following the techniques outlined in Example 1.16.

1.12. Two-Phase Flow

The mechanics and characteristics of two-phase flow systems have been of interest throughout the industry for some time. In oil production, the trend toward central location of tank batteries has resulted in longer flow lines in which more than one fluid phase is flowing. Crude stabilization and total well-stream processing in distillate fields have also emphasized the need for additional information on the effects of liquid flowing along with gas in pipe lines. Optimum pipe size for the situations described above has become an important factor.

The general problem of predicting the pressure drop for two-phase pipe flow of gas and liquid is complicated. The manifold increase in pressure drop when liquid is placed in a gas pipe line was first measured by Van Wingen.[27] When two phases flow through the same pipe, the gas usually flows faster than the liquid which accumulates in the pipe and reduces the cross-sectional area available for gas flow. The pressure loss of a fluid flowing through a pipe is inversely proportional to the fifth power of the pipe diameter, i.e., a reduction of 20 per cent in diameter would cause a threefold increase in pressure drop. Other factors causing high-pressure drops are the energy required to move the liquid through the line and the energy used in the violent rising and falling of liquid. This energy must come from a reduction in pressure.

The work of Lockhart and Martinelli[28] provided one of the first correlations proposed for horizontal two-phase flow. Four tentative flow mechanisms were defined in terms of the laminar or turbulent flow for each phase, i.e., liquid-turbulent and gas-turbulent, liquid-laminar and gas-turbulent, etc.

Alves[29] demonstrated that the different flow patterns shown in Fig. 1.17 are possible for the flow mechanisms defined by Martinelli. These patterns[29] have been described as follows: "Assume a horizontal pipe with liquid flowing so as to fill the pipe, and consider the types of flow that occur as gas is added in increasing amounts.

"1. Bubble flow: Flow in which bubbles of gas move along the upper part of the pipe at approximately the same velocity as the liquid. [This is similar to froth flow where the entire pipe is filled with a froth.]

"2. Plug flow: Flow in which alternate plugs of liquid and gas move along the upper part of the pipe.

"3. Stratified flow: Flow in which the liquid flows along the bottom of the pipe and the gas flows above, over a smooth liquid-gas interface.

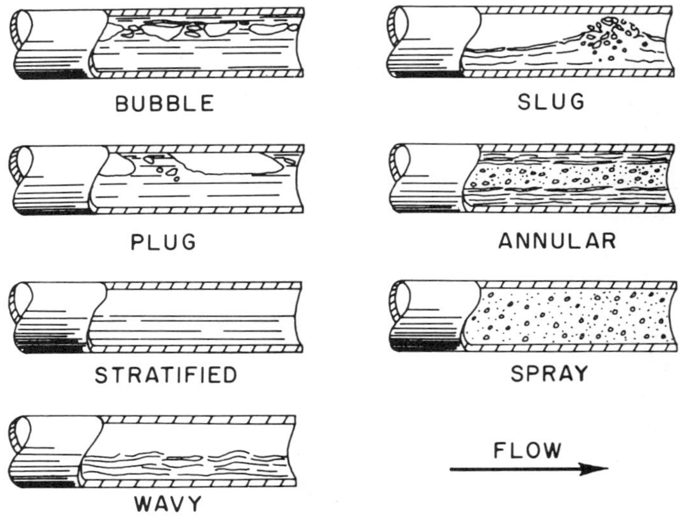

Fig. 1.17 Flow patterns. (Alves,[29] *Chemical Engineering Progress.*)

"4. Wavy flow: Flow which is similar to stratified flow except that the gas moves at a higher velocity and the interface is disturbed by waves traveling in the direction of the flow.

"5. Slug flow: Flow in which a wave is picked up periodically by the more rapidly moving gas to form a frothy slug which passes through the pipe at a much greater velocity than the average liquid velocity.

"6. Annular flow: Flow in which the liquid forms in a film around the inside wall of the pipe and the gas flows at a high velocity as a central core.

"7. Spray flow: Flow in which most or nearly all of the liquid is entrained as spray by the gas."

Baker[30] presented the results of 27 field experiments made with 8-in. and 10-in. lines. Pressure drops for large lines were found to be 40 to 60 per cent less than predicted by the Lockhart and Martinelli correlations. Using the same method of correlation, Baker correlated data for each flow pattern. Yocum[31] and Chavez[32] have also presented correlations for two-phase flow in flow lines and pipe lines. Bertuzzi, Tek, and Poettmann[33] introduced the concept of a two-phase friction factor which they called the energy dissipation function. A semiempirical approach was adopted combining ideas from the theory of single-phase flow with 267 experimental measurements on two-phase flow behavior. The two-phase friction factor was found to correlate best in terms of the Reynolds numbers of the liquid and gas phases and the gas-liquid mass ratio as presented in Fig. 1.18. The corre-

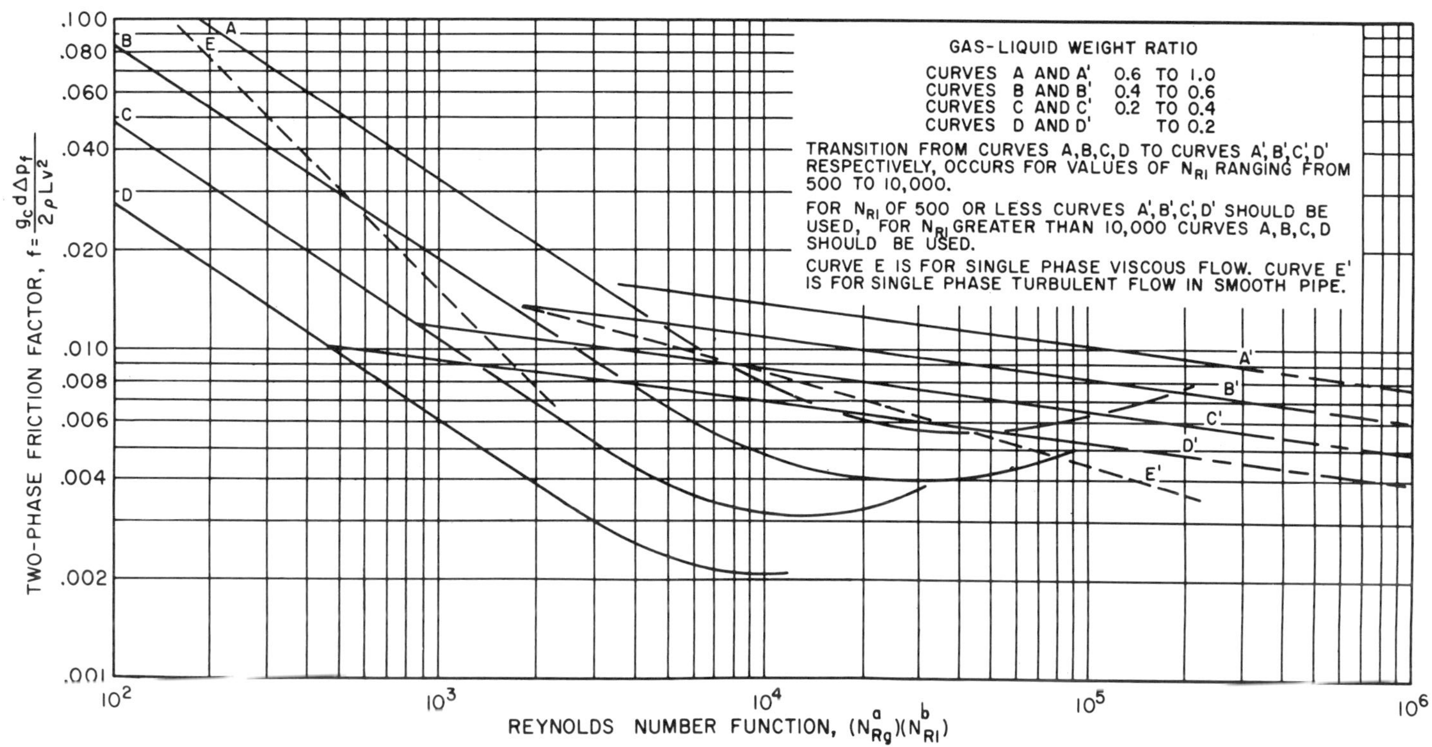

Fig. 1.18 Composite energy dissipation chart. (Bertuzzi, Tek, and Poettmann,[33] *Trans.* AIME.)

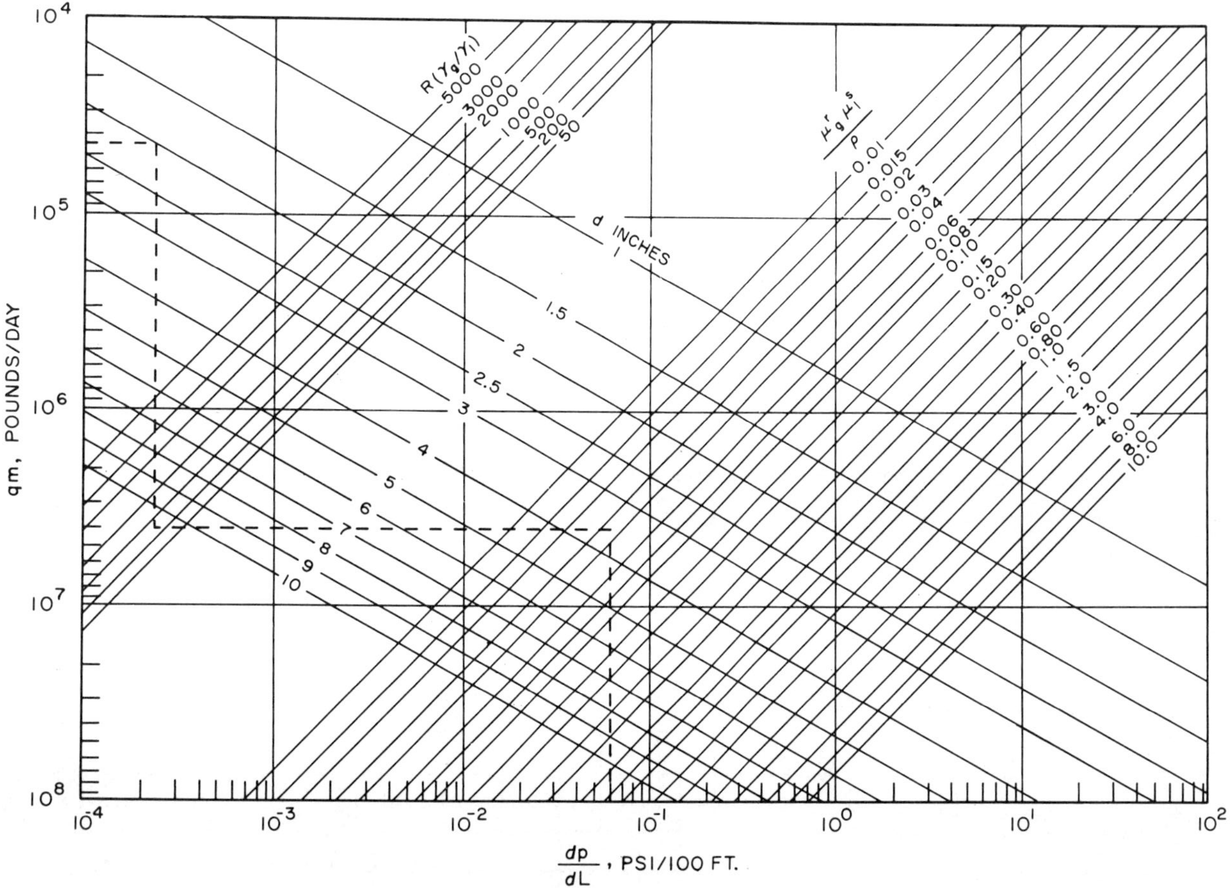

Fig. 1.19 Chart for estimating pressure drop for multiphase fluid flow in pipes. (Bertuzzi, Tek, and Poettmann,[33] *Trans.* AIME.)

lation developed by Bertuzzi[33] and co-workers has an algebraic average deviation of +0.82 per cent, and Martinelli's correlation has an algebraic average deviation of +14.4 per cent.

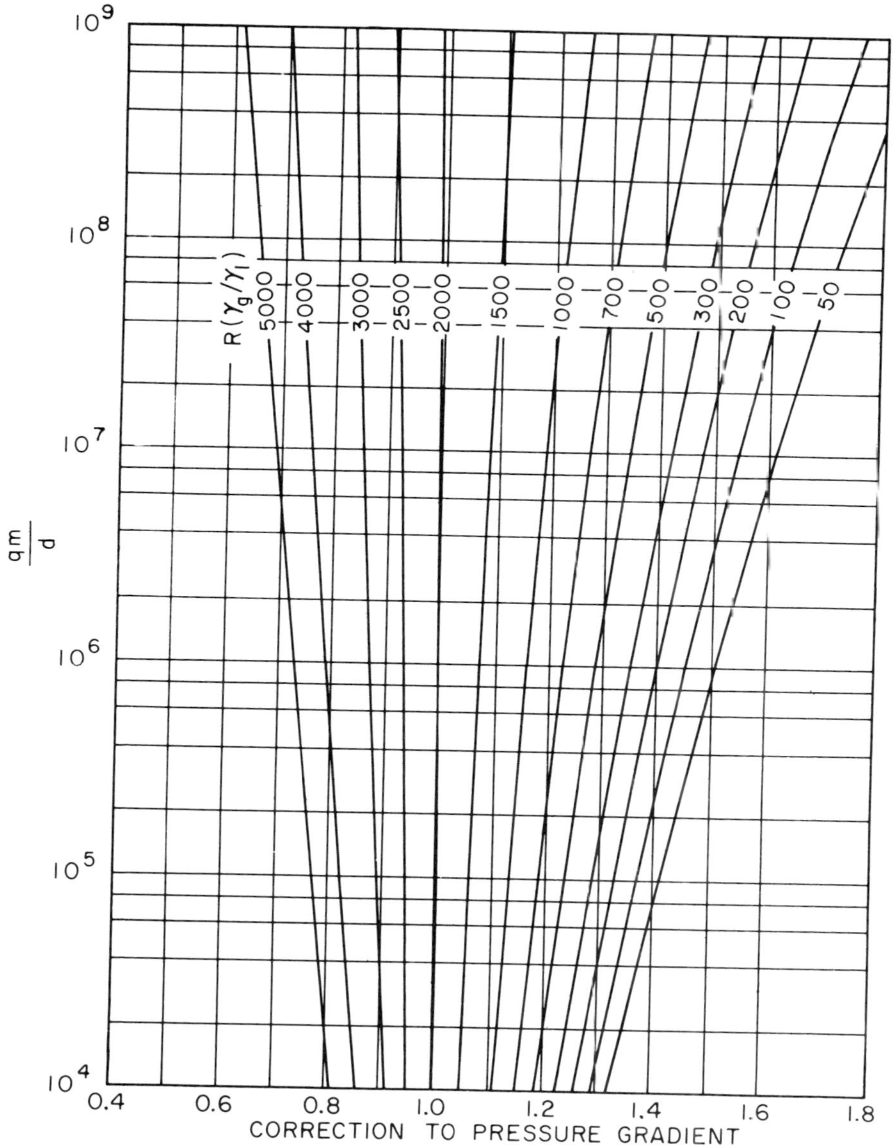

Fig. 1.20 Multiplication correction for Figure 1.19. (Bertuzzi, Tek, and Poettman,[33] *Trans.* AIME.)

To simplify the calculation of the pressure drop, the equations are placed in graphical form as presented in Figs. 1.19 and 1.20,

where q = stock-tank barrels of liquid flowing per day
m = total mass of liquid and gas associated with one stock-tank barrel of liquid, lbm/STB
qm = mass flow rate, lbm/day
R = producing gas-oil ratio, standard cubic feet per stock-tank barrel of oil
R_s = gas solubility in oil at pressure p, standard cubic feet per stock-tank barrel
γ_g = gas gravity (air = 1)
γ_l = specific gravity of liquid (water = 1)
ρ = flowing density of total fluid entering at pressure p and temperature T_{avg}, lbm/cu ft
μ_g = viscosity of gas at pressure p, lbm/ft sec
μ_l = viscosity of liquid at pressure p, lbm/ft sec
r = $0.12a$
s = $0.12b$

Figures 1.19 and 1.20 should not be used where the gas-liquid volume ratio is greater than approximately 5,000 cu ft/bbl. The procedure and equations for calculating the two-phase pressure drop in flow through a horizontal flow string are as follows:

1. The values of m and ρ are found from the following relationships:

$$m = 5.61 \times 62.4\gamma_l + 0.0764\gamma_g R \quad \text{lbm/STB} \tag{1.104}$$

where the factor 5.61 represents cu ft/bbl, 62.4 is the density of water in lbm/cu ft at 60°F, 0.0764 is the density of air in lbm/cu ft at 14.7 psia and 60°F, and the other terms are as previously defined. The density is calculated from

$$\rho = \frac{m}{V_m} \quad \text{lbm/cu ft}$$

where V_m, in cubic feet, is the volume of gas and liquid at pressure p per stock-tank barrel of liquid.

$$V_m = 5.61B_o + \frac{p_{sc}T_{avg}z}{pT_{sc}}(R - R_s) \quad \text{cu ft/STB} \tag{1.105}$$

where B_o = formation volume factor of oil at pressure p, in barrels of pipeline oil per barrel of stock-tank oil. This value should be measured at the average flow line temperature, but since this usually cannot be done, the formation volume factor at reservoir temperature is used.

p = pressure, psia
p_{sc} = base pressure, 14.7 psia
T_{avg} = arithmetic average of the inlet and outlet pipe temperatures
T_{sc} = base temperature, 520°R
z = gas compressibility factor

If the gas solubility is nearly zero, as in air-water mixtures, or if the line pressure is low, the gas solubility R_s is neglected, so that

$$V_m = 5.61 + \frac{p_{sc}T_{avg}zR}{pT_{sc}} \quad \text{cu ft/STB} \tag{1.106}$$

Generalized correlations for gas solubility and formation volume factor, liquid viscosity, gas viscosity, and the compressibility factor are found in most reservoir engineering textbooks.

2. Select the pipe diameter.
3. Calculate the gas-liquid ratio function $R(\gamma_g/\gamma_l)$.
4. Calculate the value of $(\mu_g^r \mu_l^s)/\rho$. This step establishes the pressure gradient for a given pressure. Since this function varies with pressure, it is best calculated for several pressures along the flow line, the corresponding pressure gradients (dp/dL) being determined from Fig. 1.19. As a first approximation, the fluid viscosities at outlet conditions are used. If the pressure drop is less than 100 psi, the change in fluid viscosity with pressure is neglected, and if the line pressure is low, the gas solubility is also neglected.
5. The pressure gradients (dp/dL) are corrected upon multiplication by a suitable factor obtained from Fig. 1.20.
6. The variation in pressure along the flow line is obtained by incremental integration, and a plot of p versus cumulative length L gives the pressure for any length of line.

Example 1.18 shows the design of a two-phase flow line.

Example 1.18. A well is being produced by gas lift. It is desirable to maintain a low wellhead pressure for efficient lifting; therefore, it is of interest to know the wellhead pressure when a 2-in. flow line is used. It is given that:

Oil production = 400 STB/day
Producing gas-oil ratio = 1000 SCF/STB
Flow line diameter = 2 in. (1.995 in. ID)
Length of line to separator = 1000 ft
Separator pressure = 50 psia
Average temperature = 75°F
Gas gravity = 0.65 (air = 1)
Liquid gravity = 0.80 (water = 1)
Gas viscosity = 0.014 cp
= 9.4×10^{-6} lbm/ft sec at 75°F and 50 psia

Liquid viscosity = 0.6 cp
= 4×10^{-4} lbm/ft sec at 75°F and 50 psia
Standard conditions = 14.65 psia and 60°F

SOLUTION: From Eq. (1.104), the total mass of flowing fluid per stock-tank barrel is

$$m = 5.61 \times 62.4 \times 0.80 + 0.0764 \times 0.65 \times 1000 = 329.7 \text{ lbm/STB}$$

and the mass flow rate is

$$qm = 400 \times 329.7 = 131{,}880 \text{ lbm/day}$$

This value establishes the starting point on Fig. 1.19. The gas-liquid ratio function is

$$R\left(\frac{\gamma_g}{\gamma_l}\right) = 1000 \times \frac{0.65}{0.80} = 812.5$$

The physical property function $\mu_g^r\mu_l^s/\rho$ locates another point on Fig. 1.19. The pressure drop is expected to be less than 100 psi; therefore, the change of fluid viscosity with pressure is neglected and the viscosity at the outlet conditions is used. Similarly, if the pressure level is low, the gas solubility is neglected and Eq. (1.106) is used for calculating V_m. The numerator of the physical property function is obtained as follows: From

$$a = \frac{K}{K+1}$$

where

$$K = 0.000219\, R(\gamma_g/\gamma_l) = 0.000219 \times 812.5 = 0.178$$

we have

$$a = \frac{0.178}{0.178 + 1} = 0.151$$

$$b = 1/e^{0.1K} = 1/e^{0.0178} = 0.980$$

and the values of the exponents are

$$r = 0.12a = 0.12 \times 0.151 = 0.018$$

$$s = 0.12b = 0.12 \times 0.980 = 0.118$$

whence

$$\mu_g^r = (9.4 \times 10^{-6})^{0.018} = 0.812$$

$$\mu_l^s = (4.0 \times 10^{-4})^{0.118} = 0.397$$

and

$$\mu_g^r\mu_l^s = 0.322$$

The flow density is

$$\rho = m/V_m = 329.7/V_m$$

and V_m, the cubic feet of gas and liquid flowing per stock-tank barrel of liquid, from Eq. (1.106) is

$$V_m = 5.61 + \frac{14.65 \times 535 \times 1000}{520}\frac{z}{p}$$

The values of p are assumed to be 50, 60, 70, 80, 90, and 100 psia. V_m corresponding to these pressures is given in Table 1.7. The values of the function $\mu_g^r \mu_l^s / \rho$ are

Table 1.7. CALCULATION OF ESTIMATED GRADIENTS

(1) p, psia	(2) z	(3) V_m	(4) $\rho = m/V_m$	(5) $\mu_g^r \mu_l^s$	(6) $\frac{\mu_g^r \mu_l^s}{\rho}$	(7) $\frac{dp}{dL}$, psi/100 ft
50	0.99	304.05	1.084	0.322	0.297	5.7
60	0.99	254.31	1.296	0.322	0.248	4.6
70	0.99	218.79	1.504	0.322	0.214	4.0
80	0.985	191.19	1.724	0.322	0.187	3.6
90	0.985	170.57	1.933	0.322	0.167	3.25
100	0.985	153.32	2.150	0.322	0.150	3.0

calculated and dp/dL is read from Fig. 1.19. The value of qm/d, where d is the pipe diameter in feet, is

$$\frac{qm}{d} = 131{,}880 \times \frac{12}{1.995} = 793{,}260 \text{ lbm/day ft}$$

Using a gas-liquid ratio function of 812.5, a correction factor of 1.21 is read from Fig. 1.20. The corrected values of dp/dL are given in Table 1.8, Col. (4). Table 1.8 further shows the incremental integration of dp/dL. A plot of p versus cumulative L gives a pressure of 97 psia at the wellhead. Since the separator pressure is 50 psia, the pressure drop is 47 psi.

Table 1.8. STEPWISE INTEGRATION

(1) p	(2) Δp	(3) dp/dL	(4) Correct dp/dL	(5) $(dp/dL)_{\text{avg}}$	(6) $\Delta L = (2)/(5)$	(7) L_{cum} ft
50	—	5.7	6.9	—	—	0
60	10	4.6	5.6	6.25	160	160
70	10	4.0	4.8	5.2	192	352
80	10	3.6	4.4	4.6	217	569
90	10	3.25	3.9	4.15	241	810
100	10	3.0	3.6	3.75	267	1077

PROBLEMS

1.1. (a) Complete the following table of conversion factors for volumetric flow rate. For example, 1 cu ft/sec = 448.80 gal/min.

ft^3/sec	gpm	bbl/day	cc/hr
1	448.80		
	1		
		1	
			1

Ans. (Column 1): 1, 2.23×10^{-3}, 6.50×10^{-5}, 9.81×10^{-9} ft³/sec.
(b) Prepare a similar table of conversion factors for pressure which includes the units: atm, psi, dynes/sq cm, in. Hg (0°C). *Ans.* 1 atm = 14.696 psi = 1.0133×10^6 dynes/sq cm = 29.921 in. Hg (0°C).

1.2. Develop a dimensionally homogeneous expression for the potential energy E_p imparted to a body of mass m by raising it a vertical distance z. *Ans.* $E_p = mgz/g_c$.

1.3. At sea level, a body weighs 1400 pounds. Calculate the mass and weight of the body at an altitude where $g = 28$ ft/sec². *Ans.* 1400 lbm, 1218 lbw.

1.4. Derive a factor for converting an absolute viscosity in centipoises (1 cp = 0.01 poises) to (a) absolute English units of lbm/ft sec, (b) gravitational English units of lbf sec/ft².

1.5. The terminal velocity v_t attained by a spherical particle of radius r_p and density ρ_p while falling through a viscous medium is given by Stokes' law:

$$v_t = \left(\frac{2g}{9}\right)\frac{(\rho_p - \rho_m)r_p^2}{\mu_m}$$

where ρ_m and μ_m are the density and absolute viscosity of the medium, respectively, and g is the local acceleration of gravity.
(a) Show that the above equation is dimensionally homogeneous.
(b) Modify the equation to accept directly the following (inconsistent) units: v_t = ft/min, $g = 32.17$ ft/sec², r_p = mm, μ_m = cp, and specific gravities (water = 1) instead of densities. *Ans.* $(2g/9)$ becomes 428.6.
(c) The dispersed water droplets in a crude oil emulsion have an average radius of 5 microns (1 micron = 0.001 mm) and a specific gravity of 1.08. The continuous oil phase has a specific gravity of 0.858 and a viscosity of 9.0 cp. How many hours would be required for these water droplets to settle 6 ft? *Ans.* 378 hr.

1.6. A 32.0° API oil is flowing at a steady rate of 180 bbl/hr through a 2.900-in. ID gathering line.
(a) Calculate the mass rate of flow in lbm/min. *Ans.* 909.
(b) What is the average flow velocity in ft/sec? *Ans.* 6.12.
(c) The 2.900-in. ID line joins a 3.826-in. ID line through which flows 40.0° API oil at a steady rate of 205 bbl/hr. What is the average flow velocity in the 3.826-in. ID line downstream of the junction? *Ans.* 7.52 ft/sec.

1.7. A 6830-ft gas well producing 0.60 gravity gas (air = 1) through $2\frac{1}{2}$-in. tubing (2.441 in. ID) is flowing at a steady rate of 4.08 MM SCF per day (14.7 psia and 60°F). The following data, obtained by direct measurement or from correlation charts, pertain to flowing bottom-hole and wellhead conditions.

Bottom Hole	*Wellhead*
$p_1 = 2953$ psia	$p_2 = 2490$ psia
$t_1 = 171$°F	$t_2 = 105$°F
$z_1 = 0.88$	$z_2 = 0.80$
$h_1 = 140$ Btu/lbm-mole	$h_2 = -560$ Btu/lbm-mole

For 1 lbm of gas flowing up the tubing from point 1 (well bottom) to point 2 (wellhead), calculate the following:
(a) Volumetric flow rates of gas entering and leaving the tubing. *Ans.* $q_1 = 15.05$ ft³/min, $q_2 = 14.53$ ft³/min.
(b) Average flow velocity at points 1 and 2. *Ans.* $v_1 = 7.72$ ft/sec, $v_2 = 7.45$ ft/sec.
(c) Change in kinetic energy. *Ans.* −0.06 ft lbf/lbm.
(d) Change in potential energy. *Ans.* 6830 ft lbf/lbm.
(e) Change in enthalpy. *Ans.* −31,299 ft lbf/lbm.
(f) Net heat energy transferred to or from the flowing gas. *Ans.* −24,469 ft lbf/lbm or −31.45 Btu/lbm; the negative sign implies that heat energy is transferred from the gas to the formation.

1.8. A pump is handling 45.0° API oil at 90 gpm. Suction and discharge pressures are 6 psig and 56 psig, respectively.
(a) Calculate the mass rate of flow. *Ans.* 602 lbm/min.
(b) How much mechanical energy, expressed as ft lbf/lbm, is transferred to the flowing oil? *Ans.* 144 ft lbf/lbm.
(c) At what rate is mechanical energy transferred to the oil? *Ans.* 86,688 ft lbf/min or 2.63 hp.
(d) If the mechanical efficiency of the pump is 85 per cent, what is the input horsepower to the pump? *Ans.* 3.09 hp.
(e) What additional sources of mechanical energy loss have been overlooked in the calculation of input horsepower?
(f) From the pump discharge the oil flows through a 2.469-in. ID line that discharges to atmospheric storage which is 91 ft higher than the pump. What is the mechanical energy loss along the line per pound mass of flowing oil? *Ans.* 70.2 ft lbf/lbm.
(g) Calculate the increase in potential energy per pound mass of oil that traverses the line. *Ans.* 91.0 ft lbf/lbm.
(h) In part (b) it was found that the pump supplies only 144 ft lbf/lbm to the oil, and yet from the sum of the answers to parts (f) and (g) each pound mass of oil transferred through the line requires an expenditure of 161.2 ft lbf/lbm. Explain this discrepancy.
(i) Calculate the frictional pressure loss along the line. *Ans.* 24.4 psi.

1.9. Show that in laminar pipe flow the average velocity v defined as q/A is half the maximum or center-line velocity of the flow.

1.10. If the oil in Problem 1.8 has a viscosity of 3.5 cp, calculate the Reynolds number of the flow in the 2.469-in. ID line. *Ans.* 26,397.

1.11. (a) Develop Eq. (1.17a) from Eq. (1.17).
(b) Develop Eq. (1.18a) from Eq. (1.18).

1.12. By equating (1.17a) to (1.18a) and solving for f, show that the friction factor–Reynolds number relationship for laminar pipe flow is given by $f = 16/N_R$.

1.13. What friction factor curve in Fig. 1.4 should be used for a 3.068-in. ID plastic-coated pipe which has an absolute roughness of 0.00025 in.? *Ans.* The 7.98 in. (8-in.) curve.

1.14. (a) For Reynolds numbers between 5000 and 100,000 the friction factor curves in Fig. 1.4 may be approximated by a straight line passing through the points $f = 0.0094$ at $N_R = 5000$ and $f = 0.0048$ at $N_R = 100{,}000$. From the equation for this straight line, derive an approximate friction factor–Reynolds number relationship for turbulent pipe flow. *Ans.* $f = 0.061 N_R^{-0.22}$.
(b) Plot this approximate (linear) relationship on Fig. 1.4.

1.15. A crude oil pipeline, 20 miles long, handles per day 40,000 bbl of 37.0° API crude with a viscosity of 9.5 cp. The line is constructed from 8-in. nominal diameter schedule number 80 steel pipe whose ID is 7.625 in. Calculate the following quantities:
(a) Oil density. *Ans.* 7.00 ppg.
(b) Average flow velocity. *Ans.* 8.20 ft/sec.
(c) Reynolds number of the flow. *Ans.* 42,750.
(d) Friction factor from Fig. 1.4. *Ans.* 0.0055.
(e) Frictional pressure loss along the line. *Ans.* 1390 psi.
(f) Pressure loss corrected for an assumed line efficiency of 90 per cent. *Ans.* 1716 psi.

1.16. The pipeline of Prob. 1.15 discharges to atmospheric storage which is 400 ft higher than the inlet.
(a) What is the loss in hydrostatic pressure between inlet and outlet? *Ans.* 145 psi.
(b) Calculate the pump discharge pressure, at the line inlet, required to maintain a flow rate of 40,000 bbl/day. *Ans.* 1861 psig.

1.17. The pipeline considered in the two preceding problems is switched to a standby pump whose discharge pressure is limited to 1000 psig. The reduced capacity of the system is to be determined.
(a) The frictional pressure loss along the line may not exceed what value? *Ans.* 855 psi.
(b) As the efficiency of the line was assumed to be 90 per cent, what value of Δp_f should be used in all calculations which involve the use of the friction factor or transmission factor curves of Fig. 1.4? *Ans.* 692.5 psi.
(c) Calculate the Karman number of the flow (with $\Delta p_f = 692.5$ psi). *Ans.* 2238.
(d) Determine the transmission factor for the line. *Ans.* 13.00.
(e) Calculate the average flow velocity. *Ans.* 5.58 ft/sec.
(f) What is the reduced capacity of the line when on the standby pump? *Ans.* 27,220 bbl/day.
(g) Over the range of flow rates investigated in this and the previous problem, frictional pressure loss varies with what power of flow rate? *Ans.* $\Delta p_f \propto q^{1.81}$.

1.18. A salt-water disposal well is tubed to bottom with 3500 ft of 4-in. ID cement-lined tubing. The static bottom-hole pressure is 1100 psia, and the capacity index of the well is 0.90 bbl water/hr/psi (of bottom-hole injection pressure in excess of static pressure). For an injection rate of 9000 bbl/day of 1.04 specific gravity brine, which has a viscosity of 1.12 cp, calculate the following:
(a) Bottom-hole injection pressure. *Ans.* 1517 psia.
(b) Hydrostatic pressure produced by a 3500 ft vertical column of brine. *Ans.* 1576 psi.
(c) Density of brine. *Ans.* 8.67 ppg.
(d) Average flow velocity down the tubing. *Ans.* 6.70 ft/sec.
(e) Reynolds number of the tubing flow. *Ans.* 1.925×10^5.
(f) Fanning friction factor from Fig. 1.4. *Ans.* $f = 0.0042$.
(g) Frictional pressure loss in the tubing. (Although cement-lined pipe has an absolute roughness no greater than clean steel pipe, an E factor of 0.90 should be used in design calculations as a safety factor.) *Ans.* 68 psi.
(h) Wellhead or surface injection pressure. *Ans.* 9 psia, i.e., 5.7 psi below atm or 11.6 in. Hg vacuum.

1.19. From a centralized collection, filtration, and pumping station, 9000 bbl/day of salt water is pumped 5 miles through an asbestos-cement pipeline to the disposal well considered in the previous problem. The elevation of the pump discharge is 80 ft above the wellhead, and the discharge pressure is 150 psig. Asbestos-cement pipe is available in internal diameters of 4, 5, 6, 7, 8, and 10 in., and the minimum size pipe is to be selected for the pipeline.
(a) Calculate the maximum allowable value of frictional pressure loss along the pipeline. *Ans.* 191.7 psi.
(b) As a tentative selection, calculate the average flow velocity, the Reynolds number, the friction factor, and the frictional pressure loss in a 4-in. ID line. Assume an E factor for the pipe of 0.90. *Ans.* 6.70 ft/sec, 1.925×10^5, 0.0042, 516 psi.
(c) Repeat the previous calculations for a 6-in. ID line. *Ans.* 2.98 ft/sec, 1.284×10^5, 0.0044, 71.3 psi.
(d) From a log-log plot of the two line diameters investigated versus their associated pressure loss, estimate the minimum pipe size for the line, by linear interpolation. *Ans.* minimum ID = 4.9 in., use 5-in. ID pipe.

1.20. (a) Combine the approximate friction factor–Reynolds number relationship given in Prob. 1.14 with the Fanning equation to obtain an expression, in practical engineering units, for the direct calculation of frictional pressure loss in turbulent pipe flow, having the form: $\Delta p_f = \phi\ (\mu, \rho, q, L, d)$.

Ans.
$$\Delta p_f = \frac{\mu^{0.22}\rho^{0.78}q^{1.78}\, L}{9359\ d^{4.78}}$$

(b) By the use of this last equation, recalculate part (f) of Probs. 1.15 and 1.17, and explain the discrepancy to be found in these present and previous results. *Ans.* $\Delta p_f = 1475/(0.90)^2 = 1821$ psi; $q = 762.4$ gpm or 26,140 bbl/day for $\Delta p_f = 692.5$ psi.

(c) By taking the derivative of the equation given in part (a) of this problem with respect to d, calculate the approximate percentage error in Δp_f which would result from a 2 per cent error in a measurement of d. Similarly, calculate the approximate error in Δp_f which would result from a 10 per cent error in μ. *Ans.* -9.6 per cent, 2.2 per cent.

1.21. The analysis of a complex pipe system can be simplified by introducing the concept of an equivalent line. An equivalent line is defined as that single line of uniform diameter d_e and length L_e whose flow rate–pressure loss characteristics would be the same as those of a more complex system.
(a) By use of the equation given in part (a) of Prob. 1.20, show that a series combination of two lines, having diameters d_1 and d_2 and lengths L_1 and L_2, may be represented by an equivalent line of length $(L_1 + L_2)$ and diameter

$$d_e = \left[\frac{(L_1 + L_2)\, d_1^{4.78}}{L_1 + L_2(d_1/d_2)^{4.78}}\right]^{1/4.78}$$

(b) Show that a parallel combination of two lines, having diameters d_1 and d_2 and common length L, may be represented by an equivalent line of length L and diameter

$$d_e = (d_1{}^{2.68} + d_2{}^{2.68})^{1/2.68}$$

(c) Describe a possible source of error in using the concept of an equivalent line when L_e is made larger than L, the length of the actual pipe system.

1.22. (a) Evaluate the Lamb coefficient C_L, defined in Eq. (1.22), and also the approximate annular coefficient C_s, defined in Eq. (1.25), for values of the annular geometry parameter $\alpha = 0.2$ and 0.6. *Ans.* For $\alpha = 0.2$, $C_L = 0.4257$ and $C_s = 0.4096$; for $\alpha = 0.6$, $C_L = 0.0689$ and $C_s = 0.0683$.
(b) With $\alpha = 0.6$, what is the percentage error in q as calculated from the approximate annular flow equation, Eq. (1.25)? *Ans.* -0.87 per cent.

1.23. (a) Develop the three forms of Eq. (1.21a) from Eq. (1.21).
(b) Develop Eq. (1.26a) from Eq. (1.26).

1.24. In designing a hydraulic fracture treatment for an 8000-ft well, the surface injection pressure is to be determined. A bottom-hole treating pressure of 2500 psia is desired while injecting 40 bbl/min of an 8.0 ppg fracture fluid which has an absolute viscosity of 100 cp. If the fluid is pumped through the annulus between 2-in. tubing (2.375 in. OD, 1.995 in. ID) and 7-in. casing (7 in. OD, 6.366 in. ID), calculate the following:
(a) Hydrostatic pressure at the bottom of the well produced by an 8000-ft annular column of fracture fluid. *Ans.* 3320 psi.
(b) Annular geometry factor α, and the diameter of a circular pipe equivalent to the annular section. *Ans.* $\alpha = 0.373$, $d_e/d_o = 0.81$, $d_e = 5.156$ in.
(c) Average flow velocity through the equivalent pipe. *Ans.* 25.82 ft/sec.
(d) Reynolds number of the flow through the equivalent pipe. *Ans.* 9883.
(e) Friction factor for the equivalent pipe flow. *Ans.* 0.0078.
(f) Frictional pressure loss in the annular flow. *Ans.* 2502 psi.
(g) Surface injection pressure in psig, while fracturing through the annulus. *Ans.* 1667 psig.

1.25. The surface injection pressure, as calculated in the previous problem, can be reduced by pumping the fracture fluid into a common manifold connected to both the tubing and the annulus.

(a) Although the annulus is equivalent to a circular pipe of diameter 5.156 in., the tubing-annulus combination is equivalent to two lines in parallel having internal diameters of 1.995 in. and 5.156 in. Calculate the diameter of the single line which is equivalent to this parallel combination. *Ans.* 5.303 in.

(b) For the same injection rate of 40 bbl/min, calculate the average flow velocity, the Reynolds number, the friction factor, and the frictional pressure loss in the 5.303-in. ID equivalent line. *Ans.* 24.41 ft/sec, 9610, 0.0079, 2202 psi.

(c) What surface injection pressure is required while fracturing through the tubing-annulus combination? *Ans.* 1367 psig.

(d) Although a reduction of 300 psi in surface injection pressure may appear to be of little consequence, calculate the corresponding reduction in power input to the fracture fluid from the injection pumps. *Ans.* 294 hp.

1.26. Integrate by parts the expression given in Eq. (1.40) to obtain Eq. (1.41). Recall that

$$\int_a^b U\,dV = UV\Big]_a^b - \int_a^b V\,dU$$

and proceed by setting

$$U = \int_\tau^{\tau_w} \phi(\tau)\,d\tau$$

and

$$dV = \tau\,d\tau$$

1.27. Derive Eq. (1.42) from Eq. (1.41), and show that the term $\frac{1}{3}(\tau_y/\tau_w)^4$ may be dropped from Eq. (1.42) without introducing an error greater than 2 per cent, provided that τ_w is greater than 2.5 τ_y.

1.28. (a) Develop Eq. (1.44a) from Eq. (1.44).
(b) Develop Eq. (1.45) from Eq. (1.43).
(c) Develop Eq. (1.45a) from Eq. (1.45).
(d) Develop Eq. (1.46a) from Eq. (1.46).

1.29. From Eqs. (1.46a) and (1.47a) with $N_{Re} = 2000$, derive Eq. (1.48a).

1.30. A 14.9 ppg cement slurry having a plastic viscosity of 66 cp and a yield point of 140 lbf/100 sq ft is pumped through a 200 ft length of 2.900-in. ID pipe.

(a) At what value of average flow velocity will the flow become turbulent? *Ans.* 13.37 ft/sec.

(b) At a steady flow rate of 180 gpm, calculate the average flow velocity and the frictional pressure loss along the line. *Ans.* 8.74 ft/sec, 52.0 psi.

(c) For a turbulent flow rate of 500 gpm, calculate the average flow velocity, the Reynolds number, the friction factor, and the frictional pressure loss along the line ($E = 0.90$). *Ans.* 24.28 ft/sec, 1.475×10^4, 0.0072, 209 psi.

1.31. Derive the general flow equation for laminar slot flow, as given in Eq. (1.49), in terms of the geometric parameters given in Fig. 1.10.

1.32. Show that Eq. (1.49), upon integration by parts, will reduce to Eq. (1.50).

1.33. Replace the function $k\phi(\tau)$ in Eq. (1.50) by its Newtonian equivalent $g_c\tau/\mu$, and carry out the integration of Eq. (1.50) to obtain the expression for laminar slot flow of a Newtonian fluid as given in Eq. (1.24).

1.34. (a) Develop Eq. (1.53) from Eqs. (1.51) and (1.52), but neglect the term $\frac{1}{2}(\tau_y/\tau_w)^3$.
(b) Develop Eq. (1.54a) from Eq. (1.54).

1.35. Solve Eq. (1.24) for Δp_f in terms of μ, L, v, and W, and show that upon replacing μ by the quantity μ_{es}, as given in Eq. (1.58), Eq. (1.53) is the result.

1.36. With N_{Ra} set equal to 2000, derive Eq. (1.62a) from Eqs. (1.60a) and (1.61a).

1.37. A string of $5\frac{1}{2}$-in. OD casing is being cemented inside an $8\frac{1}{2}$-in. wellbore (bit size). The physical properties of the cement slurry are as follows: density = 15.7 ppg, plastic viscosity = 55 cp, yield point = 138 lbf/100 sq ft.
(a) At what displacement rate, in cu ft/min, will the annular flow of the cement become turbulent? *Ans.* $v_{ca} = 11.06$ ft/sec, $q = 152$ ft^3/min.
(b) For a displacement rate of 100 cu ft/min, calculate the frictional pressure loss per 1000 ft of cement in the annulus. *Ans.* $v = 7.28$ ft/sec, $\Delta p_f = 274$ psi/1000 ft.

1.38. Calculate the circulating bottom-hole pressure, CBHP, for the well system considered in Sec. 1.10. *Ans.* 6516 psia.

1.39. In the same well as studied in Sec. 1.10, a slight loss in circulation (annular returns) is observed while drilling at 10,000 ft. The pumping rate is immediately reduced from 56 strokes/min (400 gpm) to 38 strokes/min (271 gpm) in an effort to reduce the pressure on the open hole. At this reduced circulation rate, calculate the following:
(a) Actual velocities inside the drill pipe and drill collars, and in the annulus around the drill collars and the drill pipe. *Ans.* $v_p = 6.05$, $v_c = 12.30$, $v_{ac} = 2.96$, $v_{ap} = 1.53$ ft/sec.
(b) Frictional pressure losses (1) in the drill pipe and surface connections, (2) in the drill collars, (3) across the bit, (4) in the annulus around the collars, (5) in the annulus around the drill pipe. *Ans.* (1) 352.6, (2) 75.1, (3) 739.9, (4) 30.1, (5) 223.5 psi.
(c) Pump discharge pressure and the CBHP. *Ans.* 1421 psig, 6498 psia.

1.40. (a) From the results of the two previous problems and those developed in Sec. 1.10, calculate the reduction in pump discharge pressure and the corresponding decrease in CBHP to be realized upon reducing the mud circulation rate from 400 to 271 gpm. *Ans.* 1291 psi in standpipe pressure, 18 psi in CBHP.
(b) Explain the causes of this apparent anomaly.

1.41. After a string of casing has been run, the hole is usually circulated with mud for about one hour to "clean up" the annulus. Though company policies may vary, a general practice is to circulate at a rate such that the annular velocity will not exceed that value which was maintained during drilling, in order to minimize the pressure on the open hole.

(a) If a string of 7-in. OD casing were run to 10,000 ft in the well studied in Sec. 1.10, what circulation rate would normally be maintained while conditioning the hole before cementing? *Ans.* 267 gpm.

(b) Calculate the CBHP while conditioning the hole and compare this value with that calculated in Prob. 1.38. *Ans.* 6710 psia, or 194 psi greater than the CBHP while drilling.

(c) Could the excess in CBHP be eliminated by reducing the mud circulation rate while conditioning the hole?

(d) Suppose that having drilled to a total depth of 10,000 ft the drilling mud were conditioned before running the casing so that, while maintaining its density at 12.0 ppg, its plastic viscosity were lowered to 35 cp and its yield point reduced to 5 lbf/100 sq ft. Calculate the CBHP with the casing on bottom while maintaining the same circulation rate of the new mud as before, 267 gpm. *Ans.* 6427 psia, or 89 psi less than the CBHP while drilling with the old mud.

1.42. (a) The mud pump on the drilling rig considered in Sec. 1.10 has an input horsepower (IHP) rating of 1000 hp and a mechanical efficiency of 85 per cent. If the IHP rating is not to be exceeded, what is the maximum discharge pressure for this pump when handling mud at (a) 766 gpm and (b) 400 gpm? *Ans.* (a) 1900 psig, (b) 3640 psig (with suction at 1 atm).

(b) The drilling contractor has found, however, that operating at pump pressures in excess of 3000 psig will greatly reduce the service life of the pump pistons (swab rubbers) and liners. If the pump pressure is not to exceed 3000 psig, how much additional hole can be drilled (below 10,000 ft) with $q = 400$ gpm before it becomes necessary to reduce the circulation rate and/or increase the size of the jet bit nozzles? *Ans.* 3146 ft.

1.43. It is desired to calculate the length of $4\frac{1}{2}$-in., 16.6-lbw drill pipe (ID = 3.826 in.) which would provide the same frictional pressure loss as the surface connections listed as combination No. 3 in Table 1.4.

(a) By use of the equation given in part (a) of Prob. 1.20, show that a given pipe of length L and diameter d may be represented by an equivalent pipe of diameter d_e and length

$$L_e = L\left(\frac{d_e}{d}\right)^{4.78}$$

(b) With reference to Fig. 1.5, the 180° bend at the top of the 4-in. ID standpipe is equivalent to what length of straight 4-in. ID pipe? *Ans.* 24 ft.

(c) What length of $4\frac{1}{2}$-in. drill pipe is equivalent to the (45 + 24) ft length of 4-in. ID standpipe? *Ans.* 55.8 ft.

(d) The 55-ft length of 3-in. ID drilling hose is equivalent to what length of $4\frac{1}{2}$-in. drill pipe? *Ans.* 175.7 ft.

(e) Although the $2\frac{1}{2}$-in. ID swivel washpipe is only 5 ft in length, in combination with the gooseneck (considered to be a 180° bend) it is equivalent to what total length of $2\frac{1}{2}$-in. ID pipe? *Ans.* 21 ft.
(f) The swivel washpipe and gooseneck is equivalent to what length of $4\frac{1}{2}$-in. drill pipe? *Ans.* 160.3 ft.
(g) What length of $4\frac{1}{2}$-in. drill pipe is equivalent to the 40-ft kelly having a bore of $3\frac{1}{4}$ in.? *Ans.* 87.2 ft.
(h) Combination No. 3 of surface connections is equivalent to what length of $4\frac{1}{2}$-in. drill pipe? *Ans.* 479 ft.

1.44. (a) Calculate the total nozzle area for a jet bit such that for a circulation rate of 800 gpm of 9.5 ppg mud the pressure drop across the bit will be 865 psi. *Ans.* $A = 0.8045$ sq in.
(b) What combination of three nozzles will provide (approximately) the required exit area? *Ans.* Two $\frac{9}{16}$-in. and one $\frac{5}{8}$-in. nozzle, $A = 0.8038$ sq in.
(c) Calculate the jet bit horsepower. *Ans.* 404 hp.
(d) Calculate the jet velocity of the mud issuing from the $\frac{9}{16}$-in. and also the $\frac{5}{8}$-in. nozzle. *Ans.* 319 ft/sec, same.

1.45. In drilling a hole of given size, while maintaining a fixed annular velocity, the use of 5-in. drill pipe in place of $4\frac{1}{2}$-in. pipe results in a substantial reduction in parasitic energy losses within the drill string.
(a) While drilling a $9\frac{5}{8}$-in. hole with annular returns maintained at 150 ft/min, calculate the reduction in circulation rate that can be realized when using 5-in. pipe in place of $4\frac{1}{2}$-in. drill pipe. *Ans.* 6.6 per cent.
(b) By use of the equation given in part (a) of Prob. 1.20, calculate the percentage decrease in frictional pressure loss within the pipe to be realized from this 6.6 per cent reduction in circulation rate and the additional reduction caused by the increase in internal diameter of 5-in. pipe (ID = 4.276 in.) over that of $4\frac{1}{2}$-in. pipe (ID = 3.826 in.). *Ans.* 47.9 per cent.
(c) The 6.6 per cent reduction in circulation rate together with a 47.9 per cent reduction in frictional pressure loss will provide what percentage decrease in parasitic energy losses within the drill pipe? *Ans.* 51.3 per cent.

1.46. (a) From the data given in Ex. 1.14, construct the absolute flow curve for the calcium carbonate suspension, i.e., prepare a Cartesian plot of shear stress (lbf/sq ft) versus shear rate (sec^{-1}).
(b) On this same graph, construct flow curves for Newtonian fluids having absolute viscosities of 1, 3, 6, 10, 15, 30, and 100 cp.
(c) Determine the values of shear stress at which the suspension will exhibit apparent viscosities of 30, 15, and 10 cp. *Ans.* 0.74, 0.95, 1.13 lbf/sq ft.
(d) With reference to the general flow curves given in Fig. 1.7, how should the flow behavior of this suspension be classified?
(e) On 5×3 cycle log-log paper, construct a logarithmic plot of shear stress versus shear rate, and determine the slope and intercept value of this curve on τ axis where shear rate equals unity. *Ans.* slope = 0.275, intercept = 0.105.
(f) Determine the equation of the curve developed in part (e). *Ans.* Eq. (1.30) with $K = 0.105$ and $n = 0.275$.

(g) Calculate K from Eq. (1.73), knowing that $K'_p = 0.121$ lbf sec$^{n'}$/ft^2 and $n = n' = 0.275$.
(h) On the logarithmic graph, construct a family of curves which show the shear stress–shear rate behavior of Newtonian fluids having absolute viscosities of 1, 2, 5, 10, 20, 50, 100, 200, 500, and 1000 cp.
(i) From the logarithmic plot, determine the values of shear stress at which the suspension exhibits apparent viscosities of 1000, 200, 50, and 10 cp. *Ans.* 0.196, 0.359, 0.610, 1.12 lbf/sq ft.

1.47. A gelled water fracturing fluid weighing 8.22 ppg has a constant flow-behavior index $n' = 0.508$ and a consistency index $K'_p = 0.0453$ lbf sec$^{n'}$/ft^2.
(a) The fluid is pumped at a steady rate of 22 gpm through $\frac{3}{4}$-in. nominal diameter (0.742-in. ID) steel pipe. Calculate the average velocity and generalized Reynolds number of the flow. *Ans.* $v = 16.32$ ft/sec, $N'_R = 1840$ (laminar).
(b) Calculate the frictional pressure loss in psi/100 ft of pipe run. *Ans.* 99.38 psi/100 ft.
(c) Calculate the shear stress at the wall of the pipe and also the corresponding rate of shear at the wall. *Ans.* $\tau_w = 2.212$ lbf/ft^2, $(-dv_r/dr)_w = 2623$ sec^{-1}.

1.48. By use of the pipe viscometer described in Ex. 1.14, the following laminar flow data were obtained for a gelled water fluid weighing 61.5 lbm/cu ft:

Flow Rate $q \times 10^3$, cu ft/min	*Pressure Drop* Δp_f, psi	*Flow Rate* $q \times 10^3$, cu ft/min	*Pressure Drop* Δp_f, psi
2.511	2.31	60.26	11.38
5.859	3.85	89.56	13.45
10.46	4.74	115.9	16.00
13.39	5.39	151.1	18.19
24.69	7.08	202.8	21.33
40.59	9.18	223.9	22.58

(a) At each flow rate, calculate the shear stress τ_w developed at the pipe wall and also the correlating parameter $(8v/d)$. *Ans.* At $q = 2.511 \times 10^{-3}$ ft^3/min, $\tau_w = 0.2783$ lbf/ft^2, $(8v/d) = 34.288$ sec^{-1}; at $q = 151.1 \times 10^{-3}$ ft^3/min, $\tau_w = 2.1919$ lbf/ft^2, $(8v/d) = 2063.3$ sec^{-1}.
(b) Prepare a log-log plot of τ_w versus $(8v/d)$ on 3×5 cycle logarithmic coordinates, and determine the flow-behavior and consistency indices, n' and K'_p, for the fluid. *Ans.* $n' = 0.508$, $K'_p = 0.0453$ lbf sec$^{n'}$/ft^2.
(c) Notice that this is the same fluid introduced in the preceding problem. Recalculate the generalized Reynolds number of the flow in part (a) of that problem by use of Eq. (1.78a) and the logarithmic plot of τ_w versus $(8v/d)$. *Ans.* $(8v/d) = 2111$ sec^{-1}, $\tau_w = 2.23$ lbf/ft^2 (from graph), $N'_R = 1830$.
(d) If this fluid were pumped at a steady rate of 15 bbl/min through $2\frac{1}{2}$-in. tubing (ID = 2.441 in.), what would be the generalized Reynolds number of the flow, the friction factor, and the frictional pressure loss per 1000 ft of tubing? *Ans.* $N'_R = 14{,}190$, $f = 0.0033$, 1050 psi/1000 ft.
(e) Calculate the equivalent consistency index K'_a for this fluid by use of the solution curves in Fig. 1.15. *Ans.* $K'_a = 0.0468$.

(f) For the analysis of annular flow systems, prepare a plot of τ_w versus $[12v/(d_o - d_i)]$ on the same logarithmic coordinates as were used in part (b).
(g) For a steady flow of 600 gpm of the gelled water through an annulus formed by 6.184-in. ID casing and 2.875-in. OD tubing, calculate the average velocity and generalized Reynolds number of the flow. *Ans.* $v = 8.18$ ft/sec, $[12\,v/(d_o - d_i)] = 356\ \text{sec}^{-1}$, $\tau_w = 0.94\ \text{lbf/ft}^2$ (from graph), $N'_{Ra} = 1630$.
(h) To continue with part (g), calculate the frictional pressure loss per 1000 ft of annulus. *Ans.* 94.7 psi/1000 ft.

1.49. Given below are the data recorded for a drilling fluid in a standard Model 35 V-G Meter ($N = 1.0$).

Cylinder speed, rpm:	600	300	200	100	6	3
Dial reading, θ:	80	51	38	24	3.5	2.5

(a) Assume that the fluid behaves as a power-law fluid, and calculate the flow-behavior index n' and the equivalent consistency index K'_v from the instrument readings at 300 and 600 rpm. *Ans.* $n' = 0.648$, $K'_v = 0.0095$ lbf $\text{sec}^{n'}/\text{ft}^2$.
(b) For each rotational speed, calculate the shear stress τ_b developed at the inner bob of the viscometer. Construct a log-log plot of τ_b versus $[2\omega_c\,\beta^2/(\beta^2 - 1)]$ on 3×5 cycle logarithmic coordinate paper, and determine the indices n' and K'_v from this plot. *Ans.* $n' = 0.670$, $K'_v = 0.0083$ lbf $\text{sec}^{n'}/\text{ft}^2$.
(c) By use of the correlation diagram given in Fig. 1.15, calculate the consistency indices K'_p and K'_a for pipe and annular flow systems. *Ans.* $K'_p = 0.0088$, $K'_a = 0.0090$ lbf $\text{sec}^{n'}/\text{ft}^2$.
(d) For the analysis of pipe and annular flow systems, prepare a plot of τ_w versus $(8v/d)$ and also a plot of τ_w versus $[12v/(d_o - d_i)]$ on the same logarithmic coordinates as were used in part (b).
(e) While circulating this drilling fluid through $4\frac{1}{2}$-in. drill pipe (ID = 3.826 in.) at a steady rate of 400 gpm, calculate the generalized Reynolds number of the flow and also the friction factor and frictional pressure loss per 1000 ft of pipe. The fluid has a density of 9.8 ppg. *Ans.* $(8v/d) = 280\ \text{sec}^{-1}$, $\tau_w = 0.37\ \text{lbf/ft}^2$(from graph), $N'_R = 6136$, $f = 0.0068$, $\Delta p_f = 84.1$ psi/1000 ft.
(f) If this fluid were treated as a Bingham plastic with $\mu_p = 29$ cp and $\tau_y = 22$ lbf/100 sq ft (based on two-point data), calculate the Reynolds number, the friction factor, and the frictional pressure loss for the pipe flow system considered in part (e). *Ans.* $N_R = 13{,}390$, $f = 0.0072$, $\Delta p_f = 89.0$ psi/1000 ft.

1.50. (a) From the data given in Ex. 1.18, what is the correction factor for a 3-in. line? *Ans.* 1.19.
(b) By use of 5-lb increments, calculate the pressure at the wellhead *Ans.* 59 psia.

REFERENCES

[1]M. H. Cullender and R. V. Smith, "Practical Solution of Gas-Flow Equations for Wells and Pipelines with Large Temperature Gradients," *Trans. AIME*, **207,** 281 (1956).

[2]L. F. Moody, "Friction Factors for Pipe Flow," *Trans. ASME*, **66**, 671 (1944).

[3]R. H. Crowe, "Internal Coating of Gas Pipelines" *The Oil and Gas Journal*, April 6, 1959, p. 107.

[4]The Crane Company, *Technical Paper No.* 409 (Chicago, 1942).

[5]Sir Horace Lamb, *Hydrodynamics*, 6th Ed. (New York: Dover Publications, 1945), p. 586.

[6]Bob C. Crittendon, "The Mechanics of Design and Interpretation of Hydraulic Fracture Treatments," *Journal of Petroleum Technology*, **11**, No. 10, 21 (1959).

[7]D. H. Caldwell and H. E. Babbit, "Flow of Muds, Sludges, and Suspensions in Circular Pipe," *Industrial and Engineering Chemistry*, **33**, 249, (1941).

[8]R. W. Beck, W. F. Nuss, and T. H. Dunn, "The Flow Properties of Drilling Muds," *Drilling and Production Practice* (API, 1947), p. 11.

[9]I. Havenarr, "The Pumpability of Clay-Water Drilling Fluids," *Trans. AIME*, **201**, 287 (1954).

[10]B. O. A. Hedstrom, "Flow of Plastic Materials in Pipes," *Industrial and Engineering Chemistry*, **44**, 651 (1952).

[11]*Recommended Practice Standard Procedure for the Evaluction of Hydraulic Fracturing Fluids*, API RP 39, First Edition (July, 1960).

[12]W. M. Laird, "Slurry and Suspension Transport," *Industrial and Engineering Chemistry*, **49**, 138 (1957).

[13]A. G. Fredrickson and R. B. Bird, "Non-Newtonian Flow in Annuli," *Industrial and Engineering Chemistry*, **50**, 347 (1958).

[14]J. C. Melrose, J. G. Savins, W. R. Foster, and E. R. Parish, "A Practical Utilization of the Theory of Bingham Plastic Flow in Stationary Pipes and Annuli," *Trans. AIME*, **213**, 316 (1958).

[15]R. A. Bobo, R. S. Hoch, G. S. Boudreaux, and R. R. Angel, *Keys to Successful Competitive Drilling* (Houston: Gulf Publishing Company, Bock Division, 1958), p. 67.

[16]J. R. Eckel and W. J. Bielstein, "Nozzle Design and its Effect on Drilling Rate and Pump Operation," *Drilling and Production Practice* (API, 1951), p. 28.

[17]*Hydraulics in Rotary Drilling*, Bulletin No. 1-C (Houston: Hughes Tool Company, 1958), p. 14.

[18]J. G. Savins and W. F. Roper, "A Direct-Indicating Viscometer for Drilling Fluids," *Drilling and Production Practice* (API, 1954), p. 7.

[19]B. Rabinowitsch, "Über die Viskosität und Elastizität von Solen," *Z. physik. Chem.*, **A145**, 20 (1929).

[20]M. Mooney, "Explicit Formulas for Slip and Fluidity," *Journal of Rheology*, **2**, 213 (1931).

[21]A. B. Metzner and J. C. Reed, "Flow of Non-Newtonian Fluids — Correlation of the Laminar, Transition, and Turbulent-flow Regions," *Journal of American Institute Chemical Engineers*, **1**, 434 (1955).

[22]L. L. Melton and C. D. Saunders, "Rheological Measurements of Non-Newtonian Fluids," *Trans. AIME*, **210**, 196 (1957).

[23]A. B. Metzner, "Non-Newtonian Flow," *Industrial and Engineering Chemistry*, **49**, 1429 (1957).

[24]D. W. Dodge and A. B. Metzner, "Turbulent Flow of Non-Newtonian Systems," *Journal of American Institute of Chemical Engineers*, **5**, 189 (1959).

[25]J. G. Savins, "Generalized Newtonian (Pseudoplastic) Flow in Stationary Pipes and Annuli," *Trans. AIME*, **213**, 325 (1958).

[26]I. M. Krieger and S. H. Maron, "Direct Determination of the Flow Curves of Non-Newtonian Fluids, III," *Journal of Applied Physics*, **25**, 72 (1954).

[27]N. Van Wingen, "Pressure Drop for Oil-Gas Mixtures in Horizontal Flow Lines," *World Oil*, **122**, 156 (1949).

[28]R. W. Lockhart and R. C. Martinelli, "Proposed Correlation of Data for Isothermal Two-Phase Two-Component Flow in Pipes," *Chemical Engineering Progress*, **45**, 39 (1949).

[29]George E. Alves, "Cocurrent Liquid-Gas Flow in a Pipe Line Contactor," *Chemical Engineering Progress*, **50**, 449 (1954).

[30]Ovid Baker, "Simultaneous Flow of Oil and Gas," *The Oil and Gas Journal*, **53**, No. 12, 185 (1954).

[31]B. T. Yocum, "A Mathematical Study of Two-Phase Flow in Well-Flowlines," *The Petroleum Engineer*, **31**, B-40 (1959); *ibid.*, **32**, B-59 (1960).

[32]Jose A. Chavez, "New Correlation for 2-Phase Pipeline Flow," *The Oil and Gas Journal*, **57**, No. 35, 100 (1959).

[33]A. F. Bertuzzi, M. R. Tek, and F. H. Poettmann, "Simultaneous Flow of Liquid and Gas through Horizontal Pipe," *Trans. AIME*, **207**, 17–24 (1956).

CHAPTER TWO

Design of Casing Strings

2.1. Introduction

As the drilling of an oil or gas well progresses, it becomes necessary to line the walls of the hole with heavy steel pipe called casing. The casing, together with the cement, performs six important functions, namely to (1) prevent caving of the hole, (2) prevent contamination of fresh water in the upper sands, (3) exclude water from the producing formation, (4) confine production to the well bore, (5) provide a means for controlling pressure, and (6) facilitate installation of the subsurface equipment required if artificial lift methods become necessary in producing the well.*[1]

The total length of casing (of uniform outside diameter) which is run in the well during a single operation is called a casing string. There are three principal types of casing strings, the classification being based on the primary function of the string. The surface string protects the fresh-water sands. The intermediate or protection string has as its primary objective the prevention of caving, and thus it facilitates drilling. The oil or production string is the member through which the well is completed, produced, and controlled. Each string of casing in a well has its upper end at the surface, and a completed well may contain several intermediate strings in addition to the surface string and the production string.

Casing itself is classified according to five properties. They are (1) the outside diameter, (2) the wall thickness, (3) the grade of the material of construction, (4) the type of joint, and (5) the length range. The outside

*References throughout the text are given at the end of each chapter.

diameter and the wall thickness determine a sixth property, i.e., unit weight. Casing is most frequently classified by a nominal weight, which is a weight adjusted from the true unit weight to compensate for threaded ends and coupling. For our purposes, a section of a casing string is defined as a continuous length of casing of uniform grade and wall thickness, and with the same type joint throughout. A casing string consisting of more than one section is called a combination string.

When casing is set in a hole, it is subjected to three significant forces. These forces result from an external pressure, an internal pressure, and a longitudinal or axial loading on the casing. External pressure tends to collapse the casing and internal pressure tends to burst the casing. Axial loading may be tension due to dead weight or compression due to buoyancy. Axial tension has two pronounced effects: it tends to pull the casing apart, and it lowers the resistance of the casing to collapse from external pressures. The problem of designing a casing string is essentially one of selecting the most economical grades and weights of casing which will withstand, without failure, the three forces to which the casing will be subjected. Since the forces under consideration are not uniform but vary with depth, it is possible to use a combination string to great advantage. Thus, the most expensive grades and weights can be reserved for the sections of the string which are exposed to the most severe forces, and a properly designed combination string results in impressive savings in money and material. Since the cost of casing is usually the greatest single item of expense in the cost of an oil well,[2] the casing program of a typical well represents an extremely large investment. Selection of casing sizes, weights, grades, and joints for a given situation, therefore, constitutes an engineering and economic problem of great importance.

2.2. Grades of Casing

Casing is graded on the basis of its minimum yield strength. The American Petroleum Institute recognizes five grades of casing (one grade being tentative). Table 2.1 lists the grades and the corresponding minimum yield

Table 2.1. GRADES OF CASING

Casing grade	Minimum yield strength, psi
F-25	25,000
H-40	40,000
J-55	55,000
N-80	80,000
P-110	110,000

strengths. The yield strength, for these purposes, is defined as the tensile stress required to produce a total elongation of 0.5 per cent of the length (except in the case of grade P-110 casing, where yield strength is defined as the tensile stress required to produce a total elongation of 0.6 per cent of the length).[3]

2.3. Casing Dimensions

Casing is designated also by the length range into which it falls. API Standard 5A[3] establishes three length ranges with limits and tolerances as shown in Table 2.2. Casing is designated also by (1) outside diameter and

Table 2.2. LENGTH RANGES OF CASING
(*After Reference 3*)

Range	Length range, ft	Minimum length,[a] ft	Maximum length variation,[a] ft
1	16–25	18	6
2	25–34	28	5
3	Over 34	36	6

[a]Tolerances given are for 95 per cent or more of carload when shipment from mill to final destination is entirely by rail.

(2) wall thickness or nominal weight. Table 2.3, taken from API Standard 5A,[3] lists 74 sizes of casing, varying in diameter from 4.5 to 20 in., in wall thickness from 0.205 to 0.595 in., and in weight from 9.50 to 94.00 lb per ft.

Table 2.3. API CASING LIST
(*After Reference 3*)

Outside diameter, in.	Wall thickness, in.	Nominal weight, lb/ft	Available grades[a]	Available threads[b]
$4\frac{1}{2}$	0.205	9.50	F,H,J	S
	0.250	11.60	J	S,L
	0.250	11.60	N,P	L
	0.290	13.50	N,P	L
	0.337	15.10	P	L
5	0.220	11.50	F,J	S
	0.253	13.00	J	S,L
	0.296	15.00	J	S,L
	0.296	15.00	N,P	L
	0.362	18.00	N,P	L

Table 2.3. (CONTINUED)

Outside diameter, in.	Wall thickness, in.	Nominal weight, lb/ft	Available grades[a]	Available threads[b]
$5\frac{1}{2}$	0.228	13.00	F	S
	0.244	14.00	H,J	S
	0.275	15.50	J	S,L
	0.304	17.00	J	S,L
	0.304	17.00	N,P	L
	0.361	20.00	N,P	L
	0.415	23.00	N,P	L
6	0.238	15.00	F	S
	0.288	18.00	H	S
	0.288	18.00	J	S,L
	0.288	18.00	N	L
	0.324	20.00	N	L
	0.380	23.00	N,P	L
	0.434	26.00	P	L
$6\frac{5}{8}$	0.245	17.00	F	S
	0.288	20.00	H	S
	0.288	20.00	J	S,L
	0.352	24.00	J	S,L
	0.352	24.00	N,P	L
	0.417	28.00	N,P	L
	0.475	32.00	N,P	L
7	0.231	17.00	F,H	S
	0.272	20.00	H,J	S
	0.317	23.00	J	S,L
	0.317	23.00	N	L
	0.362	26.00	J	S,L
	0.362	26.00	N,P	L
	0.408	29.00	N,P	L
	0.453	32.00	N,P	L
	0.498	35.00	N,P	L
	0.540	38.00	N,P	L
$7\frac{5}{8}$	0.250	20.00	F	S
	0.300	24.00	H	S
	0.328	26.40	J	S,L
	0.328	26.40	N	L
	0.375	29.70	N,P	L
	0.430	33.70	N,P	L
	0.500	39.00	N,P	L

[a]Casing grades listed here are those referred to in Sec. 2.2 and Table 2.1.
[b]"S" indicates availability in short threads, "L" in long threads.

Table 2.3. (CONTINUED)

Outside diameter, in.	Wall thickness, in.	Nominal weight, lb/ft	Available grades[a]	Available threads[b]
$8\frac{5}{8}$	0.264	24.00	F,J	S
	0.304	28.00	H	S
	0.352	32.00	H	S
	0.352	32.00	J	S,L
	0.400	36.00	J	S,L
	0.400	36.00	N	L
	0.450	40.00	N,P	L
	0.500	44.00	N,P	L
	0.557	49.00	N,P	L
$9\frac{5}{8}$	0.281	29.30	F	S
	0.312	32.30	H	S
	0.352	36.00	H	S
	0.352	36.00	J	S,L
	0.395	40.00	J	S,L
	0.395	40.00	N	L
	0.435	43.50	N,P	L
	0.472	47.00	N,P	L
	0.545	53.50	N,P	L
$10\frac{3}{4}$	0.279	32.75	F,H	S
	0.350	40.50	H,J	S
	0.400	45.50	J	S
	0.450	51.00	J,N,P	S
	0.495	55.50	N,P	S
	0.545	60.70	P	S
	0.595	65.70	P	S
$11\frac{3}{4}$	0.300	38.00	F	S
	0.333	42.00	H	S
	0.375	47.00	J	S
	0.435	54.00	J	S
	0.489	60.00	J,N	S
$13\frac{5}{8}$	0.330	48.00	F,H	S
	0.380	54.50	J	S
	0.430	61.00	J	S
	0.480	68.00	J	S
	0.514	72.00	N	S
16	0.312	55.00	F	S
	0.375	65.00	H	S
	0.438	75.00	J	S
	0.495	84.00	J	S
20	0.438	94.00	F,H	S

2.4. Casing Threads and Couplings

Individual lengths of casing are usually joined by means of threaded couplings. Couplings are graded in the same manner as casing, and the physical properties of a coupling must be at least equal to those of the casing sections it joins.[3] Couplings are classified also according to the outside diameter and the wall thickness of the casing with which they are to be used. Finally, couplings are classified as either long or short in accordance with the lengths of the threads of the casing with which they are to be used. Table 2.3 shows the couplings available for each size and grade of casing. API Standard 5A gives dimensions of long and short threads and couplings as shown in Fig. 2.1 and Tables 2.4 and 2.5. When casing

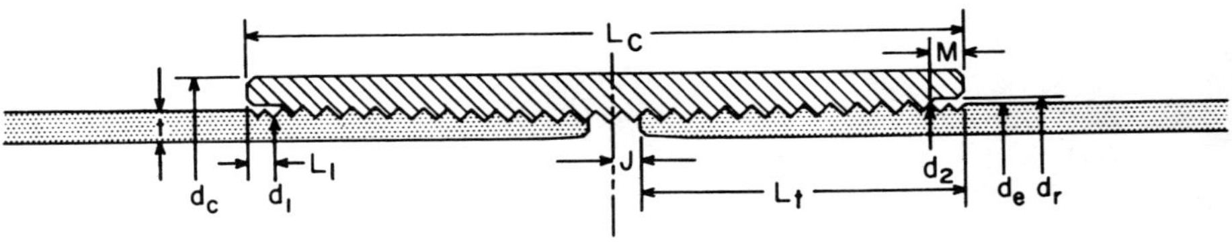

Fig. 2.1 Dimensions of threads and couplings.

d_e Outside diameter of casing.
d_c Outside diameter of couplings.
d_1 Pitch diameter at distance L_1 from vanish point.
d_2 Pitch diameter at hand-tight plane.
J Distance from end of casing to center of coupling for power-tight make-up.
L_c Coupling length.
L_t Distance from end of casing to vanish point of threads.
L_1 Distance from vanish point of casing threads to plane of d_1.
M Distance from face of coupling to hand-tight plane.
d_r Diameter of coupling recess.
t Casing wall thickness.

For numerical values of dimensions, see Tables 2.4 and 2.5.

is furnished with the threads and couplings, the length is taken to be the overall length of pipe plus attached coupling, i.e., the length as measured from the uncoupled end of the pipe to the outer face of the coupling at the opposite end, with the coupling made up power-tight.[3] When two or more lengths are made up to form a section or a string of casing, the overall length of the string is less than the sum of the individual lengths of casing

Table 2.4. DIMENSIONS OF SHORT THREADS AND COUPLINGS[a]
(After Reference 3)

d_e	t	d_1	d_2	J	L_c	L_t	L_1	M	d_r	d_c
$4\frac{1}{2}$	All	4.43175	4.40337	0.500	5	2.000	0.625	0.704	$4\frac{19}{32}$	5.000
5	0.220	4.93175	4.90337	0.750	$6\frac{1}{2}$	2.500	0.625	0.704	$5\frac{3}{32}$	5.563
5	X[b]	4.93175	4.90337	0.500	$6\frac{1}{2}$	2.750	0.625	0.704	$5\frac{3}{32}$	5.563
$5\frac{1}{2}$	0.228	5.43175	5.40337	0.750	$6\frac{3}{4}$	2.625	0.625	0.704	$5\frac{19}{32}$	6.050
$5\frac{1}{2}$	X	5.43175	5.40337	0.500	$6\frac{3}{4}$	2.875	0.625	0.704	$5\frac{19}{32}$	6.050
6	All	5.93175	5.90337	0.500	7	3.000	0.625	0.704	$6\frac{3}{32}$	6.625
$6\frac{5}{8}$	All	6.55675	6.52837	0.500	$7\frac{1}{4}$	3.125	0.625	0.704	$6\frac{23}{32}$	7.390
7	0.231	6.93175	6.90337	1.250	$7\frac{1}{4}$	2.375	0.625	0.704	$7\frac{3}{32}$	7.656
7	X	6.93175	6.90337	0.500	$7\frac{1}{4}$	3.125	0.625	0.704	$7\frac{3}{32}$	7.656
$7\frac{5}{8}$	0.250	7.55675	7.52418	0.875	$7\frac{1}{2}$	2.875	0.625	0.709	$7\frac{23}{32}$	8.500
$7\frac{5}{8}$	X	7.55675	7.52418	0.500	$7\frac{1}{2}$	3.250	0.625	0.709	$7\frac{23}{32}$	8.500
$8\frac{5}{8}$	0.264	8.55675	8.52418	0.875	$7\frac{3}{4}$	3.000	0.625	0.709	$8\frac{23}{32}$	9.625
$8\frac{5}{8}$	X	8.55675	8.52418	0.500	$7\frac{3}{4}$	3.375	0.625	0.709	$8\frac{23}{32}$	9.625
$9\frac{5}{8}$	0.281	9.55675	9.52418	0.625	$7\frac{3}{4}$	3.250	0.625	0.709	$9\frac{23}{32}$	10.625
$9\frac{5}{8}$	X	9.55675	9.52418	0.500	$7\frac{3}{4}$	3.375	0.625	0.709	$9\frac{23}{32}$	10.625
$10\frac{3}{4}$	0.279	10.68175	10.64918	1.250	8	2.750	0.625	0.709	$10\frac{27}{32}$	11.750
$10\frac{3}{4}$	X	10.68175	10.64918	0.500	8	3.500	0.625	0.709	$10\frac{27}{32}$	11.750
$11\frac{3}{4}$	0.300	11.68175	11.64918	0.750	8	3.250	0.625	0.709	$11\frac{27}{32}$	12.750
$11\frac{3}{4}$	X	11.68175	11.64918	0.500	8	3.500	0.625	0.709	$11\frac{27}{32}$	12.750
$13\frac{3}{8}$	All	13.30675	13.27418	0.500	8	3.500	0.625	0.709	$13\frac{15}{32}$	14.375
16	All	15.93175	15.89918	0.500	9	4.000	0.625	0.709	$16\frac{3}{32}$	17.000
20	All	19.93175	19.89918	0.500	9	4.000	0.625	0.709	$20\frac{3}{32}$	21.000

[a]All dimensions in inches.

[b]All thicknesses except that listed immediately above.

Table 2.5. DIMENSIONS OF LONG THREADS AND COUPLINGS[a]
(*After Reference 3*)

d_e	t	d_1	d_2	J	L_c	L_t	L_1	M	d_r	d_c
$4\frac{1}{2}$	All	4.43175	4.40337	0.500	7	3.000	0.625	0.704	$4\frac{19}{32}$	5.000
5	All	4.93175	4.90337	0.500	$7\frac{3}{4}$	3.395	0.625	0.704	$5\frac{3}{32}$	5.563
$5\frac{1}{2}$	All	5.43175	5.40337	0.500	8	3.500	0.625	0.704	$5\frac{19}{32}$	6.050
6	All	5.93175	5.90337	0.500	$8\frac{1}{2}$	3.750	0.625	0.704	$6\frac{3}{32}$	6.625
$6\frac{5}{8}$	All	6.55675	6.52837	0.500	$8\frac{3}{4}$	3.875	0.625	0.704	$6\frac{23}{32}$	7.390
7	All	6.93175	6.90337	0.500	9	4.000	0.625	0.704	$7\frac{3}{32}$	7.656
$7\frac{5}{8}$	All	7.55675	7.52418	0.500	$9\frac{1}{4}$	4.125	0.625	0.709	$7\frac{23}{32}$	8.500
$8\frac{5}{8}$	All	8.55675	8.52418	0.500	10	4.500	0.625	0.709	$8\frac{23}{32}$	9.625
$9\frac{5}{8}$	All	9.55675	9.52418	0.500	$10\frac{1}{2}$	4.750	0.625	0.709	$9\frac{23}{32}$	10.625

[a] All dimensions in inches.

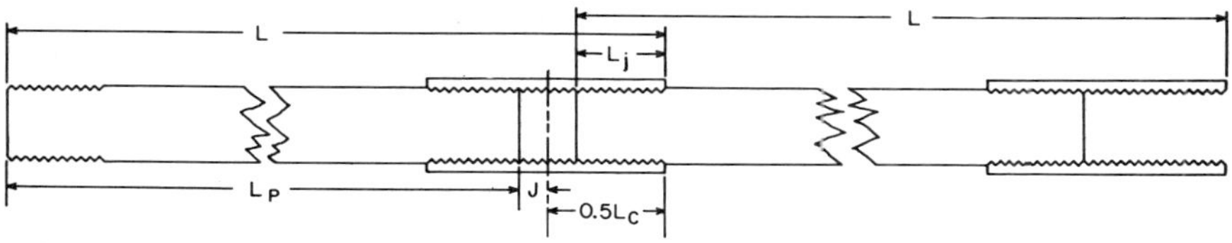

Fig. 2.2 Make-up loss per joint for casing.

L Length of casing and attached coupling.
L_c Length of coupling.
J Distance from end of casing to center of coupling.
L_j Make-up loss.
L_p Length of pipe.

For numerical values of make-up loss, see Tables 2.6 and 2.7.

because of the make-up loss at the couplings. Reference to Fig. 2.2 shows that the make-up loss per joint for a string made up power-tight is

$$L_j = 0.5L_c - J \tag{2.1}$$

where J is a dimension as defined in Fig. 2.2.

Example 2.1. Calculate the make-up loss per joint for 7-in., 26-lb/ft casing with short threads and couplings. Also, calculate the loss per thousand feet of casing (as tallied on the pipe rack) and the casing required per thousand feet of made-up string, expressing the answers in terms of L, the average length of the casing in feet.

SOLUTION: From Table 2.4,

$$L_c = 7.25 \text{ in. and } J = 0.500 \text{ in.}$$

$$L_j = 0.5L_c - J = 0.5 \times 7.25 - 0.500 = 3.125 \text{ in./joint}$$

The number of joints in 1000 ft of tallied casing is $1000/L$, and so

$$\text{Make-up loss per thousand feet} = 3.125 \times 1000/L = 3125/L \text{ in.}$$

Let L_T = feet of casing required per thousand-foot string. Then,

$$\text{Make-up loss} = (L_T/1000)(3125/L) \text{ in.} = 3.125L_T/12L \text{ ft}$$

$$L_T - (3.125L_T/12L) = 1000$$

$$L_T\,[1 - (3.125/12L)] = 1000$$

$$L_T = \frac{1000}{1 - (3.125/12L)} = \frac{1000L}{L - (3.125/12)} = \frac{1000L}{L - 0.26}$$

By use of the method illustrated by Example 2.1, and assuming the average lengths of casing in ranges 1, 2, and 3 to be, 22, 29, and 38 ft, respectively, Tables 2.6 and 2.7 were set up to show, for each length range, the make-up loss per thousand feet and the casing required per thousand feet of made-up string.

Table 2.6. MAKE-UP LOSS FOR CASING (SHORT THREADS AND COUPLINGS)

Casing OD, in.	Nominal weight, lb/ft	Loss per joint, in.	Range 1 casing A[a], ft	Range 1 casing B[b], ft	Range 2 casing A, ft	Range 2 casing B, ft	Range 3 casing A, ft	Range 3 casing B, ft
$4\frac{1}{2}$	All	2.000	7.576	1007.6	5.747	1005.8	4.386	1004.4
5	11.50	2.500	9.470	1009.6	7.184	1007.2	5.482	1005.5
5	X[c]	2.750	10.417	1010.5	7.902	1008.0	6.031	1006.1
$5\frac{1}{2}$	13.00	2.625	9.943	1010.0	7.543	1007.6	5.757	1005.8
$5\frac{1}{2}$	X	2.875	10.890	1011.0	8.261	1008.3	6.305	1006.3
6	All	3.000	11.364	1011.5	8.621	1008.7	6.579	1006.6
$6\frac{5}{8}$	All	3.125	11.837	1012.0	8.980	1009.1	6.853	1006.9
7	17.00	2.375	8.996	1009.1	6.825	1006.9	5.208	1005.2
7	X	3.125	11.837	1012.0	8.980	1009.1	6.853	1006.9
$7\frac{5}{8}$	20.00	2.875	10.890	1011.0	8.261	1008.3	6.305	1006.3
$7\frac{5}{8}$	X	3.250	12.310	1012.5	9.339	1009.4	7.127	1007.2
$8\frac{5}{8}$	24.00	3.000	11.364	1011.5	8.621	1008.7	6.579	1006.6
$8\frac{5}{8}$	X	3.375	12.784	1012.9	9.698	1009.8	7.401	1007.5
$9\frac{5}{8}$	29.30	3.250	12.310	1012.5	9.339	1009.4	7.127	1007.2
$9\frac{5}{8}$	X	3.375	12.784	1012.9	9.698	1009.8	7.401	1007.5
$10\frac{3}{4}$	32.75	2.750	10.417	1010.5	7.902	1008.0	6.031	1006.1
$10\frac{3}{4}$	X	3.500	13.258	1013.4	10.058	1010.2	7.676	1007.7
$11\frac{3}{4}$	38.00	3.250	12.310	1012.5	9.339	1009.4	7.127	1007.2
$11\frac{3}{4}$	X	3.500	13.258	1013.4	10.058	1010.2	7.676	1007.7
$13\frac{3}{8}$	All	3.500	13.258	1013.4	10.058	1010.2	7.676	1007.7

[a]Make-up loss per thousand feet of casing.
[b]Casing required per thousand feet of made-up string.
[c]All weights except that listed immediately above.

Table 2.7. MAKE-UP LOSS FOR CASING (LONG THREADS AND COUPLINGS)

Casing OD, in.	Loss per joint, in.	Range 1 casing A[a], ft	Range 1 casing B[b], ft	Range 2 casing A, ft	Range 2 casing B, ft	Range 3 casing A, ft	Range 3 casing B, ft
$4\frac{1}{2}$	3.000	11.364	1011.5	8.621	1008.7	6.579	1006.6
5	3.375	12.784	1012.9	9.698	1009.8	7.401	1007.5
$5\frac{1}{2}$	3.500	13.258	1013.4	10.058	1010.2	7.676	1007.7
6	3.750	14.205	1014.4	10.776	1010.9	8.224	1008.3
$6\frac{5}{8}$	3.875	14.678	1014.9	11.135	1011.3	8.498	1008.6
7	4.000	15.151	1015.4	11.494	1011.6	8.772	1008.8
$7\frac{5}{8}$	4.125	15.625	1015.9	11.853	1012.0	9.046	1009.1
$8\frac{5}{8}$	4.500	17.045	1017.3	12.931	1013.1	9.868	1010.0
$9\frac{5}{8}$	4.750	17.992	1018.3	13.649	1013.8	10.417	1010.5
$10\frac{3}{4}$	4.750	17.992	1018.3	13.649	1013.8	10.417	1010.5
$11\frac{3}{4}$	4.750	17.992	1018.3	13.649	1013.8	10.417	1010.5
$13\frac{3}{8}$	4.750	17.992	1018.3	13.649	1013.8	10.417	1010.5

[a]Make-up loss per thousand feet of casing.
[b]Casing required per thousand feet of made-up string.

When casing is run, the couplings forming the joints in the upper sections are in tension because of the weight of the casing suspended below them. These joints must, of course, possess sufficient strength to resist rupture or deformation under the axial stresses to which they will be subjected. Additionally, they must be leak-resistant in tension if the casing string is to perform its functions properly. Thomas and Bartok[4] have reported the results of investigations concerned with the leak resistance of joints in tension. Among the conclusions drawn were two which should be noted here, viz.:

1. Although several factors influence the leak resistance of a joint made up in the conventional manner, yield strength is apparently the controlling factor.
2. For standard API round threads, a joint in tension is leak-resistant up to the yield strength of the joint (and usually beyond), and hence the leak resistance in tension is greater than that required in service.

The axial tension load which can be supported at a casing joint is called the joint strength. Joint strength depends on the grade, size, and weight of the casing, and on the effective length of the threads. A subcommittee of the API Committee on the Standardization of Oil-Country Tubular Goods in 1939 recommended the use of the empirical equations shown below in calculating the average joint strengths for standard API round-thread casing.[5] For casing with short threads and couplings:

$$F_{ajs} = C_s A_j [33.71 - d_e][24.45 + 1/(t - h)] \tag{2.2}$$

For casing with long threads and couplings:

$$F_{ajl} = C_l A_j [25.58 - d_e][24.45 + 1/(t - h)] \tag{2.3}$$

where d_e = outside diameter of pipe, in.

t = wall thickness of pipe, in.
h = height of threads, in.
A_j = cross-sectional area of pipe wall under the last perfect thread, sq in.
C_s, C_l = constants for the grade of the casing, as shown in Table 2.8.
F_{ajs}, F_{ajl} = average joint strengths, pounds.

Table 2.8. JOINT STRENGTH CONSTANTS

Joint Grade	C_s	C_l	C
F-25	53.5	—	134
H-40	72.5	—	182
J-55	96.5	159	243
N-80	112.3	185	282
P-110	146.9	242	369

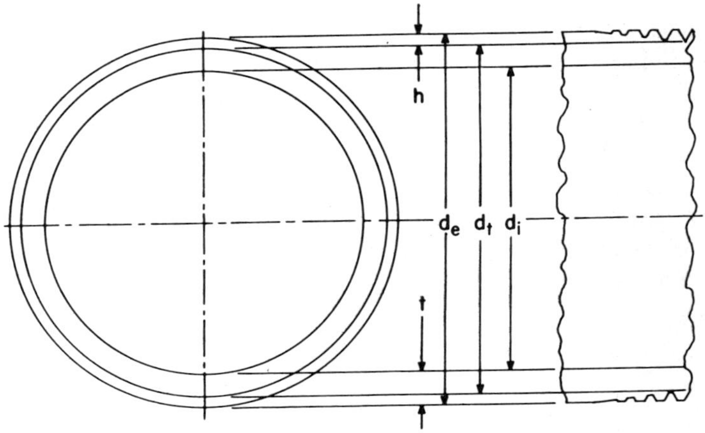

Fig. 2.3 Relationship of pipe OD, pipe ID, and diameter to base of last perfect thread.

Considering the cross-sectional area of the pipe wall under the last perfect thread, it is apparent from an inspection of Fig. 2.3 that

$$A_j = (\pi/4)d_t^2 - (\pi/4)d_i^2 \tag{2.4}$$

where d_t is the diameter of the pipe measured from the base of the thread and d_i is the internal diameter of the pipe. These dimensions are determined from

$$d_t = d_e - 2h \quad \text{and} \quad d_i = d_e - 2t \tag{2.5}$$

Substituting Eqs. (2.5) into Eq. (2.4),

$$\begin{aligned} A_j &= (\pi/4)[(d_e - 2h)^2 - (d_e - 2t)^2] \\ &= (\pi/4)[d_e^2 - 4d_eh + 4h^2 - d_e^2 + 4d_et - 4t^2] \\ &= \pi[d_e(t - h) - (t^2 - h^2)] \\ &= \pi(t - h)(d_e - t - h) \end{aligned} \tag{2.6}$$

Substituting Eq. (2.6) into Eqs. (2.2) and (2.3),

$$F_{ajs} = \pi C_s[33.71 - d_e][d_e - t - h][24.45(t - h) + 1] \tag{2.7}$$

$$F_{ajl} = \pi C_l[25.58 - d_e][d_e - t - h][24.45(t - h) + 1] \tag{2.8}$$

It is emphasized here that Eqs. (2.7) and (2.8) give *average* values of joint strength. Statistical analysis of test performance data indicates that the minimum joint strength to be expected is approximately 80 per cent of the average. Designating the minimum joint strengths by F_{js} and F_{jl},

$$F_{js} = 0.80\pi C_s[33.71 - d_e][d_e - t - h][24.45(t - h) + 1] \tag{2.9}$$

$$F_{jl} = 0.80\pi C_l[25.58 - d_e][d_e - t - h][24.45(t - h) + 1] \tag{2.10}$$

Equations (2.9) and (2.10) define the minimum strength of casing joints as given by API Bulletin 5C2, the standard reference for performance properties of casing.[6] These equations can be simplified by making use of several facts. For example, the height of threads on standard API round-thread casing is always 0.07125 in. Also, inspection of Table 2.8 shows a constant ratio between C_s and C_l such that

$$C_l = 1.647C_s \tag{2.11}$$

Finally, we can define a constant C such that

$$C = 0.80\pi C_s \tag{2.12}$$

Values of C for all grades of casing are shown in Table 2.8. Substituting Eqs. (2.11) and (2.12) into Eqs. (2.9) and (2.10),

$$F_{js} = C(33.71 - d_e)(d_e - t - 0.071)(24.45t - 0.742) \tag{2.13}$$

$$F_{jl} = 1.647C(25.58 - d_e)(d_e - t - 0.071)(24.45t - 0.742) \tag{2.14}$$

Example 2.2. Calculate the joint strength of 7-in., 26-lb/ft grade J-55 casing with long threads and couplings. Express the answer in terms of feet of 26-lb casing which could be suspended below a joint without exceeding the joint strength.

Solution: From Table 2.3, $t = 0.362$ in. From Table 2.8, $C = 242.5$. Using Eq. (2.14),

$$F_{jl} = 1.647 \times 242.5(25.58 - 7.00)(7.000 - 0.362 - 0.071)(24.45 \times 0.362 - 0.742)$$

$$= 1.647 \times 242.5 \times 18.58 \times 6.567 \times 8.109 = 395{,}000 \text{ lb}$$

$$\frac{395{,}000 \text{ lb}}{26 \text{ lb/ft}} = 15{,}200 \text{ ft of 26-lb casing}$$

Joint strengths calculated from Eqs. (2.13) and (2.14) are shown in Table 2.9.

Table 2.9. JOINT STRENGTHS OF CASING[a]
(*After Reference 6*)

Casing OD, in.	Nominal weight, lb/ft	F-25 S[b]	H-40 S	J-55 S	J-55 L[c]	N-80 S	N-80 L	P-110 S	P-110 L
$4\frac{1}{2}$	9.50	71	96	128					
	11.60			159	189		220		288
	13.50						258		337
	15.10								394
5	11.50	84		152					
	13.00			178	210				
	15.00			210	247		288		377
	18.00						354		463
$5\frac{1}{2}$	13.00	95							
	14.00		139	186					
	15.50			211	247				
	17.00			234	275		320		418
	20.00						382		500
	23.00						440		576
6	15.00	108							
	18.00		179	239	278		323		
	20.00						366		
	23.00						432		565
	26.00								646
$6\frac{5}{8}$	17.00	121							
	20.00		195	259	299				
	24.00			320	370		430		562
	28.00						511		669
	32.00						582		762
7	17.00	118	160						
	20.00		191	254					
	23.00			300	344		400		
	26.00			345	395		460		602
	29.00						520		681
	32.00						578		756
	35.00						635		831
	38.00						688		900
$7\frac{5}{8}$	20.00	138							
	24.00		227						
	26.40			333	378		439		
	29.70						505		661
	33.70						581		760
	39.00						676		885

[a]All joint strengths are given in thousands of pounds.
[b]Joint strength with short threads and couplings.
[c]Joint strength with long threads and couplings.

Table 2.9. (CONTINUED)

Casing OD, in.	Nominal weight, lb/ft	F-25 S	H-40 S	J-55 S	J-55 L	N-80 S	N-80 L	P-110 S	P-110 L
$8\frac{5}{8}$	24.00	161		288					
	28.00		252						
	32.00		295	393	437				
	36.00			448	499		581		
	40.00						655		858
	44.00						729		954
	49.00						812		1062
$9\frac{5}{8}$	23.90	185							
	32.30		279						
	36.00		318	442	462				
	40.00			477	521		606		
	43.50						670		875
	47.00						727		952
	53.50						841		1100
$10\frac{3}{4}$	32.75	196	265						
	40.50		338	450					
	45.50			518					
	51.00			585		680		890	
	55.50					750		981	
	60.70							1081	
	65.70							1180	
$11\frac{3}{4}$	38.00	222							
	42.00		236						
	47.00			507					
	54.00			593					
	60.00			668		778			
$13\frac{3}{8}$	48.00	260	352						
	54.50			545					
	61.00			613					
	68.00			695					
	72.00					868			
16	55.00	258							
	65.00		423						
	75.00			662					
	84.00			753					
20	94.00	359	487						

2.5. *Weights of Casing*

In discussing casing weights, it is necessary to differentiate between plain-end weight, average weight with threads and couplings, and nominal weight of casing. Also, it should be understood that the discussion which follows applies only to non-upset casing. (Upset tubular goods are those which have an increased wall thickness at the ends to compensate for the reduced strength which would otherwise result from the presence of threads. For upset casing, the dimensions of the upset are not covered by standards, but are governed by agreement between the casing manufacturer and the purchaser.[3])

The plain-end weight of casing is the weight of the casing without threads and couplings. It can be calculated readily by assuming the casing to be a cylindrical tube of uniform wall thickness, which in fact it is if we eliminate all consideration of upset casing. If the outside diameter of the casing is d_e inches and the wall thickness is t inches, then the internal diameter is $d_e - 2t$ inches and the mean diameter is

$$0.5[d_e + (d_e - 2t)] = d_e - t \tag{2.15}$$

The mean circumference of the casing wall is

$$\pi \times \text{mean diameter} = \pi(d_e - t) \tag{2.16}$$

and the cross-sectional wall area will be the product of wall thickness and mean circumference,

$$A = \pi t(d_e - t) \tag{2.17}$$

Considering one linear foot of casing, the volume, in cubic inches, is

$$V = 12\pi t(d_e - t) \tag{2.18}$$

Taking the density of steel to be 0.2833 lb per cu in., the plain-end weight, w_p, in pounds per linear foot, is

$$w_p = 0.2833 \times 12\pi t(d_e - t) = 10.68t(d_e - t) \tag{2.19}$$

The average weight (per foot) of a length of casing with threads at both ends and a coupling attached power-tight at one end is designated by the symbol w_{tc}. It is determined from the plain-end weight w_p, the weight of metal removed in cutting threads at each end w_t, the weight of the coupling w_c, the overall length of the casing and attached coupling L, and the actual length of pipe without coupling L_p, as follows:

$$w_{tc} = (L_p w_p - 2w_t + w_c)/L \tag{2.20}$$

Figure 2.2 shows the relationship between the lengths of the pipe with and without attached couplings to be

$$L_p = L - (0.5L_c + J)/12 \tag{2.21}$$

the factor 12 being necessary because L_p and L are normally expressed in feet, whereas L_c and J are generally given in inches, as in Tables 2.4 and 2.5. We may rewrite Eq. (2.21) in a more convenient form,

$$L_p = L - (L_c + 2J)/24 \tag{2.22}$$

The American Iron and Steel Institute has developed equations for calculating the weights of finished couplings and the weights of casing

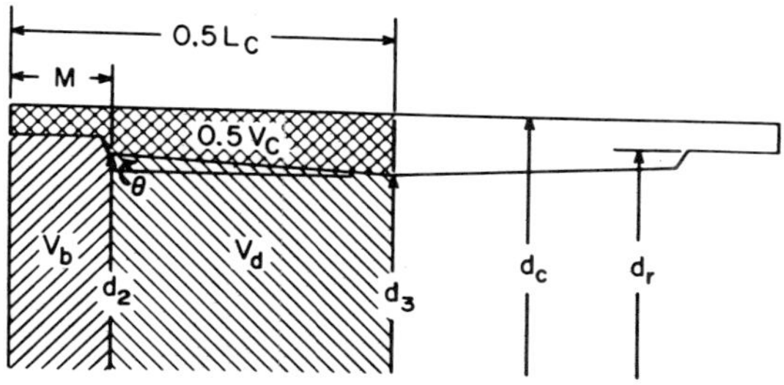

Fig. 2.4 Calculation of coupling weight.

removed in threading.[7] Considering Fig. 2.4, the total volume of a cylinder of length $0.5L_c$ and diameter d_c is given by

$$V_a = 0.5L_c(\pi/4)d_c^2 \tag{2.23}$$

The volume V_b, which is that of a cylinder of diameter d_r and length M, is

$$V_b = M(\pi/4)d_r^2 \tag{2.24}$$

The volume V_d, which is that of the frustum of a cone of altitude $0.5L_c - M$ and bases of diameter d_2 and d_3, is

$$\begin{aligned} V_d &= (1/3)[0.5L_c - M]\left[(\pi/4)d_2^2 + (\pi/4)d_3^2 + \sqrt{(\pi/4)d_2^2 \times (\pi/4)d_3^2}\right] \\ &= (\pi/12)(0.5L_c - M)(d_2^2 + d_2d_3 + d_3^2) \end{aligned} \tag{2.25}$$

The volume $V_a - V_b - V_d$ represents one half the total coupling volume V_c:

$$\begin{aligned} V_c &= 2[0.5L_c(\pi/4)d_c^2 - M(\pi/4)d_r^2 - (\pi/12)(0.5L_c - M)(d_2^2 + d_2d_3 + d_3^2)] \\ &= (\pi/12)[3L_cd_c^2 - 6Md_r^2 - (L_c - 2M)(d_2^2 + d_2d_3 + d_3^2)] \end{aligned} \tag{2.26}$$

Taking the density of steel to be 0.2833 lb per cu in.,

$$\begin{aligned} w_c &= 0.2833V_c \\ &= 0.222L_cd_c^2 - 0.445Md_r^2 \\ &\quad - 0.074(L_c - 2M)(d_2^2 + d_2d_3 + d_3^2) \end{aligned} \tag{2.27}$$

The dimensions L_c, d_c, M, d_r, and d_2 can be obtained from Tables 2.4 and 2.5. Dimension d_3 can be calculated from

$$d_3 = d_2 - (0.5L_c - M)\tan\theta \tag{2.28}$$

For API threads and couplings, the thread taper for all sizes is such that $\tan \theta = 0.0625$, and

$$d_3 = d_2 - 0.0625(0.5L_c - M) \tag{2.29}$$

The weight of pipe (at each end) which is removed in threading can be determined by considering a cylinder of length L_t and diameter d_e as shown in Fig. 2.5. L_t is the total length of the pipe threads from the end of

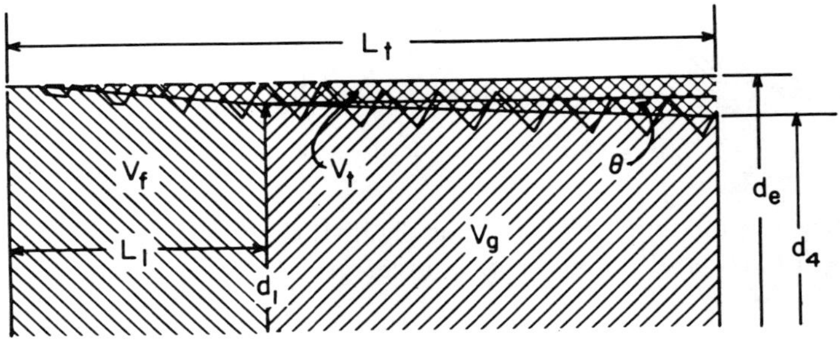

Fig. 2.5 Calculation of weight removed in threading.

the pipe to the vanishing point of the threads, and d_e is the outside diameter of the casing, as previously defined. The volume of this cylinder is

$$V_e = L_t(\pi/4)d_e^2 \tag{2.30}$$

The volume V_f, which is that of the frustum of a cone of altitude L_1 and bases of diameters d_e and d_1, is

$$\begin{aligned} V_f &= (1/3)L_1\left[(\pi/4)d_e^2 + (\pi/4)d_1^2 + \sqrt{(\pi/4)d_e^2 \times (\pi/4)d_1^2}\right] \\ &= (\pi/12)L_1(d_e^2 + d_ed_1 + d_1^2) \end{aligned} \tag{2.31}$$

The volume V_g, which is that of the frustum of a cone of altitude $L_t - L_1$ and bases of diameters d_1 and d_4, is

$$\begin{aligned} V_g &= (1/3)(L_t - L_1)\left[(\pi/4)d_1^2 + (\pi/4)d_4^2 + \sqrt{(\pi/4)d_1^2 \times (\pi/4)d_4^2}\right] \\ &= (\pi/12)(L_t - L_1)(d_1^2 + d_1d_4 + d_4^2) \end{aligned} \tag{2.32}$$

The volume removed in threading, V_t, is related to the volumes V_e, V_f, and V_g by

$$V_t = V_e - V_f - V_g \tag{2.33}$$

Combining Eqs. (2.30) through (2.33),

$$\begin{aligned} V_t &= (\pi/4)L_t d_e^2 - (\pi/12)L_1(d_e^2 + d_e d_1 + d_1^2) \\ &\quad - (\pi/12)(L_t - L_1)(d_1^2 + d_1 d_4 + d_4^2) \\ &= (\pi/12)[3L_t d_e^2 - L_1(d_e^2 + d_e d_1 + d_1^2) \\ &\quad - (L_t - L_1)(d_1^2 + d_1 d_4 + d_4^2)] \end{aligned} \tag{2.34}$$

and the weight removed in threading is

$$\begin{aligned} W_t &= 0.2833 V_t \\ &= 0.222 L_t d_e^2 - 0.074 L_1(d_e^2 + d_e d_1 + d_1^2) \\ &\quad - 0.074(L_t - L_1)(d_1^2 + d_1 d_4 + d_4^2) \end{aligned} \tag{2.35}$$

The dimensions L_t, d_e, L_1, and d_1 can be obtained from Tables 2.4 and 2.5. Dimension d_4 can be calculated from

$$\begin{aligned} d_4 &= d_1 - (L_t - L_1)\tan\theta \\ &= d_1 - 0.625(L_t - L_1) \end{aligned} \tag{2.36}$$

Example 2.3. Calculate, for the casing of Example 2.2, the average weight (in pounds per foot) for 20-ft lengths.

SOLUTION:

$$d_e = 7.000 \text{ in.},\ L = 20 \text{ ft},\ t = 0.362 \text{ in.}$$

From Table 2.5, $d_1 = 6.93175$ in., $d_2 = 6.90337$ in., $J = 0.500$ in., $L_c = 9.000$ in., $L_t = 4.000$ in., $L_1 = 0.625$ in., $M = 0.704$ in., $d_r = 7.094$ in., and $d_c = 7.656$ in. From Eq. (2.29),

$$d_3 = 6.90337 - 0.0625(0.5 \times 9.000 - 0.704) = 6.666 \text{ in.}$$

From Eq. (2.36),

$$d_4 = 6.93175 - 0.0625(4.000 - 0.625) = 6.721 \text{ in.}$$

From Eq. (2.19),

$$w_p = 10.68 \times 0.362(7.000 - 0.362) = 25.66 \text{ lb/ft}$$

From Eq. (2.21),

$$L_p = 20 - (0.5 \times 9.000 + 0.500)/12 = 19.58 \text{ ft}$$

From Eq. (2.27),

$$\begin{aligned} w_c &= 0.222 \times 9.000 \times 7.656^2 - 0.445 \times 0.704 \times 7.094^2 \\ &\quad - 0.074(9.000 - 2 \times 0.704)(6.903^2 + 6.903 \times 6.666 + 6.666^2) \\ &= 23.67 \text{ lb} \end{aligned}$$

From Eq. (2.35),

$$\begin{aligned} w_t &= 0.222 \times 9.000 \times 7.000^2 \\ &\quad - 0.074 \times 0.625(7.000^2 + 7.000 \times 6.932 + 6.932^2) \\ &\quad - 0.074(4.000 - 0.625)(6.932^2 + 6.932 \times 6.721 + 6.721^2) = 1.86 \text{ lb} \end{aligned}$$

From Eq. (2.20),

$$w_{tc} = (19.58 \times 25.66 - 2 \times 1.86 + 23.67)/20 = 26.13 \text{ lb/ft}$$

Table 2.10 gives plain-end weights, coupling weights, and weights with threads and couplings for all API sizes of casing. The weights with threads and couplings are based on 20-ft lengths.

Table 2.10. WEIGHTS OF CASING AND COUPLINGS
(*After Reference 3*)

Casing OD, in.	Nominal weight, lb/ft	Plain-end weight, lb/ft	Coupling weight		Weight threaded and coupled[a]	
			short, lb	long, lb	short, lb/ft	long, lb/ft
$4\frac{1}{2}$	9.50	9.40	6.05		9.55	
	11.60	11.35	6.05	9.07	11.47	11.54
	13.50	13.04		9.07		13.20
	15.10	14.98		9.07		15.12
5	11.50	11.23	10.18		11.50	
	13.00	12.83	10.18	12.56	13.07	13.12
	15.00	14.87	10.18	12.56	15.08	15.13
	18.00	17.93		12.56		18.14
$5\frac{1}{2}$	13.00	12.84	11.44		13.12	
	14.00	13.70	11.44		13.97	
	15.50	15.35	11.44	14.03	15.59	15.64
	17.00	16.87	11.44	14.03	17.09	17.14
	20.00	19.81		14.03		20.03
	23.00	22.54		14.03		22.70
6	15.00	14.65	14.53		15.03	
	18.00	17.57	14.53	18.29	17.91	17.99
	20.00	19.64		18.29		20.02
	23.00	22.81		18.29		23.13
	26.00	25.80		18.29		26.06
$6\frac{5}{8}$	17.00	16.69	19.97		17.29	
	20.00	19.49	19.97	24.82	20.04	20.17
	24.00	23.58	19.97	24.82	24.06	24.18
	28.00	27.65		24.82		28.16
	32.00	31.20		24.82		31.64
7	17.00	16.70	18.34		17.20	
	20.00	19.54	18.34		20.01	
	23.00	22.63	18.34	23.67	23.03	23.15
	26.00	25.66	18.34	23.67	26.02	26.13
	29.00	27.82		23.67		29.12
	32.00	31.68		23.67		32.01
	35.00	34.58		23.67		34.86
	38.00	37.26		23.67		37.48

[a]Weights with threads and couplings are based on 20-ft lengths. These (calculated) weights must be adjusted for lengths longer or shorter than 20 ft in accordance with Eq. (2.37) developed in Example 2.4.

Table 2.10. (CONTINUED)

Casing OD, in.	Nominal weight, lb/ft	Plain-end weight, lb/ft	Coupling weight		Weight threaded and coupled[a]	
			short, lb	long, lb	short, lb/ft	long, lb/ft
7⅝	20.00	19.69	26.93		20.55	
	24.00	23.47	26.93		24.26	
	26.40	25.56	26.93	34.23	36.32	26.51
	29.70	29.04		34.23		29.91
	33.70	33.04		34.23		33.83
	39.00	38.05		34.23		83.73
8⅝	24.00	23.57	35.58		24.75	
	28.00	27.02	35.58		28.13	
	32.00	31.10	35.58	47.48	32.14	32.48
	36.00	35.14	35.58	47.48	36.11	36.42
	40.00	39.29		47.48		40.48
	44.00	43.39		47.48		44.48
	49.00	48.00		47.48		48.98
9⅝	29.30	28.04	39.51		29.32	
	32.30	31.03	39.51		32.25	
	36.00	34.86	39.51	55.77	36.01	36.46
	40.00	38.94	39.51	55.77	40.01	40.44
	43.50	42.70		55.77		44.11
	47.00	46.14		55.77		47.47
	53.50	52.85		55.77		54.02
10¾	32.75	31.20	45.53		32.65	
	40.50	38.88	45.53		40.20	
	45.50	44.22	45.53		45.44	
	51.00	49.50	45.53		50.63	
	55.50	54.21	45.53		55.25	
	60.70	59.40	45.53		60.34	
	65.70	64.53	45.53		65.37	
11¾	38.00	36.69	49.61		38.22	
	42.00	40.60	49.61		42.08	
	47.00	45.56	49.61		46.94	
	54.00	52.57	49.61		53.32	
	60.00	58.81	49.61		59.94	
13⅜	48.00	45.98	56.23		47.64	
	54.50	52.74	56.23		54.28	
	61.00	59.45	56.23		60.87	
	68.00	66.11	56.23		67.40	
	72.00	70.60	56.23		71.81	
16	55.00	52.36	78.98		54.70	
	65.00	62.58	78.98		64.71	
	75.00	72.72	78.98		74.63	
	84.00	81.97	78.98		83.68	
20	94.00	91.41	98.25		93.86	

Example 2.4. Calculate, for the casing of Example 2.3, the average weight (with threads and couplings) for each length range. Assume the average lengths of casing in ranges 1, 2, and 3 to be 22, 29, and 38 ft, respectively.

SOLUTION: Let w'_{tc} be the threaded and coupled weight for casing in lengths of 20 ft and let w_{tc} be the threaded and coupled weight for the same casing in lengths of L feet. From Eqs. (2.20) and (2.21),

$$w_{tc} = \left(\frac{1}{L}\right)\left[\left(L - \frac{L_c + 2J}{24}\right)w_p - 2w_t + w_c\right]$$

$$= \left(\frac{1}{L}\right)\left[Lw_p - \frac{L_c w_p}{24} - \frac{J w_p}{12} - 2w_t + w_c\right]$$

and

$$-\frac{L_c w_p}{24} - \frac{J w_p}{12} - 2w_t + w_c = Lw_{tc} - Lw_p = L(w_{tc} - w_p)$$

$$w'_{tc} = \left(\frac{1}{20}\right)\left[\left(20 - \frac{L_c + 2J}{24}\right)w_p - 2w_t + w_c\right]$$

$$= \left(\frac{1}{20}\right)\left[20w_p - \frac{L_c w_p}{24} - \frac{J w_p}{12} - 2w_t + w_c\right]$$

and

$$-\frac{L_c w_p}{24} - \frac{J w_p}{12} - 2w_t + w_c = 20w'_{tc} - 20w_p = 20(w'_{tc} - w_p)$$

$$L(w_{tc} - w_p) = 20(w'_{tc} - w_p)$$

$$w_{tc} = (20/L)(w'_{tc} - w_p) + w_p \tag{2.37}$$

Applying Eq. (2.37) to the casing of Example 2.3,

$$w_{tc} = (20/L)(26.13 - 25.66) + 25.66 = 9.4/L + 25.66$$

For range 1, $w_{tc} = 9.4/22 + 25.66 = 26.09$ lb/ft
For range 2, $w_{tc} = 9.4/29 + 25.66 = 25.98$ lb/ft
For range 3, $w_{tc} = 9.4/38 + 25.66 = 25.91$ lb/ft

A review of the discussion up to this point will show that the weights under consideration have been weights for individual lengths of casing. To determine the average weight per foot of a made-up string, the make-up loss must be considered. If L_T is the total length of casing required for a 1000-ft string, the total weight of the string is $L_T w_{tc}$, and the average weight per foot of the string is

$$w = \frac{L_T w_{tc}}{1000} = \frac{L_T}{1000} \times \left[\frac{20}{L}(w'_{tc} - w_p) + w_p\right] \tag{2.38}$$

If L-foot lengths of casing are used, the number of joints in the string will be L_T/L and, from Eq. (2.1), the loss per joint is $(0.5L_c - J)/12$ feet. The casing required is the made-up length plus the total make-up loss, or

$$L_T = 1000 + (L_T/L)(0.5L_c - J)/12 \tag{2.39}$$

from which

$$L_T = \frac{1000L}{L - (0.5L_c - J)/12} \tag{2.40}$$

Combining Eqs. (2.38) and (2.40),

$$w = \frac{20(w'_{tc} - w_p) + Lw_p}{L - (0.5L_c - J)/12} \tag{2.41}$$

Example 2.5. Considering $9\frac{5}{8}$-in. casing with 0.395 in. wall thickness, calculate the average weights of made-up strings for each length range in both long and short threads and couplings. Make the assumptions (as to average length of pipe in each range) made in Example 2.4.

Solution: For the three length ranges, Eq. (2.41) becomes

$$w_1 = \frac{20(w'_{tc} - w_p) + 22w_p}{22 - (0.5L_c - J)/12} = \frac{240(w'_{tc} - w_p) + 264w_p}{264 - 0.5L_c + J} \tag{2.42}$$

$$w_2 = \frac{20(w'_{tc} - w_p) + 29w_p}{29 - (0.5L_c - J)/12} = \frac{240(w'_{tc} - w_p) + 348w_p}{348 - 0.5L_c + J} \tag{2.43}$$

$$w_3 = \frac{20(w'_{tc} - w_p) + 38w_p}{38 - (0.5L_c - J)/12} = \frac{240(w'_{tc} - w_p) + 456w_p}{456 - 0.5L_c - J} \tag{2.44}$$

Tables 2.4 and 2.10 give the following dimensions in the case of short threads and couplings: $L_c = 7.75$ in., $J = 0.500$ in., $w_p = 38.94$ lb/ft, $w'_{tc} = 40.01$ lb/ft. Tables 2.5 and 2.10 give the following dimensions in the case of long threads and couplings: $L_c = 10.50$ in., $J = 0.500$ in., $w_p = 38.94$ lb/ft, $w'_{tc} = 40.44$ lb/ft. Substituting these values into Eqs. (2.42)–(2.44) we obtain for short threads and couplings:

$$w_1 = 40.4 \text{ lb/ft} \quad w_2 = 39.9 \text{ lb/ft} \quad w_3 = 39.7 \text{ lb/ft}$$

and for long threads and couplings:

$$w_1 = 41.1 \text{ lb/ft} \quad w_2 = 40.4 \text{ lb/ft} \quad w_3 = 40.1 \text{ lb/ft}$$

Calculations such as those of Examples 2.4 and 2.5 will show that the nominal weights, as given in Table 2.3, are (as the word "nominal" implies) weights in name only. Nominal weights are useful for identification purposes and in cases where it is not necessary to specify the *exact* average weight per foot. The casing of Example 2.5, for instance, has a nominal weight of 40 lb per ft, and this value is accurate enough for most calculations. Conventionally, neither the length ranges of the casing nor the dimensions of the threads and couplings are considered in calculations involving the weight of a section of a casing string, since satisfactory accuracy can be obtained by use of the nominal weight corresponding to the outside diameter and the wall thickness of the casing.

2.6. Liners

A liner is an abbreviated oil string extending from the bottom of the hole upward to a point approximately 100 ft above the lower end of the protection string, where it is suspended from a liner hanger and sealed off. Its function is similar to that of an oil string. Its obvious advantage over a conventional string (which would extend from the bottom of the hole to the surface) is economy, since less pipe is needed for a liner. This distinct advantage is not realized without certain disadvantages, among them the possibility of leakage at the point at which the liner is suspended.[8]

API liners are manufactured only of grade J-55 steel; however, any grade, weight, and type of casing *may* be used as a liner. Sizes of liners are designated in the same manner as casing sizes, i.e., by outside diameter and wall thickness. The length limits and tolerances for liners are the same as for casing in ranges 2 and 3, as given in Table 2.2. Liners are manufactured with square-cut plain ends, and weights can be calculated (as in the case of plain-end casing) by use of Eq. (2.19). The tentative API liner list, with plain-end weight and inside diameter for each size, is given in Table 2.11.[3]

Table 2.11. API LINERS
(After Reference 3)

Outside diameter, in.	Wall thickness, in.	Plain-end weight, lb/ft	Inside diameter, in.
$3\frac{1}{2}$	0.289	9.91	2.922
4	0.286	11.34	3.428
$4\frac{1}{2}$	0.290	13.04	3.920
5	0.362	17.93	4.276
$5\frac{1}{2}$	0.361	19.81	4.778
$6\frac{5}{8}$	0.417	27.65	5.791

2.7. Effects of External Pressure

If casing is lowered into a hole, the pressure outside the casing may be greater than pressure inside the casing because of fluid pressure in the formation opposite the casing or because of the column of fluid standing between casing and hole. When the excess of external pressure over internal pressure is of sufficient magnitude, there is a tendency for the casing to collapse. If collapse is preceded by permanent deformation, the casing is said to have experienced plastic failure. If, on the other hand, collapse occurs under elastic deformation, failure is said to be elastic. The ability of casing to withstand external pressure without experiencing either plastic or elastic failure is called collapse resistance.

The subject of collapse resistance of casing has been investigated thoroughly from both the theoretical and experimental standpoints. (Reference 1 and References 9–15 provide particularly good background material for a detailed study of the factors which affect collapse resistance.) As a result of these investigations, it has become apparent that collapse resistance is determined by (1) the ratio of pipe diameter to wall thickness, (2) the characteristics of the material of construction, and (3) the axial tension or axial compression to which the casing is subjected. In the discussion and mathematical presentation which follow in this section, only the first two factors will be considered. The effect of axial tension is considered in Sec. 2.9, which deals with axial loading of casing.

Lévy,[16] in 1884, developed an expression for the *elastic* collapse of long cylinders under external pressure as follows:

$$p_c = \frac{2E}{1 - \mu^2} \times \frac{1}{[(d_e/t) - 1]^3} \tag{2.45}$$

where E is Young's modulus and μ is Poisson's ratio for the material of construction. In the derivation of Eq. (2.45), Lévy assumed that the external pressure acts on the mean diameter of the pipe. Clinedinst[12] showed that, in the event that external pressure acts on the outside diameter (as it obviously does in cases of practical importance), Eq. (2.45) becomes

$$p_c = \frac{2E}{1 - \mu^2} \times \frac{1}{(d_e/t)[(d_e/t) - 1]^2} \tag{2.46}$$

Tests have shown average collapse pressures to be approximately 95 per cent of those calculated from Eq. (2.46).[6] Making use of this fact, and taking $E = 30 \times 10^6$ and $\mu = 0.3$, we can write Eq. (2.46) as

$$p_c = \frac{62.6 \times 10^6}{(d_e/t)[(d_e/t) - 1]^2} \tag{2.47}$$

In considering the more practical case (from the standpoint of casing collapse) of *plastic* failure, casing is divided into two classes according to the d_e/t ratio. For pipe having d_e/t less than 14, the collapse pressure is calculated from the Lamé equation:[6]

$$p_c = 2Y_a \left[\frac{(d_e/t) - 1}{(d_e/t)^2} \right] \tag{2.48}$$

where Y_a is the average yield strength for the material of construction.

Stewart,[17] using the results of a monumental series of tests on lap-welded Bessemer steel tubes, developed empirical expressions for collapse pressure. For d_e/t ratios less than 43.5, the Stewart equation is

$$p_c = 86{,}670(t/d_e) - 1386 \tag{2.49}$$

and for d_e/t greater than 43.5, the equation is

$$p_c = 50.21 \times 10^6(t/d_e)^3 \tag{2.50}$$

Pipe having a d_e/t value greater than 43.5 would, by Eq. (2.50), have a collapse pressure of less than 610 psi, and would have limited usefulness as oil well casing. Equation (2.49), on the other hand, applies to pipe having sufficient wall thickness to withstand collapse under well conditions, and, as a matter of fact, it serves as the basis for the API equation for plastic collapse of casing with d_e/t greater than 14. The Stewart equation (2.49) can be written

$$p_c = \frac{86{,}670}{(d_e/t)} - 1386 \tag{2.51}$$

The API equation for plastic failure at d_e/t values greater than 14 is

$$p_c = Y_a\left[\frac{2.503}{(d_e/t)} - 0.046\right] \tag{2.52}$$

It can be seen that Eq. (2.52) is an expression of the Stewart form. It is empirical, being the result of collapse tests on seamless pipe having yield strengths ranging upward from 40,000 psi. Equation (2.52) also incorporates a factor to correct for deviation of the measured average d_e/t ratio from the nominal d_e/t value.[6] Values of Y_a for the various grades of steel are shown in Table 2.12.

Table 2.12. COLLAPSE RESISTANCE PROPERTIES OF CASING
(*After Reference 6*)

Casing grade	Maximum d_e/t for plastic collapse	Average yield strength, psi
H-40	40.02	50,000
J-55	30.73	65,000
N-80	23.91	85,000
P-110	18.57	123,000

Equations (2.47) through (2.52) give average collapse pressures, as determined from a large number of tests. The engineer, however, must design not on the basis of the *average* physical properties of his materials of construction, but on the basis of *minimum* properties. Analysis of test data indicates minimum collapse pressures to be no less than 75 per cent of average values except in an insignificant number of cases.[1] Accordingly, minimum collapse pressures are calculated from the following equations:

For elastic failure,

$$p_c = \frac{46.95 \times 10^6}{(d_e/t)[(d_e/t) - 1]^2} \tag{2.53}$$

For plastic failure with d_e/t less than 14,

$$p_c = 1.50 Y_a \left[\frac{(d_e/t) - 1}{(d_e/t)^2} \right] \tag{2.54}$$

For plastic failure with d_e/t greater than 14,

$$p_c = Y_a \left[\frac{1.877}{(d_e/t)} - 0.0345 \right] \tag{2.55}$$

The Stewart equation for d_e/t less than 43.5,

$$p_c = \frac{65{,}000}{(d_e/t)} - 1040 \tag{2.56}$$

The Stewart equation for d_e/t greater than 43.5,

$$p_c = \frac{37.66 \times 10^6}{(d_e/t)^3} \tag{2.57}$$

Equations (2.53) through (2.57) are used by the American Petroleum Institute to determine performance properties of API casing. The Stewart equations are used to calculate collapse pressures for grade F-25 casing only. Equations (2.53) through (2.55) are used to predict collapse pressures for casing of grades H-40, J-55, N-80, and P-110.

Example 2.6. Calculate the collapse resistances of the following: $5\frac{1}{2}$-in. 17-lb J-55 casing; 7-in. 38-lb N-80 casing; $9\frac{5}{8}$-in. 47-lb P-110 casing; 16-in. 55-lb F-25 casing.

Solution: Except for grade F-25 casing, it is necessary in each instance to determine which of Eqs. (2.53) through (2.55) is applicable. Since collapse resistance is considered to have been exceeded if either plastic or elastic failure occurs, the lower of the two values predicted by the plastic and elastic equations must be used. The key to the selection of the proper equation is the d_e/t ratio for the pipe. It is obvious that, for the casing to fail elastically,

$$\frac{46.95 \times 10^6}{(d_e/t)[(d_e/t) - 1]^2} < Y_a \left[\frac{1.877}{(d_e/t)} - 0.0345 \right]$$

if elastic failure occurs at $d_e/t > 14$, or

$$\frac{46.95 \times 10^6}{(d_e/t)[(d_e/t) - 1]^2} < 1.50 Y_a \left[\frac{(d_e/t) - 1}{(d_e/t)^2} \right]$$

if elastic failure occurs at $d_e/t < 14$. Application of these expressions to grade P-110 casing (for which $Y_a = 123{,}000$ psi) and solutions by trial and error give $d_e/t > 18.57$ in the first case and $d_e/t > 17.43$ in the second case. However, the second expression does not apply at $d_e/t > 17.43$, and hence this solution is not valid. Therefore, we can say that for grade P-110 casing failure will be elastic for pipe having d_e/t values greater than 18.57; for d_e/t values under 18.57, failure will be plastic. Similarly, it is possible to determine limiting d_e/t values for the elastic failure of grades H-40, J-55, and N-80. These values are given in Table 2.12. For the $5\frac{1}{2}$-in. 17-lb J-55 casing, $d_e/t = 5.500/0.304 = 18.09$. According to Table 2.12, failure will be plastic, and Eq. (2.55) applies:

$$p_c = Y_a\left[\frac{1.877}{(d_e/t)} - 0.0345\right] = 65{,}000\left[\frac{1.877}{18.09} - 0.0345\right]$$

$$= 4500 \text{ psi}$$

Considering the 7-in. 38-lb N-80 casing,

$$d_e/t = 7.000/0.540 = 12.96$$

Failure will be plastic, and Eq. (2.54) applies:

$$p_c = 1.50Y_a\left[\frac{(d_e/t) - 1}{(d_e/t)^2}\right] = 1.50 \times 85{,}000\left[\frac{12.96 - 1}{12.96^2}\right]$$

$$= 9080 \text{ psi}$$

Considering the $9\frac{5}{8}$-in. 47-lb P-110 casing,

$$d_e/t = 9.625/0.472 = 20.4$$

and failure will be elastic:

$$p_c = \frac{46.95 \times 10^6}{(d_e/t)[(d_e/t) - 1]^2} = \frac{46.95 \times 10^6}{20.4[20.4 - 1]^2} = 6120 \text{ psi}$$

For the 16-in. 55-lb F-25 casing,

$$d_e/t = 16.000/0.312 = 51.3$$

and Eq. (2.57) applies:

$$p_c = \frac{37.66 \times 10^6}{(d_e/t)^3} = \frac{37.66 \times 10^6}{51.3^3} = 290 \text{ psi}$$

Table 2.13 gives collapse resistances for casing of all API sizes and grades. With a few exceptions, the collapse resistances shown are calculated from Eqs. (2.53) through (2.57). For H-40 casing the collapse resistance values for sizes with d_e/t greater than 30 depart from those calculated by use of Eq. (2.55) to represent better correlation with test data.[6]

Table 2.13. Collapse resistances of five grades of casing

(After Reference 6)

Casing OD, in.	Nominal weight, lb/ft	Collapse resistance, psi				
		F-25	H-40	J-55	N-80	P-110
$4\frac{1}{2}$	9.50	1920	2550	3320		
	11.60			4540	5930	8590
	13.50				7350	10,640
	15.10					12,730
5	11.50	1820		3130		
	13.00			3930		
	15.00			4980	6520	9430
	18.00				8550	12,390
$5\frac{1}{2}$	13.00	1660				
	14.00		2440	3170		
	15.50			3860		
	17.00			4500	5890	8520
	20.00				7580	10,910
	23.00				8900	12,870
6	15.00	1540				
	18.00		2780	3620	4740	
	20.00				5690	
	23.00				7180	10,380
	26.00					12,380
$6\frac{5}{8}$	17.00	1370				
	20.00		2360	3060		
	24.00			4250	5550	7850
	28.00				7110	10,290
	32.00				8490	12,280
7	17.00	1100	1370			
	20.00		1920	2500		
	23.00			3290	4300	
	26.00			4060	5320	7220
	29.00				6370	9220
	32.00				7400	10,700
	35.00				8420	12,180
	38.00				9030	13,130
$7\frac{5}{8}$	20.00	1100				
	24.00		1970			
	26.40			3010	3930	
	29.70				4910	6180
	33.70				6070	8780
	39.00				7530	10,900

Table 2.13. (CONTINUED)

Casing OD, in.	Nominal weight, lb/ft	Collapse resistance, psi				
		F-25	H-40	J-55	N-80	P-110
$8\frac{5}{8}$	24.00	950		1430		
	28.00		1580			
	32.00		2110	2740		
	36.00			3420	4470	
	40.00				5390	7420
	44.00				6320	9140
	49.00				7370	10,660
$9\frac{5}{8}$	29.30	860				
	32.30		1320			
	36.00		1710	2220		
	40.00			2770	3530	
	43.50				4280	4760
	47.00				4900	6120
	53.50				6110	8830
$10\frac{3}{4}$	32.75	650	830			
	40.50		1340	1730		
	45.50			2300		
	51.00			2870	3750	3750
	55.50				4420	5040
	60.70					6790
	65.70					8540
$11\frac{3}{4}$	38.00	620				
	42.00		940			
	47.00			1630		
	54.00			2270		
	60.00			2840	3680	
$13\frac{3}{8}$	48.00	560	740			
	54.50			1140		
	61.00			1670		
	68.00			2140		
	72.00				2880	
16	55.00	290				
	65.00		640			
	75.00			1010		
	84.00			1480		
20	94.00	410	520			

2.8. *Effects of Internal Pressure*

During the entry of formation fluid into casing, as well as in such operations as squeezing and fracturing, casing is often subjected to high internal pressures. In the lower portions of a casing string, external pressure is normally greater than internal pressure, owing to factors noted in Sec. 2.7. In the upper portions of the string, however, external pressure is negligible, since near the surface there is neither formation pressure nor significant fluid column pressure opposite the casing. Therefore, any appreciable internal pressure (whether resulting from entry of fluid or from surface pump pressure) would cause an excess of internal pressure over external pressure in the upper portions, with a resulting tendency of the casing to fail by longitudinal splitting. The excess internal pressure at which this type of failure takes place is called *bursting pressure*.

Conventionally, bursting pressures for steel pipes are calculated from Barlow's formula,[1]

$$p_b = 2St/d_e \tag{2.58}$$

where p_b is the pressure required to cause rupture of the pipe and S is the tensile strength of steel, both quantities being expressed in pounds per square inch. For oil well casing, however, failure is considered to have occurred when applied stress exceeds yield strength, resulting in permanent deformation of the pipe.[1] This type of failure takes place at stresses considerably below those which cause actual rupture. In view of this, the effect of internal pressure on casing is best analyzed in terms of internal yield pressure of the casing by substituting minimum yield strength, Y_m, for tensile strength in Barlow's formula.

A second consideration in the application of Barlow's formula is the fact that casing wall thicknesses may vary, within API tolerances, by 12.5 per cent from nominal values. This means that, if nominal values for wall thickness are used, bursting or internal yield pressures may be as much as 12.5 per cent lower than values predicted by Eq. (2.58). Making allowance for this, the expression for minimum internal yield pressure, p_i, becomes

$$p_i = 2Y_m t(1 - 0.125)/d_e = 1.75Y_m t/d_e \tag{2.59}$$

Example 2.7. Calculate the minimum internal yield pressures for the casing of Example 2.6.

Solution:

$$p_i = \frac{1.75Y_m t}{d_e} = \frac{1.75Y_m}{d_e/t}$$

Values of the ratio d_e/t are given in the solution to Example 2.6 and in Table 2.15. Values of minimum yield strength are given in Table 2.1. These values and the internal yield pressures as calculated from the above equation are tabulated below.

Casing	Y_m, psi	d_e/t	p_i, psi
$5\frac{1}{2}$-in. 17-lb J-55	55,000	18.09	5,320
7-in. 38-lb N-80	80,000	12.96	10,800
$9\frac{5}{8}$-in. 47-lb P-110	110,000	20.4	9,440
16-in. 55-lb F-25	25,000	51.3	850

Table 2.14 gives internal yield pressures for casing of all API sizes and grades, as calculated from Eq. (2.59).[6]

Table 2.14. Internal yield pressures of five grades of casing

(After Reference 6)

Casing OD, in.	Nominal weight, lb/ft	Internal yield, psi				
		F-25	H-40	J-55	N-80	P-110
$4\frac{1}{2}$	9.50	1990	3190	4380		
	11.60			5350	7780	10,690
	13.50				9020	12,410
	15.10					14,420
5	11.50	1930		4240		
	13.00			4870		
	15.00			5700	8290	11,390
	18.00				10,140	13,940
$5\frac{1}{2}$	13.00	1810				
	14.00		3110	4270		
	15.50			4810		
	17.00			5320	7740	10,640
	20.00				9190	12,640
	23.00				10,560	14,520
6	15.00	1740				
	18.00		3360	4620	6720	
	20.00				7560	
	23.00				8870	12,190
	26.00					13,920
$6\frac{5}{8}$	17.00	1620				
	20.00		3040	4180		
	24.00			5110	7440	10,230
	28.00				8810	12,120
	32.00				10,040	13,800
7	17.00	1440	2310			
	20.00		2720	3740		
	23.00			4360	6340	
	26.00			4980	7240	9960
	29.00				8160	11,220
	32.00				9060	12,460
	35.00				9960	13,690
	38.00				10,800	14,850
$7\frac{5}{8}$	20.00	1430				
	24.00		2750			
	26.40			4140	6020	
	29.70				6890	9470
	33.70				7890	10,860
	39.00				9180	12,630

Table 2.14. (CONTINUED)

Casing OD, in.	Nominal weight, lb/ft	Internal yield, psi				
		F-25	H-40	J-55	N-80	P-110
$8\frac{5}{8}$	24.00	1340		2950		
	28.00		2470			
	32.00		2860	3930		
	36.00			4460	6490	
	40.00				7300	10,040
	44.00				8120	11,160
	49.00				9040	12,430
$9\frac{5}{8}$	29.30	1280				
	32.30		2270			
	36.00		2560	3520		
	40.00			3950	5750	
	43.50				6330	8700
	47.00				6870	9440
	53.50				7930	10,900
$10\frac{3}{4}$	32.75	1140	1820			
	40.50		2280	3130		
	45.50			3580		
	51.00			4030	5860	8060
	55.50				6450	8860
	60.70					9760
	65.70					10,660
$11\frac{3}{4}$	38.00	1120				
	42.00		1980			
	47.00			3070		
	54.00			3560		
	60.00			4010	5830	
$13\frac{3}{8}$	48.00	1080	1730			
	54.50			2730		
	61.00			3090		
	68.00			3450		
	72.00				5380	
16	55.00	850				
	65.00		1640			
	75.00			2630		
	84.00			2980		
20	94.00	960	1530			

2.9. *Effects of Axial Loading*

As stated in Sec. 2.1, the effect of axial tension is twofold. First, it tends to cause the casing to fail by longitudinal deformation or yielding, and second, it lowers the resistance of casing to collapse.* Considering the

*It should be noted that the other type of axial loading, i.e., axial compression, increases resistance to collapse. By itself, axial compression caused by buoyancy loading is not sufficiently large to create a design problem. Critical axial compression loading would not occur unless the string were dropped or it hit a bridge while being run into the hole.

first effect, the stress at which permanent deformation takes place is the minimum yield strength Y_m. If the axial load is sufficiently large to cause deformation anywhere along the pipe, the deformation will occur at the root of the last perfect thread (in the case of API threads), since the cross-sectional area of the pipe wall is least at that point. The axial load (in pounds) causing longitudinal yielding is therefore

$$F_a = Y_m A_j \tag{2.60}$$

The area at the root of the thread, A_j, is calculated from Eq. (2.6). Values of A_j for all sizes and weights of casing are given in Table 2.15.

Table 2.15. Cross-sectional dimensions of casing
(After References 3 and 6)

Casing OD, in.	Nominal weight, lb/ft	d_e/t	Drift diameter, in.	Plain-end area, sq in.	Root thread area, sq in.
$4\frac{1}{2}$	9.50	21.951	3.965	2.766	1.775
	11.60	18.000	3.875	3.338	2.347
	13.50	15.517	3.795	3.836	2.844
	15.10	13.353	3.701	4.407	3.416
5	11.50	22.727	4.435	3.304	2.200
	13.00	19.762	4.369	3.773	2.670
	15.00	16.891	4.283	4.374	3.271
	18.00	13.812	4.151	5.275	4.171
$5\frac{1}{2}$	13.00	24.122	4.919	3.776	2.561
	14.00	22.541	4.887	4.029	2.814
	15.50	20.000	4.825	4.514	3.299
	17.00	18.092	4.767	4.962	3.747
	20.00	15.235	4.653	5.828	4.613
	23.00	13.253	4.545	6.630	5.414
6	15.00	25.210	5.399	4.308	2.981
	18.00	20.833	5.299	5.168	3.841
	20.00	18.518	5.227	5.777	4.450
	23.00	15.789	5.115	6.709	5.382
	26.00	13.825	5.007	7.589	6.262
$6\frac{5}{8}$	17.00	27.040	6.010	4.911	3.444
	20.00	23.003	5.924	5.734	4.267
	24.00	18.821	5.796	6.937	5.470
	28.00	15.887	5.666	8.122	6.666
	32.00	13.947	5.550	9.177	7.710
7	17.00	30.303	6.413	4.912	3.361
	20.00	25.735	6.331	5.749	4.198
	23.00	22.082	6.241	6.656	5.105
	26.00	19.337	6.151	7.549	5.998
	29.00	17.157	6.059	8.449	6.899
	32.00	15.453	5.969	9.317	7.766
	35.00	14.056	5.879	10.173	8.622
	38.00	12.962	5.795	10.959	9.408

Table 2.15. (CONTINUED)

Casing OD, in.	Nominal weight, lb/ft	d_e/t	Drift diameter, in.	Plain-end area, sq in.	Root thread area, sq in.
$7\frac{5}{8}$	20.00	30.500	7.000	5.792	4.115
	24.00	25.417	6.900	6.904	5.213
	26.40	23.246	6.844	7.519	5.828
	29.70	20.333	6.750	8.541	6.850
	33.70	17.732	6.640	9.720	8.029
	39.00	15.250	6.500	11.192	9.501
$8\frac{5}{8}$	24.00	32.670	7.972	6.934	5.020
	28.00	28.371	7.892	7.947	6.032
	32.00	24.502	7.796	9.149	7.234
	36.00	21.562	7.700	10.336	8.421
	40.00	19.166	7.600	11.557	9.642
	44.00	17.250	7.500	12.763	10.848
	49.00	15.485	7.386	14.118	12.203
$9\frac{5}{8}$	29.30	34.252	8.907	8.249	6.110
	32.30	30.849	8.845	9.128	6.990
	36.00	27.343	8.765	10.254	8.116
	40.00	24.367	8.679	11.454	9.315
	43.50	22.126	8.599	12.559	10.421
	47.00	20.391	8.525	13.572	11.434
	53.50	17.660	8.379	15.547	13.408
$10\frac{3}{4}$	32.75	38.530	10.036	9.178	6.788
	40.50	30.714	9.894	11.435	9.045
	45.50	26.875	9.794	13.006	10.616
	51.00	23.888	9.694	14.561	12.171
	55.50	21.717	9.604	15.947	13.557
	60.70	19.725	9.504	17.473	15.082
	65.70	18.067	9.404	18.982	16.592
$11\frac{3}{4}$	38.00	29.166	10.994	10.791	8.178
	42.00	35.285	10.928	11.944	9.331
	47.00	31.333	10.844	13.402	10.788
	54.00	27.011	10.724	15.463	12.849
	60.00	24.028	10.616	17.301	14.687
$13\frac{3}{8}$	48.00	40.530	12.559	13.525	10.546
	54.50	35.197	12.459	15.513	12.535
	61.00	31.104	12.359	17.486	14.508
	68.00	27.864	12.259	19.447	16.468
	72.00	26.021	12.191	20.769	17.791
16	55.00	51.282	15.188	15.378	11.812
	65.00	42.666	15.062	18.407	14.841
	75.00	36.529	14.938	21.413	17.847
	84.00	32.323	14.822	24.112	20.546
20	94.00	45.662	18.936	26.917	22.456

Example 2.8. What loads could be carried (without failure) by the casing of Example 2.6? Compare these loads with joint strengths as given in Table 2.9.

SOLUTION: For $5\frac{1}{2}$-in. 17-lb J-55 casing, $Y_m = 55{,}000$ psi and $A_j = 3.747$ sq in.;
$F_a = 205{,}000$ lb, $F_{js} = 234{,}000$ lb, $F_{jl} = 275{,}000$ lb

For 7-in. 38-lb N-80 casing, $Y_m = 80{,}000$ psi and $A_j = 9.408$ sq in.;
$F_a = 753{,}000$ lb, $F_{jl} = 668{,}000$ lb

For $9\frac{5}{8}$-in. 47-lb P-110 casing, $Y_m = 110{,}000$ psi and $A_j = 11.434$ sq in.;
$F_a = 1{,}258{,}000$ lb, $F_{jl} = 952{,}000$ lb

For 16-in. 55-lb F-25 casing, $Y_m = 25{,}000$ psi and $A_j = 11.812$ sq in.;
$F_a = 295{,}000$ lb, $F_{js} = 258{,}000$ lb

Example 2.8 shows that casing will, in some instances, fail in tension at loads below the joint strength of the casing.

In analyzing the effect of axial loading on collapse, it should first be noted that the collapse resistances given in Table 2.13, which were calculated from Eqs. (2.53)–(2.57), apply to casing subjected to no axial stress. The presence of axial stress has no effect on the resistance of a tube to *elastic* collapse; however, it can substantially reduce resistance to *plastic* failure. If we let p_c be the (plastic) collapse pressure for casing under no axial load and p_{cc} be the collapse pressure corrected for the effect of an axial load W, then

$$p_{cc}/p_c = S_t/S_o \tag{2.61}$$

where S_o is the yield stress in pure tension and S_t is the apparent yield stress in the tangential or peripheral direction in the presence of axial stress S_z. If the cross-sectional wall area of the tube is A, then

$$S_z = -W/A \tag{2.62}$$

According to Holmquist and Nadai,[13] it can be assumed that, for a moderately thick-walled tube, plastic yielding takes place if

$$S_t^2 - S_tS_z + S_z^2 = S_o^2 \tag{2.63}$$

Dividing by S_o^2, and substituting for S_z the value given by Eq. (2.62),

$$\frac{S_t^2}{S_o^2} + \frac{S_tW}{AS_o^2} + \frac{W^2}{A^2S_o^2} = 1 \tag{2.64}$$

Substituting Eq. (2.61) into Eq. (2.64),

$$\left(\frac{p_{cc}}{p_c}\right)^2 + \left(\frac{2W}{K}\right)\left(\frac{p_{cc}}{p_c}\right) + \left(\frac{4W^2}{K^2} - 1\right) = 0 \tag{2.65}$$

where $K = 2AS_o$, a constant determined by the dimensions and the material of construction of the tube. For casing, A is the plain-end area as given by Eq. (2.17), and S_o is considered to be the average yield strength Y_a. Solving for p_{cc}/p_c,

$$\frac{p_{cc}}{p_c} = \frac{1}{2}\left[-\frac{2W}{K} \pm \sqrt{\frac{4W^2}{K^2} - 4\left(\frac{4W^2}{K^2} - 1\right)}\right] = -\frac{W}{K} \pm \sqrt{-\frac{3W^2}{K^2} + \frac{K^2}{K^2}} \tag{2.66}$$

Since the ratio p_{cc}/p_c is positive, we must eliminate the negative root in Eq. (2.66). Solving for p_{cc},

$$p_{cc} = (p_c/K)(\sqrt{K^2 - 3W^2} - W) \tag{2.67}$$

Example 2.9. Calculate the collapse resistance of 7-in. 26-lb N-80 casing if a 2000-ft section of 7-in. 29-lb N-80 casing is suspended below it.

SOLUTION: From Eq. (2.17), or from Table 2.15,

$$A = \pi t(d_e - t) = 3.1416 \times 0.362(7.000 - 0.362) = 7.549 \text{ in.}^2$$

By calculation, or from Table 2.16,

$$K = 2AY_a = 2 \times 7.549 \text{ in.}^2 \times 85{,}000 \text{ lb/in.}^2 = 1.283 \times 10^6 \text{ lb}$$

From Table 2.13,

$$p_c = 5320 \text{ psi}$$

$$W = 2000 \text{ ft} \times 29 \text{ lb/ft} = 58{,}000 \text{ lb}$$

From Table 2.15, $d_e/t = 19.337$. Reference to Table 2.12 indicates that failure would be plastic. From Eq. (2.67),

$$p_{cc} = \frac{5320}{1.283 \times 10^6}\left(\sqrt{1.646 \times 10^{12} - 0.010 \times 10^{12}} - 0.058 \times 10^6\right) = 5060 \text{ psi}$$

Table 2.16. VALUES OF K IN EQ. (2.67) FOR FOUR GRADES OF CASING

Casing OD, in.	Nominal weight, lb/ft	K (thousands of pounds)			
		H-40	J-55	N-80	P-110
$4\frac{1}{2}$	9.50	278	360		
	11.60		434	567	821
	13.50			652	944
	15.10				1084
5	11.50		430		
	13.00		490		
	15.00		569	744	1076
	18.00			897	1298
$5\frac{1}{2}$	14.00	403	524		
	15.50		587		
	17.00		645	844	1221
	20.00			991	1434
	23.00			1127	1631
6	18.00	517	672	879	
	20.00			982	
	23.00			1141	1650
	26.00				1867
$6\frac{5}{8}$	20.00	573	745		
	24.00		902	1179	1707
	28.00			1383	2001
	32.00			1560	2258

Table 2.16. (CONTINUED)

Casing OD, in.	Nominal weight, lb/ft	K (thousands of pounds)			
		H-40	J-55	N-80	P-110
7	17.00	491			
	20.00	575	747		
	23.00		865	1132	
	26.00		981	1283	1857
	29.00			1436	2078
	32.00			1584	2292
	35.00			1729	2503
	38.00			1863	2696
$7\frac{5}{8}$	24.00	690			
	26.40		977	1278	
	29.70			1452	2101
	33.70			1652	2391
	39.00			1903	2753
$8\frac{5}{8}$	24.00		901		
	28.00	795			
	32.00	915	1189		
	36.00		1344	1757	
	40.00			1965	2843
	44.00			2170	3140
	49.00			2040	3473
$9\frac{5}{8}$	32.30	913			
	36.00	1025	1333		
	40.00		1489	1947	
	43.50			2135	3090
	47.00			2307	3339
	53.50			2643	3825
$10\frac{3}{4}$	32.75	918			
	40.50	1144	1487		
	45.50		1691		
	51.00		1893	2475	3582
	55.50			2711	3923
	60.70				4298
	65.70				4670
$11\frac{3}{4}$	42.00	1194			
	47.00		1742		
	54.00		2010		
	60.00		2509	2941	
$13\frac{3}{8}$	48.00	1352			
	54.50		2017		
	61.00		2273		
	68.00		2528		
	72.00			3531	
16	65.00	1841			
	75.00		2784		
	84.00		3135		
20	94.00	2692			

In connection with the use of Eq. (2.67), two points must be emphasized. First, the value of p_c for substitution in the equation must be the pressure for *plastic* failure at zero axial stress. Second, if casing fails by elastic collapse at zero axial stress, the collapse resistance under axial loading remains constant until the load is of sufficient magnitude to cause the value found by Eq. (2.67) to be less than the elastic collapse resistance at zero axial stress. These facts are illustrated by the following problem.

Example 2.10. What is the smallest axial load at which $9\frac{5}{8}$-in. 43.5-lb P-110 casing will exhibit plastic failure?

Solution: For the casing under consideration, d_e/t = 22 162 (Table 2.15). The maximum d_e/t value for plastic collapse of P-110 casing is 18.57 (Table 2.12). Therefore, failure at zero axial stress will be elastic (4760 psi, from Table 2.13). As axial stress increases above zero, collapse resistance will remain constant until

$$(p_c/K)\left(\sqrt{K^2 - 3W^2} - W\right) < 4760$$

where $K = 3.090 \times 10^6$ lb (Table 2.16) and

$$p_c = Y_a\left[\frac{1.877}{(d_e/t)} - 0.0345\right] = 123{,}000\left[\frac{1.877}{22.162} - 0.0345\right]$$
$$= 6175 \text{ psi}$$

Substituting these values of K and p_c gives

$$\frac{6175}{3.090 \times 10^6}\left(\sqrt{9.548 \times 10^{12} - 3W^2} - W\right) < 4760$$

Solving for W, we obtain $W < 558{,}500$ lb.

Table 2.17 is a list of all API casing sizes which fail elastically at zero axial stress. Also shown are minimum loads at which the equations for plastic collapse apply. The effect of axial loading on grade H-40 casing is not considered, since this casing is not normally used in combination strings; i.e., a section of it is not normally subjected to axial stress by the suspension of another section below it.

2.10. Design Factors

It is proper at this point to review briefly some of the properties of casing discussed in Sections 2.4, 2.7, 2.8, and 2.9. Joint strength, F_j, is a measure of the resistance of casing to failure in tension at the joints. The tensile yield load, F_a, is a measure of the resistance of casing to failure in tension, based on the minimum yield strength and the cross-sectional area of the material of construction. The collapse pressure, p_{cc}, is a measure of the resistance of casing to failure by collapse under external pressure. Internal yield pressure, p_i, is a measure of the resistance of casing to failure by yielding or bursting from internal pressure. In summary, the quantities

Table 2.17. Elastic collapse of API casing[a]

Grade	OD, in.	Nominal weight, lb/ft	Minimum axial load at which plastic collapse equations apply, pounds	psi
H-40	$13\frac{3}{8}$	48.00		
	16	65.00		
	20	94.00		
J-55	$8\frac{5}{8}$	24.00	36,800	5300
	$11\frac{3}{4}$	47.00	22,500	1680
	$13\frac{3}{8}$	54.50	123,500	7960
	$13\frac{3}{8}$	61.00	9000	520
	16	75.00	201,200	9400
	16	84.00	104,800	4340
N-80	$9\frac{5}{8}$	40.00	43,500	3800
	$11\frac{3}{4}$	60.00	20,800	1200
	$13\frac{3}{8}$	72.00	309,200	14,890
P-110	$6\frac{5}{8}$	24.00	35,000	5040
	7	26.00	106,000	14,040
	$7\frac{5}{8}$	29.70	235,200	27,540
	$8\frac{5}{8}$	40.00	127,500	11,030
	$9\frac{5}{8}$	43.50	558,500	44,470
	$9\frac{5}{8}$	47.00	386,200	28,460
	$10\frac{3}{4}$	51.00	816,200	56,060
	$10\frac{3}{4}$	55.50	658,500	41,290
	$10\frac{3}{4}$	60.70	348,500	19,940

[a]All grades of casing shown fail by elastic collapse at zero axial stress.

F_j, F_a, p_{cc}, and p_i indicate maximum allowable stresses (based on minimum physical properties) to which casing can be subjected without failure.

It is seldom desirable to subject any material to its maximum allowable stress. This is particularly true for the material of which casing is constructed, since minimum physical properties are determined statistically, and the properties of any individual length of casing may deviate considerably from the statistical average. Accordingly, we must introduce the concept of the *safety factor*, which is the ratio of maximum allowable stress to actual working stress. With safety factors designated by N (with appropriate subscripts), the working loads and pressures to be sustained by casing are

$$\text{Joint load} = F_j/N_j \tag{2.68}$$

$$\text{Axial load} = F_a/N_a \tag{2.69}$$

$$\text{External pressure} = p_{cc}/N_c \tag{2.70}$$

$$\text{Internal pressure} = p_i/N_i \tag{2.71}$$

Many authorities[18] point out that the term *safety factor* is a misnomer in view of the fact that *actual* physical properties are not generally known.

They prefer to use the term *design factor*, which is more descriptive of the use made of the factor. The latter terminology appears more logical, and the ratio of allowable stress to working stress will be referred to as the *design factor* throughout this chapter.

The selection of design factors for any engineering design problem is governed largely by four basic considerations. According to Hills,[19] these are:

1. The reliability and accuracy of strength data used for design. The closer these data approach true minima, the smaller the design factor may be.
2. The degree of similarity between service conditions and the test conditions used for determining strength data. If service and test conditions impose the same type of loading (as to impact, acceleration, fatigue, etc.), smaller design factors may be used.
3. The reliability and accuracy of loading assumptions used for design. The closer these loading assumptions approach the maxima encountered in service, the smaller the design factor may be.
4. The consequences of failure. As possibilities of hazard to personnel and economic loss increase, larger design factors must be used.

Considering in turn failure by collapse, failure in tension, and failure by internal yield, we arrive at the following conclusions:

1. The reliability and accuracy of strength data are approximately the same for all three types of failure.
2. The similarity between test and service conditions is about the same for all three types.
3. In determining loading data, the worst possible conditions are generally used. For example, when considering collapse, it is common practice to assume that the pressure outside the casing results from an external fluid column extending to the surface, and that the pressure inside the casing is zero, i.e., that the pipe is empty. Thus, the excess of external pressure over internal pressure will normally be much less than that calculated. On the other hand, when internal yield is considered, it is usually assumed that the external pressure is zero and that the internal pressure is the reservoir pressure. Finally, in calculating the effects of tension, buoyant forces on the casing are generally neglected, which results in a calculated axial load approximately 12 per cent greater than the actual load.
4. In the case of collapse, failure involves no hazard to personnel. Bursting, however, might be considerably hazardous to personnel, and tension failure would involve moderate risk to personnel. The economic loss in any case would be considerable.

In 1955 the API Mid-Continent District Study Committee on Casing Programs reported the results of an investigation of design factors being applied in casing programs. The report presented a summary of replies to questionnaires sent to all members of the committee. An analysis of 50 replies received from 38 companies indicates the following practices concerning design factors:[20]

1. Design factors for collapse vary from 1.000 to 1.500. A design factor of 1.125 was used on 68 per cent of the casing strings reported.
2. Design factors for tension vary from 1.50 to 2.00, the factors 1.60 (on 29 per cent of the casing strings), 1.75 (on 21 per cent), 1.80 (on 25 per cent), and 2.00 (on 24 per cent) being most commonly used. In this connection, about one half the replies indicated use of the same tension design factor for ensuring against failure in the pipe body and in the joints. Where a distinction was made between the two types of tension failure, the design factor for failure in the pipe body was generally somewhat lower than that for joint strength. For example, many replies showed that tensile yield load design factors of 1.60 and 1.80 were used with joint strength design factors of 1.80 and 2.00 respectively.
3. Design factors for internal yield vary from 1.00 to 1.75, with a factor of 1.10 being used on 32 per cent of the casing strings reported.

Example 2.11. A combination string in a certain well is to consist of (1) from zero to 3740 ft, 7-in. 20-lb J-55 casing with short threads and couplings, and (2) from 3740 to 7390 ft, 7-in. 23-lb N-80 casing with long threads and couplings. The mud weight is to be 10.2 ppg, and the expected bottom-hole pressure is 3000 psi. Calculate the design factors for collapse, internal yield, joint strength, and yield strength (for both sections), assuming the worst possible loading conditions.

SOLUTION:

Weight of lower section = 23 lb/ft × (7390 − 3740) ft = 84,000 lb

The collapse pressures for 23-lb N-80 and 20-lb J-55 casing are, respectively, 4300 and 2500 psi (Table 2.13). The collapse pressure for the 20-lb casing (upper section) must be corrected for the effect of axial loading (the weight of the lower section suspended below it). For the 20-lb casing, K = 747,000 lb (Table 2.16), and so

$$p_{cc} = \frac{p_c}{K}\left(\sqrt{K^2 - 3W^2} - W\right)$$

$$= \frac{2500}{747,000}\left(\sqrt{53.7 \times 10^{10}} - 84,000\right) = 2170 \text{ psi}$$

The fluid column pressures outside the casing are

$$0.433(10.2/8.33) \text{ psi/ft} \times 7390 \text{ ft} = 3920 \text{ psi}$$

at the bottom of the lower section, and

$$0.433(10.2/8.33) \text{ psi/ft} \times 3740 \text{ ft} = 1980 \text{ psi}$$

at the bottom of the upper section. Therefore the design factors for collapse are:

$$N_c = 4300/3920 = 1.10 \text{ for the lower section}$$
$$N_c = 2170/1980 = 1.10 \text{ for the upper section}$$

The internal yield pressures for 23-lb N-80, and 20-lb J-55 casing are, respectively, 6340 and 3740 psi (Table 2.14), and thus the design factors for internal yield are:

$$N_i = 6340/3000 = 2.12 \text{ for the lower section}$$
$$N_i = 3740/3000 = 1.25 \text{ for the upper section}$$

The load on the top joint of the lower section is 84,000 lb, and the weight of the upper section is 20 lb/ft (3740 − 0) ft = 74,800 lb; therefore, the load on the top joint of the upper section is

$$84{,}000 + 74{,}800 = 158{,}800 \text{ lb}$$

The joint strengths of 23-lb N-80 and 20-lb J-55 casing are, respectively, 400,000 and 254,000 lb (Table 2.9). Therefore the design factors for joint strength are:

$$N_j = 400{,}000/84{,}000 = 4.76 \text{ for the lower section}$$
$$N_j = 254{,}000/158{,}800 = 1.60 \text{ for the upper section}$$

The root thread areas for 23-lb and 20-lb casing are, respectively, 5.105 and 4.198 sq in. (Table 2.15), and so the unit stresses at roots of last perfect threads are:

$$84{,}000/5.105 = 16{,}500 \text{ psi (lower section)}$$
$$158{,}000/4.198 = 37{,}800 \text{ psi (upper section)}$$

The minimum yield strengths for grades N-80 and J-55 casing are, respectively, 80,000 and 55,000 psi (Table 2.1), and hence the design factors for yield strength are:

$$N_a = 80{,}000/16{,}500 = 4.85 \text{ for the lower section}$$
$$N_a = 55{,}000/37{,}800 = 1.46 \text{ for the upper section}$$

2.11. Selection of Casing Sizes

The controlling factor in the selection of casing size is the outside diameter of the production string. As outlined by Blum,[21] considerations in the determination of this factor are as follows:

1. *Drilling cost.* As hole diameter increases, so will the cost of drilling and completion. The cost of large-diameter holes should be balanced against expected economic advantages.
2. *Methods of production.* A well may flow naturally in its early history but later may require pumping. The hole should be large enough to accommodate necessary production equipment.
3. *Production rates.* Rate is an important factor only where it is high enough to cause appreciable pressure drop in the production tubing.
4. *Possibilities of multizone completion.* Hole sizes should be large

enough to handle equipment for multizone completion, if this is a reasonable possibility.

5. *Number of intermediate strings.* If the expected drilling conditions necessitate one or more intermediate strings, the maximum size of the production string will be limited.

6. *Nature of the fluids produced.* This factor is important primarily because it affects the choice of production equipment, and in turn, the downhole equipment and accessories limit the minimum hole size.

7. *Rig limitations.* Normally, the selection of a rig depends on size and depth of the hole to be drilled. There are cases, however, when rig selection is limited in a given area. In these cases, sizes of hole and casing are determined by rig capabilities.

8. *Workovers.* If experience indicates that remedial work is commonly needed, hole size should be large enough to accommodate the necessary equipment.

9. *Availability of casing.* Shortage of casing has, in many instances, been the determining factor in establishing production string size.

10. *Common practice.* Even after careful consideration of the above factors, the experiences of others in given areas and situations should be studied before final determination of the casing size.

11. *Type of well* (*exploratory or development*). In an exploratory well, the prime purpose of drilling is to "prove up" the existence of commercial zones. Frequently, it is necessary to set casing, and in the interest of economy, slim-hole drillling and completion may be used to extend the exploratory budget. This type of drilling and completion, however, should be weighed against possible production problems at some later date.

When the outside diameter of the production string has been established, bit sizes and sizes of intermediate and surface strings can be determined. In selecting the hole size for a given casing OD, it is necessary to consider the coupling OD (for the casing) and to provide sufficient clearance (beyond the coupling OD) to allow for mud cake and also for casing appliances such as centralizers and scratchers and hole conditions such as caving formations and crookedness. Table 2.18, based on recommendations of Blum,[21] gives minimum bit sizes for all API casing sizes. The last column of Table 2.18 shows some minimum bit sizes which, although standard, are not commonly used. Table 2.19 is a list of the most frequently used bit sizes. To determine the smallest casing through which a given bit will pass, reference is made to Table 2.15, which shows, for each API casing size, the drift diameter, i.e., the diameter of the largest tool which will pass through the casing.

Example 2.12. The production string for a certain well is to consist of $5\frac{1}{2}$-in. casing. Determine minimum casing sizes for the intermediate and surface strings and bit sizes for all strings.

Table 2.18. RECOMMENDED BIT SIZES FOR RUNNING API CASING[a]
(*After Blum*[21])

Casing OD	Coupling OD	Recommended clearance	Minimum hole size	Minimum bit size
$4\frac{1}{2}$	5.000	1.000	6.000	6
5	5.563	1.250	6.813	7
$5\frac{1}{2}$	6.050	1.250	7.300	$7\frac{3}{8}$
6	6.625	1.750	8.375	$8\frac{3}{8}$
$6\frac{5}{8}$	7.390	1.750	9.140	$9\frac{1}{2}$
7	7.656	2.000	9.656	$9\frac{3}{4}$
$7\frac{5}{8}$	8.500	2.500	11.000	11
$8\frac{5}{8}$	9.625	3.000	12.625	$13\frac{3}{4}$
$9\frac{5}{8}$	10.625	3.250	13.875	$14\frac{3}{4}$
$10\frac{3}{4}$	11.750	3.250	15.000	15
$11\frac{3}{4}$	12.750	3.500	16.250	17
$13\frac{3}{8}$	14.375	3.500	17.875	18
16	17.000	3.500	20.500	$20\frac{3}{4}$
20	21.000	3.500	24.500	$25\frac{1}{2}$

[a]All dimensions in inches.

Table 2.19. COMMON BIT SIZES IN INCHES
(*After Blum*[21])

$3\frac{3}{4}$	6	$7\frac{3}{4}$	$9\frac{5}{8}$	$13\frac{3}{4}$
$3\frac{7}{8}$	$6\frac{1}{8}$	$7\frac{7}{8}$	$9\frac{7}{8}$	15
$4\frac{1}{4}$	$6\frac{1}{4}$	$8\frac{1}{2}$	$10\frac{5}{8}$	$17\frac{1}{2}$
$4\frac{5}{8}$	$6\frac{3}{4}$	$8\frac{5}{8}$	11	22
$4\frac{3}{4}$	$7\frac{3}{8}$	$8\frac{3}{4}$	12	23
$5\frac{5}{8}$	$7\frac{5}{8}$	9	$12\frac{1}{4}$	

SOLUTION: For $5\frac{1}{2}$-in. (production) string use a $7\frac{3}{8}$-in. bit (Table 2.18). The minimum size for intermediate string is the OD of the smallest casing with a drift diameter of at least $7\frac{3}{8}$ (= 7.375) in. Drift diameters for $8\frac{5}{8}$-in. casing vary from 7.386 to 7.972 in., and all casings smaller than $8\frac{5}{8}$ in. have drift diameters less than 7.375 in. (Table 2.15). Therefore, for the intermediate string, use $8\frac{5}{8}$-in. casing.

For the $8\frac{5}{8}$-in. (intermediate) string use a $13\frac{3}{4}$-in. bit (Table 2.18). The minimum size for surface string is the OD of the smallest casing with a drift diameter of at least $13\frac{3}{4}$ (= 13.750) in. Drift diameters for 16-in. casing vary from 14.822 to 15.188 in., and all casings smaller than 16 in. have drift diameters less than 13.750 in. (Table 2.15). Therefore, for the surface string use 16-in. casing

For the 16-in. (surface) string use a $20\frac{3}{4}$-in. bit (Table 2.18). The smallest commonly used bit with the minimum desirable diameter is the 22-in. bit (Table 2.19).

Summarizing, the surface string requires a 22-in. bit and 16-in. casing; the intermediate string a $13\frac{3}{4}$-in. bit and $8\frac{5}{8}$-in. casing, and the production string a $7\frac{3}{8}$-in. bit and $5\frac{1}{2}$-in. casing.

2.12. Design of a Combination String

A combination string (i.e., a casing string consisting of more than one section) is used in order to obtain a string which will satisfy the desired design factors with the least investment. Thus the starting point for a design is a statement of the weights and grades of casing available, together with the design factors to be employed. In connection with the latter, it should be noted that the physical properties almost universally considered are joint strength, collapse pressure, and internal yield. Many authorities recommend, in addition, the consideration of longitudinal yielding, although in most instances the design factor for longitudinal yielding will automatically be satisfied if the design factor for joint strength is satisfied.

Once the available casing and the design factors to be used have been determined, all grades and weights of casing which will not meet the requirements for internal yield are eliminated. It will be recalled from Section 2.10 that the worst possible conditions are used in determining loading data. In line with this, the internal pressure (for design purposes) is assumed to be full reservoir pressure, p_{ws}, and the external pressure is assumed to be zero. Thus the minimum allowable internal yield strength for the casing to be used in the string is

$$p_i = p_{ws} N_i \tag{2.72}$$

For casing which will meet the requirements for internal yield, the controlling factor in the lower portions of the string is collapse pressure, and the controlling factor in the upper portions of the string is joint strength (or, possibly, longitudinal yielding).

For purposes of investigating the setting depth limitations imposed by collapse resistance, it is assumed that the external pressure is that due to the external fluid column, and that the internal pressure is zero. Accordingly, the lowest section of the casing string will be composed of casing of the least expensive weight and grade which will satisfy the equation

$$p_c = 0.052 N_c \rho L_s \tag{2.73}$$

where L_s is the setting depth for the casing and ρ is the density (in lb per gal) of the external fluid column. The factor 0.052ρ $(= 0.433\rho/8.33)$ is the pressure gradient of the fluid column. In determining setting depths for sections other than the lowest, the effect on collapse pressure of longitudinal tension must be considered. This normally involves the use of either trial-and-error or graphical solutions.

At some point up the hole, collapse resistance ceases to be the controlling factor in casing string design. From this point to the top of the string, the primary considerations are joint strength and longitudinal yielding. In this region the casing must be designed to satisfy the equations

$$F_j = WN_j \tag{2.74}$$

and

$$Y_m A_j = WN_a \tag{2.75}$$

where W is the weight of casing suspended below the casing under consideration.

By way of illustration, the steps to be followed in the design of a combination string are shown in Example 2.13.

Example 2.13. Design a 7-in., 8000-ft combination casing string for a well where the mud weight is 12 ppg and the expected (formation) pressure gradient is 0.5 psi per ft, using the worst possible loading assumptions. All weights of API casing in grades J-55 and N-80 are available. The design factors to be satisfied are 1.125 for collapse, 2.00 for joint strength, 1.25 for yield strength, and 1.00 for internal yield.

Grade	Weight	p_i	p_c	K	F_{jl}	F_{js}	Y_m	A_j
J	20	3740	2500	747,000		254,000	55,000	4.198
J	23	4360	3290	865,000	344,000	300,000	55,000	5.105
J	26	4980	4060	981,000	395,000	345,000	55,000	5.998
N	23	6340	4300	1,132,000	400,000		80,000	5.105
N	26	7240	5320	1,283,000	460,000		80,000	5.998
N	29	8160	6370	1,436,000	520,000		80,000	6.899
N	32	9060	7400	1,584,000	578,000		80,000	7.766
N	35	9960	8420	1,729,000	635,000		80,000	8.622
N	38	10,800	9080	1,863,000	688,000		80,000	9.408

SOLUTION: The available casings (along with pertinent physical properties) are listed above. In case the reservoir pressure is not known, it is estimated by the use of a reasonable gradient; i.e.,

$$p_{ws} = 8000 \text{ ft} \times 0.5 \text{ psi/ft} = 4000 \text{ psi}$$

The minimum internal yield for any section of the string must be

$$p_i = p_{ws}N_i = 4000 \times 1.00 = 4000 \text{ psi}$$

This requirement excludes the use of 20-lb, J-55 casing (which has an internal yield pressure of 3740 psi) at any point in the string. Since all other weights and grades have internal yield pressures greater than 4000 psi, they are retained for further consideration. The lowest section of the string will consist of the least expensive casing available for which the collapse pressure is at least

$$p_c = 0.052N_c\rho L_s = 0.052 \times 1.125 \times 12 \times 8000 = 5620 \text{ psi}$$

Therefore the lowest section (which will hereafter be designated Section 1) will consist of 29-lb N-80 casing with long threads and couplings. The length of Section

1 is limited (physically) only by the axial load which can be sustained at the top joint of the section. Considering joint strength,

$$W_{\max} = F_j/N_j = 520{,}000/2.00 = 260{,}000 \text{ lb}$$

and considering yield strength,

$$W_{\max} = Y_m A_j/N_a = 80{,}000 \times 6.899/1.25 = 442{,}000 \text{ lb}$$

The maximum length for Section 1 is therefore

$$\frac{260{,}000 \text{ lb}}{29 \text{ lb/ft}} = 8970 \text{ ft}$$

which is greater than the setting depth. The next lowest section (hereafter called Section 2) will consist of the next lighter casing, namely, 26-lb N-80 casing with long threads and couplings. Neglecting the effect of axial tension (due to the weight of Section 1 suspended below it), the setting depth of Section 2 is

$$L_s = \frac{p_c}{0.052 N_c \rho} = \frac{5320}{0.052 \times 1.125 \times 12} = 7580 \text{ ft}$$

This is the first assumed setting depth of Section 2. Under this assumption, the weight of Section 1 is

$$(8000 - 7580) \text{ ft} \times 29 \text{ lb/ft} = 12{,}180 \text{ lb}$$

For this axial load, the collapse pressure of Section 2 is

$$\begin{aligned} p_{cc} &= \frac{p_c}{K}\left(\sqrt{K^2 - 3W^2} - W\right) \\ &= \frac{5320}{1{,}283{,}000}\left(\sqrt{1.646 \times 10^{12} - 0.445 \times 10^9} - 12{,}180\right) = 5270 \text{ psi} \end{aligned}$$

and the setting depth of Section 2 is

$$\frac{5270}{0.052 \times 1.125 \times 12} = 7510 \text{ ft}$$

This is the second assumed setting depth of Section 2. Under this assumption, the weight of Section 1 is

$$29 \times (8000 - 7510) = 14{,}210 \text{ lb}$$

and hence

$$\begin{aligned} p_{cc} &= \frac{5320}{1{,}283{,}000}\left(\sqrt{1.646 \times 10^{12} - 0.606 \times 10^9} - 14{,}210\right) \\ &= 5260 \text{ psi} \end{aligned}$$

for Section 2. The third assumed setting depth for Section 2 is

$$\frac{5260}{0.052 \times 1.125 \times 12} = 7490 \text{ ft}$$

The weight of Section 1 and the collapse pressure of Section 2 are, under this assumption, 14,790 lb and 5260 psi, respectively. The resulting setting depth agrees with the third assumed setting depth of 7490 ft, which is thus taken to be the correct setting depth for Section 2. Also, for Section 2 the maximum joint load is

$$\frac{F_j}{N_j} = \frac{460{,}000}{2.00} = 230{,}000 \text{ lb}$$

and the maximum yield load is

$$\frac{Y_m A_j}{N_a} = \frac{80{,}000 \times 5.998}{1.25} = 384{,}000 \text{ lb}$$

Since the weight of casing suspended below Section 2 is 14,790 lb, the maximum length of Section 2 is

$$\frac{(230{,}000 - 14{,}790) \text{ lb}}{26 \text{ lb/ft}} = 8280 \text{ ft}$$

which is greater than the setting depth. Section 3 will consist of 23-lb N-80 casing with long threads and couplings, which has an uncorrected collapse pressure of 4300 psi. Again neglecting the effect of axial tension due to the weights of Sections 1 and 2, the first assumed setting depth for Section 3 is

$$\frac{4300}{0.052 \times 1.125 \times 12} = 6130 \text{ ft}$$

On this basis, the weight of Section 2 is

$$(7490 - 6130) \text{ ft} \times 26 \text{ lb/ft} = 35{,}400 \text{ lb}$$

and the total axial load below Section 3 is

$$14{,}790 + 35{,}400 = 50{,}200 \text{ lb}$$

The corrected collapse pressure for Section 3 is

$$\frac{4300}{1{,}132{,}000}\left(\sqrt{1.281 \times 10^{12} - 0.008 \times 10^{12}} - 50{,}200\right)$$
$$= 4090 \text{ psi}$$

from which the second assumed setting depth for Section 3 is

$$\frac{4090}{0.052 \times 1.125 \times 12} = 5830 \text{ ft}$$

By continuing the trial-and-error procedure as in the case of Section 2, the setting depth for Section 3 is calculated to be 5780 ft. For this setting depth, the total weight of Sections 1 and 2 is 59,200 lb and the collapse pressure of Section 3 is 4060 psi. The maximum allowable joint load for Section 3 is

$$400{,}000/2.00 = 200{,}000 \text{ lb}$$

and the maximum yield load is

$$\frac{80{,}000 \times 5.105}{1.25} = 327{,}000 \text{ lb}$$

The maximum length of Section 3 is

$$\frac{(200{,}000 - 59{,}200) \text{ lb}}{23 \text{ lb/ft}} = 6120 \text{ ft}$$

which is again greater than the setting depth. Thus collapse pressure continues to be the controlling factor, and will determine the setting depth of Section 4. The least expensive of the remaining grades and weights is 26-lb J-55 casing with short threads and couplings, and this will constitute Section 4. The setting depth of Section 4 is found by trial-and-error to be 5310 ft, the total weight of Sections 1, 2, and 3 is 71,400 lb, and the collapse pressure of Section 4 is 3730 psi. The maximum allowable joint and yield loads for Section 4 are, respectively,

$$345{,}000/2.00 = 172{,}500 \text{ lb}$$

and

$$55{,}000 \times 5.998/1.25 = 264{,}000 \text{ lb}$$

The maximum length of Section 4 is

$$\frac{(172{,}500 - 71{,}400) \text{ lb}}{26 \text{ lb/ft}} = 3890 \text{ ft}$$

Since 3890 ft is less than the allowable setting depth of Section 4, the setting depth for Section 5 is governed not by collapse pressure but by joint strength. Section 5, composed of 26-lb J-55 casing with long threads and couplings, has a setting depth given by

$$L_s = 5310 - 3890 = 1420 \text{ ft}$$

For Section 5, maximum allowable joint and yield loads are, respectively,

$$395{,}000/2.00 = 197{,}500 \text{ lb}$$

and

$$55{,}000 \times 5.998/1.25 = 264{,}000 \text{ lb}$$

The weight of all casing below Section 5 is

$$71{,}400 + (26 \times 3890) = 172{,}500 \text{ lb}$$

and the maximum length of Section 5 is

$$\frac{197{,}500 - 172{,}500}{26} = 960 \text{ ft}$$

The maximum setting depth of Section 6 is

$$1420 - 960 = 460 \text{ ft}$$

It is obvious that Section 6 must consist of casing with a joint strength greater than that of Section 5 (i.e., greater than 395,000 lb). No weight of J-55 casing will satisfy this requirement, and we therefore must use 23-lb N-80 casing with long threads and couplings. For Section 6, allowable joint and yield loads are, respectively,

$$400{,}000/2.00 = 200{,}000 \text{ lb}$$

and

$$80{,}000 \times 5.105/1.25 = 327{,}000 \text{ lb}$$

The weight of all casing below Section 6 is

$$172{,}500 + (26 \times 960) = 197{,}500 \text{ lb}$$

the maximum length of Section 6 is

$$\frac{(200{,}000 - 197{,}500)\text{ lb}}{23\text{ lb/ft}} = 110\text{ ft}$$

and the setting depth of Section 7 is

$$460 - 110 = 350\text{ ft}$$

Section 7 must consist of casing with a joint strength greater than 400,000 lb. The obvious choice is 26-lb N-80 casing with long threads and couplings. For this casing the maximum joint and yield loads are (from calculations previously made) 230,000 and 384,000 lb, respectively. The maximum length for Section 7 is therefore

$$\frac{(230{,}000 - 200{,}000)\text{ lb}}{26\text{ lb/ft}} = 1150\text{ ft}$$

Since this is greater than the allowable setting depth of Section 7, this section can continue to the top of the hole. Summarizing,

Section	*Interval, ft*	*Length, ft*	*Grade*	*Weight*	*Coupling*
1	7490–8000	510	N-80	29	Long
2	5780–7490	1710	N-80	26	Long
3	5310–5780	470	N-80	23	Long
4	1420–5310	3890	J-55	26	Short
5	460–1420	960	J-55	26	Long
6	350–460	110	N-80	23	Long
7	0–350	350	N-80	26	Long

The calculations involving design factors are tabulated below:

Section	N_i	N_c	N_j	N_a
1	2.04	1.277	35.2	37.3
2	1.81	1.125	7.77	8.11
3	1.58	1.125	5.60	5.71
4	1.24	1.125	2.00	1.91
5	1.24	3.54	2.00	1.67
6	1.58	11.55	2.00	2.04
7	1.81	19.50	2.20	2.30

The design problem of Example 2.13 was worked out in a great deal more detail than is normally necessary. It is obvious that, once joint strength becomes the controlling factor, it is unnecessary to check for collapse resistance for each section. It is also apparent that, in most instances, the design factor for axial yielding will be satisfied if the design factor for joint strength is satisfied.

The tedious trial-and-error calculations inherent in the determination of setting depth (where collapse is the controlling factor) can often be simplified by use of charts, examples of which are shown in Figs. 2.6 through 2.8. In each case, the curved lines are plots of Eq. (2.67) and the straight lines are plots of the equation

$$p_{cc} = 0.052 N_c \rho L_s \tag{2.76}$$

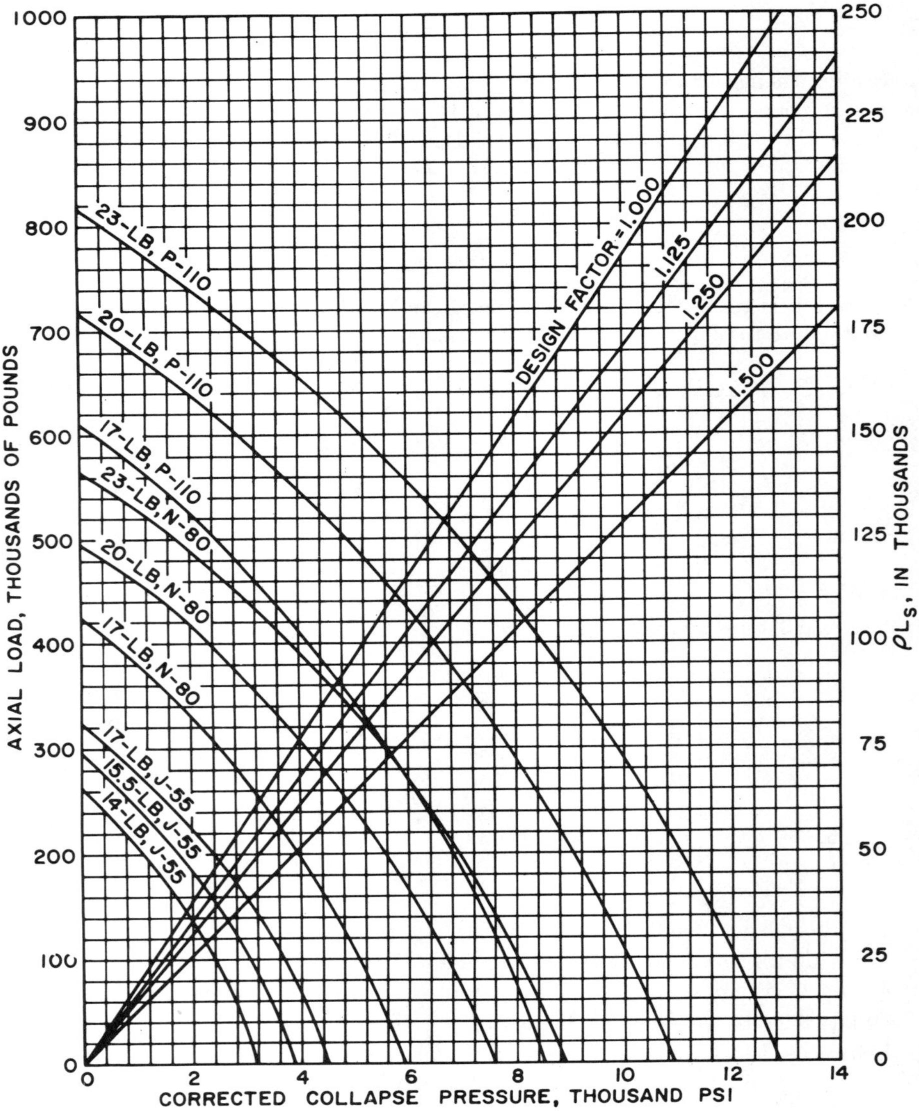

Fig. 2.6 Collapse design chart for $5\frac{1}{2}$-inch casing.

The method of using these charts is illustrated by the following example.

Example 2.14. Considering Example 2.13, determine the setting depth of Section 2 of the combination string, using the collapse design chart for 7-in. casing (Fig. 2.7).

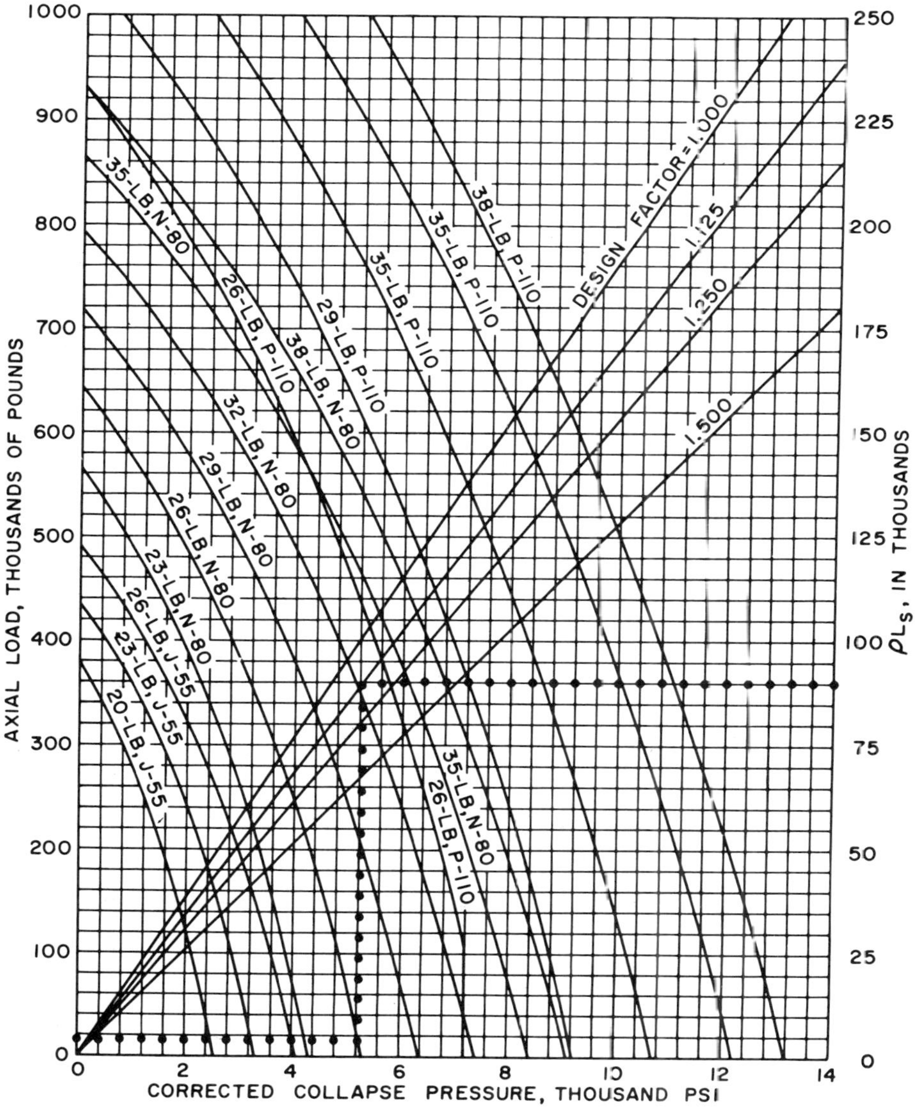

Fig. 2.7 Collapse design chart for 7-inch casing.

Solution: Section 1 consists of 29-lb N-80 casing. Section 2 consists of 26-lb N-80 casing. Neglecting the effect of axial loading, ρL_s for Section 2 is 91,000 (from Fig. 2.7). Therefore $L_s = 91{,}000/12 = 7580$ ft. This is the first assumed setting depth. On this assumption, the axial load on Section 2 is

$$29 \text{ lb/ft} \times (8000 - 7580) \text{ ft} = 12{,}180 \text{ lb}$$

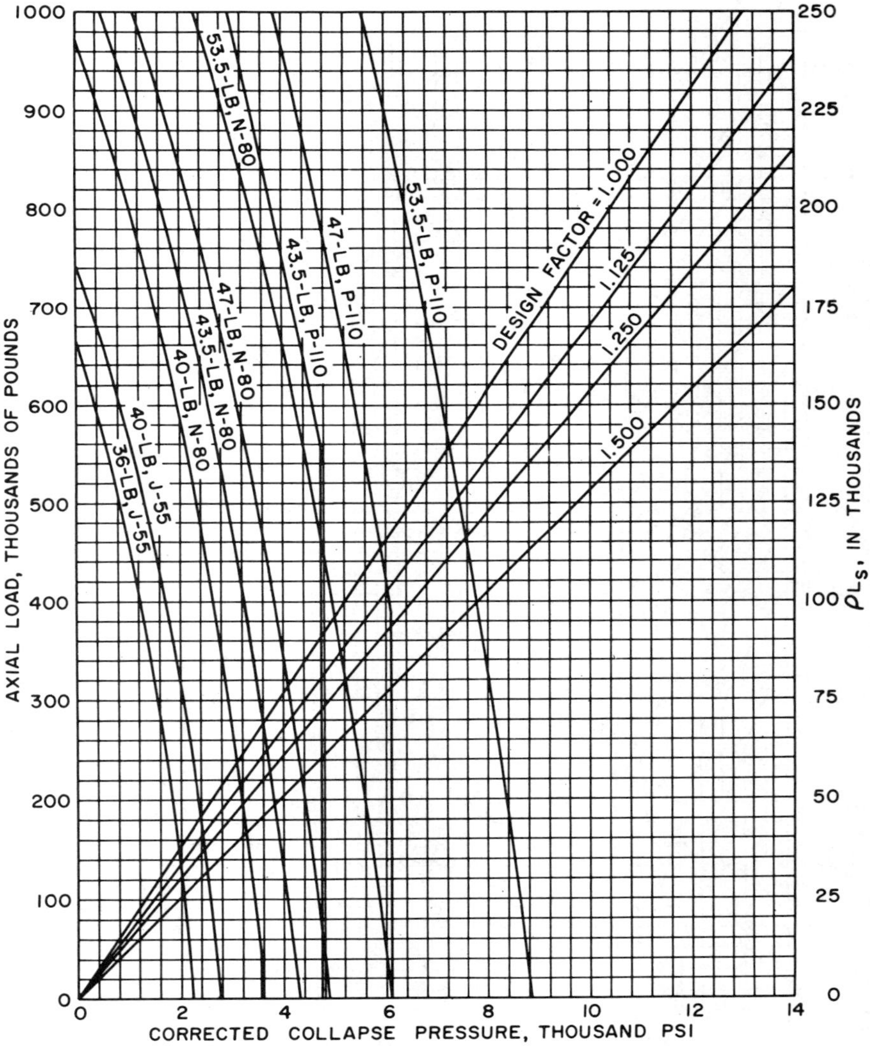

Fig. 2.8 Collapse design chart for $9\frac{5}{8}$-inch casing.

From Fig. 2.7, $\rho L_s = 90{,}000$, and the second assumed setting depth is 90,000/12 = 7500 ft. On this assumption, the axial load is

$$29(8000 - 7500) = 14{,}500 \text{ lb}$$

and within the limits to which the chart can be read, ρL_s is again 90,000. Thus the maximum setting depth of Section 2 is taken to be 7500 ft. This is in significant agreement with the value of 7490 ft, as determined by more rigorous methods.

PROBLEMS

2.1. What is the total make-up loss for a 9000-ft string of 5½-in. range 3 casing with long threads and couplings? *Ans.* 69 ft.

2.2. Calculate the minimum bit size for running the surface string of a well if the casing program calls for a surface string, an intermediate string, and a 7-in. production string. *Ans.* 20¾ in.

2.3. A 5½-in. OD combination casing string is to consist of 15.5-lb and 17-lb J-55 casing to a total depth of 6300 ft, set in 12 ppg mud. Using the collapse design chart, determine the maximum setting depth of the 15.5-lb casing for a design factor of 1.125. *Ans.* 5420 ft.

2.4. What is the total make-up loss for a 4000-ft string of 9⅝-in. 40-lb J-55 range 1 casing with short threads and couplings? *Ans.* 51 ft.

2.5. Determine the casing size for the surface string and the bit sizes for surface and production strings if the casing program calls for a surface string and a 9⅝-in. production string. *Ans.* 16 in.; 20¾ in.; 14¾ in.

2.6. Express the joint strength of 9⅝-in. 36-lb J-55 casing with short threads and couplings as equivalent length of casing for a design factor of 2.00. *Ans.* 6140 ft.

2.7. The (made-up) casing for each section of the 7-in. combination string of Example 2.13 is given in the accompanying table. Considering the make-up loss, calculate the amount of casing to be ordered on location, assuming range 2 casing. *Ans.* 520 ft of 29-lb N-80 long; 2090 ft of 26-lb N-80 long; 590 ft of 23-lb N-80 long; 3930 ft of 26-lb J-55 short; 980 ft of 26-lb J-55 long.

Section	*Grade*	*Weight*	*Coupling*	*Length*
1	N-80	29	Long	510
2	N-80	26	Long	1710
3	N-80	23	Long	470
4	J-55	26	Short	3890
5	J-55	26	Long	960
6	N-80	23	Long	110
7	N-80	26	Long	350

2.8. For which weights of $5\frac{1}{2}$-in. N-80 casing would failure occur by axial yielding; i.e., for which weights is the joint strength greater than the load for axial yielding? *Ans.* All weights.

2.9. A $5\frac{1}{2}$-in. OD combination casing string consists of 15.5-lb J-55 casing with short couplings from zero to 4000 ft and 17-lb N-80 casing with long couplings from 4000 to 6000 ft. The mud weight at time of setting is 11 ppg, and the bottom-hole pressure is 3000 psi. Calculate the following design factors: collapse (upper and lower sections); joint strength (upper section); axial yield (upper section); internal yield (upper section). *Ans.* 1.58; 1.72; 2.20; 1.89; 1.60.

2.10. A string of $9\frac{5}{8}$-in. casing is to be set to a total depth of 6500 ft (where the expected bottom-hole pressure is 3300 psi) in 11 ppg mud. Design factors to be used are 1.25 for collapse, 1.80 for joint strength, and 1.10 for internal yield. All weights of API casing in grades J and N are available. Determine the interval and the composition of each section of the combination string. *Ans.* 0–70 ft, 40-lb J-55 LTC (long threads and couplings); 70–3500 ft, 40-lb J-55 STC (short threads and couplings); 3500–4730 ft, 40-lb N-80 LTC; 4730–5920 ft, 43.5-lb N-80 LTC; 5920–6500 ft, 47-lb N-80 LTC.

REFERENCES

[1]Blaine B. Wescott, C. A. Dunlop, and E. N. Kemler, "Setting Depths for Casing," *Drilling and Production Practice* (API 1940), pp. 125–196.

[2]L. C. Uren, *Petroleum Production Engineering* (*Oil Field Development*), 3rd Ed. (New York: McGraw-Hill Book Co., Inc., 1946), p. 387.

[3]*API Specification for Casing, Tubing, and Drill Pipe* (API Standard 5A) 22nd Ed. (March, 1958).

[4]P. D. Thomas and A. W. Bartok, "Leak Resistance of Casing Joints in Tension," *Drilling and Production Practice* (API, 1941), pp. 243–273.

[5]"Report of the Chairman of the Committee on the Standardization of Oil-Country Tubular Goods," *Production Bulletin No. 223* (API 1939), p. 155.

[6]*Performance Properties of Casing and Tubing* (API Bulletin 5C2), 7th Ed. (May, 1957).

[7]*Steel Tubular Products Manual* (AISI, December, 1951), pp. 205–210.

[8]J. W. Peret, "Casing String Design, Handling, and Usage," *The Petroleum Engineer*, **28,** B33–B40 (November, 1956).

[9]T. McLean Jasper, "Importance of Compression Yield Strength in Establishing Setting Depths for Casing," *Drilling and Production Practice* (API, 1936), pp. 250–277.

[10]W. M. Frame, "Casing Setting Depths Are Not Assured By Physical Properties of the Steel," *Drilling and Production Practice* (API, 1938), pp. 322–352.

[11]J. O. Hills, "Design of Casing Programs," *Drilling and Production Practice* (API, 1939), pp. 369–382.

[12]W. O. Clinedinst, "A Rational Expression for the Critical Collapsing Pressure of Pipe Under External Pressure," *Drilling and Production Practice* (API, 1939), pp. 383–391.

[13]J. L. Holmquist and A. Nadai, "A Theoretical and Experimental Approach to the Problem of Collapse of Deep-Well Casing," *Drilling and Production Practice* (API, 1939), pp. 392–420.

[14]Walter C. Main, "Combining Bending and Hoop Stresses to Determine Collapsing Pressure of Oil-Country Tubular Goods," *Drilling and Production Practice* (API, 1939), pp. 421–431.

[15]S. H. Edwards and C. P. Miller, "Discussion of the Effect of Combined Longitudinal Loading and External Pressure on the Strength of Oil-Well Casing," *Drilling and Production Practice* (API, 1939), pp. 483–502.

[16]M. Lévy, "Problème de l'élastique," *Journal de mathématique pures et appliquées,* Ser. 3, **10,** 5–42 (1884).

[17]Reid T. Stewart, "Collapsing Pressures of Lap-Welded Steel Tubes," *Trans. ASME,* **27,** 730–822 (1906).

[18]J. E. Saye and T. W. G. Richardson, "Field Testing of Casing-String Design Factors," *Drilling and Production Practice* (API, 1954), pp. 23–28.

[19]J. O. Hills, "A Review of Casing-String Design Principles and Practice," *Drilling and Production Practice* (API, 1951), pp. 91–107.

[20]Mid-Continent District Study Committee of Casing Programs, "Survey Report on Casing-String Design Factors," *Drilling and Production Practice* (API 1955), pp. 154–164.

[21]Harold Blum, "Casing Program," *Fundamentals of Rotary Drilling* (Dallas: The Petroleum Engineer Publishing Company, 1959), pp. 111–119.

CHAPTER THREE

Cementing

3.1. Introduction

The development of portland cement started in 1824 when Joseph Aspdin*,[1,2] was granted a patent for an artificial cement made by burning a blend of limestone and clay. Aspdin called his product *portland cement* because the concrete produced from it resembled the stone quarried on the Isle of Portland off the coast of England. Portland cement is a hydraulic product, i.e., it hardens to a set mass when mixed with water. From the very first it was used for subaqueous masonry because of its ease of placement and ability to set up under water. These facts forcibly remind us of its present use in oil-well cementing. At first, cement plants were generally located where the limestones had impurities which would yield a cement of the desired composition. The selection of impure limestones as raw materials was gradually replaced by the blending of different materials to give the compounds desired in cements. Today the principal calcareous raw materials used in cement manufacture are limestone, oyster shells, and marl. The principal argillaceous materials used are clay, shale, slate, and blast-furnace slag. In the case of slow-set cements it is necessary to modify the composition by the addition of small amounts of silica from sandstone and iron ore or other iron-bearing materials. The modern cement plant operates on a mass production basis, since normally 584 lb of raw material are required to produce 1 bbl of cement weighing 376 lb.

The kiln feed, either as powder, filter cake, or slurry is fed to rotary kilns, and fired with coal, oil, or gas. The raw materials lose water as they

*References throughout the text are given at the end of each chapter.

advance in the kiln, and the carbonates (limestones) lose carbon dioxide. As they approach the hot zone of the kiln, the reaction of CaO with Al_2O_3, Fe_2O_3, and SiO_2 begins to take place. These reactions continue as the mixture progresses in the kiln and are finally completed in the hot zone where the temperatures range from 2600 to 2800° F, and where about 25 per cent of the mixture is a liquid. During its travel through the kiln, the feed rolls up into balls which may be as small as dust particles or as large as about two inches in diameter. These balls, known as cement clinker, pass from the kiln into grate or rotary air coolers and are then stored in either silos or bins before grinding. The seasoned clinker, which is really the finished cement in lump form, is then taken to the grinding mills where gypsum ($CaSO_4 \cdot 2H_2O$) is added to retard setting time and increase strength. The finished cement either is stored in silos or moved to the packing house for sacking in multiwalled, moisture-resistant paper sacks each containing 94 lb of cement. Cement is transported in bulk by hopper-bottom railroad cars, bulk trucks, or barge. Each car or shipment of cement is sampled and tested in accordance with API specifications and a portion of the sample is retained for future reference. The unit sold by the cement company to any recognized dealer or agency is the *barrel*, containing 376 lb or four 94-lb sacks. The premium cements are priced 50 cents per bbl higher than standard portland cement. For additional information on the manufacture of portland cements and the materials used in making concrete, the reader is referred to the work of Blanks and Kennedy.[3]

3.2. Composition of Portland Cement

As the constitution of portland cement became better understood, cement chemists adopted the system proposed by Bogue[4] for calculating the composition of cement from chemical analysis in terms of the compounds expected to be present in the clinker. There are four crystalline phases in portland cement clinker that are considered to be the principal cementing materials which hydrate to form, or aid in the formation of, a rigid structure. *Tricalcium silicate* ($3CaO \cdot SiO_2$ = "C_3S") is the major contributor to strength at all stages, but particularly during early stages of curing (up to 28 days). The average tricalcium silicate content is 45 to 65 per cent with a maximum of 67 per cent for high early-strength cement. *Dicalcium silicate* ($2CaO \cdot SiO_2$ = "C_2S") hydrates very slowly and is the constituent which produces long-term strength. The average dicalcium silicate content is 25 to 35 per cent, but being slow to hydrate it has no effect on the setting time. *Tricalcium aluminate* ($3CaO \cdot Al_2O_3$ = "C_3A") hydrates rapidly and contributes most to the heat of hydration, its contribution being nearly 2 cal/gm of cement over 3 days for each per cent of the compound present. To a minor extent, it also contributes to strength and produces that portion of cement which is most readily attacked by sulfate waters. The API specifications[5] permit a maximum of 15 per cent C_3A for *regular type* high early-strength cement and 3 per cent for *highly sulfate-resistant* cements. The rapid setting produced by this compound is controlled by the addition of proper amounts of gypsum ($CaSO_4 \cdot 2H_2O$) as a retarder. *Tetracalcium aluminoferrite* ($4CaO \cdot Al_2O_3 \cdot Fe_2O_3$ = "C_4AF") has little effect on the physical properties of the cement. API specifications require that the sum of the tetracalcium aluminoferrite content plus twice the tricalcium aluminate may not exceed a maximum of 24 per cent for highly sulfate-resistant cements.

Table 3.1[6] shows the oxide analysis for a standard portland, a high early-strength, and a retarded cement, and the calculated Bogue constituents.

Table 3.1. COMPOUND ANALYSIS AND CHEMICAL COMPOSITION OF PORTLAND CEMENTS

(*After Carter and Smith*[6])

	Standard portland cement	High early-strength cement	Retarded cement
Analysis by Oxides			
Lime (CaO)	65.6	66.5	64.0
Silica (SiO_2)	22.2	21.1	23.2
Alumina (Al_2O_3)	5.8	4.8	4.2
Ferric oxide (Fe_2O_3)	2.8	2.6	5.0
Magnesia (MgO)	1.9	1.2	1.2
Sulfur trioxide (SO_3)	1.8	2.7	2.0
Ignition loss	0.7	0.9	0.7
Analysis by oxide compounds			
C_3S	50	67	43
C_2S	26	10	34
C_3A	11	8	3
C_4AF	9	8	15
$CaSO_4$	3	5	3
MgO	2	1	1
Surface area, sq cm/gm	1771	2700	1520

The lower C_3S content of the retarded cement is reflected in a marked reduction in chemical activity, resulting in a slower development of strength up to 28 days. Considerable strength is gained after 28 days owing to the high C_2S content. The reduced C_3A content markedly improves the resistance of the cement to the corrosive attack of sulfate water. The equations for calculating compound compositions from oxide analysis are given in ASTM C 150-56[7] as derived by Blanks and Kennedy.[3] These equations are:

$$\text{Tricalcium silicate} = (4.07 \times \text{per cent CaO}) - (7.60 \times \text{per cent } SiO_2) - (6.72 \times \text{per cent } Al_2O_3) - (1.43 \times \text{per cent } Fe_2O_3) - (2.85 \times \text{per cent } SO_3) \quad (3.1)$$

$$\text{Dicalcium silicate} = (2.87 \times \text{per cent } SiO_2) - (0.754 \times \text{per cent } 3CaO \cdot SiO_2) \quad (3.2)$$

$$\text{Tricalcium aluminate} = (2.65 \times \text{per cent } Al_2O_3) - (1.69 \times \text{per cent } Fe_2O_3) \quad (3.3)$$

$$\text{Tetracalcium aluminoferrite} = 3.04 \times \text{per cent } Fe_2O_3 \quad (3.4)$$

$$\text{Calcium sulfate} = 1.7 \times \text{per cent } SO_3 \quad (3.5)$$

For example, each per cent of iron oxide combines with 0.64 per cent alumina,

$$\frac{Al_2O_3}{Fe_2O_3} = \frac{101.94}{159.68} = 0.64$$

and with 1.40 per cent CaO,

$$\frac{4CaO}{Fe_2O_3} = \frac{224.32}{159.68} = 1.40$$

to form 3.04 per cent tetracalcium aluminoferrite. Also, each per cent of sulfur anhydrite combines with 0.70 per cent lime to form 1.70 per cent calcium sulfate,

$$\frac{CaO}{SO_3} = \frac{56.08}{80.07} = 0.70$$

Example 3.1 illustrates the calculation of cement compounds from the oxide analysis. The standard portland cements[8] are characterized by an average

Example 3.1. Calculate the percentages of compounds in a cement from the oxide analysis as reported from the control laboratory: 21.1 per cent SiO_2, 2.6 per cent Fe_2O_3, 4.8 per cent Al_2O_3, 66. 5 per cent CaO, 1.2 per cent MgO, and 2.7 per cent SO_3.

SOLUTION: Substituting in Eqs. (3.1) through (3.5),

$$\text{Tricalcium silicate} = (4.07 \times 66.5) - (7.60 \times 21.1) - (6.72 \times 4.8) - (1.43 \times 2.6) - (2.85 \times 2.7) = 66.6 \text{ per cent}$$

$$\text{Dicalcium silicate} = (2.87 \times 21.1) - (0.754 \times 66.6) = 10.3 \text{ per cent}$$

$$\text{Tricalcium aluminate} = (2.65 \times 4.8) - (1.69 \times 2.6) = 8.3 \text{ per cent}$$

$$\text{Tetracalcium aluminoferrite} = 3.04 \times 2.6 = 7.9 \text{ per cent}$$

$$\text{Calcium sulfate} = 1.7 \times 2.7 = 4.6 \text{ per cent}$$

fineness or surface area of 1735 sq cm/gm and a composition of 56 per cent

C_3S, 19 per cent C_2S, and 12.7 per cent C_3A, as compared with high early-strength cements with an average surface area of 2436 sq cm/gm and a composition of 66 per cent C_3S, 9 per cent C_2S, and C_3A ranging from zero for the sulfate-resistant type to 12.7 per cent for the regular type.

The slow-set cements fall into one of two categories: *Unretarded* slow-set cement is obtained (theoretically) by limiting the $3CaO \cdot Al_2O_3$ content to zero. The properties of the *retarded* slow-set cement are controlled by the use of organic additives. These materials include modified starches, sugars, salts of lignin sulfonic acid, boric acid and its salts, and gums. It is believed that the retarders function mainly by changing the surface forces on the cement particles by adsorption. Ernsberger and France[9] showed that calcium lignosulfonate was adsorbed by cement. Adsorption of the lignosulfonate anions by the cement particles results in the cement particles bearing a negative charge which causes them to repel each other, i.e., it aids in dispersing the particles. The purposes of these additives are to retard the rate at which cement reacts with water and, at the same time, to maintain the viscosity of the slurry below 25 poises for a period of several hours. The slow-set cements range in fineness from 1300 to 1500 sq cm/gm, with an average fineness of 1470 and average contents of 50 per cent C_3S, 28 per cent C_2S, and C_3A ranging from zero to 4.7 per cent.

Most cement companies produce only one slow-set cement for deep wells. The retardation is adjusted to give a thickening time sufficient for placement of the cement in 14,000-ft wells, longer than is needed in shallow wells. However, this longer time has proven to be *no* disadvantage because the slurries harden rapidly after they have been placed.

Ludwig[10] has summarized the studies on the different compounds in dry cement and their hydration products at surface temperature and pressure

Table 3.2. REACTIONS OF CEMENT MINERALS WITH WATER

(*After Ludwig*[2,10])

Products in dry cement	Hydration products at surface temperature and pressure
Tricalcium silicate ($3CaO \cdot SiO_2$) Dicalcium silicate ($2CaO \cdot SiO_2$)	$1.5CaO \cdot SiO_2 \cdot (1.0 \text{ to } 2.5)H_2O$ $+ Ca(OH)_2$
Tricalcium aluminate ($3CaO \cdot Al_2O_3$) Gypsum ($CaSO_4 \cdot 2H_2O$) Alkalies [$(ssNa_2SO_4) + K_2SO_4$]	$3CaO \cdot Al_2O_3 \cdot 3CaSO_4 \cdot 31H_2O$ $+ 3CaO \cdot Al_2O_3 \cdot 6H_2O$ $+ KOH + NaOH$
Tetracalcium aluminoferrite ($4CaO \cdot Al_2O_3 \cdot Fe_2O_3$) Gypsum ($CaSO_4 \cdot 2H_2O$) Calcium hydroxide [$Ca(OH)_2$]	$3CaO \cdot R_2O_3 \cdot 3CaSO_4 \cdot 31H_2O$ $+ 3CaO \cdot R_2O_3 \cdot 6H_2O$[a]
Magnesia (MgO)	$Mg(OH)_2$
Free lime (CaO)	$Ca(OH)_2$

[a]R_2O_3 is either Al_2O_3 or Fe_2O_3 or a mixture of both.

(Table 3.2). For a better understanding of the structure of hardened cement including combined and uncombined water, consider a given volume of a mixture of cement and water which is free of air. This volume can be visualized as consisting of space filled with solids and water. One of the reactions occurring in the slurry is represented by the equation:

$$\underset{\text{solid}}{3CaO\cdot Al_2O_3} + \underset{\text{solid}}{3(CaSO_4\cdot 2H_2O)} + \underset{\text{liquid}}{25H_2O} \rightarrow$$

$$\underset{\text{solid}}{3CaO\cdot Al_2O_3\cdot 3CaSO_4\cdot 31H_2O}$$

In this reaction, two of the solid phases in the cement react with water to produce a new solid phase. This new phase occupies more space than the original two solids but less space than the original solids plus water. Some shrinkage of slurry or paste can take place while the mixture is mobile; however, void space is formed as a result of further adsorption of water by cement after initial set.

3.3. Standardization of Oil-Well Cements

The *API Recommended Practice for Testing Oil-Well Cements and Cement Additives*[11] contains the test procedures for evaluating the properties of cements used in oil and gas wells. This material is covered in the manual in the following order: (I) Sampling; (II) Preparation of Slurry; (III) Determination of Water Content of Slurry; (IV) Determination of Density of Slurry; (V) Filter-Loss Test; (VI) Strength Tests; (VII) Soundness and Fineness Tests; (VIII) Thickening-Time Tests: (a) Pressure-Temperature Thickening-Time Test, (b) Atmospheric Pressure Thickening-Time Test A, (c) Atmospheric Pressure Thickening-Time Test B.

The physical properties of viscosity, thickening time, and strength are of primary concern in cementing oil and gas wells. The viscosity and thickening time (pumpability) are given first consideration since cement or cement additives must produce a slurry that will remain *pumpable* long enough to be placed in a well at depth.

Three thickening-time testers have been developed for the measurement of viscosity and thickening time of cement slurries. As the name implies, thickening-time tests determine the length of time a cement slurry remains pumpable under given laboratory conditions.

The Standard Oil Company of California thickening-time tester, Fig. 3.1, was described by Silcox and Rule[12] and by Davis.[13] The torque on the paddles is related to the pumpability or thickness of the cement, and the elapsed time between the initial starting of the apparatus and the occurrence of a torque pull of 40 oz is considered as the thickening time of the cement.

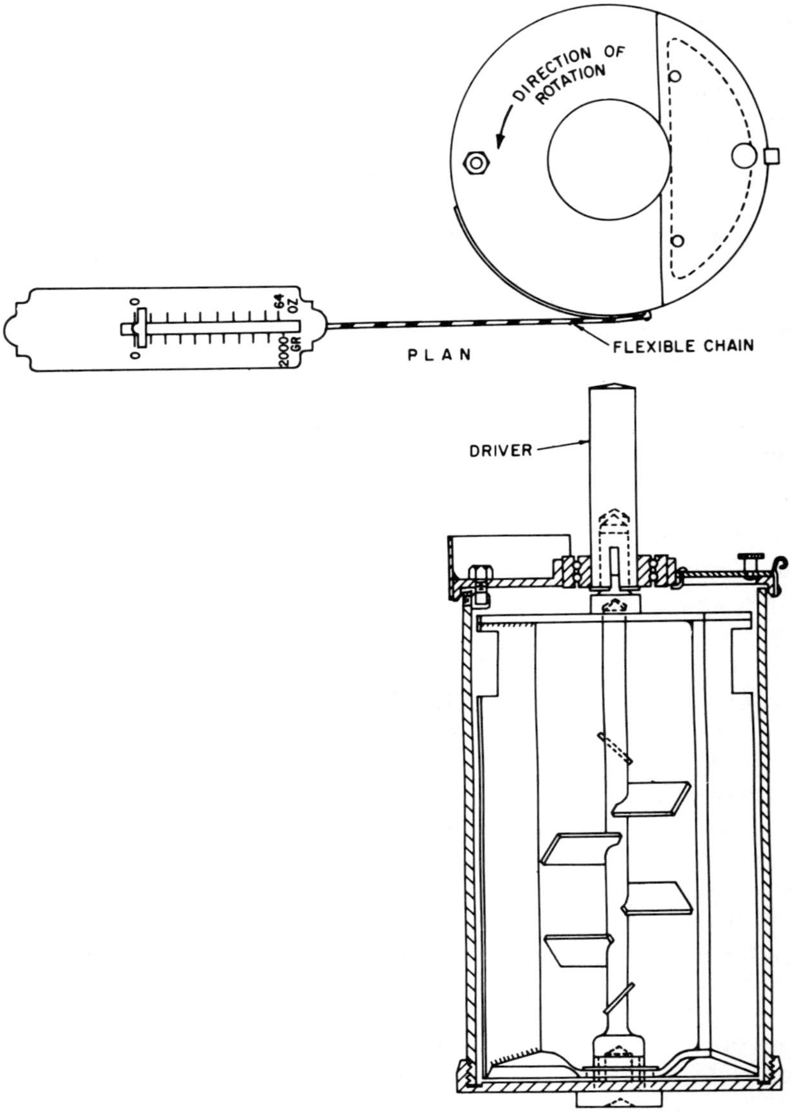

Fig. 3.1 Standard Oil Company of California thickening-time tester. (After API RP 10B.[11])

The Halliburton cement consistometer also utilizes a paddle arrangement which stirs the cement slurry and indicates resistance to the movement of the paddle through the slurry. It is built to test two samples

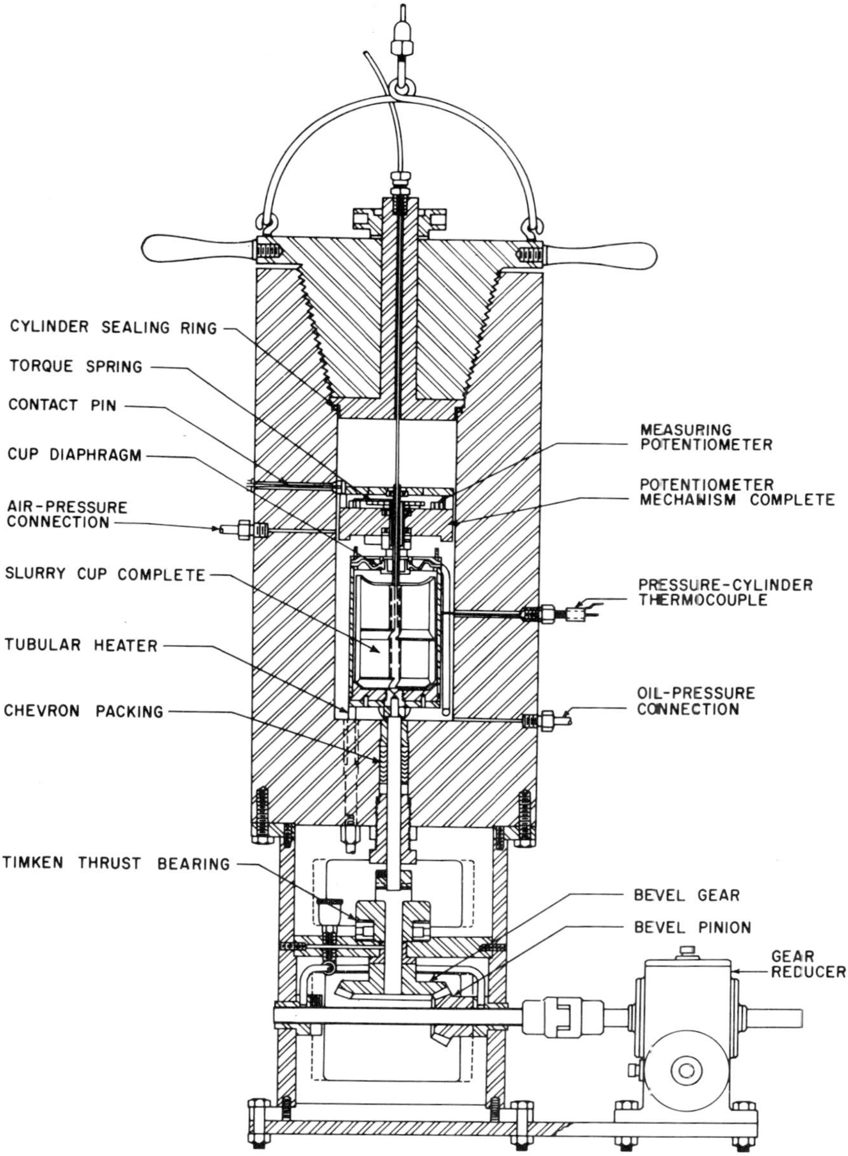

Fig. 3.2 Pan American thickening-time tester. (After API RP 10B.[11])

simultaneously for comparative purposes and to indicate the consistency in poises. A consistency of 100 poises is considered to be the limit of pumpability. The term viscosity is not applied to cement slurries since they do not behave as true fluids. The thickness or body of cement slurries is indicated by the term *consistency* and the poise is retained as the unit. The Standard Oil Company of California and Halliburton thickening-time testers operate only at atmospheric pressure and at temperatures up to 200°F.

Farris[14] described the first high-pressure consistometer (Fig. 3.2). This apparatus provided the first practical evaluation of cement slurries at elevated cementing temperatures and pressures similar to those imposed on a slurry as it flows down the casing. With the Pan American thickening-time tester, it is possible to duplicate actual well cementing temperatures and pressures in the laboratory and to obtain results that have a close relation to cement pumpability. To make a test with this instrument, the slurry is placed in the slurry container, the paddle assembly inserted, the cup placed in the pressure chamber, and the chamber filled with bath oil. After the head assembly of the pressure chamber is in place, the slurry container is set rotating, the oil-pressure pump started, and the resistance to the movement of the paddle through the slurry is recorded. The time required for the consistency to reach 100 poises is called the thickening time.

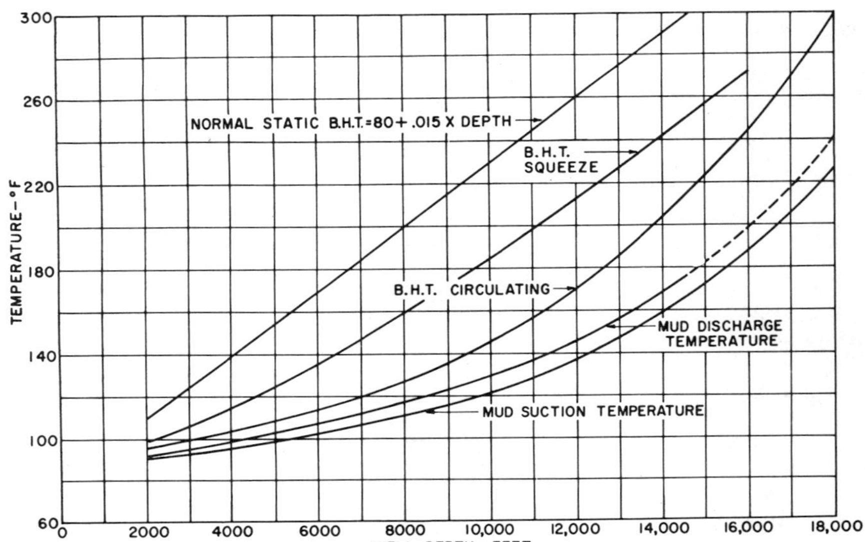

Fig. 3.3 Average temperature of Gulf Coast wells.[15]

In Fig. 3.3[15] is given the normal *static* bottom-hole temperature as calculated from Eq. (3.6):

$$T = 0.015D \text{ ft} + 80°\text{F} \tag{3.6}$$

This is taken to be the average formation temperature at various depths along the Gulf Coast. The *circulating* bottom-hole temperature curve was obtained from actual measurements. This is the temperature which cement slurry should reach during casing cementing. The vertical difference between this curve and the static bottom-hole temperature curve represents the amount of cooling a well undergoes during circulation. The mud *discharge* temperature curve is the average temperature of the mud returns at the surface. The vertical difference between this curve and the circulating bottom-hole temperature curve represents the cooling a slurry undergoes while returning to the surface from the shoe of the casing. The mud *suction* temperature curve is the average temperature of mud entering the well through the mud pump. The vertical distance between this curve and the mud discharge temperature curve represents the cooling a drilling mud undergoes when flowing through the mud pits. From Fig. 3.3, for example, the average 10,000-ft well should have a static bottom-hole temperature of 230°F. After circulation has been established and the bottom-hole temperature stabilized, the circulating temperature will be near 145°F and the mud discharge temperature near 129°F.

From field data, the American Petroleum Institute correlated temperatures and pressures with well depth and the time required to complete oil-well cementing operations. Table 3.3[11] reflects actual temperatures and pressures that may be imposed on oil-well cements during normal casing-cementing operations. API casing-cementing well-simulation *test schedules* are based on these data. These schedules simulate the *average* conditions of temperature and pressure encountered at specified depths in primary cement jobs. The pressure and temperature within the thickening-time tester are programmed to reach maximum conditions in the total cementing time. In the API test procedures, detailed schedules are provided which show the rates of temperature and pressure increase with time that will simulate normal conditions encountered in cementing wells at depths ranging from 1000 to 18,000 ft. Pressure and temperature changes to be expected when a cement slurry is pumped down an 8000-ft well are given in Table 3.4.[11]

It has been mentioned previously that several varieties of portland cement are available for use in oil and gas wells, including regular construction-type cement and special oil-well types. To provide coverage of the several types, the API system[5] provides for seven classes of portland cement based on chemical and physical requirements in regular and highly sulfate-resistant types as follows:

Table 3.3. Basis for API casing-cementing well-simulation test schedules[a]

Schedule no.	Depth, ft	Mud weight,[b] lb/gal	Surface pressure,[c] psi	Bottom-hole circulating temperature,[d] °F	Bottom-hole pressure,[e] psi	Total cementing time,[f] min
1	1000	10	500	80	1020	23
2	2000	10	500	91	1540	27
3	4000	10	500	103	2580	37
4	6000	10	750	113	3870	46
5	8000	10	1000	125	5160	55
5A	9000	11	1120	135	6270	60
6	10,000	12	1250	144	7480	65
7	12,000	14	1500	172	10,230	74
8	14,000	16	1750	206	13,390	84
9	16,000	17	2000	248	16,140	91
10	18,000	18	2000	300	18,800	100

[a]From API RP 10B.[11]

[b]Mud weights obtained from a review of field data.

[c]Surface pressure obtained from a review of field data.

[d]Bottom-hole circulating temperatures averaged from actual field tests run at various depths.

[e]Bottom-hole pressures calculated from surface pressures and mud weights.

[f]Total cementing time calculated from the following assumed conditions: casing size, 7 in. OD; size of job, 300 sacks of cement; pumping rate, 50 cu ft/min.

Table 3.4. SCHEDULE 8: 8000-FT CASING-CEMENTING WELL-SIMULATION TEST[a,b]

(After Reference 11)

Time, min	Pressure, psi	Temperature, °F
0	1000	80
2	1150	82
4	1300	83
6	1450	85
8	1600	87
10	1750	88
12	1900	90
14	2050	91
16	2200	93
18	2350	95
20	2500	96
22	2650	98
24	2800	100
26	2950	101
28	3100	103
30	3250	104
32	3400	106
34	3550	108
36	3700	109
38	3850	111
40	4000	112
42	4150	114
44	4300	116
46	4450	117
48	4600	119
50	4750	121
52	4900	122
54	5050	124
55	5160	125

[a]Final temperature and pressure should be held constant to completion of test, within ±2°F and ±10 psi respectively.

[b]Field conditions assumed: surface temperature, 80°F; surface pressure, 1000 psi; size of job, 300 sacks; mud weight, 10 lb/gal (74.8 lb/cu ft); bottom-hole temperature, 125°F; bottom-hole pressure, 5160 psi; mixing time, 18 min; total cementing time, 55 min.

Class A: Intended for use from surface to 6000-ft depth,* and when special properties are not required. Available in regular type only (similar to ASTM C 150, Type I).

Class B: Intended for use from surface to 6000-ft depth * Available in

*These depth limits are based on the conditions imposed by the casing-cementing well-simulation tests (Schedules 1–9 incl., RP 10B), and should be considered as approximate values.

regular type (similar to ASTM C 150, Type II) for conditions requiring moderate sulfate resistance, and in the highly sulfate-resistant types.

Class C: Intended for use from surface to 6000-ft depth,* and for conditions requiring high early strength. Available in the regular type (similar to ASTM C 150, Type III) and in the highly sulfate-resistant type.

Class N: Intended for use from 6000 to 9000-ft depth,* and for conditions of moderate temperature and pressure. Available in the regular (moderate sulfate resistance) and in the high sulfate-resistant type.

Class D: Intended for use from 6000 to 12,000-ft depth,* and for conditions of moderately high temperature and moderately high pressure. Available in the regular (moderate sulfate resistance) and in the high sulfate-resistant type.

Class E: Intended for use from 6000 to 14,000-ft depth,* and for conditions of high temperature and pressure. Available in the regular (moderate sulfate resistance) and in the high sulfate-resistant type.

Class F: Intended for use from 10,000 to 16,000-ft depth,* and for conditions of extremely high temperature and pressure. Available in regular (moderate) and high sulfate-resistant types.

Figure 3.4 shows the total cementing times for various quantities of cement under the stated conditions and also the minimum thickening-time requirements for each class of cement. This figure is useful in selecting the cement best suited for a particular job. The line keyed "300 sacks," for example, shows the time required to mix 300 sacks of cement at a rate of 20 sacks per min and to displace the slurry in 7-in. casing at a displacement rate of 50 cu ft/min.

On long strings of casing, the time required to mix the cement and displace the cement slurry is of great importance. If the mixing and displacement time exceeds the thickening time of the slurry, cement is left in the casing. The average mixing rate is in the range of 20 to 50 sacks per min and the displacement rate will depend on the rig pumps.

Table 3.5[16] gives the delivery volumes for duplex double-acting pumps in cubic feet per cycle, a cycle being one revolution of the crankshaft on a power pump. For triplex double-acting pumps the values shown in this table must be multiplied by 1.5. The values for this table were calculated from Eq. (3.7),

$$V = 9.09 \times 10^{-4} S(2d_l^2 - d_r^2) \tag{3.7}$$

where S (stroke length), d_l (liner diameter), and d_r (rod diameter) are in inches. Example 3.2 illustrates the method of calculating total cementing time.

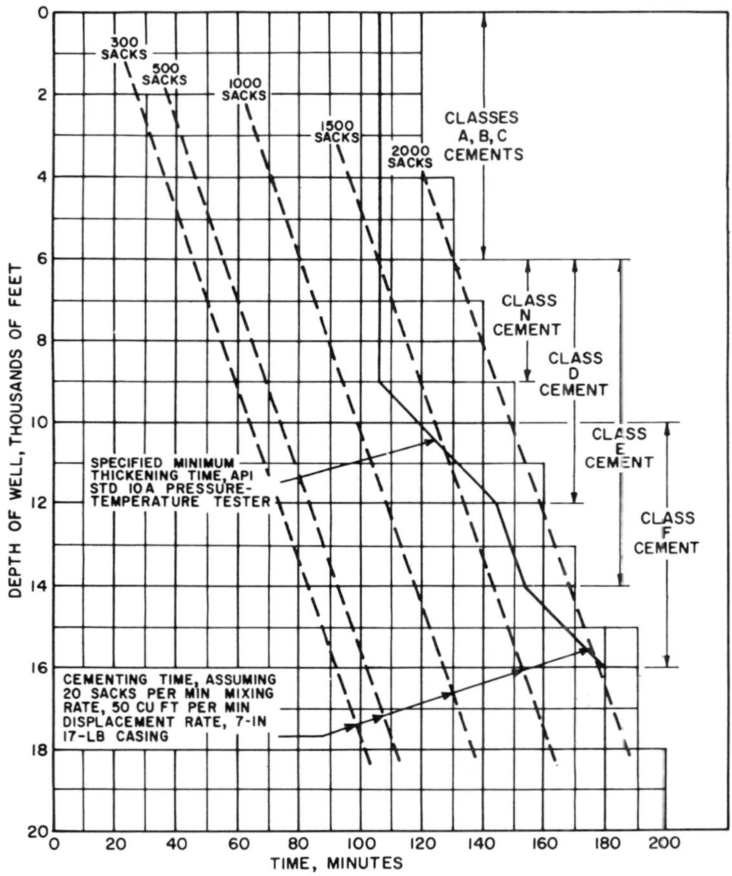

Fig. 3.4 Well depth and cementing-time relationships. (After API STD 10A.[5])

Example 3.2. Calculate the total cementing time for a long protection string of casing, given: 12,000 ft (top float collar) of 53.5 lb/ft, $9\frac{5}{8}$-in. casing (capacity = 0.3973 cu ft per lin ft) set in a $12\frac{1}{4}$-in. hole; 1000 sacks of cement and one cementing truck having a mixing capacity of 20 sacks per min; and a rig mud pump with an 18-in. stroke and $6\frac{1}{2}$-in. liners, operating at 60 rpm with 90 per cent efficiency, and with a capacity of 1.1523 cu ft per cycle (Table 3.5).

SOLUTION:

$$\text{Mixing time} = \frac{\text{No. sacks}}{\text{Truck capacity}} = \frac{1000}{20} = 50 \text{ min}$$

$$\text{Volume of casing} = 0.3973 \times 12{,}000 = 4768 \text{ cu ft}$$

$$\text{Displacing time} = \frac{\text{Volume of casing}}{\text{Pump capacity}} = \frac{4768}{60 \times 1.1523} = 69 \text{ min}$$

$$\text{Total cementing time} = 50 + 69 = 119 \text{ min}$$

Table 3.5. DUPLEX DOUBLE-ACTING PUMP DELIVERY VOLUMES[a]

(*After Reference 16*)

Pump data			Cubic feet per cycle for percentage efficiencies indicated				
Liner size, in.	Stroke, in.	Rod diameter, in.	100	95	90	85	80
4.00	6	1.5	0.1622	0.1541	0.1460	0.1379	0.1298
4.00	8	1.5	0.2163	0.2055	0.1947	0.1839	0.1730
4.00	10	1.5	0.2704	0.2569	0.2434	0.2298	0.2163
4.00	12	1.5	0.3245	0.3083	0.2921	0.2758	0.2596
4.50	8	1.5	0.2782	0.2643	0.2503	0.2364	0.2225
4.50	10	1.5	0.3477	0.3303	0.3129	0.2955	0.2782
4.50	12	1.5	0.4172	0.3964	0.3755	0.3546	0.3338
4.50	14	1.5	0.4869	0.4625	0.4380	0.4137	0.3894
5.00	10	2.0	0.4181	0.3972	0.3763	0.3554	0.3345
5.00	12	2.0	0.5017	0.4766	0.4515	0.4264	0.4014
5.00	14	2.0	0.5854	0.5561	0.5269	0.4976	0.4683
5.00	16	2.5	0.6363	0.6045	0.5727	0.5409	0.5090
5.00	18	2.5	0.7158	0.6800	0.6442	0.6084	0.5726
5.50	12	2.0	0.6163	0.5855	0.5547	0.5239	0.4930
5.50	14	2.0	0.7190	0.6831	0.6471	0.6112	0.5752
5.50	16	2.5	0.7890	0.7496	0.7101	0.6707	0.6312
5.50	18	2.5	0.8876	0.8432	0.7988	0.7545	0.7101
6.00	12	2.0	0.7417	0.7046	0.6675	0.6304	0.5934
6.00	14	2.0	0.8653	0.8220	0.7788	0.7355	0.6922
6.00	16	2.5	0.9563	0.9085	0.8607	0.8129	0.7650
6.00	18	2.5	1.0758	1.0220	0.9682	0.9144	0.8606
6.25	12	2.0	0.8085	0.7681	0.7277	0.6872	0.6468
6.25	14	2.0	0.9433	0.8961	0.8490	0.8018	0.7546
6.25	16	2.5	1.0453	0.9930	0.9408	0.8885	0.8362

Table 3.5. (CONTINUED)

6.25	18	2.5	1.1760	1.1172	1.0584	0.9996	0.9408
6.50	12	2.0	0.8781	0.8342	0.7903	0.7464	0.7025
6.50	14	2.0	1.0244	0.9732	0.9220	0.8707	0.8195
6.50	16	2.5	1.1381	1.0812	1.0243	0.9674	0.9105
6.50	18	2.5	1.2803	1.2163	1.1523	1.0883	1.0242
6.50	20	2.5	1.4226	1.3515	1.2803	1.2092	1.1381
6.75	12	2.0	0.9503	0.9028	0.8553	0.8078	0.7602
6.75	14	2.0	1.1087	1.0533	0.9978	0.9424	0.8870
6.75	16	2.5	1.2344	1.1727	1.1110	1.0492	0.9875
6.75	18	2.5	1.3887	1.3193	1.2498	1.1804	1.1110
6.75	20	2.5	1.5430	1.4659	1.3887	1.3116	1.2344
7.00	12	2.0	1.0253	0.9740	0.9228	0.8715	0.8202
7.00	14	2.0	1.1962	1.1364	1.0766	1.0168	0.9570
7.00	16	2.5	1.3344	1.2677	1.2010	1.1342	1.0675
7.00	18	2.5	1.5012	1.4261	1.3511	1.2760	1.2010
7.00	20	2.5	1.6680	1.5846	1.5012	1.4178	1.3344
7.25	14	2.5	1.2583	1.1954	1.1325	1.0696	1.0066
7.25	16	2.5	1.4380	1.3661	1.2942	1.2223	1.1504
7.25	18	2.5	1.6178	1.5369	1.4560	1.3751	1.2942
7.25	20	2.5	1.7976	1.7077	1.6178	1.5280	1.4381
7.50	14	2.5	1.3521	1.2845	1.2169	1.1493	1.0817
7.50	16	2.5	1.5453	1.4680	1.3908	1.3135	1.2362
7.50	18	2.5	1.7385	1.6516	1.5647	1.4777	1.3908
7.50	20	2.5	1.9316	1.8350	1.7384	1.6419	1.5453
7.75	14	2.5	1.4492	1.3767	1.3043	1.2318	1.1594
7.75	16	2.5	1.6562	1.5734	1.4906	1.4078	1.3250
7.75	18	2.5	1.8632	1.7700	1.6769	1.5837	1.4906
7.75	20	2.5	2.0703	1.9668	1.8633	1.7598	1.6562
7.75	22	2.5	2.2773	2.1634	2.0496	1.9357	1.8218
8.0	20	2.5	2.2134	2.1027	1.9921	1.8814	1.7707

[a]Reproduced through the courtesy of the Halliburton Oil Well Cementing Company.

The selection of the API class of cement and the additive to be used in a particular well is governed by the bottom-hole pressure, possibility of lost circulation zones up the hole, temperature range, and placement time. Table 3.6 shows the technical data[17] on one brand of Class E cement which

Table 3.6a. PRESSURE-THICKENING TIME OF A CLASS E CEMENT[a]

Well depth, ft	API schedule no.	Maximum temperature, °F	Maximum pressure, psi	Time to 100 poises, hr: min
6000	4	113	3870	5:17
10,000	6	144	7480	3:32
12,000	7	172	10,230	3:55
14,000	8	206	13,390	2:56

Table 3.6b. COMPRESSIVE STRENGTH OF A CLASS E CEMENT[a]

Well depth, ft	API schedule no.	Temperature, °F	Pressure, psi	Compressive strength, psi
		8-Hour Cure		
12,000	7-S	260	3000	2404
14,000	8-S	290	3000	3163
		24-Hour Cure		
2000	2-S	110	1600	1666
4000	3-S	140	3000	2492
6000	4-S	170	3000	5508
8000	5-S	200	3000	6006
12,000	7-S	260	3000	4175
14,000	8-S	290	3000	3667
—	—[b]	125	3000	1708
—	—[b]	150	3000	2297
—	—[b]	180	3000	3967

[a]Data of reference 17. Water-cement ratio = 40 per cent (by weight) or 4.5 gal/sack; slurry volume = 1.08 cu ft/sack; slurry weight = 16.3 lb/gal.

[b]Special schedule (non-API) to simulate actual well conditions.

at maximum water-cement ratio has a slurry weight of 15.45 lb per gal. These tests were made in accordance with API recommended practice for testing cement.[11] For example, the pressure thickening time for Schedule No. 7 is 3 hr 55 min, and the 24-hr compressive strength (minimum WOC time) is 4175 psi. For the well of Example 3.2, temperature surveys in the area indicate that the relation between static temperature and depth is given by

$$T = 0.0135D \text{ ft} + 75°\text{F} \tag{3.8}$$

Example 3.3 shows that temperatures of 238°F and 191°F may be expected in this well at depths of 12,000 and 8625 ft, respectively.

Example 3.3. Calculate the static temperature range of the cement job of Example 3.2, given:

Depth to top of float collar = 12,000 ft
1000 sacks of cement
Slurry weight = 16.3 lb/gal
Slurry volume = 1.10 cu ft/sack
Water = 4.5 gal/sack
Casing $9\frac{5}{8}$ in., 53.5 lb/ft run in a $12\frac{1}{4}$-in. hole
Float collar 60 ft above the shoe
Capacity of casing = 0.3612 sacks per lin ft
Annulus fill-up rate = 3.5125 lin ft per sack

SOLUTION:

Volume of cement left in casing = 0.3612 × 60 = 22 sacks
Fill-up in annulus = 3.5125 × 978 = 3435 ft

From Eq. (3.8), the static temperature at 12,060 ft is

$$T = 0.0135 \times 12{,}060 + 75°\text{F} = 238°\text{F}$$

At 8625 ft the static temperature is

$$T = 0.0135 \times 8625 + 75°\text{F} = 191°\text{F}$$

As shown in Sec. 3.4, the slurry used must develop at least 500 psi compressive strength in 24 hr at the cement top (8625 ft). From Table 3.6b, the compressive strength would be near 6000 psi.

3.4. Strength of Cement

The oil industry is concerned only with the performance of cement, one of the important factors in the engineering design of a cement job being the strength and density of the set slurry at the upper placement limit. Originally, it was considered good practice to use a cement of strength at least equal to that of the producing horizon; however, limited information is available on the tensile and compressive strengths of formations in any particular area. Scott,[18] who surveyed the mechanisms of formation fractures, considers that tensile strength will range from 0 to 200 psi in joints and fractures, from 200 to 1000 psi in porous sandstones and limestones, and up to 3000 psi in tight formations. Craft[19] measured the tensile and compressive strengths of eight producing horizons in the West Texas–New Mexico area. The tensile strengths ranged from zero psi for the Wolfcamp formation to as high as 1695 psi for the San Andres limestone, with an average value of 638 psi. The compressive strengths ranged from 8215 psi for a Devonian limestone to as high as 22,500 psi for the Ellenburger limestone.

Studies have been conducted by Carter and Smith[6] on the strength of cementing compositions to determine their strength behavior over curing periods of up to 180 days at elevated temperatures and pressures. The 24-hr compressive strengths of five Class E cements cured at 230°F and 3000 psi ranged from 7000 to 9400 psi, which is considerably below the compressive strengths of the formations in the West Texas area.

In the early days, operators waited as long as seven days for cement to attain strength, and for many years the waiting period was three days. Although practices vary in different states, 24 hours is the most popular waiting period at this time. "Waiting on cement" (referred to as WOC) on surface pipe and on the intermediate or protection string of casing starts when the wiper plug is down, and ends when the drilling-out of the plug is started. On production casing (oil string) the WOC interval begins when the cement is placed and ends when the well is perforated or when drilling out begins.

From an engineering standpoint the first consideration is to determine what strength a set slurry must attain to do the following: (1) secure the pipe in the hole; (2) isolate permeable zones at the well bore; (3) withstand the shock of drilling and perforating; (4) confine initial fracture pressure to the desired zone; (5) develop sufficient strength to satisfy the above requirements after mud contamination. The forces on set cement in the annulus of a well are assumed to be composed of a horizontal compressive force (formation pressure) plus a vertical or nearly vertical shearing or tensile force (weight of the casing). Farris[20] has measured the bonding strength of cement in the annulus under *ideal* conditions as the force that must be applied to $5\frac{1}{2}$-in. casing to break the cement bond and move the casing with respect to the outside $9\frac{5}{8}$-in. casing. Table 3.7 shows the increase in cement bonding strength with time and Table 3.8 shows the calculated loads and the corresponding lengths of casing which each foot of cement in the annulus will support at various cement ages. These data

Table 3.7. CEMENT BONDING STRENGTH

(*After Farris*[20])

Cement age, hr	Force to break 4-ft cement bond, lb	Cement tensile strength, psi	Remarks
1.83	400	0	Soft cement slurry
2.33	550	0	Soft cement slurry
3.08	1300	0	Initial set
3.66	4000	4[a]	Cement stiffening rapidly
4.42	18,200	8[a]	Final set
5.50	20,000+	12	Could not break bond
6.50	20,000+	20	Could not break bond

[a]Estimated value.

Table 3.8. Strength of cement

(*After Farris*[20])

Cement age, hr	Force to break 1-ft cement bond, lb	Length of pipe supported by 1 ft of cement, ft		
		$5\frac{1}{2}$-in. 17-lb	7-in. 24-lb	$13\frac{3}{8}$-in. 72-lb
1.83	100	5.8	4.1	1.3
2.33	137	8.0	5.7	1.9
3.08	325	19.1	13.5	4.5
3.66	1000	58.8	41.6	13.8
4.42	4550	267.5	189.6	63.1

show, for example, that when the cement developed an estimated tensile strength of only 8 psi or, since the ratio of tensile to compressive strength is about twelve to one, 100 psi compressive strength, each linear foot in the annulus is strong enough to support 267 ft of $5\frac{1}{2}$-in. OD, 17-lb pipe.

From laboratory data, Farris[20] found that cement in wells will attain the minimum strength of 8 psi in three times the time required for the cement to reach a consistency of 100 poises at well conditions of temperature and pressure. It was also found that *maximum* temperature development time in a cement slurry was equal to two times the 100 poise thickening time. From field tests, the shut-in casing pressure was found to reach a maximum at approximately the same time the cement down the hole reached its maximum temperature. These relationships are written in the form of equations as follows:

$$t_{8\text{ psi}} = t_{\text{min WOC}} \tag{3.9}$$

$$t_{\text{min WOC}} = t_{100\,\mu} \times 3 \tag{3.10}$$

$$t_{\max T} = t_{100\,\mu} \times 2 \tag{3.11}$$

$$t_{\max T} = t_{\max p_{cs}} \tag{3.12}$$

whence

$$t_{\text{min WOC}} = t_{\max p_{cs}} \times 1.5 \tag{3.13}$$

where $t_{8\text{ psi}}$ is the time from mixing the cement to attainment of a tensile strength of 8 psi; $t_{\text{min WOC}}$, the minimum waiting-on-cement time; $t_{100\,\mu}$, the cement well-simulation 100-poise thickening time; $t_{\max T}$, maximum temperature development time; and $t_{\max p_{cs}}$, the time to maximum static pressure on the casing. Equation (3.13) states that the *minimum* WOC time is calculated by multiplying the time from mixing the first sack of cement to attainment of maximum shut-in casing pressure by a factor of 1.5. A factor of 2 is used where mixing temperatures are below 70°F.

The cement strength required to isolate permeable zones at the well bore was investigated by Clark,[21] who has presented data which show that the ability of cement to bond to pipe can be correlated with the tensile strength of the cement. The *optimum* tensile strength, considering reduced material cost, the high fill-up obtained with light-weight slurry, and relatively little passage of fluids, is 50 psi; however, the *maximum bonding ability* occurs at 100 psi tensile strength or 1200 psi compressive strength.

The strength required to withstand the shock of drilling depends on the contractor's drilling practices. When only drill-collar weight is used on the bit in drilling the plug and there is enough hole to bury the drill collars, a compressive strength of about one hundred pounds per square inch is theoretically sufficient. However, when drill pipe weight is used on the bit with high rotary speeds, the shock loads may cause high-strength cement to fail.

In a survey made by the API Mid-Continent District Study Committee on Cementing Practices,[22] the main reasons given for the "knocking off" of the lower joints of surface casing during rotary drilling, which may result in the abandonment of the well, were (1) the use of second-hand casing, (2) the use of additives producing low-strength cement around the shoe, (3) casing set in unconsolidated formations, (4) the use of too many drill collars in drilling out immediately below the shoe, and (5) failure to use casing centralizers. It has been found by Cannon[23] that this type of cement failure will occur even though the bottom joints are welded at the collars; however, it can be eliminated by the use of wall-cleaning guides on the bottom few joints of casing to remove the mud filter cake and improve the bond between the cement and the walls of the hole. It is the general opinion that this type of cementing failure can be prevented by using Class A or Class C neat cement around the shoe, with or without an accelerator, and by placing centralizers on the bottom three joints of casing. On the other hand, if the cement is to be exposed to the high pressures encountered in hydraulic fracturing, where, according to Godbey and Hodges,[24] the apparent fracture pressure often checks closely with the overburden pressure based on an assumed value of 1 psi per ft, then the fracturing of the cement, in addition to its bonding ability, must be considered. According to Clark,[25] a cement with a tensile strength of 100 psi would be able to withstand these pressures without rupturing. On the other hand, mechanical considerations suggest that actually the problem of confining fracturing pressure or excluding extraneous fluid depends also upon the seal between the cement and the formation.

Water-cement ratio. The fineness of cement, or surface area in square centimeters per gram, is determined by the Wagner turbidimeter[26] apparatus, in which surface area measurement is based on the rate of settlement of spherical particles suspended in kerosene. The more finely a cement is ground the greater the early strength or the shorter the pumping time for any well condition. For example, the average surface areas reported by Robinson[8] are 1784, 2470, and 1395 sq cm per gm for Classes

A, C, and E, respectively. Variation in surface area causes a small change in the specific gravity of the cement.

The surface area affects the minimum and maximum slurry weights (low and high water). The *minimum* amount of water for any class of cement is defined as that amount which can be used without producing a slurry of consistency greater than 30 poises, as measured by the Halliburton consistometer described in Sec. 3.3. If less than the minimum amount of water is used, the friction in the annulus plus the hydrostatic pressure may be high enough to break down weak formations. In addition, since the water-cement ratio is low, the loss of a small amount of water at a tubing collar during squeeze cementing or opposite permeable formations will result in premature stoppage of slurry placement. On the other hand, it will be necessary to use the minimum water-cement ratio on plugback jobs where maximum strength is desired or in killing high-pressure salt-water flows where maximum slurry weight is required. The thickening or pumping time of these extremely low-water slurries is considerably shorter than that of slurries mixed with the usual amounts of water.

After cement and water are mixed to form a (*neat*) slurry, the cement immediately starts to settle since its density is approximately three times that of water. The rate of sedimentation for any class of cement depends on the surface area and the water-cement ratio. When the volume of water is small, there is little or no free water in which the cement particles can move, there are more collisions between particles, and the cement particles stay in suspension. For every surface area there is a *maximum* water-cement ratio at which the cement stays in suspension until the initial set has taken place and the set or hardened volume of the slurry is the same as the fluid volume. The *maximum* water content is defined as that quantity of water that can be mixed with cement without causing the separation of more than 2.5 ml of supernatant water when the slurry is allowed to stand for 2 hr at room temperature in a 250-ml graduate. If more than the maximum amount of water is used, settling of the cement particles in the slurry will occur to the extent that free-water pockets and low-strength cement will exist within the cement column. Perforating opposite such a section would expose all of the formation adjacent to the water pocket. Table 3.9[27]

Table 3.9. PROPERTIES OF NEAT CEMENT SLURRIES
(*After Reference 27*)

Class of cement	Maximum water[a]			Minimum water		
	Water, gal/sack	Weight, lb/gal	Volume, cu ft/sack	Water, gal/sack	Weight, lb/gal	Volume, cu ft/sack
A	5.5	15.39	1.22	3.90	16.89	1.00
C	7.9	13.92	1.54	6.32	14.80	1.33
E	4.4	16.36	1.07	3.15	17.84	0.90

[a]Allows no separation of free water.

shows the data obtained on neat cement slurries prepared with both *maximum* and *minimum* water-cement ratios. The Class C cement has two definite advantages over the Class A cement. First, the slurry can be mixed as light as 13.92 lb per gal, thus considerably reducing the hydrostatic pressure and the possibility of lost circulation, and second, the yield is 1.54 cu ft per sack as compared with 1.22 cu ft per sack for the Class A cement. The Class E cement, which has a low surface area, cannot be mixed lighter than 16.36 lb per gal without the formation of free-water pockets and a gradation in strength. On the other hand, to prevent a blow-out during cementing or after placement of the cement, the slurry should be as heavy as the mud used in drilling the well. A maximum possible slurry weight of 17.84 lb per gal may then be the major factor in the selection of this class or brand of cement for cementing a high-pressure well, the other considerations being its thickening time and compressive strength.

To provide uniformity in testing oil-well cements, the water percentage to be used with each class of cement must conform to the API standards. The water percentage by weight of cement for Classes A and B is 46 per cent (5.19 gal per sack); for Class C, 56 per cent (6.32 gal per sack); and for Classes N, D, E, and F, 40 per cent (4.51 gal per sack).

Since the addition of bentonite requires an increased amount of water, 4.5 per cent is added for each per cent bentonite in Classes A, B, and C, and 3.8 per cent water for each per cent bentonite in Classes N, D, E, and F. For example, if 3 per cent bentonite is added to a Class A cement slurry having a normal water-cement ratio of 46 per cent, the ratio is increased to 59.5 per cent. These water-cement ratios are often referred to as the "API water" for each class of cement and were used in all laboratory experiments described in this chapter unless otherwise noted.

Table 3.10 shows the (calculated) range of slurry weights and volumes that can be obtained with a Class A cement by variation of the water-cement ratio between the practical limits of 3.91 gal/sack (minimum water) and 5.50 gal/sack (maximum water)

Calculation of slurry density or "weight," usually expressed in pounds per gallon, is based on the following equation:

$$\text{Slurry weight} = \frac{\text{lb cement} + \text{lb water} + \text{lb additive}}{\text{gal cement} + \text{gal water} + \text{gal additive}} \tag{3.14}$$

wherein the (absolute) volumes of all constituents are assumed to be additive (a property assigned to ideal solutions). At this point, a clear distinction must be made between the *bulk* volume and the *absolute* volume of solid constituents usually supplied in powdered or granular form. For example, a 94-lb sack of cement contains 1 cu ft of bulk cement powder,

Table 3.10. NEAT CEMENT SLURRY WEIGHT AND VOLUME[a]

Slurry weight, lb/gal	Slurry weight, lb/cu ft	Water, gal/sack cement	Slurry volume cu ft/sack
15.4[b]	115.2[b]	5.50[b]	1.215[b]
15.5	115.9	5.37	1.198
15.6	116.7	5.25	1.182
15.7	117.4	5.13	1.166
15.8	118.2	5.01	1.150
15.9	118.9	4.91	1.136
16.0	119.7	4.79	1.120
16.1	120.4	4.69	1.107
16.2	121.2	4.59	1.094
16.3	121.9	4.49	1.080
16.4	122.7	4.38	1.066
16.5	123.4	4.29	1.053
16.6	124.2	4.19	1.040
16.7	124.9	4.10	1.028
16.8	125.7	4.00	1.015
16.9[c]	126.4[c]	3.91[c]	1.003[c]

[a]Sp. gr. of cement = 3.14.
[b]Maximum water.
[c]Minimum water.

yet the actual space occupied by the cement particles is only 0.48 cu ft — the void space (air) between the particles accounts for 0.52 cu ft of the bulk volume. Thus, when 1 cu ft of bulk cement is mixed with, say, 0.70 cu ft (5.24 gal) of water, the volume of the resulting slurry is only 1.18 cu ft, i.e., absolute volumes are additive.

Similar distinctions should be made when referring to the density of a powdered material. Cement has a bulk density of 94 lb/cu ft, an absolute density of 94/0.48 = 195.8 lb/cu ft, and an (absolute) specific gravity of 195.8/62.4 = 3.14, a handy figure to commit to memory. Returning to Eq. (3.14), the absolute volume of all solid constituents must be calculated in gallons, where

$$\text{Absolute volume, gal} = \frac{\text{lb of material}}{8.34\ \text{lb/gal} \times \text{sp. gr. of material}} \tag{3.15}$$

The volume of slurry to be realized from 1 sack of cement when mixed with a specified amount of water and possibly other additives is called the *yield* or set volume of the mix. The yield in cubic feet per sack of cement is readily calculated from the denominator of Eq. (3.14),

$$\text{Yield} = \frac{\text{gal cement} + \text{gal water} + \text{gal additive}}{7.48\ \text{gal/cu ft}} \tag{3.16}$$

provided that all entries in Eq. (3.14) are those to be associated with 1 sack or 94 lb of cement. In this connection, water-cement ratios are often expressed as a per cent mix wherein the weight of water used is given as a percentage of the weight of cement, or

$$\text{Per cent mix} = \frac{\text{lb water}}{\text{lb cement}} \times 100\% \tag{3.17}$$

Example 3.4. Calculate the weight, per cent mix, and yield or set volume of a slurry, given:

Water-cement ratio = 5.5 gal/sack
Specific gravity of cement = 3.14
One sack = 1 cu ft = 94 lb
Density of water = 8.34 lb/gal

SOLUTION: From Eqs. (3.14)–(3.17),

$$\text{Slurry weight} = \frac{94 \text{ lb/sack} + 5.5 \text{ gal/sack} \times 8.34 \text{ lb/gal}}{\dfrac{94 \text{ lb/sack}}{8.34 \text{ lb/gal} \times 3.14} + 5.5 \text{ gal/sack}}$$

$$= 15.4 \text{ lb/gal}$$

$$\text{Yield} = \frac{\dfrac{94 \text{ lb/sack}}{8.34 \text{ lb/gal} \times 3.14} + 5.5 \text{ gal/sack}}{7.48 \text{ gal/cu ft}}$$

$$= 1.215 \text{ cu ft/sack}$$

$$\text{Absolute volume} = \frac{94 \text{ lb/sack}}{8.34 \text{ lb/gal} \times 3.14} = 3.6 \text{ gal/sack}$$

$$\text{Per cent mix} = \frac{5.5 \text{ gal/sack} \times 8.34 \text{ lb/gal} \times 100}{94 \text{ lb/sack}}$$

$$= 48.8 \text{ per cent water by weight of cement}$$

Sulfate resistance. Of the various salts contained in sea waters and in natural ground waters, it has been found that sodium sulfate, magnesium

Table 3.11. WATER ANALYSES
(*After Rordam*[28])

	Oxide and Ion Contents, ppm		
	Sea water	West Texas water	South Texas water
SiO_2	—	35	56
Fe	—	0.38	10
Ca^{+2}	430	788	4000
Mg^{+2}	1330	384	446
Na^+	11,000	2912	8612
K^+	400	48	726
HCO_3^-	—	200	1630
Cl^-	19,800	4600	29,000
NO_3^-	—	7.3	—
SO_4^{-2}	2760	3160	3450

sulfate, and magnesium chloride are most destructive to cement. The water analyses in Table 3.11 show very clearly that highly corrosive waters can be expected in wells. According to Rordam,[28] these salts react with lime and alumina compounds in the cement and cause a swelling and crumbling which may lead to complete disintegration of the hardened cement. It has been established that a cement having zero to three per cent C_3A can be considered practically immune to the attack of sulfate solutions. When the C_3A content exceeds four to five per cent, failures can be expected.

To investigate the sulfate resistance of oil-well cements, an API Committee[29] made a cooperative study in which the deterioration as generally characterized by expansion was measured on cement specimens. Table 3.12[30] shows the results of accelerated laboratory tests with 5 per cent

Table 3.12. SUMMARY OF TESTS OF SULFATE RESISTANCE OF OIL-WELL CEMENTS[a]

(After Bearden[30])

API class of cement	Tricalcium aluminate content, per cent	At 100°F		At 140°F		At 180°F	
		Age, days	Expansion, per cent	Age, days	Expansion, per cent	Age, days	Expansion, per cent
A	11.5	252	F[b]	196	F	84	F
B	6.6	364	0.316	364	0.052	364	0.021
C[c]	0.0	364	0.053	364	0.005	364	0.009
D[c]	0.0	364	0.063	364	0.021	364	0.018
E	6.2	364	0.073	364	0.038	364	0.022

[a]Materials exposed to 5 per cent sodium sulfate solution under accelerated laboratory conditions.

[b]F denotes failure of specimen in indicated time interval.

[c]High sulfate-resistant type.

sodium sulfate solution. These tests indicated considerable difference in the sulfate resistance of the cements and additives, and this resistance increases as the C_3A content of the cement decreases.

3.5. *Additives for Oil-Well Cements*

The various types of cement are manufactured because of variations in well conditions such as depth, temperature, and pressure, and can be used to make slurries which may be varied within only nominal weight and thickening time limits by varying the proportions of water. Additives are used to make further variations from the normal properties of cement slurries, and have become essential in the design of oil-well cementing compositions. Most operators use some type of additive in the cement, usually for one or more of the following purposes:

1. Reducing or increasing density.
2. Increasing volume at reduced unit cost.
3. Accelerating or retarding setting time.
4. Increasing strengths.
5. Preventing lost returns.
6. Increasing or improving the durability.
7. Decreasing the water loss.

Table 3.13 shows the effect of some additives on the physical properties[31]

Table 3.13. Effects of some additives on the physical properties of cement[a]

(*After Dale*[31])

		Bentonite	Perlite	Diatomaceous earth	Pozzolan	Sand	Barite	Arsenoferrite	Calcium chloride	Sodium chloride[b]	Lignosulfonates	CMHEC[c]	Diesel oil	Low-water-loss materials[c]	Lost-circulation materials	Activated charcoal
Density	Decreased	+	+	+	x											
	Increased					+	+	+	x	x	x					
Water Requirement	Decreased										+					
	Increased	+	x	+	x	x	x	x							x	x
Viscosity	Decreased								x		+					
	Increased	x	x	x	x	x	x	x							x	x
Thickening Time	Decreased			x						x	+	+	x	+		
	Increased	x					x	x	+	+						
Setting Time	Decreased	x	x	x	x						+	+		x		
	Increased						x	x	+	+						
Early Strength	Decreased	x	x	x	x		x	x			+	+		x	x	x
	Increased								+	+						
Final Strength	Decreased	x	x	+	x		x					+		x	x	x
	Increased										x					
Durability	Decreased	x	x	x									x		x	
	Increased				+											x
Water Loss	Decreased	+									x	+	x	+	x	
	Increased		x	x												

[a]A minor effect is denoted by x; + denotes major effect and/or principal purpose for which the additive is used.

[b]Small percentages of sodium chloride accelerate thickening; large percentages may retard API Class A cement.

[c]Carboxymethyl hydroxyethyl cellulose.

Table 3.14. PHYSICAL PROPERTIES OF CEMENTING MATERIALS AND ADDITIVES

(*After Owsley*[32])

Material	Bulk density, lb/cu ft	Specific gravity	Absolute density, lb/cu ft
Portand cement	94	3.14	195.9
Pozzolans			
Fly-ash	74	2.46	153.5
Scoria	84	2.80	174.7
Pumicite	74	2.47	154.1
Bentonite	60	2.65	165.4
Activated charcoal	14	1.57	97.9
Barite	135	4.23	263.9
Nut shell (walnut)	48	1.28	79.9
Sand (Ottawa)	100	2.63	164.1
Salt	71	2.17	135.4
Calcium chloride, powder	50.5	1.96	122.3
Calcium chloride, flake	56.4	1.96	122.3
Expanded perlite	13[a]	—	—
Diatomaceous earth	16.7	2.10	131.0

[a]For 13 lb of expanded perlite use a volume of 2.32 gal at 0 psi and/or 0.71 gal for 3,000 psi.

of cement, and Table 3.14 shows the physical properties[32] of cementing materials and additives.

Bentonite. Bentonite or *gel* has been used in oil-well cements for at least 25 years. At the present time gel is the most common and frequently used additive. This clay mineral is an alteration product of volcanic rocks, the purest deposits of which occur in Wyoming. The addition of bentonite in amounts of 2 to 12 per cent permits the use of more water in the slurries and results in slurry of lower weight and lower strength, and lower unit costs. According to its proponents, the chief advantages of this type of additive are as follows:

1. Bentonite holds the cement particles in suspension.
2. Lower slurry densities give lower hydrostatic pressures, i.e., higher fill-ups behind the casing are obtained.
3. Lower strength gives better penetrability in perforating.
4. Lower cost per cubic foot of slurry.
5. Lower water loss of slurry.

The chief disadvantages are that bentonite lowers the resistance to the attack of sulfate waters, and that there is considerable loss in strength and increase in permeability at high temperatures. The permeability[6] of set cement at 290°F and 320°F is presented in Table 3.15. At 290°F all

Table 3.15. PERMEABILITY OF SET CEMENT
(*After Carter and Smith*[6])

Composition	Permeability at 290°F, md		Permeability at 320°F, md	
	3 days	7 days	3 days	7 days
API Class E cement	0.0241	0.0116	0.0967	8.0537
API Class E cement (4 per cent gel)	0.1285	0.0482	0.3503	23.3190
API Class A cement (0.7 per cent lignin retarder, 12 per cent gel)	0.4113	0.1827	0.7282	12.4322

measured values were less than one millidarcy; however, at 320°F both the Class A and Class E cements had high permeabilities.

According to Dunlap and Patchen,[33] high-temperature retrogression is eliminated in deep-well cementing by adding suitable amounts of silica flour (22 weight per cent), and a mere 0.4 weight per cent CMHEC greatly increases the thickening time of the slurry at high temperatures.

Modified cements first described by Morgan and Dumbauld,[34,35] include those slurries containing 12 to 15 per cent bentonite with 0.25 to 0.75 per cent calcium lignosulfonate per sack of cement. These cementing slurries have properties similar to those of gel cements, except that the calcium lignosulfonate acts as a retarder and dispersing agent, giving the cement slurry a longer pumping time under high-temperature well conditions. The successful use of 12.5 to 14 lb per gal slurries for placing cement columns 2000 to 4000 ft above the casing shoe on the primary cement job eliminates the time and expense of stage cementing. Table 3.16 shows the densities

Table 3.16. SLURRY WEIGHT AND VOLUME FOR 4 PER CENT GEL CEMENT

Slurry weight, lb/gal	Slurry weight, lb/cu ft	Water, gal/sack cement	Slurry volume, cu ft/sack
14.1[a]	105.5[a]	7.73[a]	1.536[a]
14.2	106.2	7.53	1.509
14.3	107.0	7.33	1.483
14.4	107.7	7.15	1.459
14.5	108.5	6.97	1.434
14.6	109.2	6.80	1.412
14.7	110.0	6.63	1.389
14.8	110.7	6.49	1.370
14.9	111.5	6.34	1.350
15.0	112.2	6.19	1.330
15.1	112.9	6.04	1.310
15.2[b]	113.7[b]	5.90[b]	1.291[b]

[a]Maximum water.
[b]Minimum water.

of 4 per cent bentonite–Class A cement slurries with *maximum* and *minimum* water.

To calculate the quantities of materials to be used on a given cement job, it is first necessary to decide from the caliper log or other data the number of sacks of normal cement required. If this amount is 1000 sacks and the normal slurry volume is 1.10 cu ft per sack, then the number of cubic feet of cement slurry required is 1000 sacks × 1.10 cu ft/sack = 1100 cu ft. One sack of 4 per cent gel cement mixed to 14.4 lb per gal has a volume of 1.459 cu ft; therefore the number of sacks of cement required are 1100/1.459 = 753, or 753 × 94 = 70,782 lb of cement. Then the amount of bentonite required is 70,782 × 0.04 = 2831.3 lb or 28.3 sacks.

The *maximum depth* at which neat cement or cement additives may be used is governed by the thickening time. The *upper limit* of depth is based on the attainment of at least 500 psi compressive strength in 24 hr. Bentonite–Class A cement mixtures have a shorter thickening time than neat cement. The thickening time of a Class A cement slurry with *maximum* water under simulated well conditions[36] is given in Table 3.17. The compressive strengths of neat cement slurries have been reported by Davis and Faulk.[37] Figure 3.5 shows the compressive strengths of a Class A *neat*

Table 3.17. THICKENING OF AN API CLASS A CEMENT[a]
(After Reference 36)

Bentonite, per cent	Thickening time at 4000-ft well conditions, hr: min	Thickening time at 8000-ft well conditions, hr: min
0	4:35	2:13
4	3:08	1:54
8	3:01	1:46
12	3:40	1:56

[a]Mixed to maximum water-cement ratio.

cement with a slurry weight of 15.7 lb per gal cured at atmospheric pressure, as compared with a Class E cement (Fig. 3.6) mixed at 16.3 lb per gal. Bentonite-cement mixtures permit the use of larger quantities of mixing water because the bentonite keeps the cement particles in suspension. This additional water results in lower slurry density and lower strength. Figures 3.7 and 3.8[31] show the compressive strengths of an API Class A cement with various percentages of bentonite, cured at 80°F and 120°F, and Fig. 3.9[37] shows the compressive strengths of a Class E cement with 4 per cent bentonite cured at various temperatures. Example 3.5 shows the amount of cement and bentonite required for a string of surface casing.

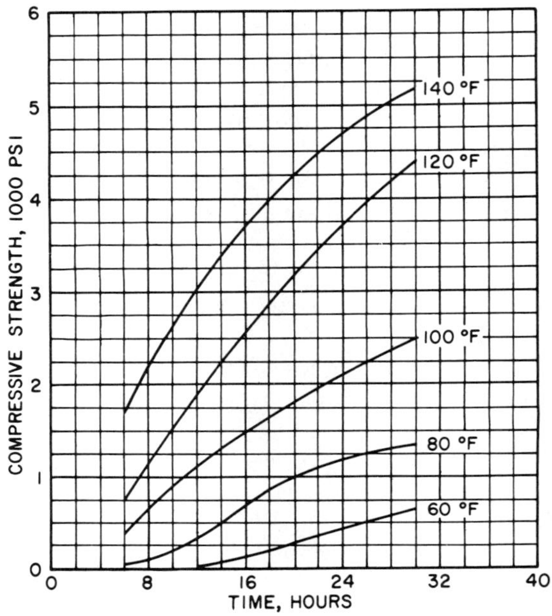

Fig. 3.5 Compressive strength of neat Class A cement. (Davis and Faulk,[37] *Drilling and Production Practice,* API.)

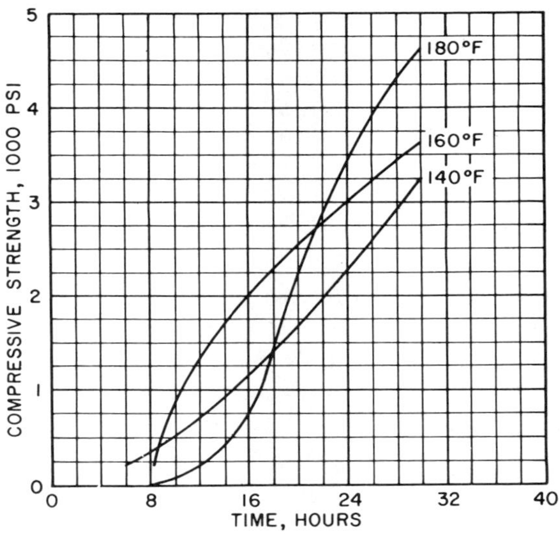

Fig. 3.6 Compressive strength of neat Class E cement. (Davis and Faulk,[37] *Drilling and Production Practice,* API.)

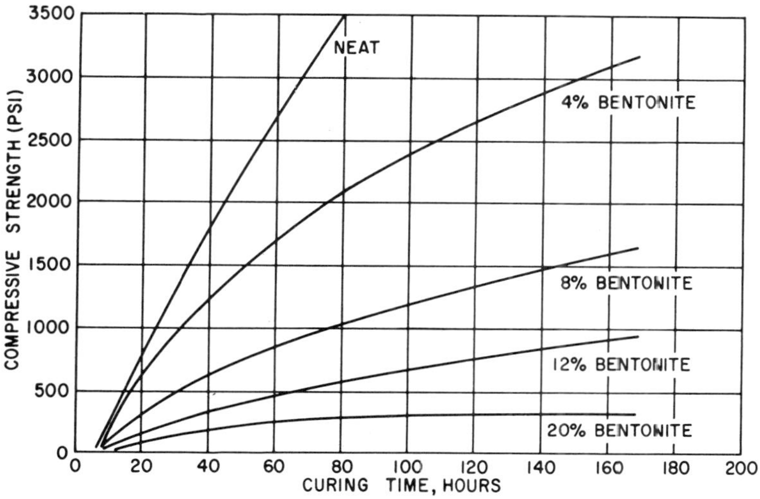

Fig. 3.7 Compressive strength of API Class A cement with various percentages of bentonite cured at 80°F. (After Dale.[31])

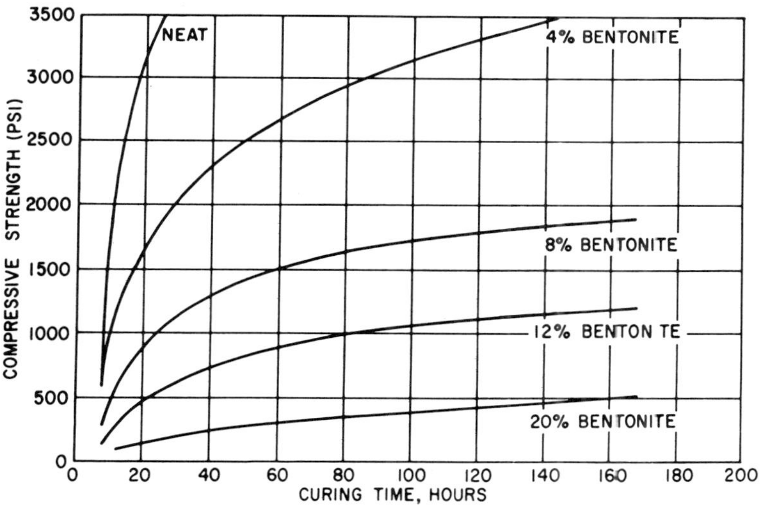

Fig. 3.8 Compressive strength of API Class A cement with various percentages of bentonite cured at 120°F. (After Dale.[31])

Example 3.5. Calculate the number of sacks of cement and bentonite required to obtain cement returns on surface casing, given:

Annular volume between $9\frac{5}{8}$-in. casing and $12\frac{1}{4}$-in. hole = 0.3469 cu ft per lin ft

Float collar placed 30 ft above the shoe

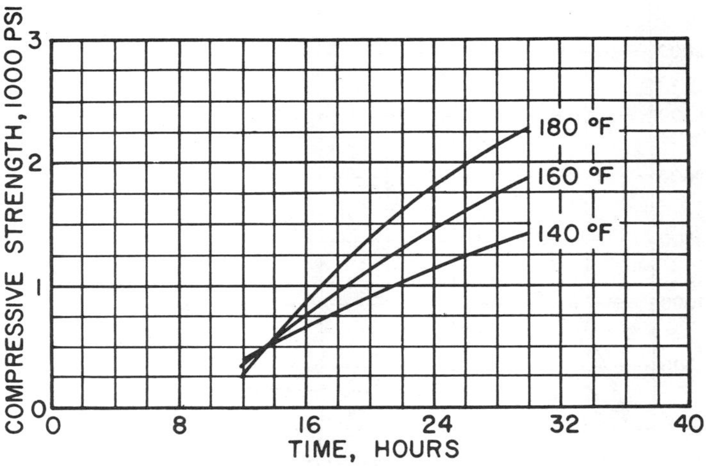

Fig. 3.9 Compressive strength of Class E cement with 4 per cent bentonite. (Davis and Faulk,[37] *Drilling and Production Practice*, API.)

Volume of $9\frac{5}{8}$-in., 40-lb/ft casing = 0.4256 cu ft per lin ft

Class A cement with 4 per cent gel

Water-cement ratio = 7.73 gal per sack

Slurry weight = 14.10 lb per gal

Slurry volume = 1.536 cu ft per sack of cement

Casing to be landed at 1400 ft

Excess cement required = 35 per cent

SOLUTION:

Cement left in casing = 30 ft × 0.4256 cu ft/ft
= 12.77 cu ft

Cement required to fill annulus
= 1400 ft × 0.3469 cu ft/ft × 1.35
= 655.64 cu ft

Total cement required = 668.41 cu ft

$$\text{Sacks of cement required} = \frac{668.41}{1.536} = 435$$

Pounds of cement = 435 sk × 94 lb/sk = 40,890

Bentonite required = 40,890 × 4 per cent
= 1636 lb or 163.6 sacks

The *range of applicability* of a neat cement or cement with an additive is described by the depth range as well as by the temperature range covered. For example, if an oil string is to be set at 6000 ft with a cement fill-up to 4000 ft, it is necessary to know the static bottom-hole temperature at 4000 ft and the circulating bottom-hole temperature at 6000 ft. If the static temperature at 4000 ft is 140°F and the circulating temperature at 6000 ft is 113°F (Table 3.3), the slurry used must develop 500 psi compressive strength in 24 hr at the upper placement limit and must remain pumpable at the lower depth conditions for a sufficient length of time to allow placement of the cement. If a Class A *neat* cement is used, the thickening time, from Table 3.22, is 2 hr 41 min as compared with an assumed placement time of 64 min. Inspection of Fig. 3.5 shows that a 500 psi compressive strength is reached in less than 7 hr at 140°F.

Expanded perlite. Perlite in its natural form is a volcanic glass with a curved fracture and contains from one to six per cent water. The raw volcanic ore is crushed and graded, and then introduced into a kiln where the temperature is raised to the fusion point of the ore. At this temperature the combined water in the ore expands, producing a cellular, thin-walled structure. It is then cooled, graded, packaged, and marketed under various brand names. Perlite, because of its structure, breaks down under pressure, allowing the slurry water to saturate the pore space. Laboratory tests have indicated that each cubic foot of perlite requires 4 to 4½ gal of water for complete saturation of the particles under pressure. If less than this minimum amount is used, the water forced into the perlite when pressure is applied will cause the slurry to become too viscous to pump.

In perlite-cement mixes, bentonite is mixed with the cement in an amount of 2 to 6 per cent. It is used to minimize water separation and to raise viscosity, which is necessary because the perlite particles tend simply to float in the mixture because their unsaturated density is less than that of the water. The investigations of Sanders and Nussbaumer[38] on the properties of perlite-cement slurries containing bentonite are presented in Table 3.18. This table gives the maximum, minimum, and recommended water ratios and the resulting physical properties for various percentages of gel. The pumping times of perlite-cement slurries cannot be measured with thickening-time testers now available, but Sanders and Nussbaumer[38] consider perlite to be chemically inert, and therefore it can be assumed that the addition of perlite does not alter the pumping time. According to the proponents of this type of additive, its chief advantages are:

1. The slurries set with less strength and thus have better gun-perforation characteristics.
2. A lower slurry weight is obtained because of the low density of the perlite and the use of additional water.

3. According to Howard and Scott,[39] the granular particles of perlite provide a bridging or sealing effect to reduce slurry losses to the formation.

Table 3.18. PHYSICAL PROPERTIES OF PERLITE–CLASS A CEMENT SLURRIES[a]

(*After Sanders and Nussbaumer*[38])

		Atmospheric pressure		3000 psi		
Bentonite, per cent	Water ratio, gal/sack	Slurry weight, lb/gal	Slurry volume, cu ft/sack	Slurry weight, lb/gal	Slurry volume, cu ft/sack	24-hr compressive strength[b]
2	9.0	12.3	2.01	13.7	1.79	1950
2	10.0	12.0	2.14	13.4	1.93	1500
2	11.0	11.8	2.27	13.0	2.06	1050
4	10.0	12.1	2.15	13.4	1.94	1600
4	11.0	11.9	2.28	13.1	2.07	1200
4	12.0	11.7	2.42	12.8	2.20	800
6	11.0	11.9	2.30	13.2	2.08	1300
6	12.0	11.7	2.43	12.9	2.22	1100
6	13.0	11.5	2.56	12.6	2.35	850

[a]One cubic foot (1 sack) cement to 1 cu ft perlite.
[b]Cured at 100°F and 3000 psi.

Morgan[40] calculated the comparative costs of a 13 lb per gal slurry of high bentonite-content cement and a bentonite-perlite cement, as shown in Table 3.19. Example 3.6 illustrates the slurry calculations for the perlite-

Table 3.19. RELATIVE COSTS OF CEMENT SLURRIES
(*After Morgan*[40])

Item of Cost	Bentonite mixture	Bentonite-perlite mixture	Neat cement (ref)
1 sack of cement	$1.00	$1.00	$1.00
10 per cent bentonite	0.375	—	—
4 per cent bentonite	—	0.125	—
1 cu ft perlite	—	1.25	—
Total cost	$1.375	$2.375	$1.00
Volume of slurry, cu ft	2.02	2.07	1.22
Cost per cu ft of slurry	$0.68	$1.15	$0.82

gel-cement mixtures of Table 3.18, and Example 3.7 shows the quantity of each material required to yield a given volume of slurry.

Example 3.6. Calculate the weight and volume of slurry for a perlite-gel-cement mixture, given:

A normal mixture containing 100 cu ft perlite, 100 sacks of cement, and 4 per cent gel by weight of cement

Dry bulk density of perlite = 13 lb/cu ft

Specific gravity of cement = 3.14

Specific gravity of gel = 2.65

Specific gravity of perlite = 0.672

Mixing water = 11 gal per sack

SOLUTION: From Eq. (3.14) the slurry weight is

$$\frac{\text{lb cement} + \text{lb water} + \text{lb perlite} + \text{lb gel}}{\text{gal cement} + \text{gal water} + \text{gal perlite} + \text{gal gel}}$$

$$= \frac{(100 \text{ sks} \times 94 \text{ lb/sk}) + (11 \text{ gal/sk} \times 100 \text{ sk} \times 8.34 \text{ lb/gal})}{\left(\dfrac{9400 \text{ lb}}{8.34 \times 3.14}\right) + 1100 \text{ gal} + \left(\dfrac{1300 \text{ lb}}{8.34 \times 0.672}\right)}$$

$$\frac{+ (100 \text{ cu ft} \times 13 \text{ lb/cu ft}) + (0.04 \times 9400 \text{ lb})}{+ \left(\dfrac{376 \text{ lb}}{8.34 \times 2.65}\right)}$$

$$= \frac{9400 + 9174 + 1300 + 376}{359 + 1100 + 232 + 17.0} = 11.86 \text{ lb per gal}$$

The yield from Eq. (3.16) is

$$\frac{\text{gal cement} + \text{gal water} + \text{gal perlite} + \text{gal gel}}{7.48 \text{ gal/cu ft}}$$

$$= \frac{\left(\dfrac{9400}{8.34 \times 3.14}\right) + 1100 + \left(\dfrac{1300}{8.34 \times 0.672}\right) + \left(\dfrac{376}{8.34 \times 2.65}\right)}{7.48}$$

$$= \frac{359 + 1100 + 232 + 17.0}{7.48}$$

$$= 228.3 \text{ cu ft/100 sacks cement, or } 2.28 \text{ cu ft/sack of cement.}$$

Example 3.7. Calculate the number of sacks of material required for a perlite-gel cement, given:

Volume of slurry required = 1000 cu ft

A 1:1 (by volume) dry mix of cement and perlite

One sack of perlite = 4 cu ft

One sack of cement = 1 cu ft dry cement

Four per cent gel by weight of cement

Solution: From Table 3.18, for a slurry weight of 11.9 lb per gal (Example 3.6), the slurry volume is 2.28 cu ft per sack of cement. Then the number of sacks of cement required is

$$\frac{1000 \text{ cu ft}}{2.28 \text{ cu ft/sack}} = 439 \text{ sacks or cu ft of cement}$$

Then, since equal dry volumes of perlite and cement are to be used,

$$\frac{439 \text{ cu ft}}{4 \text{ cu ft/sack}} = 110 \text{ sacks of perlite}$$

and

$$\frac{4 \text{ per cent} \times 439 \text{ sacks} \times 94 \text{ lb/sack}}{100 \text{ lb/sack gel}} = 16.5 \text{ sacks of gel}$$

Pozzolanic materials. Pozzolans are siliceous and aluminous mineral substances which possess no cementitious qualities themselves but which will react with the calcium hydroxide in the slurry to form compounds that *do* possess cementitious properties. According to Blanks and Kennedy,[3] the naturally occurring pozzolans include (1) clays and shales of the montmorillonite and kaolinite types that require calcination at temperatures in the range of 800 to 2000°F to make them active pozzolans; (2) the opaline materials including opaline shales, diatomaceous earths, and chert which may or may not be benefited by calcination; and (3) volcanic tuffs and pumicites of the basaltic and andesitic types, which are seldom improved by calcination.

Finely-ground pumice and an artificial pozzolan called *fly ash* (flue dust), produced in coal-burning power plants, are used extensively as additives in oil-well cements. These materials do not provide the water-carrying capacity of clays such as bentonite. Inasmuch as these pozzolans have about the same fineness as portland cement, they hold about the same amount of water. Pozzolans have specific gravities of 2.3 to 2.8, depending on the source, as compared with 3.10 to 3.20 for portland cements. Mangold[41] and Sterne[42] discussed the effects of pozzolans on the properties of cement slurries, and reviewed laboratory and field experiences with these materials. The advantages of the pozzolan–API Class A cement mixtures are:

1. Low heat of hydration.
2. Early strengths much less than those of regular cements, giving better penetrability during perforation.
3. Greater resistance to sulfate and acid-water attack.
4. Lighter slurry weight as compared with Class A cement slurries of similar consistency, resulting in an increased fill-up of cement behind the casing for a given hydrostatic pressure.
5. Savings in both cement and cost of cementing a well.

Table 3.20 shows the weight of each component for various artificial pozzolan-cement slurries.[43] Table 3.21 shows the compressive strengths of

Table 3.20. Mixtures of cement[a] and pozzolan[b]
(*After Reference 43*)

Composition, volume per cent		Component weights, lb		Total weight per equivalent sack, lb
Pozzolan	Cement	Pozzolan	Cement	
0	100	0.0	94.0	94
25	75	18.5	70.5	89
50	50	37.0	47.0	84
60	40	44.4	37.6	82
75	25	55.5	23.5	79

[a]API Class A cement, sp. gr. 3.14.
[b]"Pozmix" (Halliburton Oil Well Cementing Co.) sp. gr. 2.46, absolute density 153.5 lb/cu ft, bulk density 74 lb/cu ft.

Table 3.21. Comprehensive strength of pozzolan[a]-cement mixtures
(*After Reference 43*)

Curing time, hr	Compressive strength at indicated curing temperatures, psi				
	60°F	80°F	100°F	120°F	140°F
			0-100 Pozzolan-cement		
6	—[b]	60	225	675	1265
12	35	350	795	1820	2940
18	195	550	1465	2760	3880
24	330	1220	2020	3070	4200
48	1055	2530	4635	5830	5910
72	1620	3200	5410	6865	6975
			25-75 Pozzolan-cement		
24	220	525	970	1390	1840
48	640	1160	1920	1970	2320
72	820	1515	2165	2420	2720
			50-50 Pozzolan-cement		
6	—[b]	50	110	235	380
12	25	120	295	490	685
18	60	195	445	660	880
24	100	350	600	815	1125
48	215	610	965	1130	1685
72	375	880	1210	1450	2145

[a]"Pozmix."
[b]Did not set.

25-75 and 50-50 pozzolan-cement mixtures[43] cured for up to 72 hr as compared with the compressive strength of neat cement cured over the same time period. To complete the information required for the design of a pozzolan-cement system, Table 3.22 shows thickening times of a neat cement slurry, a 50-50 pozzolan mix, and 50-50 pozzolan mixes treated

Table 3.22. POZZOLAN[a]-CEMENT THICKENING TIME
(*After Reference 43*)

Composition, weight per cent		Calcium chloride, per cent	Thickening time at simulated well depths[b] (ft), hr:min		
Pozzolan	Cement		2000	4000	6000
0	100	0	4:45	3:36	2:41
50	50	0	4:17	3:19	2:21
50	50	2	3:35	3:04	2:04
50	50	4	2:21	1:20	1:08

[a]"Pozmix."
[b]Pan American high-pressure thickening-time tester.

respectively with 2 per cent and 4 per cent calcium chloride as an accelerator.

Diatomaceous Earth. The use of a special grade of diatomaceous earth as a light-weight additive to portland cement was first reported by Bergman, Hurley, and Shell[44] and later by Shell, Hurley, Bergman, and Fisher.[45] With the use of diatomaceous earth there is a decrease in slurry weight owing to the lower absolute density of the diatomite as compared with portland cement (2.1 versus 3.14) and the large amount of water that can be added to a slurry without separation of the solids. This dilution of cement causes some increase in thickening time and a decrease in strength. As with the other pozzolans, the silica contained in the diatomite reacts chemically with the calcium hydroxide released when portland cement sets. The pozzolanic action of diatomaceous earth produces a gel which becomes cementitious with age and temperature. Table 3.23 shows the density and volume relationships for 20 per cent and 40 per cent diatomite-cement systems,[46] and Table 3.24 shows the compressive strengths and thickening times to be used in the design of these systems.[47]

Gilsonite. Gilsonite is a naturally occurring solid hydrocarbon with a specific gravity of 1.07 and, when graded for oil field use, a bulk density of 50 lb per cu ft. It is noncellular in structure and when subjected to pressure will not absorb water from the cement slurry. The unique properties of gilsonite such as low specific gravity, resistance to corrosive fluids, chemical inertness, and low water requirements result in a low density, low lost-circulation cement slurry. According to Slagle and Carter,[48] there is *no* difference between the thickening times of slurries containing 0 or 50 lb of gilsonite per sack of cement. The compressive strength of an 11 lb per gal gilsonite-cement slurry, cured at 100°F for 24 hr, exceeds 500 psi. The relatively high compressive strength as compared with other types of light-weight cement slurries is due to the fact that the slurry weight is reduced by the solid gilsonite rather than by additional amounts of water.

Table 3.23. DENSITY AND VOLUME RELATIONSHIPS FOR DIATOMITE[a]-CEMENT SLURRIES[b]

(*After Shell*[46])

Slurry weight, lb/gal	Water, gal/sack cement	Diatomite, sacks/cu ft of slurry	Slurry volume, cu ft/sack cement
	20-80 Diatomite-cement		
12.0	15.5	0.16	2.69
12.1	14.9	0.17	2.62
12.2	14.4	0.17	2.55
12.3	14.0	0.18	2.49
12.4	13.5	0.19	2.43
12.5	13.1	0.19	2.37
	40-60 Diatomite-cement		
10.8[c]	28.1	0.17	4.52
10.9	26.8	0.17	4.34
11.0[d]	25.6	0.18	4.19
11.1[d]	24.5	0.19	4.03
11.2	23.4	0.19	3.89
11.3	22.4	0.20	3.76

[a]Sold under the trade name "Diacel D."

[b]Diatomite: 1 sack = 50 lb = 3 cu ft; bulk density = 16.67 lb/cu ft; sp. gr. = 2.10; absolute density = 2.10 × 62.4 = 131.0 lb/cu ft. Sp. gr. Class A cement = 3.15. Density of water at 80°F = 8.318 lb/gal.

[c]Recommended slurry density for most Class C-cement jobs.

[d]Recommended slurry density for most Class A-cement jobs.

Table 3.24. THICKENING TIME AND COMPRESSIVE STRENGTH FOR DIATOMITE[a]-CEMENT SLURRIES

(*After Reference 47*)

Weight, lb/gal	Calcium chloride, per cent	Compressive strength,[b] psi				API simulated thickening time, hr:min	
		80°F	100°F	120°F	140°F	6000 ft 170°F	8000 ft 200°F
		20 per cent diatomite					
12.0	0	55	160	180	500	2:16	2:00
12.0	2	150	260	450	600+	2:13	1:58
12.2	0	70	140	250	690	2:08	2:00
12.2	2	170	300	520	700+	2:26	1:46
12.4	0	95	290	510	780	2:00	1:39
12.4	2	200	380	660	800+	1:54	1:34
		40 per cent diatomite					
11.1	0	25	55	110	370	7:00+[c]	
11.1	4	45	130	300	600	7:13[c]	
11.3	0	30	85	130	580	3:22[c]	
11.3	4	70	190	520	600+	2:02[c]	

[a]Trade name "Diacel D."

[b]Developed in one day at various curing temperatures.

[c]Simulated thickening times for a depth of 10,000 ft and circulating temperature of 230°F.

Calcium chloride. The competitive nature of the petroleum industry makes it desirable for operators to conserve as much time as possible in completing various stages of the drilling program. Since Farris[20] and others have found the minimum strength required for a primary cementing job (under *ideal* conditions) for conductor and surface casing to be only 100 psi compressive strength, the next step is to obtain a more rapid development of cement strength. Calcium chloride, a chemical accelerator, is commonly used in wells having bottom-hole temperatures of less than 125°F, and at curing temperatures less than 75°F its use is considered a "must." Calcium chloride is dissolved in the mixing water or is premixed with the cement. When premixed with the dry cement, it can be stored for several days without harmful effects. It is seen from Table 3.25[49] that the addition of 4 per cent calcium chloride causes a reduction of about one half in the thickening time of Class A cement. Table 3.26[49] gives the compressive strengths of Class A cement with 0 to 4 per cent calcium chloride. Since the optimum

Table 3.25. Effect of calcium chloride on the thickening times of API Class A cement
(*After Reference 49*)

Calcium chloride, per cent	Pan American thickening time at indicated well depth (ft), hr:min		
	2000	4000	6000
0	4:00+	3:48	2:32
2	3:15	2:30	1:47
4	2:38	1:55	1:05

drill-out compressive strength is considered to be 500 psi, the use of this strength, which is reached in less than 12 hours of curing time with 2 per cent calcium chloride at 60°F, results in a substantial saving in rig time.

Sodium chloride. The operator is confronted with sodium chloride in saline water (often the most convenient source of mixing water) and as solid salt formations in wells. No cementing failures have been reported in cementing the salt strings in West Texas where the average landing depth is 4500 ft; however, in the Pine Prairie Field in South Louisiana there have been two casing failures opposite the salt section at 7500 ft. In the Beaver Lodge and Fryburg fields in North Dakota, the casing has collapsed in a number of wells opposite the Charles salt section from 7600 to 8680 ft. To prevent salt flow due to overburden pressure, the wells are drilled through the salt section with a saturated salt drilling mud and cemented with a saturated salt-brine cement. This is necessary because circulation of fresh-water slurries would dissolve the salt. The addition of salt to cement should also reduce the *skin effect*[50] (zonal damage). The dry blending of salt in a cementing composition eliminates foaming and time lost in dissolving salt in the mixing water. Table 3.27 shows the amount of salt required for

Table 3.26. Effect of calcium chloride on the compressive strength of Class A cement

(*After Reference 49*)

Curing time, hr	Calcium chloride, per cent	Compressive strength at curing temperature (°F), psi			
		60	80	100	120
6	0	—[a]	45	385	905
12	0	65	365	830	1660
18	0	185	915	1525	3060
24	0	430	1250	1805	3815
48	0	1040	1395	3490	5990
6	2	115	300	1015	1800
12	2	505	1055	2400	3260
18	2	750	1325	3075	4210
24	2	1580	2415	3910	5475
48	2	3050	4385	6340	6525
6	4	155	360	970	1445
12	4	610	1005	2090	2715
18	4	900	1395	2885	3635
24	4	1620	2385	3490	3665
48	4	2850	3715	4990	4840

[a]Did not set.

Table 3.27. Densities of sodium chloride solutions at 68°F[a]

(*After Reference 51*)

(1) Weight of Sol., lb/gal	(2) Specific gravity	(3) NaCl, lb/gal sol.	(4) NaCl, lb/bbl pure water	(5) Sol./pure water, bbl/bbl
8.4	1.0084	0.116	4.90	1.006
8.5	1.0204	0.260	11.04	1.011
8.6	1.0324	0.407	17.38	1.017
8.7	1.0444	0.555	23.84	1.023
8.8	1.0564	0.708	30.61	1.029
8.9	1.0684	0.862	37.52	1.036
9.0	1.0804	1.017	44.58	1.043
9.1	1.0924	1.175	51.88	1.051
9.2	1.1044	1.334	59.34	1.059
9.3	1.1164	1.493	66.91	1.067
9.4	1.1285	1.653	74.66	1.075
9.5	1.1405	1.816	82.70	1.084
9.6	1.1525	1.979	90.86	1.093
9.7	1.1645	2.143	99.22	1.102
9.8	1.1765	2.308	107.79	1.112
9.9	1.1885	2.473	116.51	1.122
10.0	1.2005	2.638	125.38	1.131

[a]Bulk weight of NaCl = 71 lb per cu ft; sp. gr. = 2.17.

various salt concentrations[49] in the mixing water. This table was prepared by plotting the specific gravities against salt concentrations given in Lange's Handbook[51] and reading from this plot the values given in Col. (3) corresponding to the weights in pounds per gallon in Col. (1). The number of pounds of sodium chloride required per barrel of pure water, Col. (4), was obtained by dividing the salt content of each gallon of solution, Col. (3), by the number of pounds of water in each gallon of solution, Col. (1) − Col. (3), and multiplying by the density of a barrel of water at 68°F. For example, the amount of sodium chloride required per barrel of pure water for a solution with a density of 9 lb per gal is

$$\frac{1.017 \text{ lb salt/gal sol.} \times 349.9 \text{ lb water/bbl water}}{(9.000 - 1.017) \text{ lb water/gal sol.}} = 44.58 \text{ lb salt/bbl water}$$

and the solution/pure water volume ratio, Col. (5), is

$$\frac{8.33 \text{ lb water/gal water}}{7.983 \text{ lb water/gal sol.}} = 1.043 \text{ gal sol./gal water}$$

The weight of salt required to increase the weight of gun-barrel salt water can be calculated on the basis of either a barrel of salt water or a barrel of pure water. For example, suppose it is desired to increase the weight of salt water from 8.8 to 10 lb per gal. On the basis of the pure water in the salt-water system, the salt required is (125.38 lb salt/bbl pure water) − (30.61 lb salt/bbl pure water) = 94.77 lb salt/bbl pure water. On the basis of salt water, the salt required is

$$\frac{94.77 \text{ lb salt/bbl pure water}}{1.029 \text{ bbl salt water/bbl pure water}} = 92.1 \text{ lb salt/bbl salt water}$$

To insure saturation at higher temperatures, an excess of salt, which will have no detrimental effect on the set cement, should be used. Table 3.28 shows the thickening times[2] of slurries of a Class E oil-well cement in solutions of 0 to 300,000 ppm NaCl, as measured in Halliburton and Pan American thickening-time testers.

If the safety factor for a casing cement job is 1 hr, it is seen from Table 3.28 that the addition of 10,000 ppm of NaCl to the Class E cement considerably reduces the thickening time, to less than a reasonable placement time. This table also provides a good comparison between the thickening times of a Class E slow-set cement slurry as measured with the Halliburton and Pan American thickening-time testers. Table 3.29 gives compressive strengths[2] of slurries at 24 hr. In general, the addition of salt up to 20,000 ppm increased the strengths of both classes of cements, but addition up to 300,000 ppm decreased the strengths below those obtained with 0 ppm.

Latex. Another cementing slurry which has been highly successful in plugging casing leaks and which has been recommended for "tailing in" on

Table 3.28. Effect of sodium chloride on the thickening times of Class E oil-well cement

(*After Ludwig*[2])

NaCl in mixing water, ppm	Halliburton thickening times at indicated temperature (°F), hr:min				Pan American thickening times at indicated well depth (ft), hr:min			
	140	160	180	200	8000	10,000	12,000	14,000
None	5:15	5:40	5:10	5:05	3:19	2:56	2:18	2:04
5000	4:08	4:30	4:15	4:30	2:42	2:04	1:53	1:49
10,000	3:31	4:00	3:55	4:10	2:22	1:45	1:39	1:34
20,000	2:45	3:10	3:25	3:30	2:08	1:37	1:12	1:10
30,000	2:15	2:50	3:10	3:20	2:02	1:24	1:10	1:03
100,000	1:50	2:10	2:48	2:30	1:58	1:24	1:09	1:01
200,000	3:00	3:08	—	—	2:41	1:52	1:33	1:14
300,000	5:15	4:06	—	—	5:36	3:57	2:37	2:10

Table 3.29. Effect of sodium chloride on the compressive strengths of cements

(*After Ludwig*[2])

NaCl in mixing water, ppm	Compressive strength of 2-in. cubes after 24-hr cure at indicated temperatures, psi					
	API Class A cement			API Class E cement		
	100°F	120°F	140°F	140°F	160°F	180°F
None	2108	2917	4023	2349	2715	5023
5000	2763	4030	4127	2910	4092	6183
10,000	2962	4080	4555	3350	4197	6287
20,000	3320	4417	4693	4090	4510	6978
30,000[a]	3145	4218	4632	3690	4518	6920
100,000	3067	3288	3513	4403	5253	6797
200,000	2470	2650	2960	4437	5255	5793
300,000	1484	2003	2220	1992	2208	2379

[a]Approximate maximum salt content of the Atlantic Ocean.

primary cementing jobs and low-pressure squeeze jobs is a mixture of latex and cement.[52,53] The latex compound is a formulation of latex and surfactants which are added directly to the mixing water. One of the outstanding properties[49,54] of this slurry is its low filtration rate. The fluid loss ranges from 20 to 30 cc in 30 min as compared with complete dehydration in 40 sec with neat Class A cement. When tested in sulfate brines, latex–cement suffered a strength loss of about ten per cent. The bonding strength when measured with water-wet metal was found to be 53 psi, as compared with 19 psi for Class A cement, and with water-wet limestone 92 psi, as compared with 41 psi for the Class A cement. Laboratory tests also showed the mixture to have greater resistance to contamination. For example, after 24 hr at 80°F, latex–cement mixed intimately with 20 per cent mud developed a compressive strength of 500 psi as compared with 407 psi for Class A cement.

Control of fluid loss. Low fluid loss is essential for control of the deposition of solids when a differential pressure exists between the slurry and permeable zones. During casing cementing, formations of high permeability may cause premature dehydration of the slurry. In squeeze cementing, controlled deposition facilitates filling all the perforations with cement without excessive squeeze pressures. In addition to CMHEC[33,45] another additive for controlling fluid loss has been described by Stout and Wohl[55] and McGhee,[56] a high molecular weight synthetic polymer which has no effect on the thickening times of API Class A cements but does retard the setting of most API Class D and E cements. The compressive strengths of set cements are not appreciably affected by this additive, and WOC times should be no longer than those of neat cements. In Class A cement slurries the fluid loss of 600 cc/30 min is reduced to 46 cc by the addition of 1.0 per cent by weight of this synthetic organic additive, and fluid loss in Class E cements is reduced from 900 cc to 65 cc/30 min. This additive provides greater fill-up behind the pipe and controlled dehydration during squeeze operations.

Lost circulation additives. The control and prevention of lost circulation of cement slurries is a problem frequently encountered during the cementing of oil and gas wells. Lost circulation or lost returns may be defined as the loss, to formation voids, of the drilling fluids or cement slurries used in rotary drilling and well completion. This loss may vary from a gradual lowering of the mud level in the pits to a complete loss of returns. The loss of drilling mud or cement results in loss of drilling time, plugging of productive formations, and blow-outs. A study of the literature shows that the types of formations or subsurface conditions which are responsible for lost circulation fall into four categories:

1. Natural fractures.
2. Induced or created fractures.
3. Cavernous formations.
4. Unconsolidated or highly permeable formations.

Howard and Scott,[39] from a laboratory study under simulated well conditions, reported that the concentration of lost circulation material has a critical effect on the ability of material to seal fractures, and that granular materials with a variety of particle sizes permit the imposition of pressure higher than is normally required to fracture the formation. Figure 3.10 is a summary of the evaluation tests made by Howard and Scott,[39] listing the additives in order of effectiveness as determined by the size of the fracture

MATERIAL	TYPE	DESCRIPTION	CONCENTRATION LBS/BBL
NUT SHELL	GRANULAR	50%–3/16+10 MESH 50%–10+100 MESH	20
PLASTIC	GRANULAR	SAME AS ABOVE	20
LIMESTONE	GRANULAR	SAME AS ABOVE	40
SULPHUR	GRANULAR	SAME AS ABOVE	120
NUT SHELL	GRANULAR	50%–10+16 MESH 50%–30+100 MESH	20
EXPANDED PERLITE	GRANULAR	50%–3/16+10 MESH 50%–10+100 MESH	60
CELLOPHANE	LAMELLATED	3/4 IN. FLAKES	8
SAWDUST	FIBROUS	1/4 IN. PARTICLES	10
PRAIRIE HAY	FIBROUS	1/2 IN. FIBERS	10
BARK	FIBROUS	3/8 IN. FIBERS	10
COTTON SEED HULLS	GRANULAR	FINE	10
PRARIE HAY	FIBROUS	3/8 IN. PARTICLES	12
CELLOPHANE	LAMELLATED	1/2 IN. FLAKES	8
SHREDDED WOOD	FIBROUS	1/4 IN. FIBERS	8
SAWDUST	FIBROUS	1/16 IN. PARTICLES	20

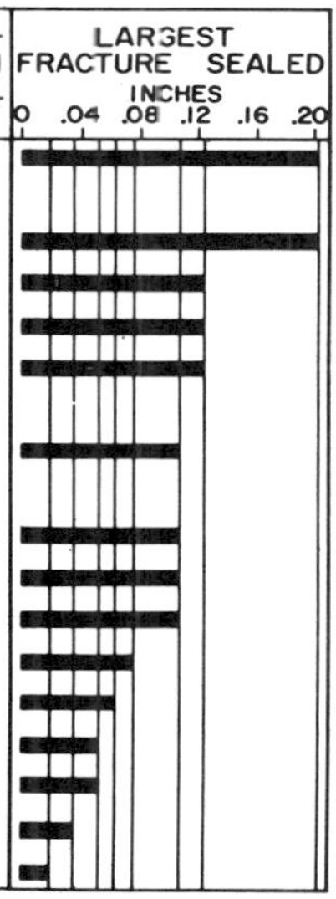

Fig. 3.10 Summary of lost circulation tests. (Howard and Scott,[39] *Trans.* AIME.)

sealed. This compilation indicates that the materials most effective for plugging fractures and withstanding high pressure differentials are of the granular type. A special blend of graded, ground walnut shell is used as an additive in cement slurries in concentrations of one to four pounds per sack of cement. In a controlled laboratory experiment,[57] one pound per sack of cement sealed a fracture 0.10 in. wide against a pressure of 3000 psi, and 2.5 lb per sack sealed a fracture 0.22 in. wide against 1000 psi. The use of a cellulose filler in cement slurry for the successful sealing-off of thief formations was first described by Leach.[58] The fillers strain out on the walls of the hole and form an impervious cake having sufficient strength to resist further loss of fluid. The filler is a chemically inert organic product in the form of very thin pieces (0.001 in. thick) with surface dimensions varying from $\frac{1}{2}$ to 1 in. Since the material has a density approximating that of drilling mud, it will remain suspended indefinitely, and the crinkled surface of the flakes prevents them from adhering to each other whether wet or dry. For serious cases, the filler is added in quantities of as much as $1\frac{1}{4}$ lb per sack of cement.

Weighting materials. Sometimes it is necessary to design a slurry that is heavier than can be obtained with neat cement. It is desirable that the slurry weight should be at least equal to that of the mud, thus reducing the tendency for the cement to channel and minimizing blow-out hazards if a high column is to be placed. Barite (barium sulfate, $BaSO_4$), ilmenite

($FeTiO_3$), and sand are used to increase the density of cement slurry where mud densities of 17 lb per gal or greater are required to control high-pressure oil and gas zones during drilling. The specific gravity of barite is 4.23, of ilmenite 4.7, and of sand 2.65. Table 3.30 shows the amounts of barite[49]

Table 3.30. Effect of barite on the properties of slurries of a Class E cement

(*After Reference 49*)

Cement,[a] lb	Barite,[b] lb	Water, gal/sack cement	Slurry wt., lb/gal	Slurry vol., cu ft/sack cement
94	—	4.5	16.26	1.08
94	22	4.9	17.22	1.22
94	37	5.3	17.63	1.33
94	55	5.8	18.02	1.46
94	76	6.4	18.40	1.62
94	108	7.3	18.84	1.87

[a]Sp. gr. 3.14.
[b]Sp. gr. 4.23, bulk weight 135 lb per cu ft.

added to a Class E oil-well cement to yield various slurry weights, and Table 3.31 gives the thickening time of barite-cement slurries.[49] Since the

Table 3.31. Effect of barite on the thickening times of API Class E cements

(*After Reference 49*)

Slurry weight, lb/gal	Pan American thickening times at 14,000 ft simulated well depth, hr:min		
	Cement A	Cement B	Cement C
16.25	3:20	3:30	3:23
17.00	2:49	3:00+	3:00+
18.00	3:05	2:28	3:00+
19.00	2:35	1:45	3:00+

addition of barite shortens the thickening time, a small amount of retarder added to the cement will provide a safe pumping time even during the cementing of casing in very deep wells.

Ottawa sand is also used as a weighting material for cement, and to obtain a *hard cement plug* in an open hole as a base for a whipstock. Table 3.32 shows the slurry properties of a Class A and a Class E cement, mixed with various amounts of sand.

Table 3.32. PROPERTIES OF OTTAWA SAND–CEMENT SLURRIES
(After Reference 49)

Cement, lb	Sand, lb	Water, gal/sack cement	Slurry wt., lb/gal	Slurry vol., cu ft/sack cement
		API Class A cement		
94	—	5.2	15.63	1.18
94	32	5.2	16.52	1.37
94	56	5.2	17.05	1.52
94	85	5.2	17.55	1.69
94	123	5.2	18.08	1.93
		API Class E cement		
94	—	4.5	16.26	1.08
94	10	4.5	16.55	1.14
94	28	4.5	17.03	1.25
94	51	4.5	17.52	1.39
94	79	4.5	18.01	1.56

Fig. 3.11 Typical casing-cementing heads. (Courtesy of Halliburton Oil Well Cementing Company.)

3.6. Surface and Subsurface Cementing Equipment

Many types of surface and subsurface equipment are used in oil-well cementing today. This equipment has been developed through the years by both oil-field operators and equipment manufacturers to assist the operator in setting casing safely at the desired depth.

Casing-Cementing heads. Various forms of cementing heads (Fig. 3.11) are employed to provide a connection from cementing and rig-pump lines to a casing string, and to provide access for insertion of the cementing plugs. One type is the quick-opening design with a coarse-thread cap. The *plug container* is designed to hold one or more plugs which are loaded prior to the actual cementing operation and from which the plugs are selectively released into the casing.

Guide and float shoes. There are two types of guide shoes in use today. These are the *plain guide shoe* and the *combination float and guide shoe.* The plain guide shoe (Fig. 3.12) is run on the bottom joint of casing and is intended only to guide the casing past any side-wall irregularities. Figure 3.12

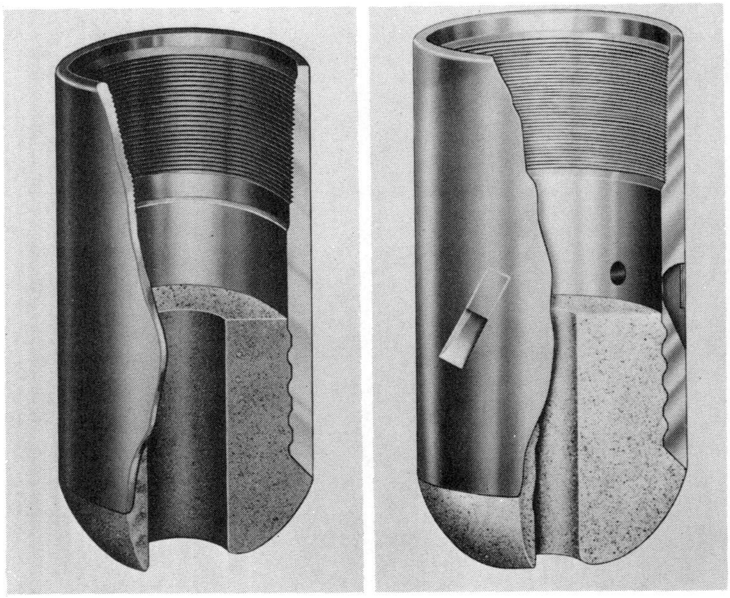

Fig. 3.12 Bottom and side-discharge guide shoes. (Courtesy of Baker Oil Tools, Inc.)

also illustrates the side-discharge type of shoe which gives more agitation to the slurry.

The combination float and guide shoe shown in Fig. 3.13 includes a back-pressure valve as an integral part of the equipment, and a side discharge. The back-pressure valve, which is composed of drillable materials such as

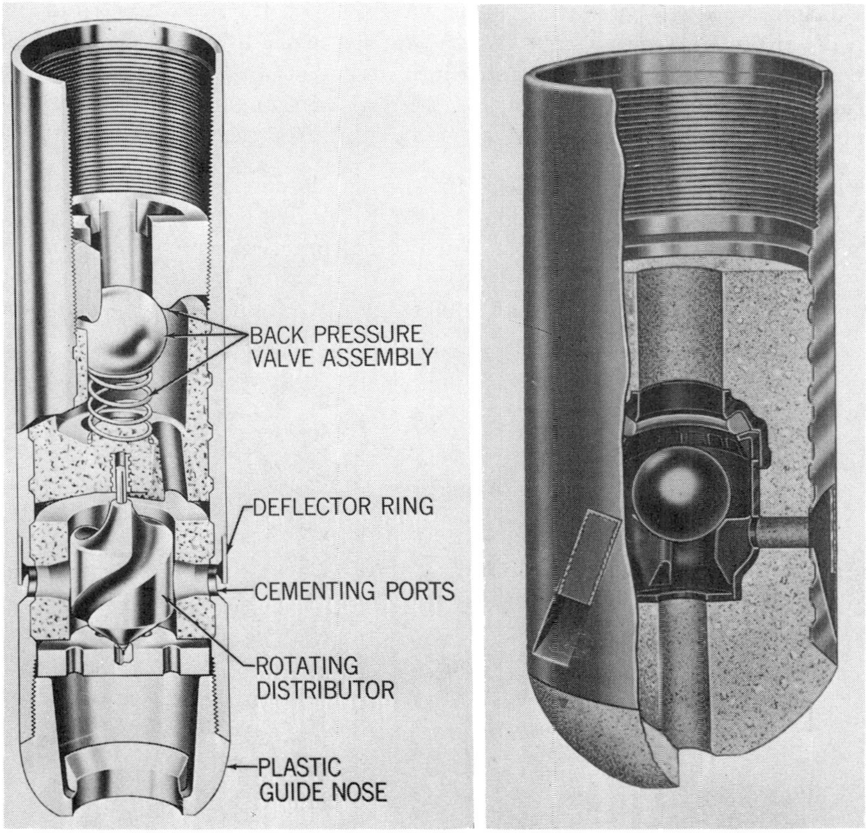

Fig. 3.13 Side-outlet combination float and guide shoes. (Courtesy of Baker Oil Tools, Inc., and Halliburton Oil Well Cementing Company.)

cement and plastic, is closed by the pressure of the outside fluid column, preventing entrance of well fluids while the casing is being lowered into the hole. The casing rides or floats down the hole to the desired depth, being only partially suspended by the elevators. In oil-well cementing operations, failure to obtain satisfactory results frequently is ascribed to the presence of a *mud cake* between the set cement and the formation, *channeling* of the cement slurry, and *off-center casing.* Thick mud filter-cake deposits occur on the wall of the hole and provide regions of weakness that ultimately permit water to pass around the cement. The hydraulic removal of the filter cake around the shoe of the casing was investigated by Jones and Berdine.[59] Down-whirler shoes were constructed with four side ports which directed 60 per cent of the fluid tangentially downward and outward with a whirling motion. The down-whirler shoe stripped the mud cake from the walls of the test holes 6 in. above the highest levels to which the shoe had been raised. Later, Howard and Clark[60] found it was possible to remove the filter cake by jetting water downward at a 45° angle with a pressure of 500 psi.

Float collars. The float collar (Fig. 3.14) is run one or more joints above the combination float and guide shoe and contains a back-pressure valve similar to that of the float shoe. The current practice, which makes use of only the top wiper plug, is to place the float collar one joint above the shoe on casing set at depths less than 6000 ft, and two joints above the shoe for

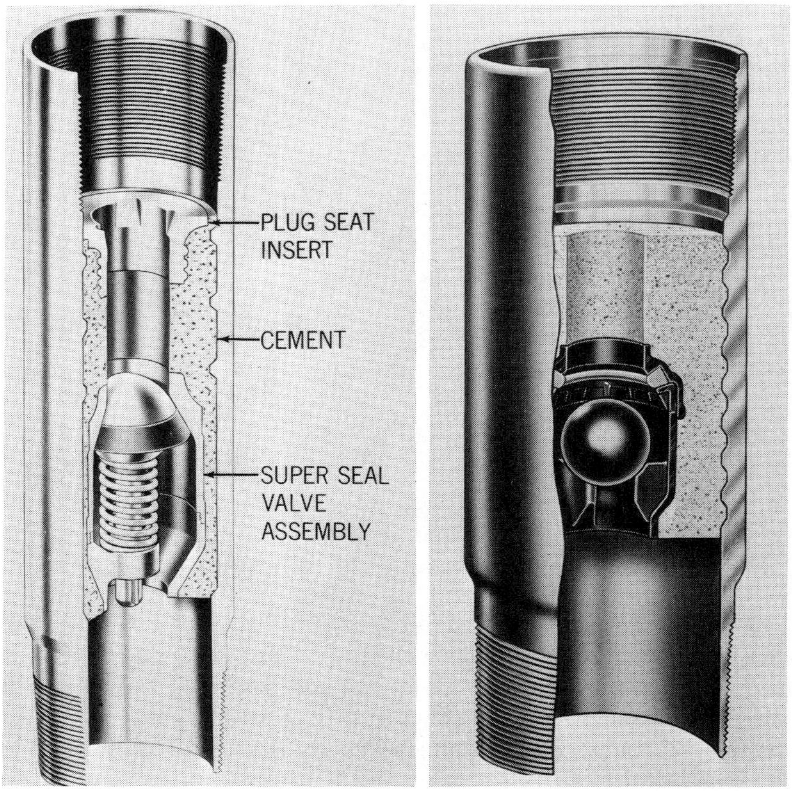

Fig. 3.14 Float collars. (Courtesy of Baker Oil Tools, Inc., and Halliburton Oil Well Cementing Company.)

longer strings. This provides for a section of contaminated cement below the float collar. The function of the float collar is not only to serve as floating equipment but also to act as a seat for the wiper plugs, thus indicating when cement placement is complete. The spacing of the float collar above the shoe therefore controls the amount of cement left inside the casing.

Stage collars. The multistage cementing tool (Fig. 3.15) is used to cement a long section in two or more stages. This assures less contamination

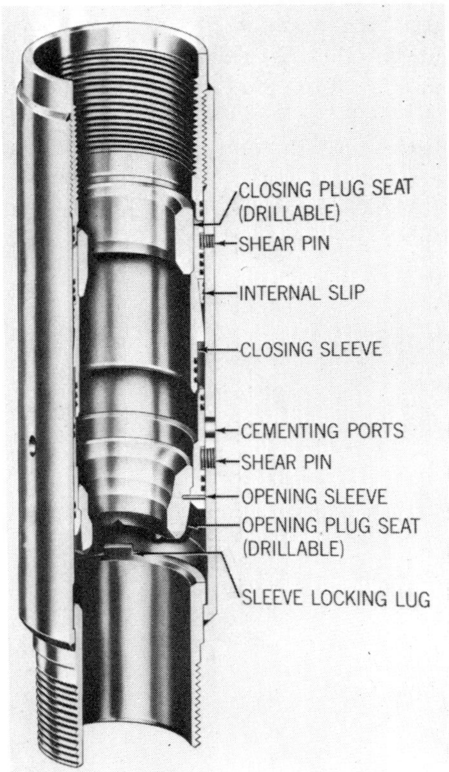

Fig. 3.15 Stage-cementing collar. (Courtesy of Halliburton Oil Well Cementing Company.)

throughout the cementing area and reduces the possibility of lost circulation due to excessive hydrostatic pressure. The tool is installed at the desired place in the casing string as it is being run and, after the cement has been placed around the bottom of the casing, the stage cementing collar is opened by dropping through the displacing fluid a bomb-type trip plug which is stopped by the stage collar. Application of pump pressure then actuates a sliding sleeve and opens the stage collar ports, permitting circulation through the collar and up the annulus. The upper-stage cement is then mixed. A collar-closing plug follows the cement as it is being displaced, the plug engaging a sleeve in the collar to close the cementing ports.

Wiper plugs. The original Perkins cementing method included the use of two plugs, one ahead of and one behind the cement slurry in the casing. Subsequently, bottom plugs became available, but their use was almost discontinued, owing to the opinion that the old-type wood plugs with a canvas wiper would plug the float collar, and to the mistaken idea that the slurry would clean away all the mud from the wall of the casing. It is now well known that the cement slurry does not remove the thin film of mud on the inner wall of the casing, as shown in Fig. 3.16, but that the top plug wipes away the boundary layer of mud, depositing it below the float collar and perhaps in the annulus. Owsley[61] has pointed out that a layer of

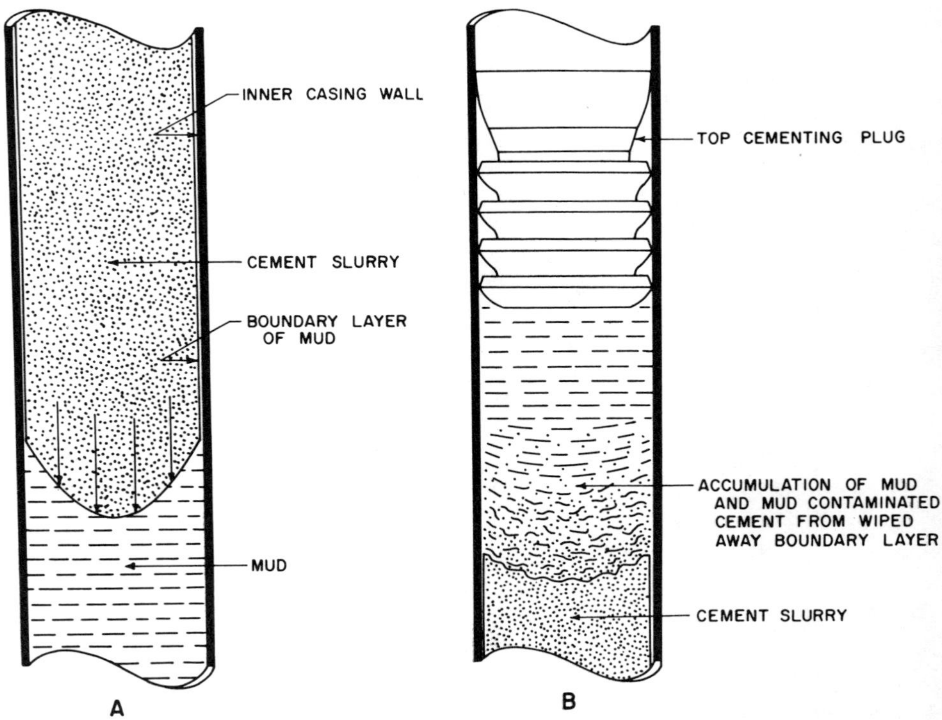

Fig. 3.16 Mud accumulation below top cementing plug. (Owsley,[61] *The Oil and Gas Journal.*)

mud 0.03 in. thick (Fig. 3.16a) on 100 ft of 7-in., 24-lb casing, when wiped off completely and collected, equals 4.2 cu ft. This will fill a 19.19-ft length of the same casing or a 16.7-ft length of the annulus if the casing is set in a $9\frac{3}{4}$-in. hole. It is not surprising that 15 to 20 ft of contaminated cement is often found below the float collar when a bottom plug is not used (Fig. 3.16b).

The top and bottom plugs presently used are of molded rubber and cast aluminum or plastic and are designed (1) to wipe the casing free of mud and cement; (2) to separate mud from cement inside the casing (the bottom plug being constructed to allow fluid circulation by means of a diaphragm, as shown in Fig. 3.17, which is ruptured by pump pressure when the plug seats

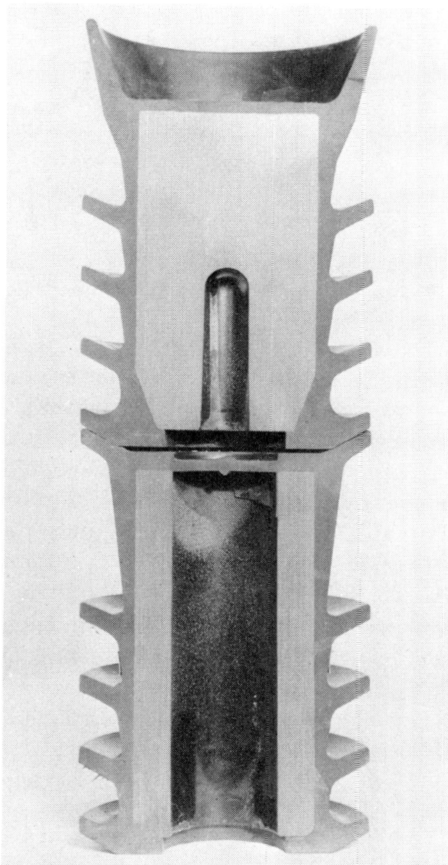

Fig. 3.17 Bottom and top wiper plugs. (Courtesy of Halliburton Oil Well Cementing Company.)

on the float collar); (3) to completely shut off circulation when the top plug reaches the bottom plug.

Casing centralizers. The ultimate shape of the collar of cement has a great deal to do with its effectiveness. In most cases the string of casing will not hang concentrically in the hole, since neither the hole nor, for that matter, the casing will be absolutely straight; consequently, at many points the casing will be in contact with the wall of the hole or the mud cake, as shown in Fig. 3.18.

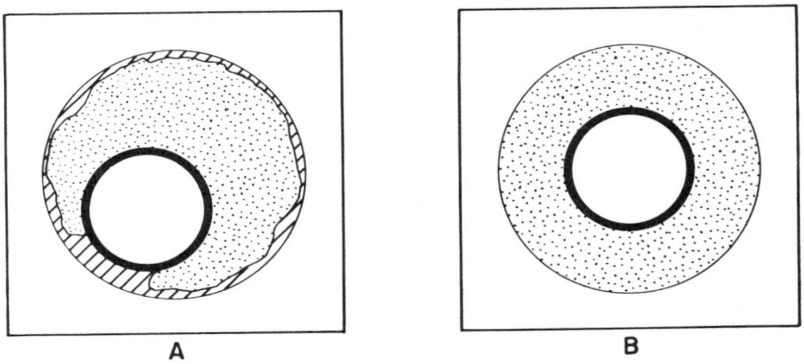

Fig. 3.18 (A) casing is improperly centered and mud cake not removed. (B) complete mud removal and proper centering with scratchers and centralizers.

Helmick and Longley[62] investigated the sticking of drill pipe due to *pressure differential.* Laboratory investigations showed that the force acting to hold the pipe against a permeable bed is proportional to the pressure differential across the drill pipe and to the area of the pipe isolated from the hydrostatic pressure by a thickening mud cake. For example, Helmick and Longley[62] found that with 2.375-in. drill pipe in a 3.0-in. hole, and a differential pressure of 500 psi, at the end of 1.5 min the pull-out force was 600 lb per ft. The mechanism of differential sticking is illustrated in Fig. 3.19. It is readily seen that the pipe/hole diameter ratio affects the initial area of pipe isolated from the hydrostatic pressure, and with additional time for continued filtration, the isolated area increases. This mechanism is operative also during the running of casing or liners in directionally-drilled holes.

Centralizers, when properly installed, assist the cementing operation by:

1. Centering the casing in the well bore.
2. Permitting equal hydrostatic pressure in the annulus, thereby preventing differential-pressure sticking.
3. Tending to hold the casing out of key seats.
4. Assisting in wall-cake removal and in breaking up cement channeling.

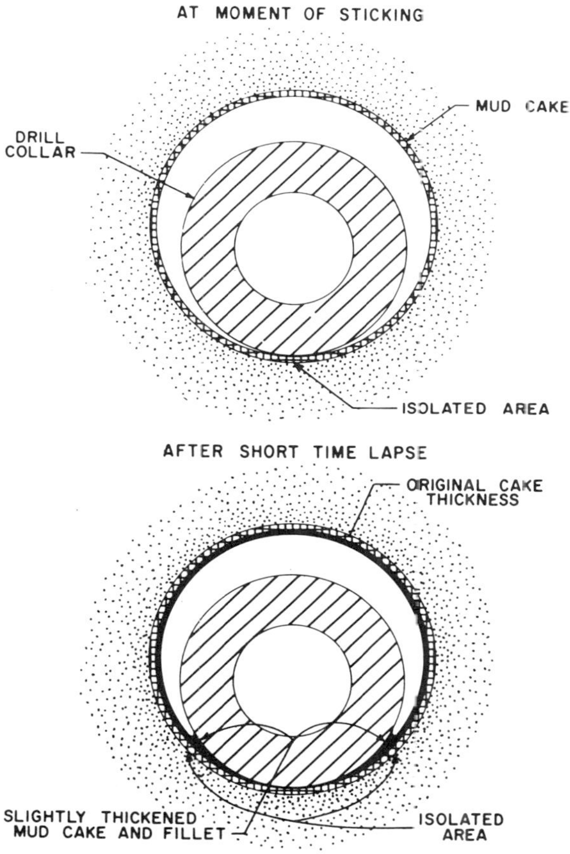

Fig. 3.19 Mechanism of differential-pressure sticking. (Helmick and Longley,[62] *Drilling and Production Pratice,* API.)

Some of the many types of centralizers are shown in Fig. 3.20. The normal recommended spacing in relatively straight holes is 90 ft in the area to be covered by cement. In crooked or directionally-drilled holes the spacing depends on hole conditions. The data in Table 3.33 are based on the condition of at least $\sim \frac{3}{4}$ in. of radial annular clearance between the pipe and wall of the hole.[63]

Scratchers. A scratcher is defined as a *mechanical wall-cleaning* device. Its purpose is to remove all wall cake from the walls of the well bore in the area covered *before* and *during* the cementing operation. This will allow good bonding of the cement to the formation and will prevent contamination and channeling during placement. Jones and Berdine[59] described the beneficial effects of abrading devices as determined from surface cementing

Table 3.33. Recommended spacing of centralizers in deviated holes[a]

(*After Hilton*[63])

Deviation, degrees	Spacing of centralizers on $5\frac{1}{2}$-in. Casing, ft		Spacing of centralizers on 7-in. Casing, ft	
	$7\frac{7}{8}$-in. hole	$8\frac{3}{4}$-in. hole	$8\frac{3}{4}$-in. hole	$9\frac{7}{8}$-in. hole
2	92	102	62	110
4	78	88	50	91
6	71	78	43	82
8	66	72	38	75
10	62	68	35	70
20	52	56	26	55
30	46	50	22	45
40	42	47	19	43

[a]Calculated on the basis of a minimum radial, annular clearance of $\sim\frac{3}{4}$ in.

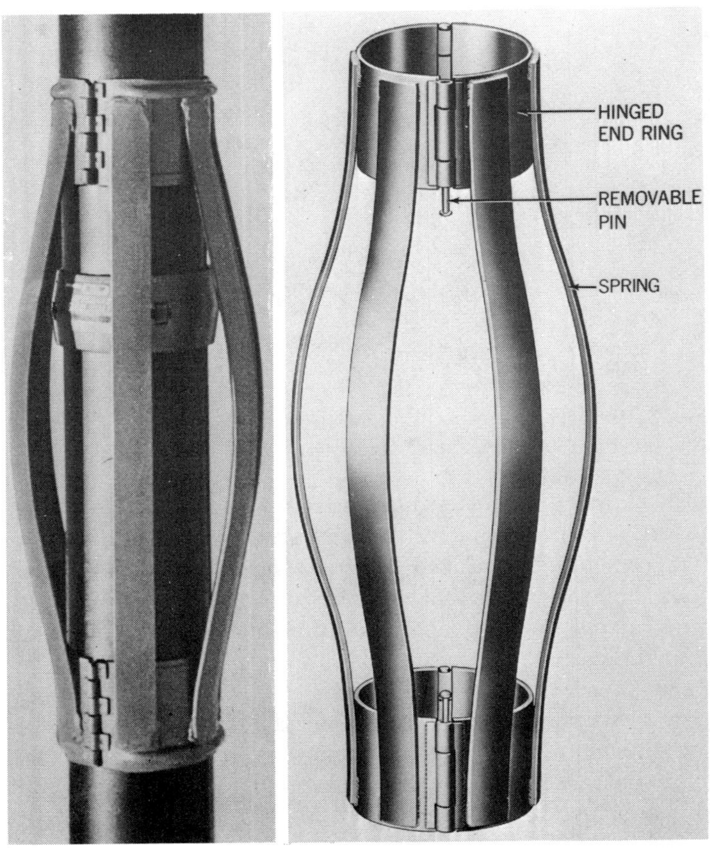

Fig. 3.20 Casing centralizers. (Courtesy of B and W Incorporated and Halliburton Oil Well Cementing Company.)

experiments intended to simulate well conditions. In no case did water circulation effect complete removal of the mud cake, and in most cases very little difference in mud-cake thickness was observed. On the other hand, over the range of travel, wall scratchers effected almost complete removal of the mud cake. In a later investigation, Teplitz and Hassebrock[64] found that the majority of cementing failures arose from incomplete displacement of the drilling mud by the cement slurry. Their field work showed that channeling or by-passing could be virtually eliminated by the proper application of scratchers and centralizers combined with a reciprocating movement of the casing during cementing. Cannon[23] described the first rotary-type wall cleaner. With this it is possible to remove the mud cake at the desired points with a minimum of disturbance at other points. The use of this type of scratcher in one area reduced the proportion of remedial squeeze jobs on oil strings from 58 per cent to 16 per cent.

Two types of scratchers are in use, i.e., the reciprocating and the rotating types. The reciprocating type (Fig. 3.21) is made of metal bands 4 to 6 in. wide with numerous wire bristles, cables, or rubber fingers attached to the band. Three types are manufactured, i.e., the solid-body type, the hinged type, and the split type. Limit rings, placed on the casing above and below the scratchers about two feet apart, allow the casing to be picked up without disturbing the filter cake while running casing. The reciprocating-type scratcher cleans the wall as the casing is reciprocated over a sufficient distance (normally 20 ft with 15-ft spacing) to ensure that each scratcher works on an area overlapping that covered by the scratcher above.

The rotating-type of scratcher is composed of a metal strip with wire bristles. This type of scratcher is installed by welding or clamping metal strips to the casing in a continuous pattern so as to locate it opposite the area covered by cement.

Howard and Clark[60] reported that both types of scratchers removed the filter cake, provided the effective inside diameter of the well bore was *less* than the effective outside diameter of the scratchers. While there was a difference in the mud pattern left by the scratchers, comparative by-pass tests indicated little or no difference in their abilities to effect a seal of the cement to the formation.

3.7. Casing-Cementing Program

In each state there is a regulatory body empowered by the state legislature to require drilling, casing, and cementing to be done in such a manner as to prevent the escape of oil or gas from one strata to another, the intrusion of water into oil or gas sands, or the pollution of fresh water sands by oil, gas, or salt water. There are also rules establishing minimum quantities of cement to be placed around the surface, protection, and production casings.

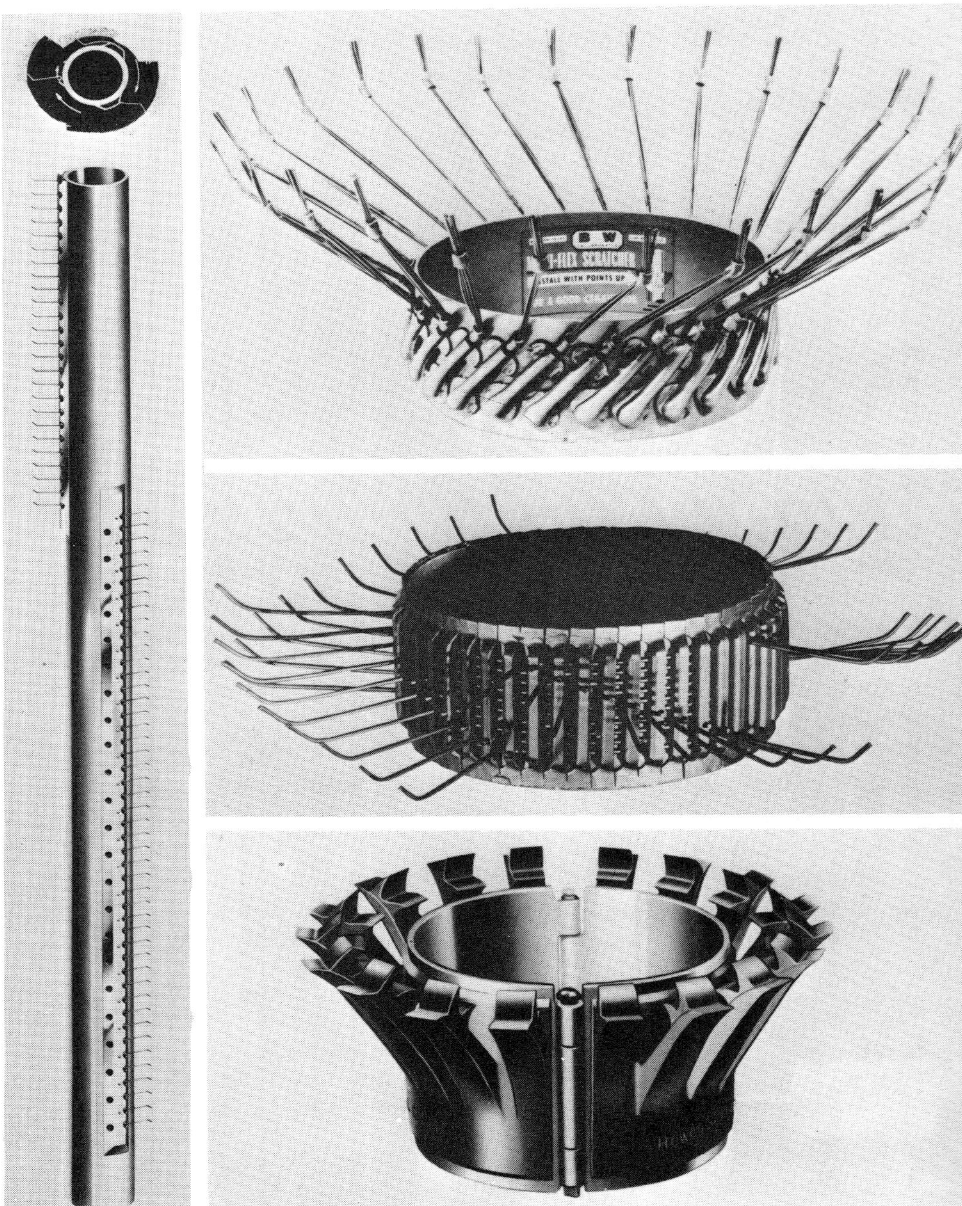

Fig. 3.21 Rotating and reciprocating scratchers. (Courtesy of B and W Incorporated and Halliburton Oil Well Cementing Company.)

As the cost of steel increases, the design of casing strings becomes more important. As a result, the proper selection of casing sizes and weights and the use of combination casing design (Chap. 2) is emphasized. Recent improvements and developments in completion technology have made it practical and economically attractive to utilize *conventional tubing* as casing in completing and servicing wells.[65] The trend has been to reduce hole sizes[66] where possible, and thereby utilize smaller casing sizes. In the selection of a casing seat, operators prefer a competent formation, either sand or shale. The selection of casing seats is generally based on the following:

1. State laws governing such selection.
2. The lithology of the formation.
3. The formation pressures.
4. The type of completion (open or cased hole).
5. The open hole allowable below the larger string of casing.

The job of running a long string of casing into the hole, landing it at the desired casing point, and successfully cementing it requires careful planning. Failure to get the pipe on bottom usually results in additional expense and may result in loss of the hole. All operators condition the hole by circulation before running casing. Some operators circulate mud through the hole once, others circulate until the returns are satisfactory, while on the Gulf Coast two circulations from bottom are usually preferred to remove the partially dehydrated mud. In most cases, no major change in the chemical treatment of the mud is made during the circulation period. Usually an effort is made to adjust the mud properties to the conditions that existed during the up-hole drilling, which requires normal, chemical, viscosity-reducing treatment.

Incomplete distribution of cement both in the annulus cross section and vertically along the casing is usually referred to as *channeling*, i.e., patches or channels of mud left within the cement column. Jones and Berdine[59] pointed out that channeling would be expected to occur near the top of the cement column. The top of the cement column is located by temperature surveys (made from 8 to 24 hr after the cement job), by radioactive tracer techniques, velocity logs, and gamma ray density logs. These surveys usually show 50 to 75 ft of contaminated cement or channeling at the top of the cement column. It has been shown in the successive displacement of fluids that when one fluid is displaced through a pipe by another fluid (the fluids being in laminar flow) and the *displacing* fluid is the more viscous, a short zone of mixing occurs, while if the *displaced* fluid is the more viscous, a *long* zone of channeling results. The rate of flow at which the cement slurry or water spacer displaces the mud greatly affects the percentages of mud displaced. Howard and Clark,[60] in an investigation of the flow characteristics of cement slurries (Fig. 3.22), found that slurries behave as plastic fluids and that different pumping rates result in three regimes of flow, i.e., plug, laminar, and turbulent. For cement slurry displacement in plug flow, where the *annular* velocities are less than one ft per sec, approximately 60 per cent of the circulatable

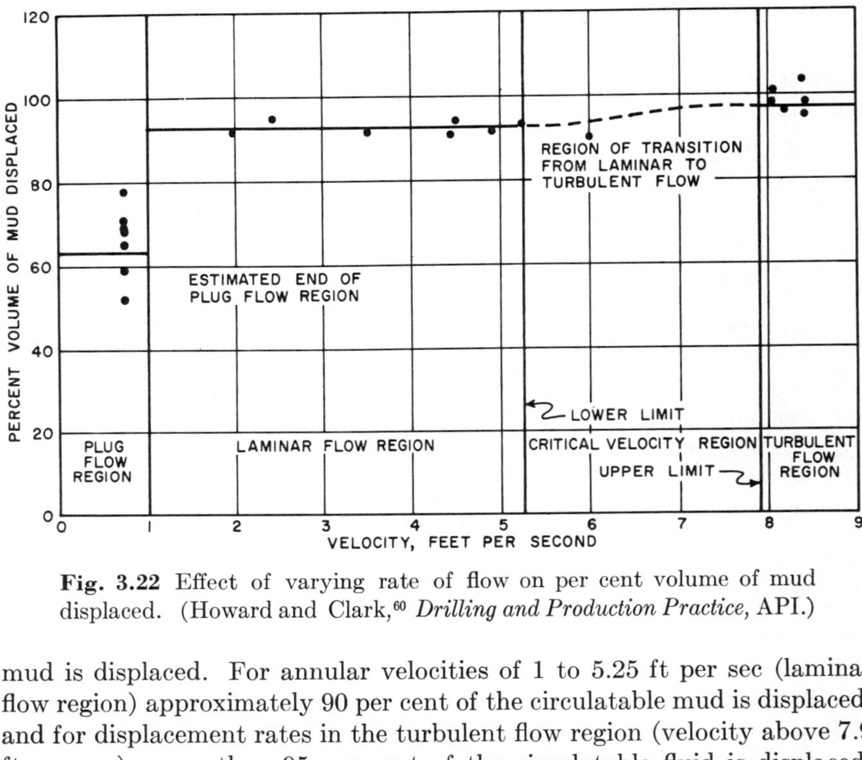

Fig. 3.22 Effect of varying rate of flow on per cent volume of mud displaced. (Howard and Clark,[60] *Drilling and Production Practice,* API.)

mud is displaced. For annular velocities of 1 to 5.25 ft per sec (laminar flow region) approximately 90 per cent of the circulatable mud is displaced, and for displacement rates in the turbulent flow region (velocity above 7.9 ft per sec), more than 95 per cent of the circulatable fluid is displaced. With the current cementing practice of using only the rig pump, or at best the rig pump plus one cementing truck, to displace cement, the displacement velocity in most cement jobs is in the laminar flow region; however, two mud pumps can be connected to the plug container. The average velocity inside drill pipe, casing, or tubing (in feet per second) is given by

$$v_i = 3.056\frac{q}{d_i^2} \tag{3.18}$$

where d_i is the inside diameter of the pipe in inches and q is the flow rate in cubic feet per minute. The average velocity in the annulus is

$$v_a = 3.056\frac{q}{(d_h^2 - d_e^2)} \tag{3.19}$$

where d_h is the diameter of the hole and d_e is the outside diameter of the casing, in inches. Example 3.8 illustrates the method of calculating annular velocity.

Example 3.8. Calculate the velocity in the annulus during displacement, given:

Hole diameter = $8\frac{1}{2}$ in.

Casing diameter = $5\frac{1}{2}$ in.

Rig pump = 8 by 18 in. with $6\frac{1}{2}$-in. liners

Pump efficiency = 90 per cent

Pump volume = 1.1523 cu ft/stroke (Table 3.5)

Pump speed = 60 rpm

SOLUTION: From Eq. (3.19),

$$v_a = 3.056 \frac{1.1523 \times 60}{(8.5)^2 - (5.5)^2} = 5.02 \text{ ft per sec}$$

Reference to Fig. 3.22 indicates that a velocity of 5.02 ft per sec is in the range of laminar flow.

Running casing. The speed at which casing is lowered depends on a number of variables, including the danger of *lost circulation*, the presence of bridges or key seats, crooked-hole conditions, and the clearance between pipe and hole. Under ideal conditions casing can be run at from 1000 ft per hr on the Gulf Coast to 2000 ft per hr in the hard-rock country.

As early as 1934 the effect of pipe motion on down-hole hydraulics was given as a possible explanation for lost circulation.[67] More recently, other authors [68–70] have presented a means of predicting mathematically the magnitude of these mud pressure surges. Figure 3.23 shows a bottom-hole pressure chart[71] of a normal casing and cementing job as it would appear on a pressure recorder at the bottom of a well containing an 11.8 lb per gal mud. The bottom-hole pressures are expressed as equivalent mud weights, or instantaneous mud weights which would be required to equal the normal static mud weight plus the surge pressure. Also shown are the upper and lower mud weights within which these pressures must be maintained. If the equivalent pressure falls below 10.5 lb per gal, formation fluids will enter the well bore. If pressures rise above the 15.4 lb per gal level, the formation will break down and lost circulation will occur. The cycle begins as the crew starts to run pipe into the hole. Surges are insignificant at first, but soon begin to increase as the annular path of the return mud increases with each joint added, and also because the rate at which each joint can be run increases as additional weight of the string overcomes the friction of the blocks, lines, and draw works. At the point marked "Hydromatic Engaged" the hydromatic brake is applied, reducing

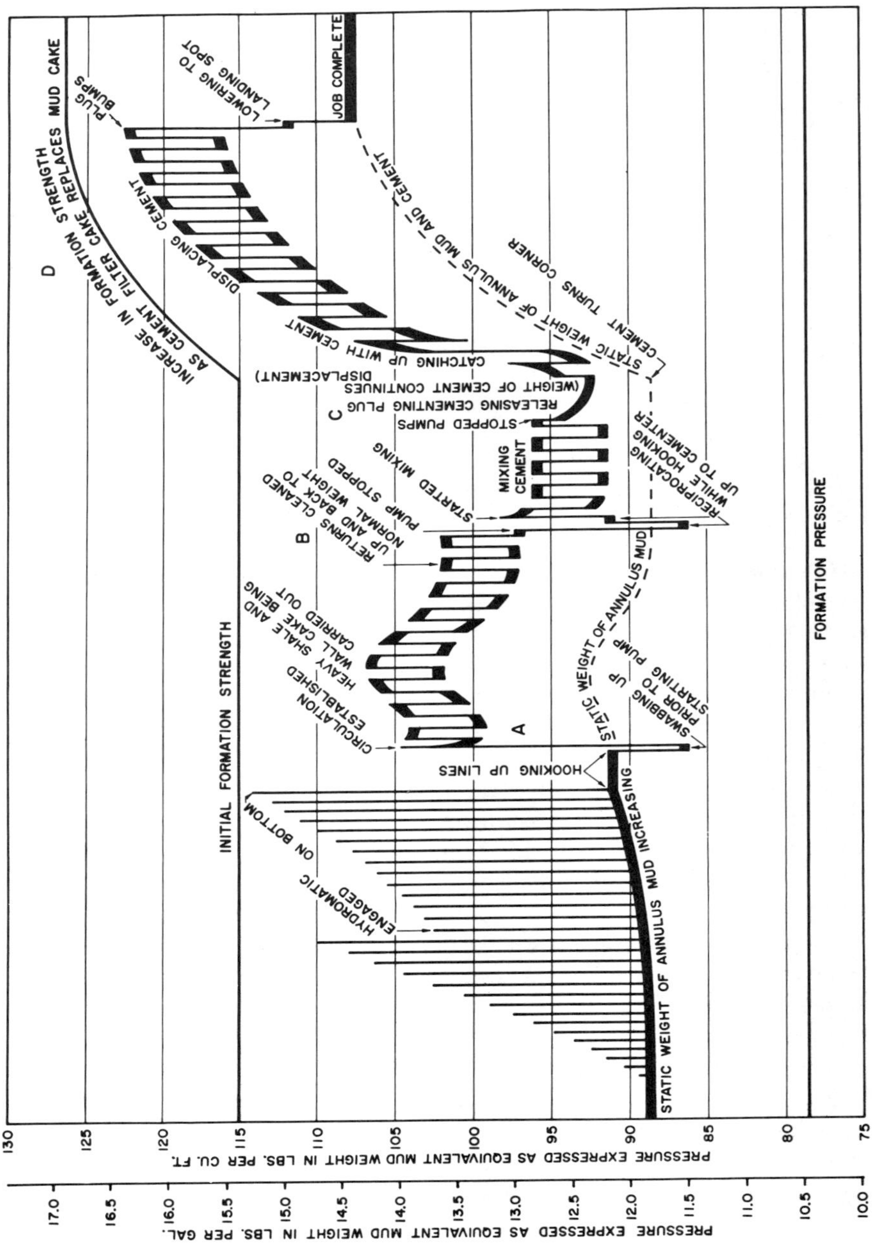

Fig. 3.23 Bottom-hole pressure chart for normal running and cementing of casing. (After Clark,[71] *Drilling and Production Practice*, API.)

the pipe-lowering speed. If the floating surges encountered during the latter portion of the downward trip exceed the strength of the formation, the fluid will fracture the formation and be forced into the fractures. As the pipe is being run a certain amount of wall cake is removed, thus increasing the mud weight, the amount of increase depending on the design of the scratching equpment. Next, after hooking up the plug container or cementing head, there is a swabbing upward of the casing prior to breaking circulation. Since the mud may have developed considerable gel strength, in establishing circulation the mud pump should be run at 10 to 20 rpm until there are full returns. Following the starting circulation (area A, Fig. 3.23), the pressure will fall slowly as the gels are broken, but it will begin to rise again as the scratchers and centralizers remove the filter cake. The stepped character of the pressure line is due to slow reciprocation of the casing. The displacing pressures during the next stage, mixing (area B), are usually lower because cement trucks have only one pump on the hole, i.e., the circulation rate is less than with the rig pump. When the mixing is complete (area C), pumping and reciprocation are stopped to release the cementing plug and change back to the rig pump. Because of the unbalanced fluid column, i.e., the heavier weight of the internal cement column, movement of the annular fluid will continue, maintaining a displacing pressure against the formation. In this cement job, cement leaves the shoe as this phase of the operation is being completed and the static weight of the cement begins to add to the static weight of the mud in the annulus. When circulation and reciprocation are resumed, a short period is necessary for the wiper plug to catch up with the cement before the displacement rate returns to normal. Pressure continues to increase as additional cement enters the annulus. The reciprocation and displacement are continued until the *plug bumps* the float collar and some operators continue to work the casing until it starts to take weight. This technique further dehydrates the cement opposite the porous formations. Clark[71] has pointed out that an increase in formation strength (area D) occurs in the annulus since the cement filter cake is replacing the mud cake. Binkley, Dumbauld, and Collins[72] showed by mathematical analysis and laboratory investigations that the factors controlling the deposition of cement filter cake are the properties of the cement slurry and the filtration time. It is this *strengthening* and *squeezing* which make it possible to raise long columns of cement without loss of returns.

Figure 3.24 shows the bottom-hole pressure of a cement job which was completed even though lost circulation occurred as a result of excessive pressure surges. As the casing was run into the well, the floating-in surges during the latter portion of the downward trip were greater than the strength of the formation, to the extent that the fluid found it easier to fracture the formation than to return up the annulus. As the pipe was run in the hole,

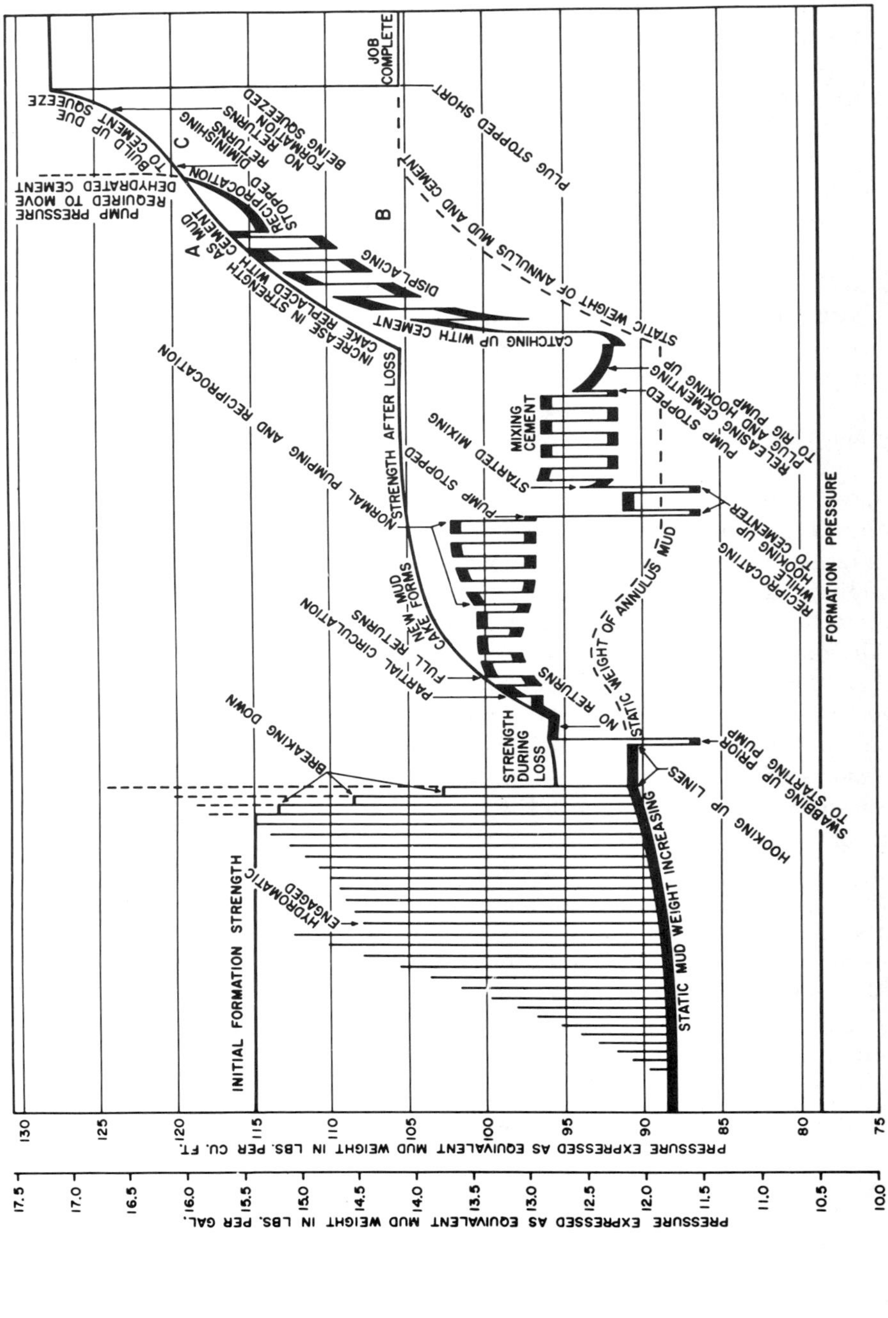

Fig. 3.24 Bottom-hole pressure chart for lost circulation caused by excessive pressure surges while running casing, and by dehydration of cement during displacement. (After Clark,[71] *Drilling and Production Practice*, API.)

higher surges further broke down the formation. With the casing on bottom, a *low* circulation rate and the addition of lost circulation material resulted in full returns. The cement was mixed and, after mixing, pumps and reciprocation were stopped to release the cementing plugs and change to the rig pumps. The cement "turned the corner" at this point and the pressure started to increase. The cement can be subjected to considerable dehydration when the pressures approach the strength of the formation (area A). Weight-indicator readings during picking up and lowering of casing before and after cementing can be used to establish a pattern for determining the amount of dehydration. In area B reciprocation was stopped, but increased circulation pressure led to even further dehydration, and surface returns diminished. At this point (area C) two things could happen, viz., the formation could break down again and take up the cement slurry, in which case the annular column could fall back, or the dehydrated cement could hold the pressure above the lost circulation point in the annulus. In the case being considered, the squeeze-pressure rose to the limit of the pumps, and displacement was stopped even though the plug was not bumped.

Amount and type of cement. The amount of cement to be used in casing-cementing depends on the total volume of slurry required and on the volume of slurry per sack of cement. The total volume of slurry required is dictated by the casing dimensions and the volume of the hole. The hole volume is determined by the bit size and the hole enlargement. The extent of hole enlargement depends on local field conditions, the type of drilling fluid, and the cement additives used. If estimated without benefit of a caliper survey, the hole enlargement is generally taken to be 50 to 100 per cent of the annular volume as calculated from the bit diameter. On the other hand, if a caliper survey is available, the enlargement can be estimated to within 10 to 15 per cent. Example 3.9 illustrates a method of using a caliper survey in volumetric calculations.[73]

Example 3.9. Calculate the depth to the top of the cement column in the annulus, given:

A string of 7-in., 17-lb/ft casing is to be set in an $8\frac{3}{4}$-in. hole at a depth of 8000 ft, using 800 sacks of cement and leaving 60 ft of cement in the casing. Caliper survey measurements are as follows:

Well depth, ft	*Hole volume, cu ft*	*Average diameter, in.*
7000–8000	454	$9\frac{1}{8}$
6000–7000	467	$9\frac{1}{4}$
5000–6000	505	$9\frac{5}{8}$
4000–5000	479	$9\frac{3}{8}$

Compare the calculated depths to the top of the cement column using experience factors of 1.15 (based on other calipered holes in the area) and 1.50 (based on bit size).

SOLUTION: Assuming the slurry volume to be 1.10 cu ft per sack,

$$\text{Slurry vol.} = 1.10 \times 800 = 880 \text{ cu ft}$$

From Table A.1 (see Appendix),

$$\text{Capacity of casing} = 0.2331 \text{ cu ft per lin ft}$$

$$\text{Slurry inside casing} = 0.2331 \times 60 = 14 \text{ cu ft}$$

$$\text{Slurry in annulus} = 880 - 14 = 866 \text{ cu ft}$$

The theoretical annular volume in any 1000-ft increment is determined from Eq. (3.22) where $d_e = 7$ in. and d_h is the average hole diameter for the increment as determined by the caliper survey. The corrected annular volume is the product of the theoretical annular volume and the experience factor 1.15.

Interval, ft	*Theoretical annular volume, cu ft*	*Corrected annular volume, cu ft*	*Cumulative annular volume, cu ft*
7000–8000	187	215	215
6000–7000	199	229	444
5000–6000	238	274	718
4000–5000	212	244	962

Unfilled annular volume below 4000 ft

$$= 962 - 866 = 96 \text{ cu ft}$$

Distance of cement top below 4000 ft

$$= \frac{96 \text{ cu ft}}{244 \text{ cu ft}/1000 \text{ ft}} = 393 \text{ ft}$$

$$\text{Depth to top of cement} = 4000 + 393 = 4393 \text{ ft}$$

If no caliper data are available, the theoretical annular volume is considered to be that given in Table A.2 in the Appendix, i.e., 0.1503 cu ft per ft. Applying an experience factor of 1.50, the corrected volume is 0.225 cu ft per ft. The height of the cement in the annulus is

$$\frac{866 \text{ cu ft}}{0.225 \text{ cu ft/ft}} = 3849 \text{ ft}$$

and the depth to the cement top is

$$8000 - 3849 = 4151 \text{ ft}$$

Temperature surveys. There are two methods for locating the top of the cement column behind the casing, viz., temperature surveys, radioactive-tracer surveys, sonic logs, and gamma ray density logs. A temperature survey consists of running a recording thermometer down the casing following the cementing operation. In the process of setting, the cement gives off heat of hydration which ranges from 63 cal per g for Class E cement to 71 cal per g for Class A cement. This heat increases the temperature of the adjacent fluid in the casing by several degrees, as shown in Fig. 3.25. The curve rises markedly below 5300 ft, indicating that heat is being given off by the setting cement. Further analysis of this survey indicates the existence of a *contaminated zone* or *channeling* up from 5350 to 5500 ft. The approximate time after cementing for running temperature surveys when different additives are used is given in Table 3.34.[49] The *time to perforate* with a minimum of shattering is obtained by doubling the temperature survey time at any well temperature. For example, the time to run a temperature survey with Class A cement at 140°F (well temperature) is 6 to 9 hr after cementing and the time to perforate is 12 to 18 hr.

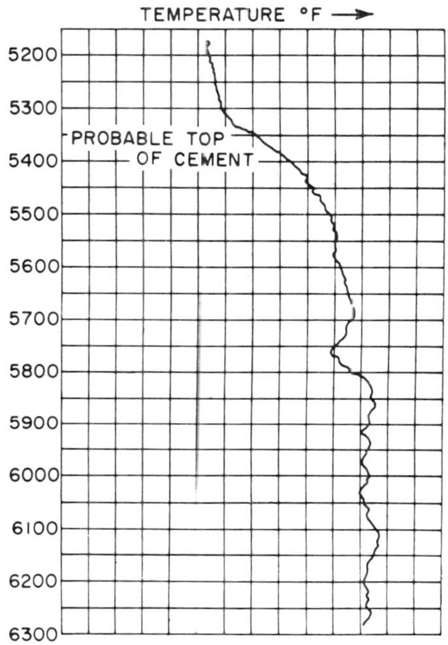

Fig. 3.25 Typical temperature survey. (Courtesy of Weatherford Oil Tool Company, Inc.)

Casing cementing. On the Gulf Coast and in coastal waters, conductor casing is set at a depth of 100 to 150 ft for the purpose of supporting unconsolidated surface deposits. Either $18\frac{5}{8}$-in. casing is run in a 22-in. hole, or $13\frac{3}{8}$-in. casing is run in a $17\frac{1}{2}$-in. hole. These strings are cemented with neat slurry made from 100 to 150 sacks of either Class A, B, or C cement. Two per cent calcium chloride is used as an accelerator when the WOC time can be reduced; usually, however, 12 to 14 hr are required to finish rigging up. The service company truck displaces the cement slurry, leaving 10 ft of cement inside the casing to allow for contamination. The equation for the internal capacity of casing in cubic feet per hundred feet is

$$V_i = 0.5454\, d_i^2 \tag{3.20}$$

and in barrels per hundred feet is

$$V_i = 0.09714\, d_i^2 \tag{3.21}$$

Table 3.34. TIME LAPSE BETWEEN CEMENTING AND TEMPERATURE SURVEY

(*After Reference 49*)

	Additive	Additive content, per cent	Time lapse for indicated well temperatures, hr			
			100°F	120°F	140°F	160°F
Class A Cement	Gel	0	8–12	8–12	6–9	4–8
		4	8–12	8–12	6–9	4–8
		8	9–12	9–12	6–9	6–9
		12	9–12	9–12	9–12	9–12
Pozzolan Cement (50-50)	Gel	0	8–12	8–12	6–9	4–8
		2	8–12	8–12	8–12	6–9
		4	8–12	8–12	8–12	6–9
Class A Cement	Diatomite	10	10–14	10–14	9–12	9–12
		20	10–14	10–14	9–12	9–12
		40	12–16	12–16	10–14	10–14
Class E Cement			12–18[a]	9–12[b]	16–24	16–24

[a]At 180°F.
[b]At 220°F.

where d_i is the inside diameter of the pipe in inches. The annular volume in cubic feet per hundred feet is

$$V_a = 0.5454(d_h^2 - d_e^2) \tag{3.22}$$

and in barrels per hundred feet is

$$V_a = 0.09714(d_h^2 - d_e^2) \tag{3.23}$$

where d_h is the diameter of the hole in inches and d_e is the outside diameter of the casing in inches. Example 3.10 shows the calculations for cementing the conductor pipe.

Example 3.10. Calculate the fill-up and displacement for cementing conductor pipe, given:

A string of $13\frac{3}{8}$-in., 48.00-lb/ft casing to be set at 100 ft in a $17\frac{1}{2}$-in. hole, using 100 sacks of Class A cement.

Maximum water-cement ratio = 5.5 gal per sack

Slurry weight = 15.40 lb per gal

Slurry volume = 1.22 cu ft per sack

On the basis of caliper surveys and experience in the area, an excess of 35 per cent cement should be used. (It is assumed that this excess will allow for the 10 ft of cement left in the casing.)

SOLUTION: From Eq. (3.20) or Table A.1, the volume of the casing is 88.18

cu ft/100 ft, and from Eq. (3.22) or Table A.2 the volume of the annulus is 69.46 cu ft/100 ft. The volume of slurry required is

$$69.46 \times 1.35 = 93.77 \text{ cu ft/100 ft}$$

Since there are 122 cu ft of slurry, there will be cement returns. The displacement is

$$0.8818 \text{ cu ft/ft} \times 90 \text{ ft} = 79.36 \text{ cu ft or } 14.12 \text{ bbl}$$

Surface casing is set from the surface to a depth below the lowest known fresh-water zone. It serves as a guide for deeper drilling, prevents caving of surface sands and gravels, and protects surface waters. Since the blow-out preventers are attached to this string, a good bond between pipe and hole is necessary for adequate protection. The bit and casing program for surface pipe is usually either to drill a 15-in. hole and set $10\frac{3}{4}$-in. casing or to drill a $12\frac{1}{4}$-in. hole and set $9\frac{5}{8}$-in. casing. The setting depth for surface casing will range from 100 ft in shallow areas to 4500 ft on offshore wells. The floating equipment consists of a float collar placed one joint from the bottom and a guide or whirler-type shoe. The general practice is to use from four to eight centralizers on the bottom joints, placed 90 ft apart. Casing scratchers, if used, are placed 15 ft apart on the bottom two or three joints to obtain a better bond between the cement and the wall of the hole. To reduce hydrostatic pressure, many operators use Class A cement with 4 to 10 per cent gel (bentonite). The number of sacks of cement varies with the amount of casing set, but usually for wells deeper than 2500 ft, either 500 sacks or the cement calculated to fill the theoretical annular space plus 10 per cent excess is used. Table A.1 gives the capacities of various sizes of casing, and Table A.2 shows the volume between tubing or casing and open hole.

The standpipe pressure gage is useful in indicating the progress of the slurry during the cement job. Figure 3.26[74] shows an idealized pressure-

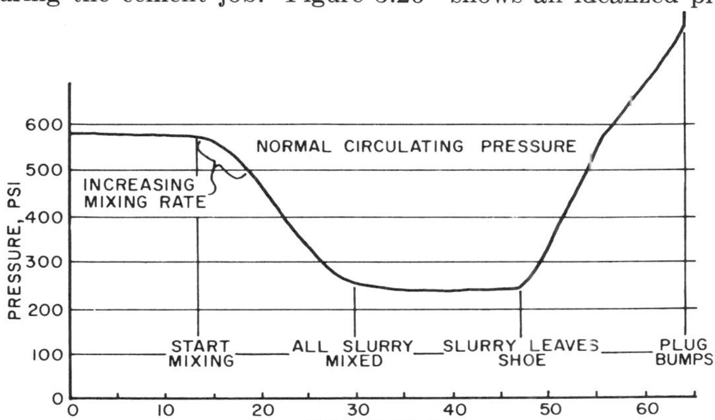

Fig. 3.26 Idealized pressure-time chart of casing cement job. (Courtesy of B. J. Service.)

time record of a casing cement job. After the pipe is on bottom, mud circulation is established with the rig pumps and is maintained until the fluid columns are equalized or returns are obtained from bottom. As the slurry enters the casing, since it is heavier than the drilling mud, the increased weight of fluid inside the pipe lowers the circulating pressure. When the slurry leaves the shoe and starts toward the surface, pump pressure begins to increase, and when the slurry height is equalized (balanced) inside and outside the casing, the pump pressure equals the normal circulating pressure. Continued displacement unbalances the columns and causes a steady increase in pump pressure as the heavier slurry accumulates outside the casing. The pressure rise continues until the plug bumps the float collar. An abnormal increase in pressure indicates either that the shoe is plugged or that the annulus is plugged by caving. On the other hand, if the pressure decreases during placement, the casing has split or parted.

The function of protection casing is to support the hole to permit deeper drilling where cavities are known to exist, or where excessive mud weights are necessary to control deeper high-pressure formations, i.e., where there is a danger of lost circulation in the shallow zones. The volume of cement used around such casing ranges from 500 to 1000 sacks with 4 to 6 per cent bentonite, usually with 100 sacks of neat cement around the shoe. Were it not for excessive hydrostatic pressures, most operators would use larger volumes of cement, as protection against the possibility of holes being worn in the casing during drilling, and some cementing programs call for a stage collar at 3500 ft to protect the casing at shallow depths. As in all cement jobs, the circulating bottom-hole temperature is an important factor. Consider, for example, a string to be set (in Southwest Texas) at 10,000 ft with cement fill-up to 7000 ft. Temperature surveys (Fig. 3.27) indicate *static* bottom-hole temperatures of 192°F at 7000 ft and 240°F at 10,000 ft. The cement selected must, therefore, develop 500 psi compressive strength in 24 hr at the upper placement limit (i.e., at 192°F) and must remain pumpable at the lower depth long enough to allow placement.

There is no set program for selecting the setting depth of production strings. Some operators may set above the productive zone and complete in open hole or with preperforated or slotted liners, whereas others may set casing through the producing sand and perforate for completion. Most operators in the Gulf Coast area now prefer cased completions with the casing set in *tension* no more than 6 ft off bottom or high enough to compensate for casing elongation during slurry placement. If the string is set below 6000 ft, a Class E cement is used and if there are sulfate waters in the producing horizon, the cement should be sulfate-resistant. Again, if it is desired to reduce density and increase the slurry volume, an additive

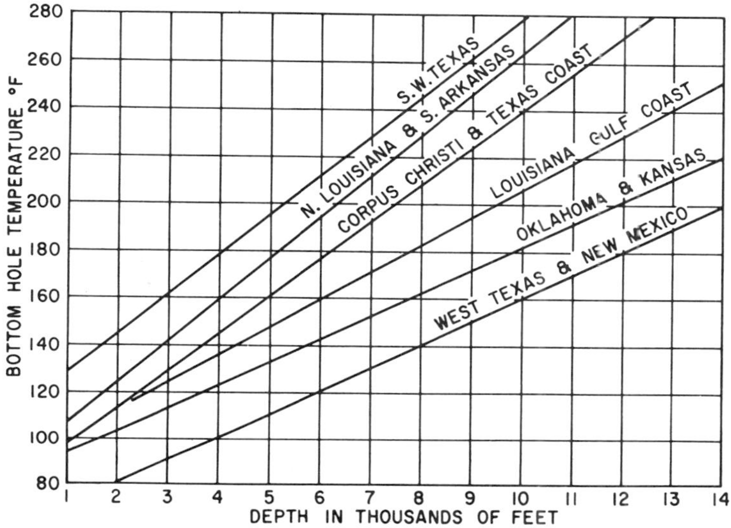

Fig. 3.27 Bottom-hole temperatures in various areas. (Davis and Faulk,[37] *Drilling and Production Practice*, API.)

such as bentonite or perlite is used. When barite or mixing water containing high concentrations of salt is used, the effect on thickening time should be determined *prior* to the job. The volume of slurry required will depend on the shallowest zone of interest. Most operators generally place the cement 500 ft above the shallowest prospective horizon.

3.8. Squeeze Cementing

Squeeze cementing is an operation in a well whereby a cement slurry is forced (squeezed) under pressure (1) into a formation (via an open hole), (2) into a channel behind the casing, or (3) through holes purposely placed in the casing.[75–80] Squeeze-cementing operations are performed either during drilling and completion or in workover operations for the following reasons:

1. To stop lost circulation in the open hole during drilling.
2. To supplement a faulty primary cementing job.
3. To reduce water-oil, water-gas, or gas-oil ratios.
4. To abandon a nonproductive or depleted zone.
5. To isolate a zone before perforating for production (block squeeze).
6. To repair defects such as joint leaks, split casing, parted casing, corrosion, etc.
7. To supplement primary cementing around casing or liners when the cement fails to reach the desired height.

When pressure is applied to a cement slurry in squeeze cementing or recementing, water is forced from the slurry if it is in contact with a permeable surface, the solid particles remaining to form a filter cake. The API filtration rates of neat-cement slurries have been reported to range from 600 to 2500 cc in 30 min. Since water is removed in the process, filtration with cement slurries is termed *dehydration* and it is not synonymous with setting of cement. When a cement has been dehydrated, the resultant filter cake is too plastic to be pumped; however, it is still soft and may be removed by either mechanical action or fluid jetting.

If the casing is perforated with drilling mud in the hole or if drilling mud is used as a workover fluid, the perforations will be filled with drilling mud and subsequently plugged with a mud filter cake. A squeeze packer is generally set 10 to 15 ft above the perforations with tubing or drill pipe. Although the cement is preceded and followed by a water spacer, the mud below the squeeze packer and the mud in the perforations must be *forced* into the formation before the cement will come in contact with the well bore. This requires a high break-down pressure to fracture the formation or to break the bond between the primary cement and formation. To push the mud and filter cake from *all* perforations, it is necessary that a fracture be developed at *each* perforation if the primary cement job filled the annular space.[81] The possibility of establishing a fracture at each perforation over a long perforated interval is extremely rare and therefore a single-stage squeeze cementing job is usually unsuccessful. However, once a number of perforations have been filled with cement, the possibility of creating fractures or opening up the remaining perforations is greatly increased and subsequent stages should result in a successful job.

Howard and Fast,[78] in *shallow* well experiments, showed that the cement is squeezed out in a horizontal sheet or pancake $1\frac{1}{2}$ to 2 in. thick, parallel to the bedding planes or existing zones of weakness, and that the lateral spread of cement varied from 45° to 270° per perforation. These tests also showed that when formations were broken down with water prior to squeezing, the cement bonded firmly and set hard. Ninety-eight per cent of the perforations broken down with water and squeezed with cement were sealed, as compared with approximately 80 per cent shut-off in the tests where the formations were broken down with mud. Further tests showed that a cement will not bond to a cement sheath or sand face covered with mud.

Harrison, Kieschnick, and McGuire[82] have pointed out that pressure-induced well-bore fractures may be horizontal when the bottom-hole fracturing pressure has exceeded the overburden load of about one psi per foot of depth. This seldom occurs below 3000 to 4000 ft of depth except possibly in the infrequent case of abnormally high horizontal stress. Field studies by Howard and Fast[78] showed that the average formation breakdown pressure was less than the theoretical overburden pressure gradient of 1 psi per ft. Cores taken from the rat hole after a squeeze job on a 9530-ft well showed that the cement was squeezed out in a *vertical* fracture. The formation of a horizontal and effective cement layer is therefore very rare except in shallow wells.

The conventional squeeze cementing of a *channel* (fluid production between sands in the annulus) involves killing the well with drilling mud and thereby filling the clean channel with mud before squeeze cementing. It is then necessary to create a fracture in the zone producing salt water or gas, and the chances of creating such a fracture at the end of the channel are rather poor since fractures are more likely to occur opposite the perforated interval than in the offending zone at the end of the channel. The use of *salt water* or other *penetrating fluids* instead of drilling mud would permit this fluid to be forced into the offending zone, allowing cement to enter the channel.

In the event that the offending zone fractures and allows the drilling mud to be displaced from the channel, the displacing cement must, nevertheless, cross the permeable producing zone. If the channel is small, the dehydration of cement opposite the producing zone would effectively block further entry of cement into the channel. Most final pressures on squeeze-cement jobs are *not* pressures exerted against the formation but are only an indication of the pressure in the casing between the perforations and the squeeze packer. Even when there is no channel and the perforations have been cemented (and the well reperforated higher or lower), the exclusion of salt water or gas still depends upon low vertical permeability or the presence of a shale barrier.

Squeeze packers. A squeeze packer or cement retainer is a tool that *confines* the *surface pressure* to the tubing or drill pipe and gives closer control of the cement slurry during squeezing.[83] It is usually necessary to maintain some pressure on the casing (1000 to 2000 psi) above the squeeze tool to withstand possible collapse from pressure outside the casing. It is also important not to have any leaks in the squeeze string. A leak will permit rapid local dehydration of the slurry and consequent plugging of the squeeze string, giving the false impression of a flash set of the cement. Therefore it is important to pressure test the casing and squeeze string before mixing the cement.

Squeeze tools are available in *drillable* and *retrievable* types. The drillable squeeze packer (Fig. 3.28) is run on tubing, drill pipe, or an electrical cable. Once set it becomes a part of the casing string, but since it is made of drillable cast iron or magnesium alloy, it can be drilled out if desired. The cast-iron retainer is resistant to salt water, acids, and caustics; the magnesium retainer is susceptible to attack by all these fluids. The two factors

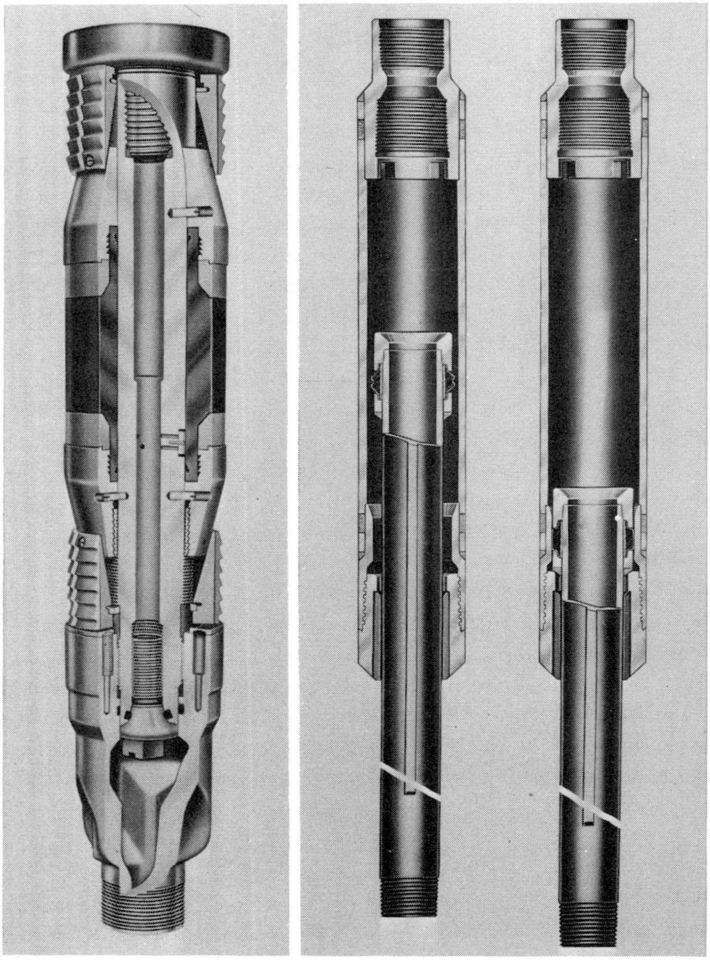

Fig. 3.28 Drillable squeeze packer. (Courtesy of Baker Oil Tools, Inc.)

to be considered in selecting a drillable packer, therefore, are *durability* and *drillability*. All drillable retainers are basically of the same construction, consisting of a retainer body on which is mounted a rubber packing element backed up by support rings which prevent extrusion of the packing. Above and below the support rings are an upper and a lower cone, on which the upper and lower slips are positioned. A back-pressure valve in the bottom of the retainer prevents backflow of the cement slurry. When tubing or drill pipe is used as the running-in string, the top slips in one type of retainer are set by rotation of the run-in string; the top slips in the other are set by first dropping a plastic ball which seats in the bottom of the retainer, permitting a pressure to build up in the tool. The pump pressure first expands the rubber packing element against the casing. Upward movement of the upper cone shears the pins in the cone and slips, forcing the slips against the wall of the casing, and the bottom slips are set by picking up on the run-in string.

The sequence of operations involving the use of a drillable squeeze packer is as follows:

1. The tool is attached to the circulation valve.
2. The run-in string is made up on the circulation valve; the tool is run to the desired depth and set.
3. The run-in string and casing are tested and formation breakdown is obtained. The *breakdown pressure* is the surface pump pressure at which the formation initially takes fluid.
4. With the circulation valve open above the retainer, a water spacer is displaced into the run-in string and is followed by the cement slurry, by a second water spacer, and, finally, by a volume of mud sufficient to displace *half* the first water spacer into the annulus.
5. The circulation valve or tool is closed and the formation is squeezed. The pumping rate is adjusted by the pressure buildup so as to obtain the desired final pressure with the available quantity of slurry. For example, with a 90-sack job, 30 sacks are pumped into the formation at a rate of two to three barrels per minute; the next 30 sacks are displaced at 1 bbl per min; and if the desired pressure has not been obtained, the last 30 sacks are squeezed at a rate of $\frac{1}{4}$ to $\frac{1}{2}$ bbl per min or slower; however, the cement slurry should be kept *moving*.
6. When the desired squeeze pressure is obtained, the pressure is held for 1 min. If the formation does not break down, the pressure is released from the casing, the circulating valve opened, and the excess slurry reversed out. The returns are the displacing fluid, the second water spacer, the excess slurry, and half of the first water spacer.
7. If the excess cement slurry cannot be reversed, the run-in string is pulled from the well rather than circulating the slurry the long way.

Retrievable squeeze packers are designed to be run in the well, used for the cementing operation, and then retrieved. They are run on tubing or drill pipe and are classified by the manner in which they are set, i.e., hydraulically, mechanically, or by a combination of the two methods. Figure 3.29 illustrates one type of retrievable squeeze tool. The sequence of operations with this tool is essentially the same as with the drillable packer. If the cement slurry has followed a channel above the perforations, rapid dehydration of the slurry is depended upon to prevent backflow. Retrievable-type squeeze tools have many advantages over drillable retainers.

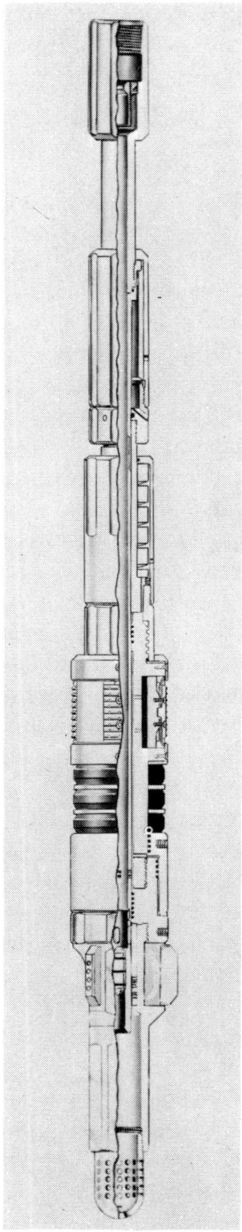

Fig. 3.29 Retrievable (hydraulic) squeeze tool. (Courtesy of Halliburton Oil Well Cementing Company.)

They can be set and released many times, they are cheaper to run, and less rig time is required if the cement is to be drilled out.

Amount of cement. The volume of cement required on squeeze jobs ranges from a few sacks to several hundred sacks when squeezing above the top of the primary cement job. According to Hodges,[83] the various criteria used to estimate the volume of cement slurry to be mixed are:

1. Two sacks of cement per ft of perforations, with a 50-sack minimum.
2. One hundred sacks minimum if an injection rate of 2 bbl per min can be obtained after breakdown.
3. The volume of slurry should not exceed the capacity of the run-in string.
4. The volume of slurry should not result in a column that cannot be reversed out.

The volume of cement squeezed depends upon:

1. Rate of pumping.
2. Breakdown pressure.
3. Filtrate loss of the cement slurry.
4. Formation permeability. (High permeability permits rapid dehydration, bridging, and pressure build-up.)
5. Length of perforated section.
6. Condition of primary cement job and top with relation to squeeze point.

Squeeze pressure. Some operators believe in *light* squeezes with low pressures, whereas others limit the squeeze pressures to those the equipment can withstand. With permanent-type completions, and in conventional wells with salt water in the hole, Tausch[81] reported that squeeze pressures need be no greater than 500 psi above reverse circulating pressures. Some of the recommended maximum squeeze pressures are those necessary to meet the following conditions:

1. Maximum final *surface* squeeze pressure at least 1000 psi higher than the break-down pressure.
2. Maximum bottom-hole pressure equal to 1 psi per ft of depth.
3. Bottom-hole pressure equal to the well depth $\times$ 0.6 + 500 psi (usually in shallow wells).
4. Maximum squeeze pressure in excess of any subsequent pressure to be put on the formation during acidizing or fracturing.

Example 3.11 shows the calculations for a typical squeeze-cement job.

Example 3.11. Make the calculations necessary for a squeeze cement job when perforations at 9215 to 9222 ft are to be squeezed with a retrievable tool set at 9200 ft and run on $2\frac{1}{2}$ in. tubing, given:

Size of casing = $5\frac{1}{2}$ in., 20 lb/ft, 0.02222 bbl/ft (Table A.1)

Size of tubing = $2\frac{1}{2}$ in. (2.875 in. OD), 0.00579 bbl/ft (Table A.4)

100 sacks of Class E cement, 1.20 cu ft per sack

10 bbl of water are spaced ahead of and 5 bbl behind the cement

Capacity of pump and discharge lines = 0.7 bbl

Solution: The slurry volume is

$$100 \times 1.20 = 120 \text{ cu ft} \quad \text{or} \quad \frac{120}{5.615} = 21.4 \text{ bbl}$$

The capacity of the system to the top of the perforations is the sum of:

Volume of pump and discharge line = 0.7 bbl

Volume of casing below retainer

$$= 15 \text{ ft} \times 0.02222 \text{ bbl/ft} = 0.3 \text{ bbl}$$

Volume of tubing = 9200 ft × 0.00579 bbl/ft = 53.3 bbl

Total volume = 54.3 bbl

The volume of the tubing and discharge line alone is 54.0 bbl. Then the by-pass (circulation joint) is closed when the displacement by the cementing truck is

$$54.0 \text{ bbl} - 5 \text{ bbl water}$$
$$- 21.4 \text{ bbl cement} - 5 \text{ bbl water} = 22.6 \text{ bbl mud}$$

The additional displacement necessary to clear the cement slurry from the tool is 5 + 21.4 = 26.4 bbl mud. If the desired squeeze pressure is not obtained, the perforations are cleared of cement by a displacement of 54.3 bbl of mud. The displacement of the pump and discharge line is usually neglected, except in the case of plugback jobs, unless the volume exceeds 1.25 bbl.

3.9. Liner Cementing

The term *liner* denotes casing which is used to case off a section of open hole below existing casing, i.e., a liner extends from the setting depth up into another string of casing. Liners are run into the well with a run-in string of drill pipe or tubing and may be set on bottom or hung from the bottom of existing casing. Some of the more important liner applications are:[84]

1. To case off open hole below a long intermediate casing string either upon initial completion or as a result of deepening the well.
2. To case off open hole which is the result of casing being stuck off bottom.
3. To case off a previous open-hole completion to control water, gas, or sand.
4. To case off zones of lost circulation or zones of high pressure encountered during drilling.

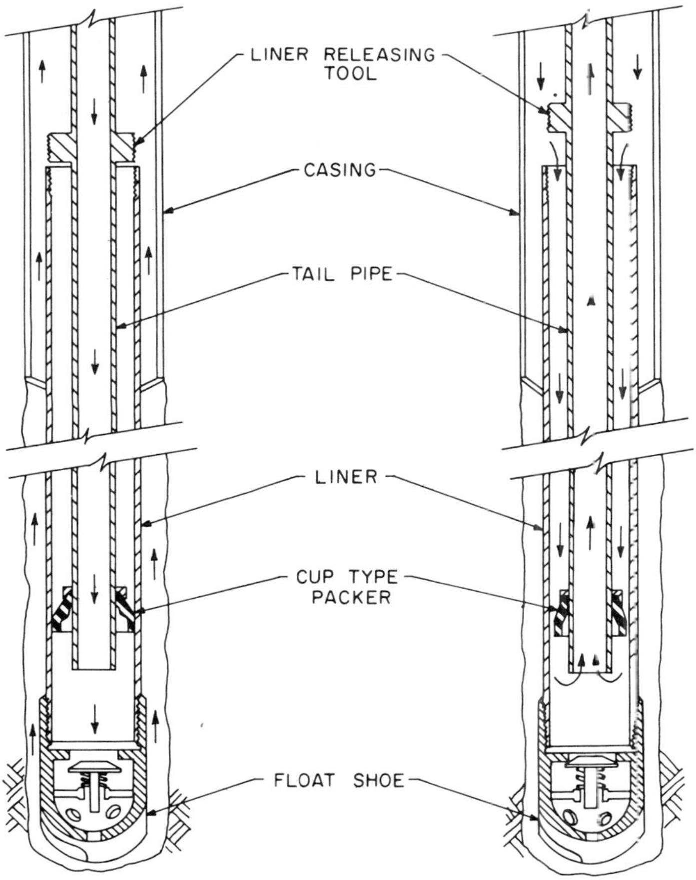

Fig. 3.30 Simple liner cementing operation. (After Davis,[84] *Oil-Well Cementing Practices In The United States*, API.)

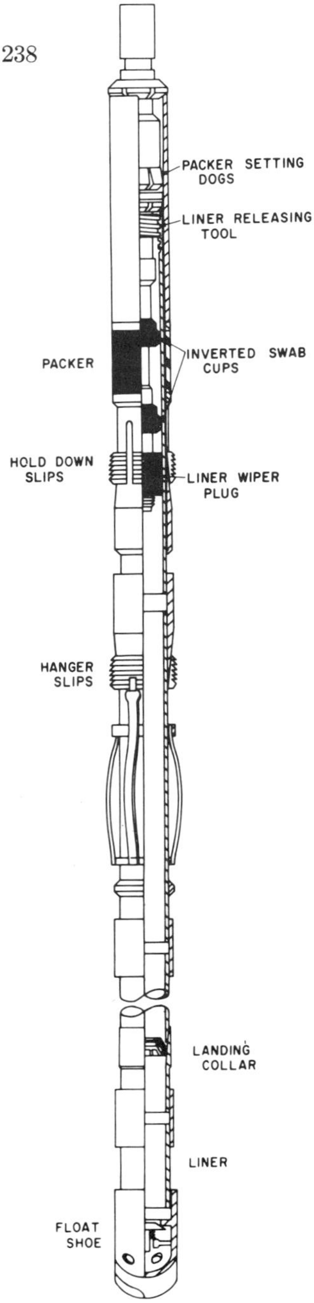

Fig. 3.31 Liner assembly for long liners. (After Davis,[84] *Oil-Well Cementing Practices in the United States*, API.)

In general, liner-setting assemblies can be classified as either short or long. Short liners are defined as those which can be set on bottom without buckling under their own weight, and long liners are those which because of their weight must be hung on slips from the existing casing. There are many different combinations of liner-setting tools; however, from a cementing standpoint they are essentially similar, their basic functions being cementing through the drill pipe and reversing out the excess cement. Figure 3.30[84] shows schematic drawings of a simple liner cementing operation. Once the liner has been set, cement is circulated down the drill pipe, out the liner shoe, and up the annulus. The displacement includes the volume of the *drill pipe* and the volume of the *tailpipe*. On completion of the displacement, the excess cement is reversed out and the drill pipe pulled out of the hole. In this case the cement seals both the annulus between liner and hole and the overlapping area between liner and casing (Fig. 3.30).

The liner assembly shown in Fig. 3.31 is designed for hanging and releasing the liner *before* cementing and then collapsing the packer, and has the added advantage of using both drill-pipe and liner-wiper plugs. The liner is run to the bottom and hung. It is released by right-hand rotation and the liner-releasing tool is lifted up 18 in. After circulation until returns are obtained from the bottom, the liner is cemented through the drill pipe and, as the last cement is mixed, a drill-pipe wiper plug is released from the cementing head. This plug wipes the drill pipe and seats in the liner-wiper plug. An increase in pump pressure shears the pin which retains the liner-wiper plug, allowing the plug to wipe the liner and seat either in the float shoe or on the landing collar. The setting tool is then raised until the packer-setting dogs are released outward. Lowering the drill pipe causes the dogs to engage the top of the liner and collapse the packer. Example 3.12 shows the calculations for cementing a liner.

Example 3.12. Calculate the number of sacks of cement and the displacement for cementing 1000 ft of 5-in. OD, 18-lb/ft liner in a 6-in. hole. The running-in string is $3\frac{1}{2}$-in., 13.30-lb/ft drill pipe. The following data are given:

Top of liner = 8600 ft

Length of liner including float shoe, hanger, and packer = 1000 ft

Length of drill pipe (including setting tool) to wiper plug = 8615 ft

Space between liner-wiper plug and landing collar = 935 ft

Volume of liner = 0.0178 bbl per ft (Table A.1)

Volume of annulus = 0.0545 sacks per ft (Table A.2)

Volume of drill pipe = 0.00742 bbl per ft (Table A.5)

Experience in the area has shown that a 40 per cent excess of cement should be used.

SOLUTION: The number of sacks of neat Class E cement required is

$$1000 \times 0.0545 \times 1.40 = 76.3 \text{ sacks}$$

The displacement to the liner plug is

$$8615 \times 0.00742 = 63.9 \text{ bbl}$$

The displacement from liner plug to the landing collar is

$$935 \times 0.0178 = 16.6 \text{ bbl}$$

Several factors make a good liner-cementing job difficult to achieve. The first is lack of adequate annular clearance which results in a very thin collar of cement; however, high annular velocities considerably improve the displacement efficiency. A second factor in the case of long liners is that operators prefer to have the drill pipe mechanically detached from the liner before cementing. This prevents casing movement and obviates the use of scratchers during placement of the cement. In general, most operators use no mechanical aids other than centralizers.

3.10. Open-Hole Plugback Cementing

A well may be plugged back from a given depth to a shallower depth for any of a number of reasons. The most important of these are:

1. *Abandonment.* Each state has extensive rules on plugging and abandoning wells, and cement plugs are normally used for this purpose. For example, if the producing string is set through the producing horizon, a cement plug not less than 15 ft long is placed

above each producing zone. Also, a plug is set at the bottom of the surface casing, about half in the casing and half in the open hole.

2. *Whipstocking.* An Ottawa sand–cement plug is used to plug off a section of a crooked hole or that part of a hole containing a fish. The plug provides a seat for a whipstock to sidetrack the plugged part of the hole.
3. *Lost Circulation.* A cement plug is placed opposite a zone of lost circulation with the hope that the cement will penetrate and seal the fractures some distance from the well bore. If pressure is applied after the slurry is in place, this is a low-pressure squeeze job.
4. *Shutting Off Water.* Plugs may be necessary in open-hole completions to shut off bottom water. These plugs are often placed with dump bailers.

Cement plugs are usually placed with open-end drill pipe or tubing, and the success of these jobs depends on preventing *contamination* of the cement and on allowing it to set without agitation. Cement plugs are placed in a well by *balancing* the fluid columns, i.e., placing the same height of slurry and displacing fluid inside and outside the drill pipe or tubing. The pump pressure[85] during the job (Fig. 3.32) is an indication of how the columns are

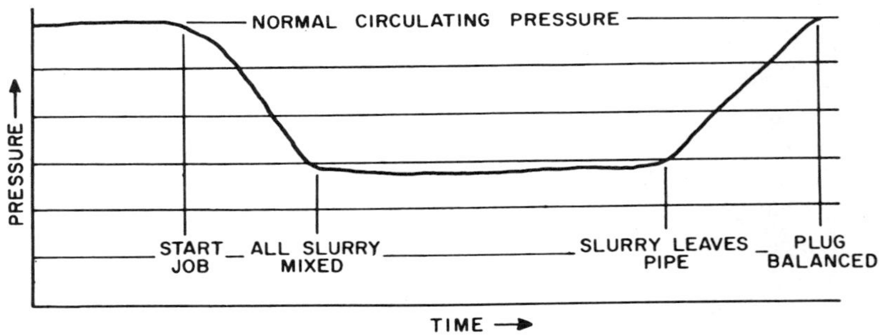

Fig. 3.32 Idealized pressure-time chart during open-hole plugback operations. (Courtesy of B. J. Service.)

balancing. At the start, the mixing pressure is equal to the normal circulating pressure. After the slurry enters the drill pipe, the pump pressure drops because the heavy slurry unbalances the fluid columns. When the slurry leaves the pipe, the pressure begins to increase and returns to normal when the cement columns are balanced, i.e., when displacement is complete. The capacities of open holes are given in Table A.3, the capacity of tubing is given in Table A.4, and the capacity of drill pipe is given in Table A.5. Example 3.13 shows the amount of cement and the barrels of displacement required for a typical plugback cement job, and Fig. 3.33 shows the balanced fluid columns after displacement. Since the displacement is made with the service company truck, the velocity in the annulus is near the plug-flow region, where approximately 60 per cent of the circulatable mud is displaced.

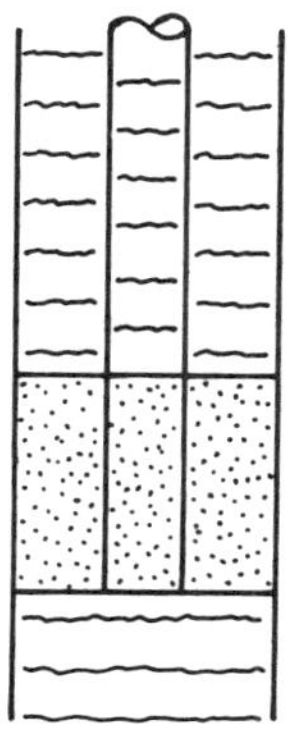

Fig. 3.33 Cement columns balanced after displacement.

Example 3.13. Calculate the cement required when an $8\frac{1}{2}$-in. hole is plugged back from 5500 to 5400 ft using $4\frac{1}{2}$-in. drill pipe, given:

Volume of $8\frac{1}{2}$-in. hole = 0.3583 sacks per ft (Table A.3)

Volume of $4\frac{1}{2}$-in., 16.6-lb/ft drill pipe = 0.01422 bbl per ft (Table A.5)

Solution: The number of sacks of cement required is

$$100 \times 0.3583 = 35.8 \text{ sacks}$$

The displacement is

$$5400 \times 0.01422 = 76.8 \text{ bbl}$$

Doherty[86] and Goins[87] have described a technique for a successful plug-back job. Mud contamination is minimized in the same manner as on a casing job, i.e., by pumping water ahead of and behind the cement. The method of balancing the columns, so that the cement will stay in position until the drill pipe is pulled, is illustrated in Fig. 3.34 and in Example 3.14. In this technique the open-end drill pipe, with a back-pressure valve in the drill string to prevent backflow, is run to the depth of the plug. After displacement at a maximum rate (exceeding the plug-flow velocity), the kelly is broken off, the drill pipe is pulled very slowly to the top of the plug, and the excess cement is reversed out.

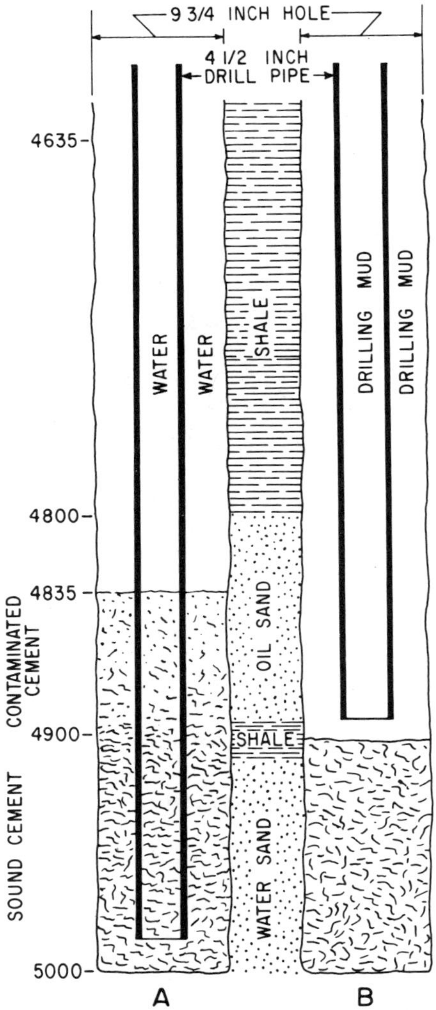

Fig. 3.34 Plugback cementing in the open hole. (After Doherty,[86] *Production Bull.*, API.)

Example 3.14. Make the necessary calculations for plugging back a $9\frac{3}{4}$-in. hole from 5000 to 4800 ft with neat Class E cement, given:

Slurry volume = 1.20 cu ft per sack

Volume of a $9\frac{3}{4}$-in. hole = 0.4714 sacks of cement per lin ft or 0.0924 bbl per lin ft (Table A.3)

Volume between $4\frac{1}{2}$-in. drill pipe and hole = 0.3709 sacks of cement per lin ft or 0.0727 bbl per lin ft (Table A.2)

Capacity of $4\frac{1}{2}$-in., 16.6-lb/ft drill pipe = 0.0726 sacks of cement per lin ft or 0.01422 bbl per lin ft (Table A.5)

Spacer behind cement = 1 bbl water

SOLUTION: The number of sacks of cement required for the plug is

$$(\text{ft of plug}) \times (\text{sacks/ft} \times K)$$

where K is a conversion factor which is used whenever the slurry volume is other than 1.10 cu ft per sack, i.e.,

$$K = \frac{1.10}{\text{Slurry volume (cu ft/sack)}}$$

Then, the cement required is 200 × 0.4714 × 1.10/1.20 or 86.4 sacks. Based on an estimated 50 per cent washout, the amount of cement used is 86.4 × 1.5 = 130 sacks. The height of the column of cement (the balanced plug) with drill pipe in the hole is

$$\frac{86.4 \text{ sacks}}{(0.3400 + 0.0666) \text{ sacks/ft}} = 212 \text{ ft}$$

where 0.3400 (= 0.3709 × 1.1/1.2) is the sacks of cement per foot in the annulus and 0.0666 (= 0.0726 × 1.1/1.2) is the sacks of cement per foot in the drill pipe. If 1 bbl of water is used as a spacer behind the cement, the spacer to be used ahead of the cement to give the same height of water (spacer) inside and outside drill pipe is

$$V_s = \frac{\text{sacks/ft in annulus}}{\text{sacks/ft in drill pipe}} \times \text{bbl of spacer in drill pipe}$$

$$= \frac{0.3709}{0.0726} \times 1 = 5 \text{ bbl}$$

The volume of the displacing fluid is

$$V_d = (\text{length of drill pipe} - \text{height of plug}) \times \text{capacity of drill pipe}$$

$$= (5000 - 212) \text{ ft} \times 0.01422 \text{ bbl/ft} = 68 \text{ bbl}$$

Since 1 bbl of the displacing fluid is water spacer, 67 bbl of mud are required. To prevent backflow, mud displacement should be less than 67 bbl, i.e., 66 bbl.

3.11. *Cementing Hydraulics*

Howard and Clark[60] first investigated the flow characteristics of cement slurries, and found that they behaved essentially as *plastic* fluids. As discussed in Sec. 1.7–1.9, the flow properties of a plastic fluid differ from those of a Newtonian fluid in that plastic materials can withstand a certain limiting shear stress indefinitely without flow. If this limit is exceeded, flow results and increases with the magnitude of the stress.

Rheology of cement slurries. Rheological measurements on a plastic fluid can easily be made with a commercially available rotational-type viscometer such as the Model 35 Fann V-G Meter. The rheological properties of cement slurries were first investigated with this viscometer by Anderson.[89] Burdyn[90] obtained relatively consistent readings when the maximum viscometer speed was limited to 300 rpm. Holden and Leidner[91] reported that the plastic viscosty of all slurries tested remained essentially constant at 56 cp regardless of bentonite content or aging temperature. However, depending on aging temperature, increasing the percentage of bentonite in a slurry from 0 to 4 per cent caused a two- to threefold increase in yield point. As a consequence, although a bentonite slurry will produce lower hydrostatic pressure gradients than a neat slurry, it will yield little, if any, reduction in flowing pressure gradients.

PROBLEMS

3.1. (a) What is the temperature range in a cement kiln?
(b) What is clinker?
(c) What is the unit of cement sold by the cement company?

3.2. (a) Name the four crystalline phases in portland cement and give the formula of each.
(b) What compound has low sulfate resistance, and what is the maximum percentage of this compound for high sulfate-resistant cements?
(c) How does the compound composition of high early-strength cement compare with that of retarded cement?
(d) What are the two types of slow-set cement?
(e) Write the reaction that takes place when tricalcium aluminate ($3CaO \cdot Al_2O_3$) combines with gypsum and water.

3.3. The oxide analysis of a cement is as follows:

Oxide	*Per cent*	*Oxide*	*Per cent*
SiO_2	21.9	SO_3	2.4
Al_2O_3	3.1	Loss	1.6
Fe_2O_3	4.2	*Total*	*99.8*
CaO	66.1	Na_2O	0.11
MgO	0.5	K_2O	0.14

(a) Calculate the Bogue constituents. *Ans.* $C_3S = 69$, $C_2S = 11$, $C_3A = 1$, $C_4AF = 13$, $CaSO_4 = 4$ per cent.
(b) Is this cement sulfate-resistant?
(c) What is the API class of this cement?

3.4. (a) What are the three types of thickening-time testers?
(b) At a depth of 12,000 ft, how does the circulating bottom-hole temperature compare with the static bottom-hole temperature?

(c) On a 14,000-ft well, what is the schedule number, mud weight, surface pressure, bottom-hole circulating temperature, bottom-hole pressure, and total cementing time on a 300-sack job?
(d) What are the seven API classes of cement and what are the depths at which they are to be used?
(e) For Schedule 8, what is the mixing time and total cementing time when 300 sacks of cement are used?
(f) What is the API cementing time when 500 sacks of cement are used on a 10,000-ft string of 7-in. casing?

3.5. Show that the constant in Eq. (3.7) is 9.09.

3.6. (a) Cement strength is based on what five functions?
(b) What is the minimum compressive strength for support of casing?
(c) What is the relationship between consistency and minimum strength?

3.7. (a) The thickening time of a Class E cement (Table 3.6), when tested using API Schedule No. 8, was 2 hr 56 min. What is the minimum WOC time? *Ans.* 8.80 hr.
(b) What will be the time for development of maximum temperature? *Ans.* 5.87 hr.

3.8. (a) What were the five reasons given for "knocking off" the lower joints of surface pipe?
(b) How can this type of failure be prevented?
(c) What is the unit of surface area of cement and how is it measured?
(d) What are the average surface areas for Class A, C, and E cements?
(e) Define minimum water in a slurry; how is it measured?
(f) Define maximum water.
(g) What are the maximum and minimum slurry weights for the three classes of cement given in Table 3.9?

3.9. What conclusions can be drawn from the sulfate-resistance tests shown in Table 3.12?

3.10. (a) Give seven reasons why additives are used in the design of cementing compositions.
(b) List the additives given in Table 3.13.

3.11. (a) What are the advantages and disadvantages of bentonite cements?
(b) What are modified cements?
(c) Is it advisable to use neat Class E cements at temperatures higher than 290°F?
(d) What effect does bentonite have on the permeability at temperatures higher than 290°F?
(e) Did bentonite appreciably affect the thickening times given in Table 3.17?
(f) From Figs. 3.5, 3.6, and 3.9, compare the 24-hr strengths at 140°F of Class A, Class E, and Class E cements with 4 per cent bentonite.
(g) Compare the 20-hr strengths of 4 per cent bentonite cements cured at 80°F and at 120°F.

3.12. (a) What is perlite?
(b) What are the chief advantages of this type of additive?
(c) How do the costs of a bentonite mixture and a perlit mixture compare?

3.13. (a) What are pozzolanic materials?
(b) What are the advantages of pozzolan cements?
(c) At 100°F and 24-hr curing time, how does the compressive strength of 50-50 pozzolan cement compare with that of neat cement?
(d) How do pozzolans affect the thickening time?

3.14. How does the 24-hr compressive strength of 20 per cent diatomite cement weighing 12 lb per gal and cured at 100°F compare with that of neat cement and of 50-50 pozzolan cement cured under the same conditions?

3.15. (a) How does the thickening time of a neat cement slurry at 2000 ft (given in Table 3.25) compare with that of a slurry containing 4 per cent calcium chloride?
(b) At a temperature of 80°F, what is the 12-hr compressive strength of slurry with various percentages of calcium chloride?

3.16. (a) At a salt concentration of 30,000 ppm and a depth of 10,000 ft, what is the thickening time as measured with the Pan American tester?
(b) Does sodium chloride in the mixing water have any effect on the compressive strength?

3.17. (a) What three weighting materials are used to increase slurry weight?
(b) What effect does barite have on the thickening time?

3.18. Describe briefly the applications of the following equipment: casing-cementing heads, guide shoes and float shoes, float collars, stage collars, wiper plugs, centralizers, and scratchers.

3.19. (a) What effect would the use of water ahead of the cement have on displacement efficiency?
(b) If you were running a 10,000-ft string of $5\frac{1}{2}$-in. casing and the casing started to take weight (stick) 2000 ft off bottom, what would you do?

3.20. A 9000-ft string of 7-in., 28-lb/ft casing is to be run in a $9\frac{3}{4}$-in. hole with the float collar placed 60 ft above the shoe, and 750 sacks of cement are to be mixed at a slurry weight of 15.5 lb per gal, with a yield of 1.2 cu ft per sack. The service company truck has a mixing capacity of 30 sacks per min. The rig pump has an 18-in. stroke with 6.50-in. liners, and operates at 60 rpm and 90 per cent efficiency.
(a) What will be the mixing time? *Ans.* 25 min.
(b) What is the displacement time? *Ans.* 27.2 min.
(c) What is the total cementing time? *Ans.* 52.2 min.
(d) What is the API minimum thickening time at the lower depth limit for each API class of cement as measured with a Pan American thickening-time tester? *Ans.* 106, 106, 106, 106, 144, 154, and 180 min.
(e) Does the Class E cement (Table 3.6) meet the above minimum thickening time on Schedule No. 8?

3.21. (a) For the conditions given in Problem 3.20, calculate the volume of cement left in the casing. *Ans.* 11.5 sacks.

(b) Assuming the hole is to gage, calculate the fill-up in the annulus. *Ans.* 3233 ft.

(c) Using Eq. (3.6), calculate the static bottom-hole temperature. *Ans.* 215°F.

(d) Using Fig. 3.3, find the circulating bottom-hole temperature, mud discharge temperature, and mud suction temperature when drilling was stopped at 9000 ft. *Ans.* 138°F, 122.5°F, 116°F.

(e) Calculate the temperature at the upper placement limit. *Ans.* 166.5°F.

(f) From Fig. 3.6, if a Class E cement is used, how many hours will it take to attain a minimum compressive strength of 500 psi at the upper placement limit? *Ans.* 11 hr.

(g) What is the approximate time after cementing to run a temperature survey? To perforate?

3.22. (a) Cement has a bulk volume of 1.00 cu ft per sack and a specific gravity of 3.14. What is its absolute volume? *Ans.* 0.480 cu ft.

(b) If the water-cement ratio is 5.01 gal per sack, what is the slurry weight? *Ans.* 15.8 lb/gal.

(c) Calculate the yield and per cent mix. *Ans.* 1.15 cu ft/sack, 44.4 per cent.

3.23. (a) The *maximum* water for a 4 per cent gel cement is 7.73 gal per sack. If the specific gravity of the cement is 3.14 and that of the bentonite is 2.65, what will be the slurry weight? *Ans.* 14.1 lb/gal.

(b) What is the yield? *Ans.* 1.54 cu ft/sack.

3.24. A string of $10\frac{3}{4}$-in., 40.5-lb/ft surface casing is to be set at a depth of 2000 ft in a 15-in. hole with the float collar placed 25 ft above the shoe. An 8 per cent gel cement is to be used, mixed at a slurry weight of 13.3 lb per gal using 9.8 gal of water per sack (yield 1.836 cu ft/sack). Experience in the area has shown that an excess of 40 per cent should be used.

(a) Calculate the number of sacks of Class A cement and bentonite required to get cement returns. *Ans.* 918 sacks cement, 69 sacks gel.

(b) From Table 3.17, the thickening time is in excess of how many hours? *Ans.* 3 hr.

(c) From Fig. 3.7, in how many hours will the set cement attain a compressive strength of 500 psi? *Ans.* 30 hr.

(d) Had a neat cement slurry containing 4 per cent $CaCl_2$ been used, what would have been the thickening time? *Ans.* 2 hr 38 min.

(e) At 80°F, in how many hours would the cement of part (d) have a compressive strength of 360 psi? *Ans.* 6 hr.

3.25. A well is to be cemented with a 50-50 pozzolan cement containing 2 per cent bentonite by weight of the mixture and mixed with 5.75 gal of water per sack. The specific gravity of the cement is 3.14; of pozzolan, 2.46; and of bentonite, 2.65.

(a) Calculate the pounds of each component per sack of the mixture. *Ans.* 47 lb cement, 37 lb pozzolan, 1.68 lb bentonite, and 47.96 lb water.

(b) Calculate the slurry density. *Ans.* 14.19 lb/gal.

(c) What is the slurry volume? *Ans.* 1.26 cu ft/sack.

(d) From Table 3.21, what will be the 24-hr compressive strength of this mixture at 120°F? *Ans.* 815 psi.

3.26. Gun-barrel salt water on an oil lease weighs 8.60 lb per gal. It is desired to increase the weight to 10 lb per gal. Calculate the pounds of sodium chloride required per barrel of original salt water. *Ans.* 106.2 lb/bbl.

3.27. A string of $5\frac{1}{2}$-in., 23-lb/ft casing is to be set in an $8\frac{3}{4}$-in. hole at a depth of 10,000 ft with the float collar 60 ft above the shoe. The caliper survey measurements are as follows:

Well depth, ft	*Hole volume, cu ft*	*Average diameter, in.*
9000–10,000	421	$8\frac{3}{4}$
8000–9000	434	$8\frac{7}{8}$
7000–8000	477	$9\frac{3}{8}$
6000–7000	499	$9\frac{1}{2}$
5000–6000	505	$9\frac{5}{8}$
4000–5000	530	$9\frac{7}{8}$

The experience factor for calipered holes in this field is 1.10.

(a) Calculate the theoretical annular volume (in cubic feet) for each 1000-ft interval below 4000 ft. *Ans.* 253 cu ft from 9000 to 10,000 ft.

(b) For each 1000-ft interval, calculate the corrected annular volume. *Ans.* 278 cu ft from 9000 to 10,000 ft.

(c) The well is to be cemented with 700 sacks of Class E cement yielding 1.10 cu ft of slurry per sack. Estimate the depth to the top of the cement, using caliper data. *Ans.* 7441 ft.

(d) If caliper survey data are not available and the experience factor for the field is 1.50, estimate the depth to the top of the cement. *Ans.* 7989 ft.

3.28. (a) A 14,000-ft well is drilled with 18 lb per gal mud. The casing is to be cemented with a barite-cement slurry weighing 18.02 lb per gal. If 55 lb of barite and 5.8 gal of water are used per sack of cement, how many sacks of cement and barite are required to mix 1000 cu ft of slurry? *Ans.* 683 sacks of cement, 376 sacks of barite.

(b) What is the approximate thickening time of this slurry? *Ans.* 2.85 hr.

3.29. (a) If the bottom-hole pressure chart of Fig. 3.23 was taken at 10,000 ft, what was the initial static mud pressure? *Ans.* 6126 psi.

(b) What was the initial formation strength? *Ans.* 7980 psi.

(c) What was the maximum surge pressure? *Ans.* 7910 psi.

(d) Why was the bottom-hole pressure less during mixing than during normal circulation?

(e) When the job was complete what was the formation strength? *Ans.* 8778 psi.

3.30. In the well of Fig. 3.24, if the pressures are taken at 10,000 ft, how much greater is the maximum surge pressure than the initial formation strength? *Ans.* 694 psi.

3.31. A well is completed with $5\frac{1}{2}$-in., 20-lb/ft casing at a depth of 8612 ft. It is desired to squeeze perforations from 8555 to 8565 ft with 100 sacks of Class E cement mixed to a slurry weight of 16.3 lb per gal (1.10 cu ft per sack). A retrievable retainer is to be set at 8540 ft with $2\frac{1}{2}$-in. tubing, and 5 bbl of water (spacer) are to be placed ahead of and 5 bbl behind the neat cement slurry.

(a) Calculate the total displacement, neglecting the pump and discharge line. *Ans.* 49.7 bbl.

(b) What is the displacement by the cementing truck when the circulation valve is closed? *Ans.* 22.1 bbl.

(c) What is the displacement to wash (clear) the perforations if the desired final squeeze pressure is not attained? *Ans.* 49.7 bbl.

(d) What would have been the effect if an extra 30 sacks of cement had been mixed by mistake?

(e) Would the reversing pressure have been higher or lower if 10 bbl of water had been used as spacer behind the cement?

3.32. A 5-in. OD, 18-lb/ft liner is to be run in a $6\frac{1}{4}$-in. hole on $2\frac{7}{8}$-in., 10.40-lb/ft drill pipe to a depth of 12,000 ft. The length of the drill pipe and setting tool to the liner-wiper plug is 9820 ft. The length of the liner, including hanger and float shoe, is 2200 ft, i.e., the top of the liner is at 9800 ft. The liner landing collar is at 11,940 ft. Experience has shown that a 50 per cent excess of cement should be used.

(a) How many sacks of Class E cement (1.10 cu ft per sack) are required? *Ans.* 230 sacks.

(b) What is the displacement to the liner plug? *Ans.* 44.2 bbl.

(c) Calculate the displacement from the liner plug to the landing collar. *Ans.* 37.7 bbl.

3.33. An $8\frac{3}{4}$-in. open hole is to be plugged back from 9000 to 8800 ft through $4\frac{1}{2}$-in. 16.6-lb/ft drill pipe, with an Ottawa sand–cement slurry weighing 18 lb per gal (Table 3.32) by use of the balanced plug technique. One barrel of water spacer will be used behind the cement.

(a) Using a 10 per cent excess of slurry, how many sacks of Class E cement and how many pounds of Ottawa sand are required? *Ans.* 59 sacks, 4661 lb sand.

(b) How many barrels of mud will be displaced before the drill pipe pulls dry? *Ans.* 123 bbl.

3.34. Plug jobs are often done in wells which are not full to the surface. The fluid level may be low because the well has been bailed down or because of low bottom-hole pressure. If the fluid in the well of Example 3.13 stands 1000 ft from the surface, what will be the amount of mud displaced to balance the columns? *Ans.* 63 bbl.

3.35. A string of 7-in. OD, 35-lb/ft casing is run in a $9\frac{7}{8}$-in. hole. The 16-in. mud pump has 6.50-in. liners, and operates at 60 rpm and 90 per cent efficiency with a flooded suction. The Class A neat cement slurry weighs 15.4 lb per gal, and has a plastic viscosity of 50 cp and a yield point of 100 lb per 100 sq ft.

(a) What is the flow velocity in the casing? *Ans.* 5.21 ft/sec.

(b) Is the cement flow in the casing laminar or turbulent?
(c) What is the velocity in the annulus? *Ans.* 3.87 ft/sec.
(d) Is the cement flow in the annulus in the laminar region?
(e) If 550 cu ft of slurry are being displaced in the casing, what is the frictional pressure loss due to the cement? *Ans.* 221 psi.
(f) What will be the frictional loss in the annulus due to the cement when 50 cu ft of slurry are left in the casing? *Ans.* 373 psi.

3.36. Continuing with Problem 3.33, the Ottawa sand–cement slurry is displaced with the service company truck at a rate of 30 cu ft per min.
(a) What is the velocity in the annulus? *Ans.* 1.6 ft/sec.
(b) Is this in the laminar region?
(c) From Fig. 3.22, would you expect to find a hard plug?

3.37. A 20-in. OD gas pipe line is to be run through a marsh. The wall thickness is $\frac{1}{2}$ in., the pipe weight 104.13 lb per ft, and the pipe is to be wrapped with $\frac{1}{8}$ in. of coal tar and fiber glass plus $2\frac{3}{4}$ in. of concrete. The weight of the coal tar and fiber glass coating is 4.66 lb per ft on 20-in. pipe and the concrete specifications call for a mix weighing 140 lb per cu ft.
(a) What is the weight of the concrete per linear foot? *Ans.* 193.2 lb/ft.
(b) What is the total weight per foot of pipe? *Ans.* 301.99 lb/ft.
(c) Calculate the weight of fresh water displaced per linear foot. *Ans.* 225.67 lb/ft.
(d) What is the specific gravity of the pipe? *Ans.* 1.34.
(e) What is the buoyancy? Is it positive or negative? *Ans.* −76.32 lb/ft.

REFERENCES

[1]Joseph Aspdin, "An Improvement in the Modes of Producing Artificial Stone," British Patent 5022 (1824).

[2]N. C. Ludwig, "Portland Cements and Their Application in the Oil Industry," *Drilling and Production Practice* (API, 1953), pp. 153–209.

[3]Robert F. Blanks and Henry B. Kennedy, *The Technology of Cement and Concrete* (New York: John Wiley & Sons, Inc., 1955), pp. 1–64.

[4]R. H. Bogue, "Calculation of the Compounds in Portland Cement," *Industrial and Engineering Chemistry,* **1,** No. 4, 192–196 (1929).

[5]*API Specifications for Oil-Well Cements and Cement Additives,* API STD 10A (Dallas: American Petroleum Institute, 1960), pp. 4–7.

[6]Greg Carter and Dwight K. Smith, "Properties of Cementing Compositions at Elevated Temperatures and Pressures," *Trans. AIME,* **213,** 20–27 (1958).

[7]*ASTM Standards on Cement, Specification for Portland Cement,* C 150-56 (Philadelphia: American Society for Testing Materials, 1958), p. 2.

[8]W. W. Robinson, "Cement for Oil Wells: Status of Testing Methods and Summary of Properties," *Drilling and Production Practice* (API, 1938), pp. 567–588.

[9]Fred M. Ernsberger and Wesley G. France, "Portland Cement Dispersion by Adsorption of Calcium Lignosulfonate," *Industrial and Engineering Chemistry*, **37**, 598 (June, 1945).

[10]N. C. Ludwig, "Chemistry of Portland Cement Used in Cil Wells," *Oil-Well Cementing Practices In The United States* (New York: American Petroleum Institute, 1959), pp. 27–33.

[11]*API Recommended Practice for Testing Oil-Well Cements and Cement Additives*, API RP 10B (Dallas: API Division of Production, 1960), pp. 2–23.

[12]D. E. Silcox and R. B. Rule, "Cement for Oil Wells," *Oil Weekly*, **78**, No. 7, 21–22 24, 26 (1935).

[13]Eugene L. Davis, "Specifications for Oil-Well Cements," *Drilling and Production Practice* (API, 1938), p. 372.

[14]R. Floyd Farris, "A Practical Evaluation of Cements for Oil Wells," *Drilling and Production Practice* (API, 1941), p. 283.

[15]Data supplied by the API Special Subcommittee On Oil Well Cement. Also see Reference 43.

[16]*Howco Cementing Tables*, Sec. 230 (Duncan: Halliburton Oil Well Cementing Company, 1958), p. 42–43.

[17]*Trinity Cement Handbook* (Dallas: Trinity Portland Cement Division, General Portland Cement Company, 1958), p. 18.

[18]J. O. Scott, "The Why and How of Fracturing," *The Oil and Gas Journal*, **55**, No. 2, 91–107 (1957).

[19]B. C. Craft, "Tensile and Compressive Strength of Formations," *The Oil and Gas Journal*, **55**, No. 26, 138 (1957).

[20]R. Floyd Farris, "Method for Determining Minimum Waiting-on-Cement Time," *Trans. AIME*, **165**, 175–188 (1946).

[21]Roscoe C. Clark, Jr., "Requirements of Casing Cement for Segregating Fluid Bearing Formations," *The Oil and Gas Journal*, **51**, No. 50, 173 (1953).

[22]F. A. Anderson, "A Study of Surface Casing and Open-Hole Plug-Back Cementing Practices in the Mid-Continent District," *Drilling and Production Practice* (API, 1955), pp. 312–322.

[23]George E. Cannon, "Improvements in Cementing Practices and the Need for Uniform Cementing Regulations," *Drilling and Production Practice* (API, 1948), p. 126.

[24]J. K. Godbey and H. D. Hodges, "Pressure Measurements During Formation Fracturing Operations," *Trans. AIME*, **213**, 65 (1958).

[25]Roscoe C. Clark, Jr., of The Western Company, personal communication.

[26]*ASTM Standards on Cement,* C 115-58 (Philadelphia: American Society for Testing Materials, 1958), p. 147.

[27]*Bentonite-Cement Information* (Duncan: Halliburton Oil Well Cementing Company, 1952), pp. 13, 33.

[28]Svend Rordam, "Sulfate-Resisting Cement," *Trans. ASME,* **60,** 233 (1938).

[29]*Report of Cooperative Tests on Sulfate Resistance of Cements and Additives,* API Mid-Continent District Study Committee on Cementing Practices and Testing of Oil-Well Cements (API 1955), pp. 34–52.

[30]William G. Bearden, "Effect of Temperature and Pressure on the Physical Properties of Cement," *Oil-Well Cementing Practices in the United States* (New York: American Petroleum Institute, 1959), p. 56.

[31]Orie O. Dale, "The Effects of Some Additives on the Physical Properties of Portland Cement," *Oil-Well Cementing Practices in the United States* (New York: American Petroleum Institute, 1959), p. 62.

[32]Wm. D. Owsley, "Surface Cementing Equipment and Supplies," *Oil-Well Cementing Practices in the United States* (New York: American Petroleum Institute, 1959), p. 98.

[33]I. R. Dunlap and F. D. Patchen, "A High-Temperature Oil Well Cement," *The Petroleum Engineer,* **29,** No. 12, B-60 (1957).

[34]B. E. Morgan and G. K. Dumbauld, "A Modified Low-Strength Cement," *Trans. AIME,* **192,** 165 (1951).

[35]B. E. Morgan and G. K. Dumbauld, "Recent Developments in the Use of Bentonite Cements," *Drilling and Production Practice* (API, 1953), p. 163.

[36]*Bentonite-Cement Information* (Duncan: Halliburton Oil Well Cementing Company, 1952), p. 19.

[37]S. H. Davis and J. H. Faulk, "Have Waiting-on-Cement Practices Kept Pace With Technology?" *Drilling and Production Practice* (API, 1957), p. 180.

[38]C. D. Saunders and F. W. Nussbaumer, "Trends in the Use of Low-Weight Cement Slurries," *Drilling and Production Practice* (API, 1952), p. 189.

[39]George C. Howard and P. P. Scott, Jr., "An Analysis and Control of Lost Circulation," *Trans. AIME,* **192,** 178 (1951).

[40]B. E. Morgan, discussion of paper by C. D. Saunders and F. W. Nussbaumer, "Trends in the Use of Low-Weight Cement Slurries," *Drilling and Production Practice* (API, 1952), pp. 198–199.

[41]George B. Mangold, "Minimizing Oil Well Cementing Failures," *World Oil,* **134,** No. 6, 112 (1952).

[42]William P. Sterne, "Pozzolans for Oil-Well Cements," *The Oil and Gas Journal,* **51,** No. 9, 72 (1952).

[43]*Pozmix Cement and Pozmix 140* (Duncan: Halliburton Oil Well Cementing Company, 1957), p. 12.

[44]W. E. Bergman, J. R. Hurley, and F. J. Shell, "Low-Water Loss, Low-Density Cement," *The Oil and Gas Journal*, **54**, No. 19, 107 (September 12, 1955).

[45]F. J. Shell, J. R. Hurley, W. E. Bergman, and H. B. Fisher, "Low Density Oil Well Cements," *World Oil*, **143**, No. 4, 131–134 (September, 1956).

[46]F. J. Shell, "DE — An Aid to Cementing Casing Through Salt Zones," *The Petroleum Engineer*, **29**, No. 3, 100–102 (1957).

[47]*Technical Data on Diacel Cement Systems* (Bartlesville, Oklahoma: Drilling Specialities Company, 1959), Bull. D-18, D-19, D-21, D-22.

[48]Knox A. Slagle and L. Gregory Carter, "Gilsonite — A Unique Additive for Oil-Well Cements," Paper No. 926-4-D, *Division of Production*, API (1959), p. 4.

[49]*Howco Cementing Tables*, Sec. 230 (Duncan: Halliburton Oil Well Cementing Company, 1959), pp. 2–29.

[50]B. C. Craft and M. F. Hawkins, *Applied Petroleum Reservoir Engineering* (Englewood Cliffs, New Jersey: Prentice-Hall, Inc., 1959), p. 319.

[51]Norbert Adolph Lange, *Handbook of Chemistry*, 6th Ed. (Sandusky: Handbook Publishers, Inc., 1946), p. 1301.

[52]"Casing Leak is Patched With a Special Cement," *Drilling*, **18**, No. 2, 95 (December 20, 1956).

[53]J. T. Rollins and R. D. Davidson, "New Latex Cement Solves Special Well Problems," *The Petroleum Engineer*, **29**, No. 2, 13–48 (1957).

[54]Wayne F. Hower and Phil C. Montgomery, "New Squeeze Cementing Mixture," *The Oil and Gas Journal*, **52**, No. 24, 136 (October 19, 1953).

[55]C. M. Stout and W. W. Wahl, "A New Organic Fluid-Loss Control Additive For Oilwell Cements," *Journal of Petroleum Technology*, AIME, **12**, No. 9, 20 (1960).

[56]Ed McGhee, "New Cement Additive Cuts Water Loss," *The Oil and Gas Journal*, **58**, No. 31, 119 (August 1, 1960).

[57]*A New Report on Lost Circulation Materials* (Duncan: Halliburton Oil Well Cementing Company, 1958), p. 2.

[58]Gilbert L. Leach, "Loss of Cement Slurry Combatted by Use of Cellulose Filler," *The Petroleum Engineer*, **12**, No. 3, 94 (December, 1940).

[59]P. H. Jones and Denis Berdine, "Oil-Well Cementing," *Drilling and Production Practice* (API, 1940), p. 45.

[60]George C. Howard and J. B. Clark, "Factors To Be Considered in Obtaining Proper Cementing of Casing," *Drilling and Production Practice* (API, 1948), p. 257.

[61]W. D. Owsley, "Improved Casing Cementing Practices in the United States," *The Oil and Gas Journal*, **48**, No. 32, 76 (December 15, 1949).

[62]W. E. Helmick and A. J. Longley, "Pressure-Differential Sticking of Drill Pipe and How It Can Be Avoided or Relieved," *Drilling and Production Practice* (API, 1957), p. 55.

[63]A. G. Hilton, "Mechanical Aids and Practices for Improvement of Primary Cementing," *Oil-Well Cementing Practices in the United States* (New York: American Petroleum Institute, 1959), p. 123.

[64]A. J. Teplitz and W. E. Hassebrock, "An Investigation of Oil Well Cementing," *Drilling and Production Practice* (API, 1946), p. 76.

[65]C. B. Corley, Jr., and J. L. Rike, "Tubingless Completions," Paper 926-4-G, *Division of Production*, API (1959), pp. 1–6.

[66]*Engineering Data* (Pittsburgh: Spang-Chalfant Division of the National Supply Company, 1956), Sec. 10, pp. 231–235.

[67]George E. Cannon, "Changes in Hydrostatic Pressure Due to Withdrawing Drill Pipe from the Hole," *Drilling and Production Practice* (API, 1934), p. 42.

[68]W. C. Goins, Jr., J. P. Weichent, J. L. Burba, Jr., D. D. Dawson, Jr., and A. J. Teplitz, "Down-the-Hole Pressure Surges and Their Effect on Loss of Circulation," *Drilling and Production Practice* (API, 1951), p. 125.

[69]W. T. Cardwell, Jr., "Pressure Changes in Drilling Wells Caused by Pipe Movement," *Drilling and Production Practice* (API, 1953), p. 97.

[70]George S. Ormsby, "Calculation and Control of Mud Pressures in Drilling and Completion Operations," *Drilling and Production Practice* (API, 1954), p. 44.

[71]E. H. Clark, Jr., "A Graphic View of Pressure Surges and Lost Circulation," *Drilling and Production Practice* (API, 1956), pp. 424–438.

[72]G. W. Binkley, G. K. Dumbauld, and R. E. Collins, "Factors Affecting the Rate of Deposition of Cement in Unfractured Perforations," *Trans. AIME*, **213**, 51 (1958).

[73]L. W. Folmar, "Methods of Detecting Top of Cement Behind Casing," *Oil-Well Cementing Practices in the United States* (New York: American Petroleum Institute, 1959), pp. 134–135.

[74]*Cementing Manual* (Long Beach, Calif.: BJ Service, Inc., 1955), p. 29.

[75]Paul D. Torrey, "Selective Exclusion of Fluids from Wells," *Drilling and Production Practice* (API, 1939), p. 205.

[76]Glen M. Stearns, "Theory of Squeeze Cementing," *The Oil and Gas Journal*, **14**, No. 14, 99 (August 12, 1944).

[77]A. W. Walker, "Squeeze Cementing," *World Oil*, **129**, No. 6, 87 (September, 1949).

[78]George C. Howard and C. P. Fast, "Squeeze Cementing," *Trans. AIME*, **189,** 53 (1950).

[79]B. E. Morgan and G. K. Dumbauld, "Measurement of Permeability of Set Cement," *Trans. AIME*, **195,** 323 (1952).

[80]T. A. Huber, G. H. Tausch, and J. R. Dublin, III, "A Simplified Cementing Technique for Recompletion Operations," *Trans. AIME*, **201,** 1 (1954).

[81]G. H. Tausch, "Squeeze Cementing with Permanent-Type Completions," *Oil-Well Cementing Practices in the United States* (New York: American Petroleum Institute, 1959), p. 162.

[82]Eugene Harrison, W. F. Kieschnick, Jr., and W. J. McGuire, "The Mechanics of Fracture Induction and Extension," *Trans. AIME*, **201,** 252 (1954).

[83]J. W. Hodges, "Squeeze Cementing Methods and Materials," *Oil-Well Cementing Practices in the United States* (New York: American Petroleum Institute, 1959), pp. 149–159.

[84]S. H. Davis, "Cementing Liners," *Oil-Well Cementing Practices in the United States* (New York: American Petroleum Institute, 1959), pp. 187–188.

[85]*Cementing Manual* (Long Beach, Calif.: BJ Service, Inc., 1955), p. 33.

[86]W. T. Doherty, "Oil-Well Cementing in the Gulf Coast Area," *Production Bull.* (API), **212,** Sec. 4, 60 (1933).

[87]W. C. Goins, Jr., "Open-Hole Plugback Operations," *Oil-Well Cementing Practices in the United States* (New York: American Petroleum Institute, 1959), pp. 193–197.

[88]Eugene C. Bingham, *Fluidity and Plasticity* (New York: McGraw-Hill Book Company, Inc., 1922), pp. 223–225.

[89]F. M. Anderson, "Mixing and Rheological Properties of Cement Slurries," *Standardization Bull. No. 118* (API, 1958), pp. 1137, 1144, 1145.

[90]R. F. Burdyn, of Socony Mobil Oil Company, Inc., personal communication.

[91]William R. Holden and Ernst J. Leidner, "Rheological Properties of Gel Cements," Paper No. 926-5-J, *Division of Production*, API (March 16, 1960), p. 4.

CHAPTER FOUR

Casing Landing

4.1. General Landing Practices

Casing landing practice is concerned with the transferral of casing weight to the well head or casing hanger after cementing. The operator is primarily interested in obtaining maximum protection and performance from API casing in overcoming the various active stresses that are present during completion and future production of the well. These stresses consist primarily of tension, collapse, burst, and longitudinal compression. Casing is designed to withstand three principal mechanical forces, viz., internal pressure, external pressure, and longitudinal or axial tensions.

An API Committee on Casing Landing Practices*,[1] reviewed the records of 3700 wells ranging in depth from 2000 to 14,000 ft. Each of the casing strings in these wells was landed by one of the four methods to be discussed.

1. Landing as cemented. The casing is considered to be in full tension while it is being cemented, its only movement being that necessary to transfer the weight of the casing to the well head or the casing hanger. If the pipe is hanging freely in fluid while being run in the well, it is subjected to *stretch* due to its weight, to the buoyant force on the pipe, and to the radial and tangential stresses due to fluid pressures inside and outside the pipe.

*References throughout the text are given at the end of each chapter.

The stretch (in inches) due to the *weight* of the pipe, e_1, is

$$e_1 = \int_0^L \frac{l\,dW}{EA} = \int_0^L \frac{l\rho_s\,dV}{EA} = \int_0^L \frac{l\rho_s A\,dl}{EA}$$

$$= \frac{\rho_s}{E}\int_0^L l\,dl = \frac{\rho_s}{E} \times \frac{1}{2} L^2 \tag{4.1}$$

where L = total length of the casing string
dW = weight of casing of length dl
E = Young's modulus
A = cross-sectional *wall* area of the casing
ρ_s = density of steel
dV = volume of steel in casing of length dl

Using conventional engineering units for length and density,

$$e_1 = \frac{\rho_s L^2}{2E} \frac{\text{lb/in.}^3 \times \text{ft}^2}{\text{lb/in.}^2} \times \frac{144 \text{ in.}^2}{\text{ft}^2} = \frac{72\rho_s L^2}{E} \text{ in.} \tag{4.2}$$

The axial compression due to the *buoyant* force acting on the cross-sectional wall area A of the casing, e_2, is related to Young's modulus by Hooke's law as follows:

$$E = \frac{\text{Stress}}{\text{Strain}} = \frac{\sigma}{\epsilon} = \frac{F/A}{e/L} = \frac{FL}{eA} \tag{4.3}$$

With F representing the buoyant force, then

$$e_2 = \frac{FL}{EA} = \frac{\rho_f V_f L}{EA} = \frac{\rho_f L}{EA} \times LA \tag{4.4}$$

where ρ_f is the density and V_f is the volume of the fluid displaced by the casing. The volume of displaced fluid is also the volume of the casing wall,

$$e_2 = \frac{\rho_f L^2}{E} \frac{\text{lb/in.}^3 \times \text{ft}^2}{\text{lb/in.}^2} \times \frac{144 \text{ in.}^2}{\text{ft}^2}$$

$$= \frac{144\rho_f L^2}{E} \text{ in.} \tag{4.5}$$

The stretch due to radial and tangential stresses, e_3, is derived in the following manner. The radial stress, σ_r, at a distance r from the center of a thick-walled cylinder[2] (due to pressure acting on the walls) is given by *Lamé's equation*:

$$\sigma_r = \frac{d_i^2 p_i - d_e^2 p_e}{(d_e^2 - d_i^2)} - \frac{d_i^2 d_e^2 (p_i - p_e)}{(d_e^2 - d_i^2) r^2} \tag{4.6}$$

where d_i and d_e are inside and outside diameters, respectively, p_i and p_e are internal and external pressures, respectively, and where positive values of σ_r indicate tension (negative values indicate compression).

Since the internal and external pressures are equal, $p_i = p_e$, the second term becomes zero. Substituting p for p_i and p_e in the first term and clearing;

$$\sigma_r = \frac{p(d_i^2 - d_e^2)}{(d_e^2 - d_i^2)} = -p \tag{4.7}$$

The general expression for the tangential stress,[2] σ_t, at distance r is

$$\sigma_t = \frac{d_i^2 p_i - d_e^2 p_e}{(d_e^2 - d_i^2)} + \frac{d_i^2 d_e^2 (p_i - p_e)}{(d_e^2 - d_i^2) r^2}$$

$$= p\frac{(d_i^2 - d_e^2)}{(d_e^2 - d_i^2)} = -p \tag{4.8}$$

where, again, a negative value indicates compression.

The general expression for axial *strain*, ϵ_z, resulting from radial, tangential, and longitudinal stresses σ_r, σ_t, and σ_z is

$$\epsilon_z = \frac{1}{E}[\sigma_z - \mu(\sigma_r + \sigma_t)] \tag{4.9}$$

where μ is Poisson's ratio. Making use of Eqs. (4.7) and (4.8),

$$\epsilon_z = \frac{1}{E}[\sigma_z - \mu(-2p)] = \frac{1}{E}[\sigma_z + 2\mu p] \tag{4.10}$$

Clearing,

$$\epsilon_z = \frac{\sigma_z}{E} + \frac{2\mu p}{E} = \frac{\sigma_z}{E} + \frac{2\mu}{E} \times 0.5\, \rho_f L$$

$$= \frac{\sigma_z}{E} + \frac{\mu \rho_f L}{E} \tag{4.11}$$

where p is considered to be the average pressure acting over length L. The axial elongation, e_z, is related to the axial strain, ϵ_z, by

$$e_z = \epsilon_z L = \frac{\sigma_z L}{E} + \frac{\mu \rho_f L^2}{E} \tag{4.12}$$

From Hooke's law,[2]

$$E = \frac{\sigma_z}{e_1/L} \quad \text{or} \quad e_1 = \frac{\sigma_z L}{E} \tag{4.13}$$

Then,

$$e_z = e_1 + \frac{\mu \rho_f L^2}{E} \tag{4.14}$$

If buoyancy is neglected, $e_z = e_1 + e_3$, and

$$e_3 = \frac{\mu \rho_f L^2}{E} \frac{\text{lb/in.}^3 \times \text{ft}^2}{\text{lb/in.}^2} \times 144 \frac{\text{in.}^2}{\text{ft}^2}$$

$$= \frac{144 \mu \rho_f L^2}{E} \text{ in.} \tag{4.15}$$

When buoyancy is considered, the total axial elongation is

$$e_z = e_1 - e_2 + e_3 \tag{4.16}$$

Substitution leads to

$$e_z = \frac{72\rho_s L^2}{E} - \frac{144\rho_f L^2}{E} + \frac{144\mu\rho_f L^2}{E}$$

$$= \frac{72L^2}{E}[\rho_s - 2\rho_f + 2\mu\rho_f]$$

$$= \frac{72L^2}{E}[\rho_s - 2\rho_f(1 - \mu)] \text{ in.} \tag{4.17}$$

Equation (4.17) can also be derived directly from Eq. (4.9). From Eqs. (4.7) and (4.8),

$$\sigma_r = \sigma_t = -p = -\rho_f\ (\mathrm{L}/2) \tag{4.18}$$

Assuming the axial stress to be a result of the weight of the casing in fluid (i.e., the casing weight in air minus the buoyant force) the *average* axial stress over length L is

$$\sigma_z = \frac{W/2}{A} - \frac{\rho_f V}{A} = \frac{\rho_s LA}{2A} - \frac{\rho_f LA}{A} = L\left(\frac{\rho_s}{2} - \rho_f\right) \tag{4.19}$$

where W is the total weight of the casing string. Substituting Eqs. (4.18) and (4.19) in Eq. (4.9),

$$\epsilon_z = \frac{1}{E}\left[L\left(\frac{\rho_s}{2} - \rho_f\right) + 2\mu\rho_f\left(\frac{L}{2}\right)\right] = \frac{L}{E}\left[\frac{\rho_s}{2} - \rho_f + \mu\rho_f\right] \tag{4.20}$$

Substituting Eq. (4.20) in Eq. (4.12),

$$e_z = \epsilon_z L = \frac{L^2}{E}\left[\frac{\rho_s}{2} - \rho_f(1 - \mu)\right]$$

$$= \frac{L^2}{2E}[\rho_s - 2\rho_f(1 - \mu)] \tag{4.21}$$

Stating Eq. (4.21) in conventional engineering units,

$$e_z = \frac{L^2\,\mathrm{ft}^2}{2E\ \mathrm{lb/in.}^2}[\rho_s - 2\rho_f(1 - \mu)]\frac{\mathrm{lb}}{\mathrm{in.}^3} \times 144\frac{\mathrm{in.}^2}{\mathrm{ft}^2}$$

$$= \frac{72L^2}{E}[\rho_s - 2\rho_f(1 - \mu)] \text{ in.} \tag{4.22}$$

2. Landing in tension at the freeze point. The freeze point for casing is usually considered to be at the top of the set cement. Where it is indicated to be up the hole, the point is established by stretch curves, by Eq. (4.23) (derived by Hayward[3]), or by other techniques:

$$L = \frac{Eew}{40.8F_d} \tag{4.23}$$

where L = length of free pipe (i.e., length of pipe above the freeze point) in feet

E = Young's modulus = 30×10^6 psi for steel

e = stretch of pipe (in inches) resulting from a differential pull of F_d pounds

w = weight of the pipe in pounds per foot

Example 4.1. Calculate the length of free pipe in a frozen string of casing, given a 9000-ft string of 20-lb casing frozen in the hole and assumed to be stuck near bottom.

SOLUTION: It is found that the minimum pull required to remove all the slack is

$$20 \times 9000 = 180{,}000 \text{ lb}$$

The casing is stretched and marked at two pull values not exceeding the strength of the casing. For this example these pull values are 200,000 and 240,000 lb and therefore the differential pull is 40,000 lb. The differential elongation corresponding to this pull is 12 in. Substituting in Eq. (4.23),

$$L = \frac{30 \times 10^6 \times 12 \times 20}{40.8 \times 40{,}000} = 4412 \text{ ft of free pipe}$$

3. Unstressed landing at the freeze point. In this practice, sufficient weight is slacked off so that the casing is in neither tension nor compression at the freeze point.

4. Landing in compression at the freeze point. In this landing method the weight slacked off ranges from 25 to 80 per cent, with an average of about one third of the total weight of the casing in mud.

In completing a well, many operations are necessary which change the magnitudes of the acting stresses to lower or higher values. According to Peret,[4] each condition or operation considered singly affects the acting stresses as shown in Table 4.1. Assuming that the casing is landed in *tension* at the freeze point, and that conditions 1, 3, 6, 7, and 10 in Table 4.1 will occur in subsequent operations, it is entirely possible that the casing will be undesirably overstressed in axial tension. The increase in axial tension will reduce resistance to collapse, requiring heavier weights and higher grades of casing.

Assuming the casing to be *neutral* or in *compression* at the freeze point, conditions 2, 3, 6, 7, and 10 in Table 4.1 may occur and the tendency to buckle is materially increased. Buckling is to be avoided since it leads to increased casing wear during well deepening, failure by thread breakage (especially in enlarged sections), and restriction of tools. The API Study Committee[1] concluded its work with the recommendation that casing be landed, *as cemented*, at the casing head exactly where it was hanging on the hook when the cement plug hit bottom. This recommendation applies to all wells where mud weights do not exceed 12.5 lb per gal, where standard

Table 4.1. EFFECTS OF COMPLETION OPERATIONS AND CHANGES IN CONDITIONS ON ACTING STRESSES[a]
(*After Peret*[4])

	Tension loading	Collapse loading	Burst loading	Buckling tendency
1. Drop in average temperature	+			−
2. Rise in average temperature	−			+
3. Rise of internal pressure	+		+	+
4. Drop of internal pressure	−		−	−
5. Rise of external pressure	−	+		−
6. Drop of external pressure	+	−		−
7. Replacement of internal fluid by a heavier fluid	+		+	−
8. Replacement of internal fluid by a lighter fluid	−		−	−
9. Replacement of external fluid by a heavier fluid	−	+		−
10. Replacement of external fluid by a lighter fluid	+	−		+
11. Swab down	−		−	−

[a] +, increase; −, decrease.

design factors are used in tension and compression, where wellhead equipment is available to support the wellhead load without damage to the casing, and where the joint strength in compression of the top section of the surface pipe is sufficient to withstand the loads imposed by the weight of the casing plus the weight of the tubing plus induced loads brought about by future operations. In wells where the design factors are low, where extreme pressures are encountered, or where excessive mud weights are necessary, the API Committee recommended that the landing practice be based on theoretical considerations developed by Lubinski.[5]

4.2. Buckling Criteria

Buckling occurs when a structural column is subjected to a load of sufficient magnitude to cause deflection of the column from its vertical alignment. Depending on the intensity of the load, the column may or may not return to its original alignment when the load is removed. For example, one type of buckling occurs when drill pipe and tubing are racked in the derrick. Here there is a deflection, but, since failure has not occurred, the pipe will return to its original alignment when picked up in the derrick. Similarly, casing which has buckled down the hole will return to its original alignment after buckling forces are removed if yield stress is not exceeded. On the other hand, if there is a large cavity opposite the buckled section, the casing will bow into the enlargement and may be permanently distorted. The

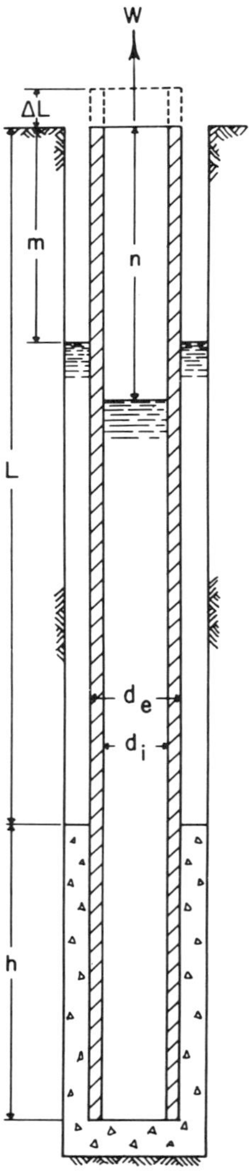

W = wellhead load, lb.
ΔS = pickup when hanging casing, lb ($-\Delta S$ indicates slack-off).
ΔL = pickup when hanging casing, in. ($-\Delta L$ indicates slack-off).
p_e = surface pressure outside casing when cement sets, psi.
Δp_e = change in surface pressure outside casing after cement sets, psi.
p_i = surface pressure inside casing when cement sets, psi.
Δp_i = change in surface pressure inside casing after cement sets, psi.
m = fluid-level drop outside casing after cement sets, ft.
n = fluid-level drop inside casing after cement sets, ft.
Δt = average change in casing temperature above cement top after cement sets, deg F.
ρ_e = fluid weight outside casing when cement sets, lb per gal.
$\Delta \rho_e$ = change in fluid weight outside casing after cement sets, lb per gal.
ρ_i = fluid weight inside casing when cement sets, lb per gal.
$\Delta \rho_i$ = change in fluid weight inside casing after cement sets, lb per gal.
d_e = outside diameter of casing, in.
d_i = inside diameter of casing, in.
w = weight of casing, lb per ft.
ρ_c = cement slurry weight, lb per gal.
L = distance from cement top to casing hanger, ft.
h = distance from casing shoe to cement top, ft.

Fig. 4.1 Parameters of the buckling-criteria and wellhead-load equations. (Cox,[8] *Drilling and Production Practice*, API.)

buckling of drill pipe, casing, and tubing has received increased attention as a result of several studies on the subject.[5–10]

The buckling tendency of casing can be expressed as the sum of seven terms, each term representing effects on buckling of down-the-hole variables discussed in detail below. The derivations which follow are based essentially on concepts developed by Cox,[8] who correlated and expanded the previous investigations of Lubinski[5] and Bouman.[7] The buckling equations are derived for single-weight casing strings. A study of equations for combination-weight strings[8] indicates that no significant error is introduced when the single-weight equations are used for combination-weight strings.

Force due to average temperature change **(B_1).** If, after the cement sets, the average temperature above the cement rises by an amount Δt, in casing assumed free at the cement top, an elongation will occur. The relation between the metal cross-sectional area A and the weight of the casing above the cement top, W, is

$$A = (A_e - A_i) = \frac{V}{L} = \frac{W/\rho_s}{L} = \frac{wL}{\rho_s L}$$

$$= \frac{w \text{ lb/ft}}{490 \text{ lb/ft}^3} \times 144 \text{ in.}^2/\text{ft}^2 = 0.294w \text{ in.}^2 \tag{4.24}$$

Designating the thermal coefficient of linear expansion by λ, the change in length ΔL of casing of length L due to a temperature change Δt is

$$\Delta L = L\lambda \, \Delta t \tag{4.25}$$

Taking the value of λ for steel to be 6.9×10^{-6} per°F,

$$\Delta L/L = \lambda \, \Delta t = 6.9 \times 10^{-6} \, \Delta t \tag{4.26}$$

From Hooke's law,

$$E = \frac{\sigma_z}{\epsilon_z} = \frac{B_1/A}{\Delta L/L} = \frac{B_1}{0.294w \times 6.9 \times 10^{-6} \, \Delta t}$$

$$= 30 \times 10^6 \text{ lb/in.}^2 \tag{4.27}$$

Solving for B_1,

$$B_1 = 0.294w \times 6.9 \times 10^{-6} \, \Delta t \times 30 \times 10^6$$

$$= +60.9w \, \Delta t \tag{4.28}$$

The positive sign in Eq. (4.28) indicates that B_1 is a force in the upward direction, tending to place the casing in compression, and thereby increasing the buckling tendency (Table 4.1). A negative sign would indicate a force which would decrease the buckling tendency. Equation (4.28) can also be written in the form

$$B_1 = A \times 6.9 \times 10^{-6} \, \Delta t \times 30 \times 10^6 = +207A \, \Delta t \tag{4.29}$$

Casing is not a continuous and uniform tube but consists of sections joined by couplings which introduce complicated stress pictures at the couplings and in the vicinity of the couplings. It is reasonable to analyze casing stresses by ignoring the effect of couplings and assuming a uniform and continuous tube. However, if the cross-sectional area is to be expressed in weight per foot of casing body, then it should be noted that the API classifies casing by its *nominal* weight per foot and not by the *true* weight per foot of the pipe body. The average value of the factor for converting nominal weight to plain-end area of casing is 3.464 lb per ft per sq in:

$$A = \frac{w}{3.464} = 0.289w \tag{4.30}$$

Substituting Eq. (4.30) in Eq. (4.29),

$$B_1 = 0.289w \times 6.9 \times 10^{-6}\,\Delta t \times 30 \times 10^6 = +59.8w\,\Delta t \tag{4.31}$$

Force due to weight of casing below cement top (B_2). If h is the distance from the casing shoe to the cement top, and w is the average unit weight of casing in the cemented zone, the force due to the weight of the casing is a downward force given by

$$B_2 = -hw \tag{4.32}$$

Force due to cement slurry weight (B_3). If ρ_c is the density of the cement slurry in pounds per gallon, the pressure due to a column of cement of height h feet is

$$\frac{\rho_c}{8.34} \times 0.434 \text{ psi/ft} \times h \text{ ft} = 0.052\rho_c h \text{ psi} \tag{4.33}$$

and the resulting upward force on the casing is

$$B_3 = 0.052\rho_c h A_e = 0.052 \times 0.785 d_e^2 h \rho_c$$

$$= +\,0.0408 d_e^2 h \rho_c \tag{4.34}$$

where A_e is the external area of the casing $(A_e = 0.785d_e^2)$.

Force due to mud weight inside casing within the cemented section (B_4). The effect of mud inside the casing is to reduce the buckling tendency. If ρ_i is the density of the mud inside the casing, then

$$B_4 = -0.052\rho_i h A_i = -0.052 \times 0.785 d_i^2 \rho_i h$$

$$= -0.0408 d_i^2 \rho_i h \tag{4.35}$$

where A_i is the internal area of the casing $(A_i = 0.785d_i^2)$.

Force due to change in mud weight and drop of fluid level outside the casing (B_5). A buckling tendency will result from tangential and radial stresses caused by a *change* in external fluid weight and fluid-level drop. The elongation which would occur as a result of these stresses if the casing were *not* cemented will be designated ΔL. The buckling force B_5 will therefore be a force equal in magnitude but opposite in direction to the force necessary to cause the hypothetical elongation ΔL. Considering an element of length dz at a vertical distance z from the top of the casing, the elongation would be

$$\mathrm{d}\Delta L = \epsilon_z\,\mathrm{d}z \tag{4.36}$$

where ϵ_z is the strain experienced by element dz. The total elongation is

$$\Delta L = \int_0^L \epsilon_z \, dz \tag{4.37}$$

where L is the total length of the casing string to the top of the cement. Considering the total axial strain resulting from radial, tangential, and axial stresses,

$$\epsilon_z = \frac{1}{E}[\sigma_z - \mu(\sigma_r + \sigma_t)] \tag{4.9}$$

Combining Eqs. (4.37) and (4.9),

$$\Delta L = \int_0^L \frac{1}{E}[\sigma_z - \mu(\sigma_r + \sigma_t)] \, dz = \frac{1}{E}\int_0^L \sigma_z \, dz - \frac{\mu}{E}\int_0^L (\sigma_r + \sigma_t) \, dz \tag{4.38}$$

Let ΔL_z be the component of ΔL resulting from axial stress, and let ΔL_{rt} be the component of ΔL resulting from radial and tangential stresses. Then Eq. (4.38) can be written

$$\Delta L = \Delta L_z + \Delta L_{rt} \tag{4.39}$$

where

$$\Delta L_z = \frac{1}{E}\int_0^L \sigma_z \, dz \tag{4.40}$$

and

$$\Delta L_{rt} = -\frac{\mu}{E}\int_0^L (\sigma_r + \sigma_t) \, dz \tag{4.41}$$

The change in pressure at depth L owing to the change in mud density $\Delta\rho_e$ is $L\Delta\rho_e$. The change in pressure at depth L owing to the drop in mud level m is $m(\rho_e + \Delta\rho_e)$, where ρ_e is the original mud density. The net change in pressure is therefore $L\Delta\rho_e - m(\rho_e + \Delta\rho_e)$, and the net change in force on the bottom of a closed tube of area A_e at depth L is $A_e[L\Delta\rho_e - m(\rho_e + \Delta\rho_e)]$. If the cross-sectional wall area of the tubing is A, the axial stress resulting from the net change in force is

$$\begin{aligned}\sigma_z &= -\frac{A_e[L\Delta\rho_e - m(\rho_e + \Delta\rho_e)]}{A} \\ &= \frac{A_e}{A}(m\rho_e + m\Delta\rho_e - L\Delta\rho_e)\end{aligned} \tag{4.42}$$

Equation (4.42) shows the axial stress σ_z to be independent of z, and therefore Eq. (4.40) can be written

$$\Delta L_z = \frac{\sigma_z}{E}\int_0^L dz = \frac{\sigma_z L}{E} \tag{4.43}$$

Combining Eqs. (4.42) and (4.43),

$$\Delta L_z = \frac{LA_e}{EA}(m\rho_e + m\Delta\rho_e - L\Delta\rho_e) \tag{4.44}$$

Taking Lamé's expressions for radial and tangential stresses as given by Eqs. (4.6) and (4.8),

$$(\sigma_r + \sigma_t)_o = 2\,\frac{d_i^2 p_i - d_e^2 p_e}{d_e^2 - d_i^2} = 2\,\frac{A_i p_i - A_e p_e}{A_e - A_i} \tag{4.45}$$

Taking the cross-sectional wall area A to be $(A_e - A_i)$,

$$(\sigma_r + \sigma_t)_o = \frac{2}{A}\,(A_i p_i - A_e p_e) \tag{4.46}$$

This is an expression for the sum of radial and tangential stresses at any point in the casing wall before the mud weight change and fluid-level drop outside the casing. A comparable expression for the sum of these stresses after the weight change and fluid-level drop is

$$(\sigma_r + \sigma_t)_f = \frac{2}{A}\,[A_i p_i - A_e(p_e + \Delta p_e)] \tag{4.47}$$

where p_i is the internal mud-column pressure, p_e is the original external mud-column pressure, and Δp_e is the change in external mud-column pressure at distance z below the top of the casing. It follows that the change in the sum of radial and tangential stresses at distance z is

$$\sigma_r + \sigma_t = (\sigma_r + \sigma_t)_f - (\sigma_r + \sigma_t)_o = -\,\frac{2A_e \Delta p_e}{A} \tag{4.48}$$

Substituting Eq. (4.48) in Eq. (4.41),

$$\Delta L_{rt} = -\,\frac{\mu}{E}\int_0^L \left(\frac{-\,2A_e\Delta p_e}{A}\right) dz = \frac{2\mu A_e}{EA}\int_0^L \Delta p_e\, dz \tag{4.49}$$

Using $\mu = 0.3$ for steel, Eq. (4.49) can be written

$$\Delta L_{rt} = \frac{0.6A_e}{EA}\left[\int_0^m \Delta p_e\, dz + \int_m^L \Delta p_e\, dz\right] \tag{4.50}$$

From 0 to m the change in external pressure results entirely from drop in fluid level:

$$\int_0^m \Delta p_e\, dz = \int_0^m (-\,z\rho_e)\, dz = -\,\rho_e \int_0^m z\, dz = -\,\frac{\rho_e m^2}{2} \tag{4.51}$$

From m to L the change in external pressure results from the drop in fluid level and from the change in fluid density as well:

$$\begin{aligned}\int_m^L \Delta p_e\, dz &= \int_m^L [z\Delta\rho_e - m(\rho_e + \Delta\rho_e)]\, dz \\ &= \int_m^L z\Delta\rho_e\, dz - \int_m^L m(\rho_e + \Delta\rho_e)\, dz \\ &= \Delta\rho_e \int_m^L z\, dz - m(\rho_e + \Delta\rho_e)\int_m^L dz \\ &= \frac{\Delta\rho_e(L^2 - m^2)}{2} - m(\rho_e + \Delta\rho_e)(L - m)\end{aligned} \tag{4.52}$$

Combining Eqs. (4.50), (4.51), and (4.52),

$$\Delta L_{rt} = \frac{0.6A_e}{EA}\left(-\frac{\rho_e m^2}{2} + \frac{L^2\Delta\rho_e - m^2\Delta\rho_e}{2} - mL\rho_e + m^2\rho_e - mL\Delta\rho_e + m^2\Delta\rho_e\right) \tag{4.53}$$

By Hooke's law,

$$E = \frac{B_5/A}{\Delta L/L} = \frac{B_5L}{A\Delta L} = \frac{B_5L}{A(\Delta L_z + \Delta L_{rt})} \tag{4.54}$$

Solving for B_5,

$$B_5 = \frac{EA(\Delta L_z + \Delta L_{rt})}{L} = \frac{EA\Delta L_z}{L} + \frac{EA\Delta L_{rt}}{L} \tag{4.55}$$

Substituting Eqs. (4.44) and (4.53) into (4.55),

$$B_5 = A_e\left[-0.7L\Delta\rho_e + (\rho_e + \Delta\rho_e)\left(0.4m + \frac{0.3m^2}{L}\right)\right] \tag{4.56}$$

If L and m are in feet and ρ_e and $\Delta\rho_e$ are in pounds per gallon, the units of B_5 are ft lb in.²/gal. If it is desired to express B_5 in pounds, the right member of Eq. (4.56) must be multiplied by

$$\frac{7.48 \text{ gal/ft}^3}{144 \text{ in.}^2/\text{ft}^2} = 0.0519$$

Rewriting Eq. (4.56) in terms of d_e,

$$B_5 = +0.0408\, d_e^2\left[-0.7L\Delta\rho_e + (\rho_e + \Delta\rho_e)\left(0.4m + \frac{0.3m^2}{L}\right)\right] \tag{4.57}$$

Force due to change in mud weight and drop of fluid level inside the casing **(B_6).** A tendency to buckle will result from tangential and radial stresses caused by changes in weight and level of the *internal* fluid. ΔL will again represent the elongation that would result from these stresses if the casing were not cemented. The buckling force B_6 is the force equal in magnitude but opposite in direction to the force necessary to cause elongation ΔL. Again, considering an element of length dz at vertical distance z from the top of the casing,

$$\mathrm{d}\Delta L = \epsilon_z\, \mathrm{d}z \tag{4.36}$$

and

$$\Delta L = \int_0^L \epsilon_z\, \mathrm{d}z \tag{4.37}$$

By the line of reasoning previously followed, it can be shown that the change in pressure at depth L owing to a drop n in mud level is $n(\rho_i - \Delta\rho_i)$ where ρ_i is the original mud density inside the casing and $\Delta\rho_i$ is the change in mud density inside the casing. The net change in force on the bottom of a closed tube of area A_i at depth L is

$$A_i[L\Delta\rho_i - n(\rho_i + \Delta\rho_i)]$$

The axial stress resulting from this net change in force is

$$\begin{aligned}\sigma_z &= \frac{A_i[L\Delta\rho_i - n(\rho_i + \Delta\rho_i)]}{A} \\ &= \frac{A_i}{A}(L\Delta\rho_i - n\rho_i - n\Delta\rho_i) \qquad (4.58)\end{aligned}$$

Letting ΔL_z be the component of ΔL resulting from axial stress,

$$\Delta L_z = \frac{LA_i}{EA}(L\Delta\rho_i - n\rho_i - n\Delta\rho_i) \qquad (4.59)$$

Using Lamé's expressions for radial and tangential stresses as given by Eqs. (4.6) and (4.8),

$$\sigma_r + \sigma_t = \frac{2A_i\Delta p_i}{A} \qquad (4.60)$$

Letting ΔL_{rt} be the component of ΔL resulting from radial and tangential stresses,

$$\begin{aligned}\Delta L_{rt} = -\frac{0.6A_i}{EA}\Bigg[&-\frac{\rho_i n^2}{2} + \frac{L^2\Delta\rho_i - n^2\Delta\rho_i}{2} \\ &- nL\rho_i + n^2\rho_i - nL\Delta\rho_i + n^2\Delta\rho_i\Bigg] \qquad (4.61)\end{aligned}$$

Using Hooke's law,

$$\begin{aligned}B_6 &= \frac{EA(\Delta L_z + \Delta L_{rt})}{L} \\ &= A_i\left[0.7L\Delta\rho_i - (\rho_i + \Delta\rho_i)\left(0.4n + \frac{0.3n^2}{L}\right)\right] \qquad (4.62)\end{aligned}$$

Expressing B_6 in pounds, L and n in feet, and ρ_i and $\Delta\rho_i$ in pounds per gallon, Eq. (4.62) becomes

$$B_6 = +\,0.0408\,d_i^2\left[0.7L\Delta\rho_i - (\rho_i + \Delta\rho_i)\left(0.4n + \frac{0.3n^2}{L}\right)\right] \qquad (4.63)$$

Force due to change of surface pressure inside and outside casing (B_7). From Lamé's equations the radial and tangential stresses can be defined as follows:

$$\sigma_{ro} = \frac{A_ip_i - A_ep_e}{A} - \frac{A_iA_e(p_i - p_e)}{Ar^2} \qquad (4.64)$$

and

$$\sigma_{to} = \frac{A_ip_i - A_ep_e}{A} + \frac{A_iA_e(p_i - p_e)}{Ar^2} \qquad (4.65)$$

where p_i and p_e are now original internal and external surface pressures, σ_{ro} and σ_{to} are the resulting original radial and tangential stresses at radius r from the center of the casing, and the other terms are as previously defined. If internal and external surface pressures are increased by Δp_i and Δp_ϵ, the final radial and tangential stresses, σ_{rf} and σ_{tf}, are given by

$$\sigma_{rf} \mathrel{\hat{=}} \frac{A_i(p_i + \Delta p_i) - A_e(p_e + \Delta p_e)}{A} - \frac{A_iA_e(p_i + \Delta p_i - p_e - \Delta p_e)}{Ar^2} \tag{4.66}$$

and

$$\sigma_{tf} = \frac{A_i(p_i + \Delta p_i) - A_e(p_e + \Delta p_e)}{A} + \frac{A_iA_e(p_i + \Delta p_i - p_e - \Delta p_e)}{Ar^2} \tag{4.67}$$

If σ_r and σ_t are stress changes resulting from pressure changes Δp_i and Δp_e,

$$\begin{aligned} \sigma_r + \sigma_t &= (\sigma_{rf} + \sigma_{tf}) - (\sigma_{ro} + \sigma_{to}) \\ &= (2/A)(A_i\Delta p_i - A_e\Delta p_e) \end{aligned} \tag{4.68}$$

Considering the casing as a closed tube, the change in internal surface pressure results in a downward force $A_i\Delta p_i$ on the closed (bottom) end. The change in external surface pressure results in an upward force $A_e\Delta p_e$ on the closed end. The net change in force is $A_i\Delta p_i - A_e\Delta p_e$ and the resultant change in axial stress is

$$\sigma_z = \frac{A_i\Delta p_i - A_e\Delta p_e}{A} \tag{4.69}$$

Substituting Eqs. (4.68) and (4.69) into Eq. (4.9) and taking $\mu = 0.3$,

$$\begin{aligned} \epsilon_z &= \frac{1}{E}\left[\frac{A_i\Delta p_i - A_e\Delta p_e}{A} - \frac{2 \times 0.3(A_i\Delta p_i - A_e\Delta p_e)}{A}\right] \\ &= \frac{0.4(A_i\Delta p_i - A_e\Delta p_e)}{EA} \end{aligned} \tag{4.70}$$

Applying Hooke's law,

$$E = \frac{B_7/A}{\epsilon_z} = \frac{B_7}{A\epsilon_z} \tag{4.71}$$

Solving for B_7,

$$B_7 = EA\epsilon_z = 0.4(A_i\Delta p_i - A_e\Delta p_e) \tag{4.72}$$

In terms of the internal and external diameters,

$$\begin{aligned} B_7 &= 0.4(0.785d_i^2\Delta p_i - 0.785d_e^2\Delta p_e) \\ &= +\,0.314(d_i^2\Delta p_i - d_e^2\Delta p_e) \end{aligned} \tag{4.73}$$

$B_1, B_2, \ldots, B_7$ represent forces affecting the buckling tendency. If ΔS is the pickup before landing, then to prevent buckling we must have

$$\Delta S > B_1 + B_2 + \ldots + B_7 \tag{4.74}$$

4.3. Wellhead Loads

The *hanging load* selected for the casing should be consistent with the buckling requirements, the strength of the casing, the strength of the outer casing strings, and the load-carrying capacity of the casing-hanging assembly. Wellhead load equations can be used to evaluate the load at any time in the well history. They are therefore valuable aids in casing design as well as in the study of buckling. The total wellhead load at the time the cement sets can be expressed as the sum of twelve terms, one term representing the effect of casing-landing practice, and the other eleven terms representing the effects on wellhead load of down-the-hole variables. The parameters in the wellhead load equations are identical with those in the buckling-criterion equations previously derived.

Force due to pickup during hanging of casing **(W_1).** If the pickup is ΔS, then

$$W_1 = +\Delta S \tag{4.75}$$

Force due to average temperature change **(W_2).** The wellhead load W_2 is equal in magnitude but opposite in direction to the buckling force B_1. Reference to Eqs. (4.24) through (4.31) shows that

$$W_2 = -59.8w\,\Delta t \tag{4.76}$$

Force due to weight of casing string **(W_3).** If w is the *average* nominal weight of the casing string of length $h + L$, then

$$W_3 = +w(h + L) \tag{4.77}$$

Force due to cement slurry weight **(W_4).** Considering the casing to be a closed tube, the pressure on the closed end resulting from the cement slurry column outside the casing is $0.052\rho_c h$. The resulting force is

$$\begin{aligned} W_4 = -0.052\rho_c h A_e &= -0.052\rho_c h \times 0.785 d_e^2 \\ &= -0.0408 d_e^2 h \rho_c \end{aligned} \tag{4.78}$$

Force due to weight of mud outside casing **(W_5).** By the line of reasoning used in Eq. (4.78),

$$W_5 = -0.0408 d_e^2 L \rho_e \tag{4.79}$$

Force due to weight of mud inside casing **(W_6).** The mud column of height $h + L$ and density ρ_i inside the casing (considered as a closed tube) results in the load

$$W_6 = +\,0.0408 d_i^2 (h + L)\rho_i \tag{4.80}$$

Force due to change in mud weight and drop of fluid level outside casing (W_7). A review of Eqs. (4.36) through (4.38) will show that the change in length, ΔL, associated with the force W_7 is

$$\Delta L = \int_0^L \epsilon_z \, dz = \int_0^L \frac{1}{E}[\sigma_z - \mu(\sigma_r + \sigma_t)] \, dz$$

$$= \frac{1}{E}\int_0^L \sigma_z \, dz - \frac{\mu}{E}\int_0^L (\sigma_r + \sigma_t) \, dz \tag{4.38}$$

Further review of previous derivations shows that

$$\Delta L_z = \frac{1}{E}\int_0^L \sigma_z \, dz = \frac{\sigma_z}{E}\int_0^L dz = \frac{\sigma_z L}{E} \tag{4.43}$$

and

$$\Delta L_{rt} = \frac{0.6A_e}{EA}\left[-\frac{\rho_e m^2}{2} + \frac{L^2\Delta\rho_e - m^2\Delta\rho_e}{2} - mL\rho_e + m^2\rho_e - mL\Delta\rho_e + m^2\Delta\rho_e \right] \tag{4.53}$$

After the cement sets, there is no change in the pipe length, that is, $\Delta L = 0$:

$$\Delta L_z + \Delta L_{rt} = 0 \tag{4.81}$$

Combining Eqs. (4.53), (4.43), and (4.81),

$$\frac{\sigma_z L}{E} = \frac{0.6A_e}{EA}[-0.5\rho_e m^2 - 0.5\Delta\rho_e m^2 - 0.5L^2\Delta\rho_e + mL\rho_e - mL\Delta\rho_e] \tag{4.82}$$

Solving for σ_z,

$$\sigma_z = \frac{0.6A_e}{A}\left[(\rho_e + \Delta\rho_e)\left(m - \frac{0.5m^2}{L}\right) - 0.5L\Delta\rho_e \right] \tag{4.83}$$

In Eq. (4.83), σ_z is the axial stress set up in the casing as a result of the change in fluid density and drop in fluid level. σ_z is related to the axial load W_7 by

$$W_7 = A\sigma_z = 0.6A_e\left[-0.5L\Delta\rho_e + (\rho_e + \Delta\rho_e)\left(m - \frac{0.5m^2}{L}\right)\right] \tag{4.84}$$

To express W_7 in pounds when L and m are in feet and ρ_e and $\Delta\rho_e$ are in pounds per gallon, Eq. (4.84) is multiplied by the conversion factor 0.0519. Using this conversion factor and substituting $0.785d_e^2$ for A_e, Eq. (4.84) becomes

$$W_7 = -0.0122d_e^2 L\Delta\rho_e + 0.0245d_e^2(\rho_e + \Delta\rho_e)\left(m - \frac{0.5m^2}{L}\right) \tag{4.85}$$

Force due to change in mud weight and drop of fluid level inside casing (W_8). By methods similar to those used above, it can be shown that

$$W_8 = +\,0.0122 d_i^2 L \Delta\rho_i - 0.0245 d_i^2 (\rho_i + \Delta\rho_i)\left(n - \frac{0.5n^2}{L}\right) \quad (4.86)$$

Force due to surface pressure outside casing (W_9). Considering casing to be a closed tube, the surface pressure outside the casing when the cement sets causes a force which reduces the wellhead load:

$$W_9 = -A_e p_e = -0.785 d_e^2 p_e \quad (4.87)$$

Force due to surface pressure inside casing (W_{10}). The surface pressure inside the casing when the cement sets causes a force which increases the wellhead load:

$$W_{10} = A_i p_i = +0.785 d_i^2 p_i \quad (4.88)$$

Force due to change of surface pressure outside casing (W_{11}). A pressure change outside the casing would tend to cause an elongation ΔL. Using Eq. (4.48),

$$\Delta L = \epsilon_z L = \frac{L}{E}[\sigma_z - \mu(\sigma_r + \sigma_t)] = \frac{L}{E}\left[\sigma_z + \frac{2\mu A_e \Delta p_e}{A}\right] \quad (4.89)$$

After the cement sets, $\Delta L = 0$, and since $\sigma_z = W_{11}/A$,

$$\frac{L}{E}\left[\frac{W_{11}}{A} + \frac{2\mu A_e \Delta p_e}{A}\right] = 0 \quad (4.90)$$

Solving for W_{11}, and taking $\mu = 0.3$,

$$W_{11} = -0.6 A_e \Delta p_e = -0.6 \times 0.785 d_e^2 \Delta p_e = -0.471 d_e^2 \Delta p_e \quad (4.91)$$

Force due to change of surface pressure inside casing (W_{12}). For the pressure change inside casing, Eq. (4.90) is written

$$\frac{L}{E}\left[\frac{W_{12}}{A} - \frac{2\mu A_i \Delta p_i}{A}\right] = 0 \quad (4.92)$$

from which

$$W_{12} = +\,0.471 d_i^2 \Delta p_i \quad (4.93)$$

4.4. Application

The application of buckling and wellhead load equations to combination strings under four conditions is illustrated by Cox,[8] who selected a 10,000-ft string with parameters as given in Example 4.2.

Example 4.2. Calculate the buckling and wellhead loads for various well conditions, given a casing string with dimensions as shown in Fig. 4.2 and other parameters as listed below:

$p_e = 0$ psi $\quad \rho_e = 13$ lb per gal
$p_i = 0$ psi $\quad \rho_i = 13$ lb per gal
$\rho_c = 15$ lb per gal

For deepening:

$\Delta t = +35°F$
$\Delta \rho_i = +5$ lb per gal

For testing liner:

$\Delta p_i = +2000$ psi
$\Delta \rho_i = +5$ lb per gal

For gas lift:

$n = 5000$ ft
$\Delta p_i = +900$ psi
$\Delta \rho_i = -4.3$ lb per gal

All variables not listed above are assumed to be zero.

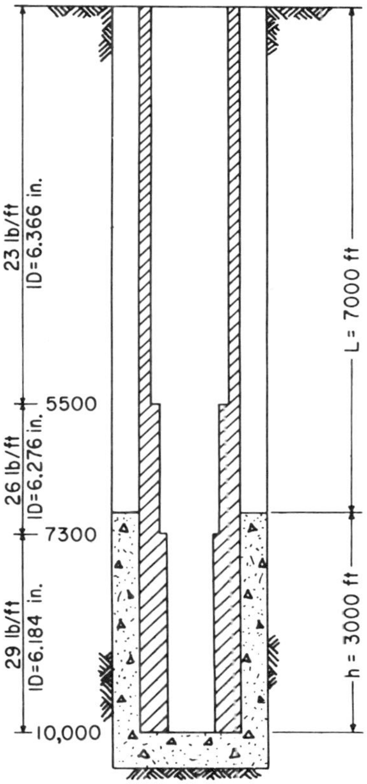

Fig. 4.2 A 7-in. combination casing string. (After Cox,[8] *Drilling and Production Practice*, API.)

Solution:

Condition when cement sets. Since no significant changes in temperature, fluid level, or fluid density occur when cement sets, the only buckling terms which apply are B_2, B_3, and B_4, and the only wellhead load terms which apply are W_3 through W_6.

$$B_2 = -hw = -(300 \times 26) - (2700 \times 29) = -86{,}100 \text{ lb}$$

$$B_3 = +0.0408 d_e^2 h \rho_c = +0.0408 \times 7.000^2 \times 3000 \times 15 = +90{,}000 \text{ lb}$$

$$\begin{aligned} B_4 &= -0.0408 d_i^2 h \rho_i \\ &= -(0.0408 \times 6.184^2 \times 2700 \times 13) \\ &\quad -(0.0408 \times 6.276^2 \times 300 \times 13) \\ &= -61{,}000 \text{ lb} \end{aligned}$$

To prevent buckling,

$$\Delta S > B_2 + B_3 + B_4 \quad \text{or} \quad \Delta S > -57{,}100 \text{ lb}$$

This indicates that, when the cement sets, the casing can be slacked off 57,100 lb without buckling resulting.

$$W_3 = +w(h + L) = +(23 \times 5500) + (26 \times 1800) + (29 \times 2700)$$
$$= +251{,}600 \text{ lb}$$
$$W_4 = -0.0408 d_e^2 h \rho_c = -0.0408 \times 7.000^2 \times 3000 \times 15 = -90{,}000 \text{ lb}$$
$$W_5 = -0.0408 d_e^2 L \rho_e = -0.0408 \times 7.000^2 \times 7000 \times 13 = -181{,}900 \text{ lb}$$
$$W_6 = +0.0408 d_i^2 (h + L)\rho_i = +(0.0408 \times 6.366^2 \times 5500 \times 13)$$
$$+ (0.0408 \times 6.276^2 \times 1800 \times 13) + (0.0408 \times 6.184^2 \times 2700 \times 13)$$
$$= +212{,}000 \text{ lb}$$

When the cement sets, the *hanging* load is

$$W_3 + W_4 + W_5 + W_6 = +191{,}700 \text{ lb}$$

Conditions for deepening. After the casing is set at 10,000 ft, the hole is deepened; the mud weight is increased from 13 to 18 lb per gal; and, since the circulated mud is at a higher temperature due to the greater depth, it is assumed that the average temperature of the casing above the cement top increases by 35°F. These changes necessitate use of the buckling terms B_1 and B_6.

The average weight of casing above the cement top is

$$w = \frac{23 \times 5500 + 26 \times 1500}{5500 + 1500} = 23.6 \text{ lb per ft}$$

hence

$$B_1 = +59.8 w \Delta t = +59.8 \times 23.6 \times 35 = +49{,}400 \text{ lb}$$

Since $n = 0$,

$$B_6 = +0.0408 d_i^2 (0.7 L \Delta\rho_i) = +0.0286 d_i^2 L \Delta\rho_i$$
$$= +(0.0286 \times 6.366^2 \times 5500 \times 5)$$
$$+ (0.0286 \times 6.276^2 \times 1500 \times 5)$$
$$= +40{,}300 \text{ lb}$$

To offset the effect on *buckling* of the increases in temperature and mud weight, the casing must be picked up 89,700 lb. Since the casing could have been slacked off 57,100 lb after cementing, the casing should be picked up 32,600 lb from the as-cemented position.

The changes in hanging load are given by terms W_2 and W_8:

$$W_2 = -59.8 w \Delta t = -59.8 \times 23.6 \times 35 = -49{,}400 \text{ lb}$$
$$W_8 = +0.0122 d_i^2 L \Delta\rho_i$$
$$= +(0.0122 \times 6.266^2 \times 5500 \times 5)$$
$$+ (0.0122 \times 6.276^2 \times 1500 \times 5)$$
$$= +17{,}200 \text{ lb}$$

In deepening, the hanging load is decreased by 32,200 lb. Since the casing was hung as cemented, the wellhead load for drilling ahead is 191,700 − 32,200 = 159,500 lb.

Conditions for testing a liner. If a liner is set and casing is pressured to 2000 psi with 18 lb per gal mud in the hole, other changes will occur. Since there is no temperature change above the top of the cement (i.e., the normal static temperature gradient exists), the only buckling terms which apply are B_6 and B_7, and the only applicable wellhead load equations are W_8 and W_{12}.

$$B_6 = +40{,}300 \text{ lb}$$

$$B_7 = +0.314(d_i^2 \Delta p_i - d_e^2 \Delta p_e)$$

In evaluating B_7, three assumptions are made, namely, that (1) the change in external pressure is zero ($\Delta p_e = 0$), (2) the change in internal pressure is effective over an area equivalent to the internal cross-sectional area of the top section ($d_i = 6.366$ in.), and (3) the casing pressure was zero during the WOC time ($\Delta p_i =$ 2000 psi). Therefore,

$$B_7 = +0.314 \times 6.366^2 \times 2000 = +25{,}500 \text{ lb}$$

When the casing is hung, the *net* pickup required to prevent buckling is

$$-\ 57{,}100 + 40{,}300 + 25{,}500 = +\ 8700 \text{ lb}$$

Finally,

$$W_8 = +17{,}200 \text{ lb}$$

$$W_{12} = +0.471 d_i^2 \Delta p_i = +0.471 \times 6.366^2 \times 2000$$

$$= +38{,}200 \text{ lb}$$

If the casing is hung as cemented, the hanging load during testing of the liner is

$$+\ 191{,}700 + 17{,}200 + 38{,}200 = +\ 247{,}100 \text{ lb}$$

Conditions during production on gas lift. Under this condition, it is assumed that the fluid level in the casing is 5000 ft, the fluid weight is 8.7 lb per gal, and the temperature change from the as-cemented value is zero. Gas is injected into the annulus between the tubing and the casing at 900 psi. The resulting changes in environment introduce the following changes in buckling tendency and wellhead load. The weighted average internal diameter squared is

$$d_i^2 = \frac{6.366^2 \times 5500 + 6.276^2 \times 1500}{7000} = 40.28$$

Therefore

$$B_6 = +0.0408 d_i^2 \left[0.7L\Delta\rho_i - (\rho_i + \Delta\rho_i)\left(0.4n + \frac{0.3n^2}{L}\right)\right]$$

$$= +0.0408 \times 40.28 \left[-0.7 \times 7000 \times 4.3 - (13 - 4.3)\left(0.4 \times 5000 + \frac{0.3 \times 5000^2}{7000}\right)\right]$$

$$= -78{,}500 \text{ lb}$$

$$B_7 = +0.314(d_i^2 \Delta p_i - d_e^2 \Delta p_e)$$

$$= +0.314(6.366^2 \times 900 - 0) = +11{,}500 \text{ lb}$$

Thus, when the casing is hung, the maximum slack-off without buckling at the time the well is on gas lift is

$$-57{,}100 - 78{,}500 + 11{,}500 = -124{,}100 \text{ lb}$$

Finally, for this condition

$$\begin{aligned} W_8 &= +0.0122d_i^2 L\Delta\rho_i - 0.0245d_i^2(\rho_i + \Delta\rho_i)\left(n - \frac{0.5n^2}{L}\right) \\ &= -0.0122 \times 40.28 \times 7000 \times 4.3 \\ &\quad -0.0245 \times 40.28 \times 8.7\left(5000 - \frac{0.5 \times 5000^2}{7000}\right) \\ &= -42{,}400 \text{ lb} \\ W_{12} &= +0.471d_i^2\Delta p_i = +0.471 \times 6.366^2 \times 900 \\ &= +17{,}200 \text{ lb} \end{aligned}$$

If the casing is hung as cemented when the well is on gas lift, the hanging load is

$$+191{,}700 - 42{,}400 + 17{,}200 = +166{,}500 \text{ lb}$$

Of the four conditions examined, the maximum required pickup (+32,600 lb) occurs during well deepening. If this pickup is applied during landing of the casing, the resulting wellhead load for each condition except deepening will be greater than that required to prevent buckling. However, if the pickup during landing is less than this value, the resulting wellhead load for deepening will be less than that required to prevent buckling. If a pickup of 32,600 lb is applied during landing, the resulting wellhead loads for the conditions investigated are:

(1) 191,700 + 32,600 = 224,300 lb as cemented

(2) 159,500 + 32,600 = 192,100 lb for deepening

(3) 247,100 + 32,600 = 279,700 lb for testing liner

(4) 166,500 + 32,600 = 199,100 lb on gas lift

If ΔS is the pickup in pounds and ΔL is the corresponding pickup in inches, ΔS and ΔL are related by Hooke's law, or

$$E = \frac{\Delta S/A}{\Delta L/12L} = \frac{12L\Delta S}{A\Delta L} \tag{4.94}$$

The factor 12 is necessary because L is expressed in feet, and ΔL and L must be in the same units. Solving for ΔL and taking the value of A as given by Eq. (4.30),

$$\Delta L = \frac{12L\Delta S}{EA} = \frac{12L\Delta S}{30 \times 10^6 \times 0.289w} = \frac{1.38 \times 10^{-6}L\Delta S}{w} \tag{4.95}$$

Example 4.3. Given that the required pickup is 32,600 lb for the well of Example 4.2, calculate the corresponding pickup in inches.

SOLUTION:

$$w = \frac{23 \times 5500 + 26 \times 1500}{5500 + 1500} = 23.6 \text{ lb per ft}$$

$$\Delta L = \frac{1.38 \times 10^{-6} \times 7000 \times 32{,}600}{23.6} = 13.3 \text{ in.}$$

PROBLEMS

4.1. What is the stretch of 11,000 ft of $5\frac{1}{2}$-in. OD 17-lb N-80 LTC casing run in 12 lb per gal mud? *Ans.* 61.3 in.

4.2. Continuing with Problem 4.1, if the average temperature of the casing on the pipe rack is 80°F and the bottom-hole temperature is 190°F, what is the elongation due to the change in temperature when the casing is run? *Ans.* 50.1 in.

4.3. Show that the constant in Eq. (4.23) is 40.8.

4.4. When coming out of the hole, 10,000 ft of $4\frac{1}{2}$-in., 16.6-lb drill pipe is stuck in a key seat. A differential pull of 50,000 lb causes an elongation of 18 in. What is the depth to the key seat (freeze point)? *Ans.* 4394 ft.

4.5. What is the distinguishing characteristic of each of the four methods of landing casing?

4.6. (a) A 12,000-ft combination string of $5\frac{1}{2}$-in. OD casing is to be cemented in an $8\frac{3}{4}$-in. hole. The string consists of 2000 ft of 17-lb 4.892-in. ID N-80 XL (no couplings); 8000 ft of 17-lb 4.892-in. ID N-80 LTC; and 2000 ft of 20-lb 4.778-in. ID N-80 LTC (on bottom). At the time of running the casing, the mud weight will be 13 lb per gal and the cement slurry weight 16 lb per gal. It is estimated that the cement top will be at 10,000 ft. When the cement sets, how many pounds can be slacked off without buckling the string?
Ans. −24,700 lb.
(b) What will be the hanging load when the cement sets? *Ans.* 161,200 lb.
(c) If the plug is bumped with 1600 psi and the pressure on the plug container is bled down to 500 psi, what will be the hanging load when the cement sets? *Ans.* 170,600 lb.

4.7. (a) After setting the $5\frac{1}{2}$-in. casing of Problem 4.6 at 12,000 ft and testing two sands, it is decided to deepen the hole by 1500 ft and to increase the mud weight from 13 to 16 lb per gal. From the temperature gradients in the area, it is known that the circulated mud will increase the average temperature of the casing above the cement top by 20°F. In order to offset the effect on buckling of the increases in temperature and mud weight, what additional pickup is required? *Ans.* +40,800 lb.
(b) Considering the allowable slack-off of Problem 4.6, part (a), and the required pickup of Problem 4.7, part (a), what is the necessary *net* pickup from the as-cemented position (when drilling ahead) to prevent buckling? *Ans.* 16,100 lb.
(c) When drilling ahead, what is the decrease in wellhead load? *Ans.* 11,500 lb.
(d) If the casing is hung as cemented, what will be the wellhead load when drilling ahead? *Ans.* 149,700 lb.

4.8. (a) With additional oil shows in the lower sands, a liner is set in the well of Problem 4.7. In squeeze-cementing the liner, the $5\frac{1}{2}$-in. casing is pressured to 2000 psi with 16 lb per gal mud in the hole, and the average temperature of the $5\frac{1}{2}$-in. casing above its cement top returns to the as-cemented value. What is the buckling tendency due to the pressure increase during squeezing? *Ans.* +15,000 lb.

(b) What is the *net* pickup required to prevent buckling? *Ans.* 10,800 lb.

(c) What is the increase in wellhead load due to the pressure increase? *Ans.* 22,500 lb.

(d) If the casing is hung as cemented, what will be the wellhead load during squeezing? *Ans.* 192,500 lb.

4.9. (a) The well of Problem 4.8 is placed on gas lift, with a working fluid level of 5000 ft, 8.5 lb per gal fluid in the annulus between casing and tubing, a gas injection pressure of 500 psi, and no temperature change from the as-cemented value. What is the change in the buckling tendency (from the as-cemented position)? *Ans.* −49,800 lb.

(b) When the casing is hung, what is the maximum allowable slack-off without causing buckling at the time the well is on gas lift? *Ans.* 74,500 lb.

(c) When the well is on gas lift, how much less is the wellhead load than the hanging load at the time the cement set? *Ans.* 26,200 lb.

(d) If the casing is hung as cemented, what will be the wellhead load on gas lift? *Ans.* 135,000 lb.

4.10. (a) Concerning the four operations considered in Problems 4.6 through 4.9, what is the maximum required pickup (in pounds and in inches) to prevent buckling? *Ans.* 16,100 lb; 13.1 in.

(b) If at the time of hanging casing, a *temporary* freeze point exists at the shoe of the surface pipe set at 2000 ft, what pickup (in pounds) is required to obtain an elongation of 13.1 in.? *Ans.* 80,700 lb.

(c) As a result of the pickup of part (b), what is the hanging load? *Ans.* 241,900 lb.

(d) The joint strength of $5\frac{1}{2}$-in. 17-lb N-80 XL casing is 457,000 lb. What is the safety factor against pullout tension? *Ans.* 1.89.

(e) What is the hanging load when the casing adjusts down to the permanent freeze point at the cement top? *Ans.* 177,300 lb.

(f) If the shoe of the surface casing is the permanent freeze point, what will be the load on the casing hanger when the liner is squeeze-cemented? *Ans.* 266,150 lb.

(g) What is the safety factor against pullout under the conditions of part (f)? *Ans.* 1.72.

(h) With the temporary freeze point at 2000 ft, what stretch of the casing will result from a pickup of 16,100 lb? *Ans.* 2.6 in.

(i) If the pickup is 16,100 lb with the temporary freeze point at 2000 ft, what is the effective pickup when the freeze point moves to the cement top? *Ans.* 3200 lb.

(j) Under which of the four conditions investigated will buckling occur if the pickup is 16,100 lb (with the temporary freeze point at 2000 ft)?

REFERENCES

[1]*Casing Landing Recommendations* (API Bull. D7)(Dallas: American Petroleum Institute, 1955), pp. 4–7.

[2]Ferdinand L. Singer, *Strength of Materials* (New York: Harper and Brothers, 1951), pp. 15, 17, 414–417.

[3]J. T. Hayward, "Methods for Determining How Much of a Frozen or Cemented Column of Pipe is Free," *Drilling and Production Practice*, (API, 1935), pp. 16–17.

[4]J. W. Peret, "Comparison of Casing Landing Methods," *The Petroleum Engineer*, **25,** No. 11, B-101 (October, 1953).

[5]Arthur Lubinski, "Influence of Tension and Compression on Straightness and Buckling of Tubular Goods in Oil Wells," API Prod. Bull. 237, Sec. IV (1951), p. 31.

[6]A. Klinkenberg, "The Neutral Zones in Drill Pipe and Casing and Their Significance in Relation to Buckling and Collapse," *Drilling and Production Practice* (API, 1951), p. 64.

[7]C. A. Bouman, "Buckling of Oil Well Piping," *The Petroleum Engineer*, **28,** No. 6, B-60 (June, 1956).

[8]W. R. Cox, "Key Factors Affecting Landing of Casing," *Drilling and Production Practice* (API, 1957), pp. 225–237.

[9]C. H. Oberg and R. W. Masters, "The Determination of Stresses in Oil-Well Casing in Place," *Drilling and Production Practice* (API, 1947), pp. 257–267.

[10]Arthur Lubinski and K. A. Blenkarn, "Buckling of Tubing in Pumping Wells, Its Effects and Means for Controlling It," *Trans. AIME*, **210,** 87 (1957).

CHAPTER FIVE

Sucker Rod Pumping

5.1. Introduction

According to a recent survey, there are, in the United States, approximately 574,000 producing oil wells, of which some 68,000 are flowing.[*,1] Of the 506,000 wells on artificial lift, a vast majority (80 to 85 per cent) are being produced by sucker rod pumping. Proper design and maintenance of sucker rod pumping installations can, therefore, be recognized as of major importance in the field of oil production. In this chapter we propose first to introduce certain theoretical expressions used in pumping calculations, then to present a method for the design of pumping installations, and finally to discuss methods of analyzing pumping installations to obtain the most efficient performance with existing equipment.

A necessary prerequisite for solving problems of design and analysis is a good understanding of the component parts of a pumping installation. These parts are (1) the prime mover, (2) the surface pumping equipment, (3) the sucker rod string, and (4) the subsurface pump. Although each component will be treated (in the descriptions which follow) as a separate unit, it cannot be emphasized too strongly that the installation is in reality a very complex linkage of the components and that no single component can be designed independently of the others. Thus, the complete design of a pumping installation becomes a difficult, trial-and-error affair, and even with the theoretical equations as guideposts, it is often necessary to resort to empirical equations and techniques based on experience.

*References throughout the text are given at the end of each chapter.

5.2. The Subsurface Pump

The functions of the pump are to admit fluid from the formation into the producing string and (primarily) to lift the fluid thus admitted to the surface. To accomplish this, any pump must contain four essential elements. As shown in Fig. 5.1 these elements are the working barrel, the

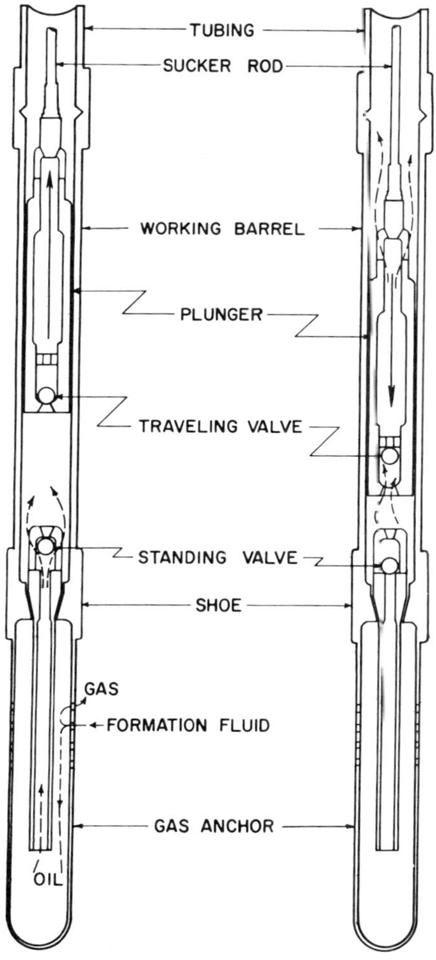

Fig. 5.1 The subsurface pump, showing locations of standing and traveling valves on upstroke (left) and downstroke (right).

plunger, the standing valve, and the traveling valve. On the upstroke part of the pumping cycle the fluid load on the plunger causes the traveling valve to close, and fluid is displaced from the tubing at the surface. Also, the motion of the plunger causes a pressure decrease above the standing valve, which opens and admits fluid from the formation. At the beginning of the downstroke the plunger begins to fall through the fluid in the tubing. Consequently, the traveling valve opens, the standing valve closes, and the fluid load is transferred from the plunger to the tubing. This transfer of fluid load is a significant factor in determining the effective plunger stroke. It should be noted that most of the production is accomplished on the upstroke, the only fluid removed from the tubing during the downstroke being that which is displaced by the falling rods.

Oil well pumps may be classed as either tubing or rod pumps. In the case of tubing pumps, the barrel is run on the tubing and the plunger is run on the rod string. In the case of rod pumps, the complete unit is run on the rod string. Rod pumps have the advantages of being more easily removed for servicing and less susceptible to damage in running, but they offer less working area for the plunger, since the maximum bore of a rod pump is necessarily less than the maximum bore of a tubing pump for the same size tubing.

Pumps may be further classed by the nature of the barrel. Liner barrels are made up of jackets with sectional or one-piece liners held in position by collars. Full barrels are one-piece tubes threaded at each end.[2]

Tubing pumps may be equipped at the lower end with regular shoes or with extension shoes and nipples. The latter arrangement permits a longer plunger stroke for a given length pump barrel. In the nonextension style, however, there is less space around the standing valve for accumulation of sand deposits. Rod pumps are of either stationary barrel type with top or bottom hold-down, or traveling barrel type. Stationary barrel pumps have reciprocating plungers actuated by valve rods. In traveling barrel pumps the barrel reciprocates over the plunger assembly.[2]

The American Petroleum Institute has adopted a classification system based on the above characteristics. Figure 5.2, which shows the five classifications, is taken from API Standard 11A, which is also the reference for all specifications pertaining to oil well pumps, and for recommended field practices pertaining to care and use of pumps. Class 1 is a tubing type with regular shoe. Class 2 is a tubing type with extension shoe and nipple. Class 3 is a rod type, stationary barrel, with top hold-down. Class 4 is a rod type, stationary barrel, with bottom hold-down. Class 5 is a rod type, traveling barrel. In Fig. 5.2, the first letter of the abbreviated designation is either "T" indicating tubing type pump or "R" for rod type pump. The second letter is either "W." referring to full barrel or "L" referring to liner barrel. The third letter may be "E" indicating use of an extension shoe and nipple, "A" indicating stationary barrel with top hold-down, "B" for stationary barrel with bottom hold-down, or "T" for traveling barrel.

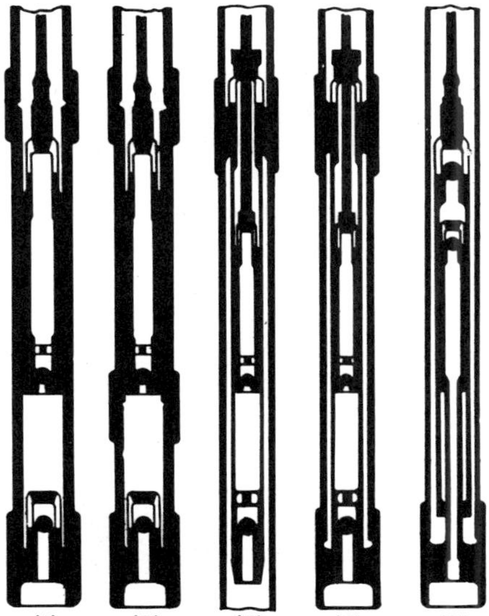

Fig. 5.2 API classification of sucker rod pumps. (API Standard 11A[2].)

5.3. The Sucker Rod String

Energy is transmitted from the surface equipment to the pump by means of a string of steel sucker rods. Sucker rods are available in five standard diameters,[3] and the problem of designing a sucker rod string is essentially a matter of determining the lightest (and therefore the most economical) string that can be used without exceeding the working stress for the rods. The maximum working stress for sucker rods has been the subject of considerable debate. It depends on the chemical composition and mechanical properties of the rods and on the nature of the fluid (i.e., whether corrosive or noncorrosive). As a rule, it is desirable to keep rod stresses under 30,000 psi; however, experience in a given area may indicate a much lower limit.

When pumps are set at depths greater than approximately 3500 feet, it is usually desirable to use a tapered rod string, i.e., a string consisting of rods of different sizes. The smallest-diameter rod is placed at the bottom of the string, immediately above the plunger, since the rod load is least at that point. At lesser depths, where rod load is greater, larger rods are used. This arrangement results in a smaller load on the surface equipment than could be obtained with an untapered string, and it represents a distinct saving in cost of sucker rods.

5.4. *The Surface Pumping Equipment*

The surface equipment transfers energy for pumping the well from the prime mover to the sucker rod string. In doing this, it must change the rotary motion of the prime mover to reciprocating motion for the sucker rods, and it must reduce the speed of the prime mover to suitable pumping speeds. Figure 5.3 shows the essential components of the surface pumping

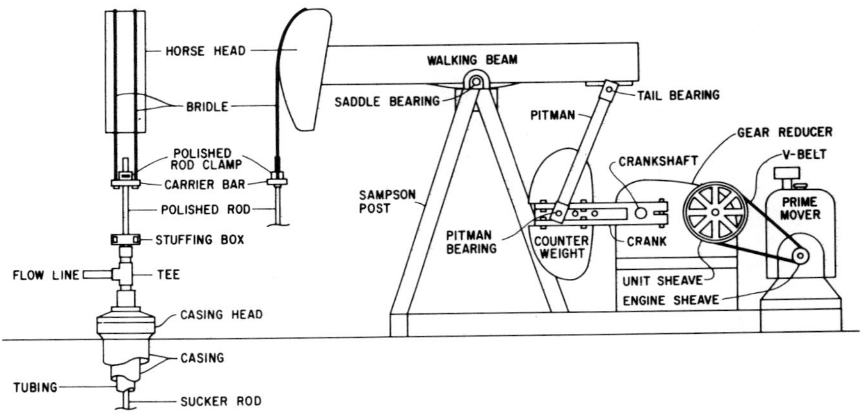

Fig. 5.3 The surface equipment of a sucker rod pumping installation.

equipment. Speed reduction is accomplished at the gear reducer, and the rest of the surface equipment is concerned with change from rotary to reciprocating motion and with counterbalance.

The direct linkage between the sucker rod string and the surface equipment is the polished rod. Polished rods are available in three sizes,[4] the size for any particular installation depending on the tubing size and the diameter of the sucker rods at the top of the rod string. The polished rod passes through a stuffing box, and fluid which has been lifted passes to the flow line through the pumping tee.

Near the upper end of the polished rod is a polished rod clamp, which is supported by the carrier bar. The carrier bar, in turn, is supported by flexible cables (called the bridle) which pass over the horse head at the end of the walking beam. Proper design of these components assures vertical travel of the polished rod, thus reducing unnecessary wear and friction losses at the stuffing box.

The walking beam is supported near its center of gravity by the sampson post. Motion is transmitted to the walking beam by the pitman, to which motion is in turn transmitted by the crank. The distance from the crank shaft to the pitman bearing defines the polished rod stroke length. Most units have three or more possible locations for the pitman bearing along the crank and, therefore, a corresponding number of possible stroke lengths.

One of the most important aspects of pumping installation design is proper selection of counterbalance. Practically all of the work of lifting fluid by a sucker rod pump is performed during the upstroke part of the pumping cycle. During this phase of the cycle, the full load of oil and sucker rods must be accelerated upward from zero velocity, and must be lifted the length of the pump stroke. On the downstroke, with fluid load transferred to the tubing, the falling rods act as a driving force for the installation. If some means were not provided for equalizing these upstroke and downstroke loads, conditions detrimental to the surface pumping equipment would result. From the standpoint of the prime mover, unnecessarily high demand for power would occur during the upstroke, with the prime mover being driven during the downstroke. (This would cause the prime mover speed to be greater on the downstroke than on the upstroke.) From the standpoint of power transmission, unnecessarily high torque would be imposed during the upstroke. Counterbalancing provides for even distribution of loads and for reduction of peak torque requirements. The counterbalance weights, in effect, store energy during the downstroke when power demand is low, and release energy during the upstroke, performing part of the work of lifting fluid and rods.[5]

To avoid a possible misinterpretation of terminology, it is necessary to differentiate between counterbalance effect (which is the net contribution of the counterbalance system *at the polished rod*) and counterbalance or counterweight (which is the weight or system of weights used to obtain the counterbalance effect). The counterbalance effect is determined by the actual weight of the counterweight, by its position, and by the geometry of the surface unit. (Pumping speed is also a factor in the case of beam counterbalance.) A given effect can be obtained by placing counterweights on the walking beam, on the pitman, or on the crank. In some cases, particularly on larger units, air pressure is used to obtain the desired counterbalance effect. The method of determining desired counterbalance effect is not influenced by the type of counterbalance used, and in the design and analysis examples which follow in this chapter, only crank counterbalance will be considered. In this instance, various counterbalance effects can be obtained by changing the position of the counterweight along the crank.

Another member of the surface pumping unit is the gear reducer. It is necessary to determine quite accurately the probable peak torque to which the gear reducer will be subjected, since the American Petroleum Institute uses peak torque as the basis for rating pumping units. The API designation for a unit is simply the maximum permissible torque (on the gear reducer) in thousands of inch-pounds. For example, an API Size 114 unit has a peak torque rating of 114,000 in.-lb. The API has standardized on 16 peak torque ratings varying from 6.4 to 1824 thousand in.-lb.[6]

The unit sheave is the member which receives power from the prime mover (through a V-belt drive). The ratio of unit sheave diameter to engine sheave diameter and the speed reduction at the gear reducer determine the overall speed reduction from prime mover to polished rod. The available unit sheave sizes and the number and type of V-belts to be used can be determined from the pumping unit manufacturer's literature.

5.5. *The Prime Mover*

The function of the prime mover is to supply to the installation mechanical energy which is eventually transmitted to the pump and used to lift fluid. The prime mover selected for a given installation must, of course, have sufficient power output to lift fluid at the desired rate from the working fluid level in the well. The prime mover may be an internal combustion engine (generally utilizing natural gas as the fuel) or it may be an electric motor.

A comparison of the advantages and disadvantages of internal combustion engines and electric motors cannot be made without considering such factors as the availability of gas and relative costs of gas and electric power. However, the use of electric motors greatly facilitates analyses of pumping unit performance in two respects. Ability to measure accurately the electrical energy supplied to a unit permits accurate calculations of efficiencies and costs. Also, counterbalance analysis of the unit can be accomplished (without interfering with the well's operation) by using a recording tachometer to measure variation in motor speed over a complete pumping cycle.

5.6. *Theoretical Analysis of Rod Motion*

If sucker rods were suspended statically from a polished rod or if they were rising or falling at constant velocity, the force on the polished rod would be the weight, W_r, of the sucker rods. However, when the sucker rods are in acceleration there will be on the polished rod an additional acceleration load, $(W_r/g)a$. The acceleration factor, or the factor by which the dead weight of the rods must be multiplied in order to obtain the maximum acceleration load, is given by

$$\alpha = a/g \tag{5.1}$$

where a is the maximum acceleration experienced by the sucker rod string.

A study of the manner in which motion is transmitted from prime mover to sucker rods shows that the rods are approximately in simple harmonic motion. The acceleration of a body in simple harmonic motion can be conveniently investigated by considering the body to be the projection (on the diameter of a reference circle) of a particle moving with uniform speed around the reference circle. If the body under consideration is the sucker rod system, the diameter of the reference circle is equal to the polished rod stroke length, and the time for one revolution of the particle around the circle is equal to the time for one complete pumping cycle. The maximum acceleration of the rod system takes place at the beginning of the upstroke and the beginning of the downstroke, i.e., when the projection has its greatest displacement from the center of the reference circle. At this time the acceleration of the projection (and thus of the rod system) is the same as that of the particle moving around the circle, which is

$$a = v_p^2/r_c \tag{5.2}$$

where v_p is the velocity of the particle, and r_c is the radius of the circle. If the time for one revolution of the particle is τ, then

$$v_p = 2\pi r_c/\tau \tag{5.3}$$

and if N is the number of revolutions per unit time,

$$v_p = 2\pi r_c N \tag{5.4}$$

since $N = 1/\tau$. Substituting into Eq. (5.1),

$$\alpha = \frac{v_p^2}{r_c g} = \frac{4\pi^2 r_c N^2}{g} \tag{5.5}$$

For a pumping well, N is the pumping speed, and r_c is related to the polished rod stroke length, S, by

$$r_c = S/2 \tag{5.6}$$

Equation (5.5) then becomes

$$\alpha = \frac{2\pi^2 SN^2}{g} \tag{5.7}$$

The polished rod stroke length is normally stated in inches and the pumping speed is given in strokes per minute. Then

$$\alpha = \frac{2\pi^2 SN^2}{32.2} \frac{\text{in./min}^2}{\text{ft/sec}^2} \times \frac{1 \text{ ft}}{12 \text{ in.}} \times \frac{1 \text{ min}^2}{3600 \text{ sec}^2} = \frac{SN^2}{70{,}500} \tag{5.8}$$

Example 5.1. One-inch sucker rods weigh 2.88 lb per ft. Calculate the maximum polished rod load resulting from 2000 ft of one-inch sucker rods if the pumping speed is 18 spm (strokes per min) and the polished rod stroke length is 74 in.

SOLUTION:

$$W_r = 2.88 \text{ lb/ft} \times 2000 \text{ ft} = 5760 \text{ lb}$$

$$\alpha = 74 \times 18^2/70{,}500 = 0.340$$

$$\text{Maximum load} = \text{weight of rods} + \text{acceleration load}$$
$$= 5760 + 5760 \times 0.340 = 7720 \text{ lb}$$

5.7. The Effective Plunger Stroke

The volume of oil handled during each stroke of the pump plunger depends not on the length of the polished rod stroke but on the movement of the plunger relative to the working barrel. This movement is called the net or effective plunger stroke, and it may differ significantly from the polished rod stroke.

Fundamentally, plunger and polished rod strokes differ because of rod and tubing stretch and because of plunger overtravel resulting from acceleration. Of a number of theoretical expressions which have been developed in an attempt to predict rod and tubing stretch and overtravel, the simplest and probably the most universally reliable result from the work of Marsh and Coberly.[7,8] The derivations which follow are based on those in the references cited, but contain some modifications resulting from more recent work.

It will be recalled from Sec. 5.2 that, as the traveling and standing valve of the subsurface pump open and close during the pumping cycle, the fluid load is transferred alternately to the tubing and to the rod string. This results in periodic elastic deformations (of rod and tubing) which are out of phase with each other by 180°. Consider the system during the downstroke, when the standing valve is closed and the traveling valve is open. At this time fluid load is on the tubing, causing a certain elongation of that member. At the beginning of the upstroke, the traveling valve closes, causing elongation of the rods. Opening of the standing valve allows the stretch to come out of the tubing. Restoration of the tubing to its original length causes the working barrel to move upward, and elongation of the rods causes the plunger to move downward. The effective plunger stroke is, therefore, decreased by an amount equal to the sum of rod and tubing elongations resulting from fluid load.

For an elastic deformation, there is a constant ratio between the stress applied to a body and the resulting strain,

$$E = \text{Stress/Strain} \tag{5.9}$$

E, the modulus of elasticity, is a characteristic of the material to which the stress is applied. Stress is applied force per unit area,

$$\text{Stress} = F/A \tag{5.10}$$

and strain is fractional change in length.

$$\text{Strain} = e/l \tag{5.11}$$

Conventionally, force, F, is expressed in pounds and cross-sectional area, A, of the member under stress is expressed in square inches. Elongation, e, and original length, l, of the member must be given in the same length units. Normally, magnitudes of elongation and length are such that these quantities are most logically expressed in inches and feet, respectively, in which case Eq. (5.11) is written

$$\text{Strain} = e/12L \tag{5.12}$$

Substitution into Eq. (5.9) gives

$$E = \frac{F/A}{e/12L} = \frac{12FL}{eA} \tag{5.13}$$

and the elongation of the member is

$$e = 12FL/EA \tag{5.14}$$

The force due to fluid load results from the pressure differential across the plunger (with the pump set at depth L) acting over the full plunger area, A_p.

$$F = \Delta p \times A_p \tag{5.15}$$

If it is assumed that the pump is set at the working fluid level in the well, the pressure differential is the pressure at depth L in a column of fluid of specific gravity G:

$$\Delta p = 0.433GL \tag{5.16}$$

For the more general case, where working fluid level is at depth D, the pressure (under the plunger) due to a column of fluid of height $(L - D)$ in the casing must be considered:

$$\Delta p = 0.433GL - 0.433G(L - D) = 0.433GD \tag{5.17}$$

From Eq. (5.14),

$$e = 12 \times 0.433GDA_pL/EA = 5.20GDA_pL/EA \tag{5.18}$$

Equation (5.18) is perfectly general, in that it gives the elongation of any member having cross-sectional area A. Specifically, the elongation of the tubing is

$$e_t = 5.20GDA_pL/EA_t \tag{5.19}$$

where A_t is the cross-sectional area of the tubing wall. For the rod string,

$$e_r = 5.20GDA_pL/EA_r \tag{5.20}$$

where A_r is the cross-sectional area of the rods. In the case of a tapered rod string, Eq. (5.18) must be applied to each section, i.e.,

$$e_1 = 5.20GDA_pL_1/EA_1, \qquad e_2 = 5.20GDA_pL_2/EA_2, \text{ etc.} \tag{5.21}$$

where e_1 is elongation of the L_1-foot section of rods of cross-sectional area A_1, e_2 is elongation of the L_2-foot section of rods of cross-sectional area A_2, etc. Then total rod stretch is

$$e_r = (5.20GDA_p/E)(L_1/A_1 + L_2/A_2 + \ldots) \tag{5.22}$$

Example 5.2.[9] A pumping well in the Brookhaven Field (Mississippi) has a 2-in. pump plunger set in $2\frac{1}{2}$-in. tubing on a tapered sucker rod string consisting of 1675 ft of 1-in. rods, 2100 ft of $\frac{7}{8}$-in. rods, and 2275 ft of $\frac{3}{4}$-in. rods. The specific gravity of the produced fluid is 0.91, and the pumping fluid level of the well is 4357 ft. Assuming the modulus of elasticity for steel to be 30×10^6 psi, calculate the loss in effective plunger stroke resulting from rod and tubing stretch.

SOLUTION:

$G = 0.91, \quad D = 4357 \text{ ft}, \quad E = 30 \times 10^6 \text{ psi}$

$L_1 = 2275 \text{ ft}, \quad L_2 = 2100 \text{ ft}, \quad L_3 = 1675 \text{ ft}$

$A_1 = 0.442$ sq in., $A_2 = 0.601$ sq in., $A_3 = 0.785$ sq in. (from Table 5.1)

$A_p = 3.142$ sq in. (from Table 5.2)

$A_t = 1.812$ sq in. (from Table 5.3)

$L = 2275 + 2100 + 1675 = 6050 \text{ ft}$

$$e_t = \frac{5.20 \times 0.91 \times 4357 \times 3.142 \times 6050}{30 \times 10^6 \times 1.812} = 7.2 \text{ in.}$$

$$e_r = \frac{5.20 \times 0.91 \times 4357 \times 3.142}{30 \times 10^6} \times \left[\frac{2275}{0.442} + \frac{2100}{0.601} + \frac{1675}{0.785}\right]$$
$$= 23.4 \text{ in.}$$

Loss in stroke $= 7.2 + 23.4 = 30.6$ in.

Table 5.1. SUCKER ROD DATA

Rod size, in.	Area, sq in.	Weight, lb/ft
$\frac{5}{8}$	0.307	1.16
$\frac{3}{4}$	0.442	1.63
$\frac{7}{8}$	0.601	2.16
1	0.785	2.88
$1\frac{1}{8}$	0.994	3.64

Table 5.2. Pump plunger data

Diameter, in.	Area, sq in.	Pump constant, bbl/day/in./spm
1	0.785	0.116
$1\frac{1}{16}$	0.886	0.131
$1\frac{1}{4}$	1.227	0.182
$1\frac{1}{2}$	1.767	0.262
$1\frac{3}{4}$	2.405	0.357
$1\frac{25}{32}$	2.488	0.369
2	3.142	0.466
$2\frac{1}{4}$	3.976	0.590
$2\frac{1}{2}$	4.909	0.728
$2\frac{3}{4}$	5.940	0.881
$3\frac{3}{4}$	11.045	1.639
$4\frac{3}{4}$	17.721	2.630

Table 5.3. Tubing data

Nominal size, in.	Outside diameter, in.	Weight, lb/ft	Wall area, sq in.
$1\frac{1}{2}$	1.900	2.90	0.800
2	2.375	4.70	1.304
$2\frac{1}{2}$	2.875	6.50	1.812
3	3.500	9.30	2.590
$3\frac{1}{2}$	4.000	11.00	3.077
4	4.500	12.75	3.601

In addition to rod stretch caused by fluid load, rod elongations also result from rod load, which consists of the dead weight of rods plus acceleration load. For an untapered string, the weight of rods suspended below any element of the string varies uniformly from zero at the bottom of the string to W_r at the top of the string. On the average, the weight of rods tending to cause elongation of an element is $W_r/2$. This is equivalent to considering the entire mass of the sucker rod string to be concentrated at $L/2$. In any event, the elongation of the rods (resulting from rod load) at the end of the downstroke will be, from Eq. (5.14),

$$e_d = \frac{12(W_r + W_r\alpha)L/2}{EA_r} \tag{5.23}$$

and the elongation of the rods at the end of the upstroke will be

$$e_u = \frac{12(W_r - W_r\alpha)L/2}{EA_r} \tag{5.24}$$

since the acceleration loads will be in opposite directions. The plunger overtravel, i.e., the net elongation resulting from acceleration, is

$$e_p = e_d - e_u = 12W_r\alpha L/EA_r \tag{5.25}$$

The weight of the rod string is

$$W_r = \rho_r LA_r/144 \tag{5.26}$$

where ρ_r is the density of the rod string in pounds per cubic foot. The density of steel is approximately 490 lb per cu ft. Substituting Eq. (5.26) in Eq. (5.25),

$$e_p = \frac{12L\alpha}{EA_r} \times \frac{490LA_r}{144} = \frac{40.8L^2\alpha}{E} \tag{5.27}$$

In the derivation of Eq. (5.27), the geometry of the surface equipment and the possibility of a tapered rod string were ignored. If these factors are considered, it is obvious that any generally applicable equation for plunger overtravel would be of a more complex form. Some authorities prefer use of

$$e_p = 32.8L^2\alpha/E \tag{5.28}$$

since the constant 32.8 is more empirically correct for a unit of "average" geometry than is the constant 40.8 in Eq. (5.27). Although Eqs. (5.27) and (5.28) give values of plunger overtravel which differ by some 25 per cent, the difference has little effect when the information is put to its ultimate use, i.e., calculation of effective plunger stroke.

Example 5.3. For the well of Example 5.2, the polished rod stroke length is 51 in. Estimate the plunger overtravel when the well is being pumped at 20 spm.

SOLUTION: $L = 6050$ ft $\quad S = 51$ in.

$E = 30 \times 10^6$ psi $\quad N = 20$ spm

$$e_p = \frac{40.8 \times 6050 \times 6050}{30 \times 10^6} \times \frac{51 \times 20 \times 20}{70{,}500} = 14.4 \text{ in.}$$

The effective plunger stroke is now seen to be the polished rod stroke decreased by the effects of rod and tubing stretch resulting from fluid load and increased by the effect of plunger overtravel. Then the effective plunger stroke is

$$S_p = S + e_p - (e_t + e_r) \tag{5.29}$$

Combining Eqs. (5.19), (5.22), (5.27), and (5.29),

$$S_p = S + \frac{40.8L^2\alpha}{E} - \frac{5.20GDA_p}{E}\left[\frac{L}{A_t} + \frac{L_1}{A_1} + \frac{L_2}{A_2} + \cdots\right] \tag{5.30}$$

In the case of an untapered rod string, Eq. (5.30) reduces to

$$S_p = S + \frac{40.8L^2\alpha}{E} - \frac{5.20GDA_pL}{E}\left[\frac{1}{A_t} + \frac{1}{A_r}\right] \tag{5.31}$$

As a final note on the subject of effective plunger stroke, it should be pointed out that there can be no tubing stretch in the case of anchored tubing. This case can best be handled by neglecting the terms involving A_t in Eqs. (5.30) and (5.31).

Example 5.4. For a well in the Santa Fe Springs Field (California) a pump with a $1\frac{3}{4}$-in. plunger is set on 4560 ft of $\frac{3}{4}$-in. rods. The fluid level is known to be low, and the tubing is anchored. With pumping at 20.5 spm and a 64-in. stroke length, production is 355 bbl per day of fluid of specific gravity 0.87. Calculate the effective plunger stroke.

SOLUTION: Assuming the pump to be set at the working fluid level, $D = L =$ 4560 ft. The sucker rod string is untapered, and Eq. (5.31) applies. Since tubing is anchored, the term L/A_t is neglected. From Tables 5.1 and 5.2, $A_r = 0.442$ sq in. and $A_p = 2.405$ sq in. From Eq. (5.8),

$$\alpha = \frac{SN^2}{70{,}500} = \frac{64 \times 20.5 \times 20.5}{70{,}500} = 0.381$$

$$S_p = S + \frac{L^2}{E}\left[40.8\alpha - 5.20\,\frac{GA_p}{A_r}\right] = 64 + \frac{20.8 \times 10^6}{30 \times 10^6}\,[15.55 - 24.60]$$

$$= 64 - 0.693 \times 9.05 = 64 - 6.3 = 57.7 \text{ in.}$$

5.8. Calculation of Polished Rod Loads

The selection of surface equipment for a pumping installation is influenced to a great extent by the anticipated maximum or peak polished rod load. The preliminary estimation of counterbalance required is based on the anticipated maximum and minimum polished rod loads. These quantities must, therefore, be determined to as great a degree of accuracy as available data permit. Several expressions have been proposed for the calculation of polished rod loads.[10] The ones presented here are believed to be reliable over the widest range of pumping speeds.

At one time or another during the pumping cycle, five factors contribute to the net polished rod load. They are (1) fluid load, (2) dead weight of sucker rods, (3) acceleration load of sucker rods, (4) buoyant force on sucker rods submerged in fluid, and (5) frictional forces. Vibration loads, which are discussed later, will not be considered here. Any acceleration load associated with the fluid being lifted will be neglected. Experimentally this seems justified, probably because of the relative compressibility of the fluid.

The weight of a tapered rod string is given by

$$W_r = M_1L_1 + M_2L_2 + \ldots \tag{5.32}$$

where M_1 is the weight, in pounds per foot, and L_1 is the length, in feet, of Section 1 of a tapered string, M_2 is the weight and L_2 is the length of

Section 2, etc. The maximum and minimum acceleration loads of the rods are given by

$$\text{Maximum acceleration load} = W_r\alpha \tag{5.33}$$

$$\text{Minimum acceleration load} = -W_r\alpha \tag{5.34}$$

Assuming the density of the rods to be 490 lb per cu ft, the volume of the rod string, and, consequently, the volume of fluid displaced by the rod string, is

$$\text{Volume} = \text{Weight/density} = W_r/490 \text{ cu ft} \tag{5.35}$$

The density of the displaced fluid (of specific gravity G) is $62.4G$ lb per cu ft. The buoyant force on the rods, which is the weight of the displaced fluid, is

$$\text{Buoyant force} = -(W_r/490)62.4G = -0.127W_rG \tag{5.36}$$

The negative sign in Eq. (5.36) is necessary because the buoyant force is always upward.

The fluid load to be used in calculating polished rod loads is the weight of the fluid column supported by the plunger. The volume of a column having as its base the plunger and as its height the sucker rod string would be

$$\text{Volume} = LA_p/144 \text{ cu ft} \tag{5.37}$$

The volume of fluid is obtained by subtracting from Eq. (5.37) the volume occupied by rods, as given by Eq. (5.35):

$$\text{Volume of fluid} = (LA_p/144) - (W_r/490) \tag{5.38}$$

Then the fluid load is

$$\begin{aligned} W_f &= 62.4G[(LA_p/144) - (W_r/490)] \\ &= 0.433G(LA_p - 0.294W_r) \end{aligned} \tag{5.39}$$

It should be recalled at this point that the fluid load is on the polished rod only during the upstroke.

The friction load cannot be predicted mathematically. For installations already in operation it can be estimated from dynamometer tests. For original design problems it is handled in the manner explained below. The friction load is here designated as $+F$ on the upstroke and $-F$ on the downstroke, since the frictional forces on a body act in a direction opposite the direction of motion of the body.

In summary, the maximum or peak polished rod load, which occurs on the upstroke, is

$$W_{\max} = W_f + W_r + W_r\alpha + F \tag{5.40}$$

and the minimum polished rod load, which occurs on the downstroke, is

$$W_{\min} = W_r - W_r\alpha - 0.127W_rG - F \tag{5.41}$$

When Eq. (5.40) is used to calculate peak load, the last term is generally neglected, as the friction load cannot be calculated with reasonable certainty. In this case

$$W_{\max} = W_f + W_r(1 + \alpha) \tag{5.42}$$

Equation (5.42) is known as the Mills equation.[10]

In the case of minimum load, buoyancy and friction are in the same direction. Here, only friction is neglected, resulting in Eq. (5.43), which, although not entirely satisfactory, is probably as good as any available:

$$W_{\min} = W_r(1 - \alpha - 0.127G) \tag{5.43}$$

Example 5.5. Calculate, for the well of Example 5.4, the maximum and minimum polished rod loads.

SOLUTION: For $\frac{3}{4}$-in. rods, $M = 1.63$ lb/ft. Then

$$W_r = 4560 \text{ ft} \times 1.63 \text{ lb/ft} = 7433 \text{ lb}$$

$$\begin{aligned} W_f &= 0.433G\,(LA_p - 0.294W_r) \\ &= 0.433 \times 0.87\,(4560 \times 2.405 - 0.294 \times 7433) \\ &= 3313 \text{ lb} \end{aligned}$$

$$\begin{aligned} W_{\max} &= W_f + W_r(1 + \alpha) = 3320 + 7440(1 + 0.381) \\ &= 3311 + 10{,}265 = 13{,}600 \text{ lb} \end{aligned}$$

$$\begin{aligned} W_{\min} &= W_r(1 - \alpha - 0.127G) = 7433(1 - 0.381 - 0.127 \times 0.87) \\ &= 7433 \times 0.508 = 3780 \text{ lb} \end{aligned}$$

5.9. Design of the Sucker Rod String

There are two possible approaches to the design of a tapered sucker rod string. Lengths of individual sections may be selected in such a manner as to make the unit stress at the top of each section the maximum permissible working stress, or lengths may be selected so as to make the unit stresses at the tops of the sections equal. The second method is more generally used, and its application will be outlined here.

In determining the stress at any point in a rod string, the force causing the stress is considered to result from fluid load on the plunger and weight of rods below the point under consideration. Four simplifying assumptions are made, namely, (1) static conditions, i.e., no acceleration loads are considered, (2) the specific gravity of the well fluid is 1.00, (3) the fluid load acts over the full plunger area, and (4) the pump is set at the working fluid level. Under these conditions the fluid load is $0.433LA_p$ pounds.

Consider first a two-way taper consisting of L_1 feet of rods of cross-sectional area A_1 square inches and weight M_1 pounds per foot, and L_2 feet of rods of cross-sectional area A_2 square inches and weight M_2 pounds per foot. Let the fractional parts (of the total rod length) represented by the sections be designated as R_1 and R_2, i.e.,

$$R_1 = L_1/L \quad \text{and} \quad R_2 = L_2/L \tag{5.44}$$

where $L = L_1 + L_2$. The stress at the top of the lower section is then

$$\frac{0.433LA_p + L_1M_1}{A_1} = \frac{0.433LA_p + LR_1M_1}{A_1} \tag{5.45}$$

Likewise, the stress at the top of the upper section is

$$\frac{0.433LA_p + L_1M_1 + L_2M_2}{A_2} = \frac{0.433LA_p + LR_1M_1 + LR_2M_2}{A_2} \tag{5.46}$$

If stresses at the tops of the sections are equal,

$$\frac{0.433A_p + R_1M_1}{A_1} = \frac{0.433A_p + R_1M_1 + R_2M_2}{A_2} \tag{5.47}$$

For selected rod and plunger sizes, the length of each section can be determined by making use of Eq. (5.47) plus the fact that

$$R_1 + R_2 = 1 \tag{5.48}$$

For a tapered string consisting of more than two sections, the analysis would differ from that presented above only in degree of complexity. For a four-way taper, for example, Eq. (5.47) must be supplemented by

$$\frac{0.433A_p + R_1M_1}{A_1} = \frac{0.433A_p + R_1M_1 + R_2M_2 + R_3M_3}{A_3} \tag{5.49}$$

and

$$\frac{0.433A_p + R_1M_1}{A_1} = \frac{0.433A_p + R_1M_1 + R_2M_2 + R_3M_3 + R_4M_4}{A_4} \tag{5.50}$$

and instead of Eq. (5.48) we must use

$$R_1 + R_2 + R_3 + R_4 = 1 \tag{5.51}$$

Equations (5.47), (5.49), (5.50), and (5.51) apply to any tapered string with up to four sections. For a two-way taper R_3 and R_4 are zero, and for a three-way taper R_4 is zero.

Example 5.6. For the sucker rod string of a certain pumping well it is desired to use a two-way taper consisting of $\frac{5}{8}$-in. and $\frac{3}{4}$-in. rods. Derive expressions for R_1 and R_2 in terms of the pump plunger area A_p.

SOLUTION: From Table 5.1,

$$A_1 = 0.307 \text{ sq in.} \qquad A_2 = 0.442 \text{ sq in.}$$

$$M_1 = 1.16 \text{ lb/ft} \qquad M_2 = 1.63 \text{ lb/ft}$$

Using Eq. (5.47),

$$\frac{0.433A_p + 1.16R_1}{0.307} = \frac{0.433A_p + 1.16R_1 + 1.63R_2}{0.442}$$

$$1.416A_p + 3.79R_1 = 0.980A_p + 2.62R_1 + 3.69R_2$$

$$1.17R_1 = 3.69R_2 - 0.436A_p$$

Using Eq. (5.48),

$$R_2 = 1 - R_1$$

so that

$$1.17R_1 = 3.69 - 3.69R_1 - 0.436A_p$$

$$4.86R_1 = 3.69 - 0.436A_p$$

and thus

$$R_1 = 0.759 - 0.0896A_p$$

$$R_2 = 0.241 + 0.0896A_p$$

By calculations similar to those of Example 5.6, it is possible to determine values of R for all probable combinations of sucker rod sizes. The results are shown in Table 5.4.

Table 5.4 DATA FOR DESIGN OF TAPERED SUCKER ROD STRING

Rod sizes in string, in.	Values of R^a
$\frac{5}{8}$–$\frac{3}{4}$	$R_1 = 0.759 - 0.0896A_p$ $R_2 = 0.241 + 0.0896A_p$
$\frac{3}{4}$–$\frac{7}{8}$	$R_1 = 0.786 - 0.0566A_p$ $R_2 = 0.214 + 0.0566A_p$
$\frac{7}{8}$–1	$R_1 = 0.814 - 0.0375A_p$ $R_2 = 0.186 + 0.0375A_p$
$\frac{5}{8}$–$\frac{3}{4}$–$\frac{7}{8}$	$R_1 = 0.627 - 0.1393A_p$ $R_2 = 0.199 + 0.0737A_p$ $R_3 = 0.175 + 0.0655A_p$
$\frac{3}{4}$–$\frac{7}{8}$–1	$R_1 = 0.664 - 0.0894A_p$ $R_2 = 0.181 + 0.0478A_p$ $R_3 = 0.155 + 0.0416A_p$
$\frac{3}{4}$–$\frac{7}{8}$–1–1$\frac{1}{8}$	$R_1 = 0.582 - 0.1110A_p$ $R_2 = 0.158 + 0.0421A_p$ $R_3 = 0.137 + 0.0364A_p$ $R_4 = 0.123 + 0.0325A_p$

[a] R_1 refers to lowest section (smallest rods), R_2 to next lowest section, etc.

Example 5.7. A pump with a 2-in. plunger is to be set at 6050 ft using a three-way taper consisting of $\frac{3}{4}$-in., $\frac{7}{8}$-in., and 1-in. rods. Sucker rods are available in 25-ft lengths. Determine the length of each section of the tapered rod string.

SOLUTION:

$$
\begin{aligned}
A_p &= 3.142 \text{ sq in.} \quad \text{(from Table 5.2)} \\
R_1 &= 0.664 - 0.0894 \times 3.142 = 0.383 \\
R_2 &= 0.181 + 0.0478 \times 3.142 = 0.331 \\
R_3 &= 0.155 + 0.0416 \times 3.142 = 0.286 \\
L_1 &= LR_1 = 6050 \times 0.383 = 2317 \text{ ft} \\
L_2 &= LR_2 = 6050 \times 0.331 = 2003 \text{ ft} \\
L_3 &= LR_3 = 6050 \times 0.286 = 1730 \text{ ft}
\end{aligned}
$$

In 25-ft increments,

$$
\begin{aligned}
L_1 &= 2325 \text{ ft} \\
L_2 &= 2000 \text{ ft} \\
L_3 &= 1725 \text{ ft}
\end{aligned}
$$

As a final step in the design of the rod string, it is necessary to check maximum anticipated stress against allowable working stress for the rods being used. This can be done by calculating maximum stress at the top of the rod string, which is the peak polished rod load divided by the cross-sectional area of the top section of rods. If the rod string has been designed as outlined above, the maximum stress at the top of the string will also be the maximum stress which can possibly be encountered at any point below the top.

Example 5.8. The sucker rods being used in the well of Example 5.7 are type 4620, of nickel-molybdenum alloy steel. The presence of hydrogen sulfide in the well fluid causes a severe corrosive condition which limits maximum working stress to 22,000 psi, as recommended by the manufacturer. Surface pumping conditions are expected to be 18 spm at a stroke length of 54 in. The well fluid has a specific gravity of 0.89. Compare the recommended maximum stress with the maximum stress to be expected under the above operating conditions.

SOLUTION: From Table 5.1,

$$
\begin{aligned}
A_3 &= 0.785 \text{ sq in.} \qquad M_1 = 1.63 \text{ lb/ft} \\
M_2 &= 2.16 \text{ lb/ft} \qquad M_3 = 2.88 \text{ lb/ft}
\end{aligned}
$$

$$
\begin{aligned}
W_r &= M_1L_1 + M_2L_2 + M_3L_3 \\
&= 1.63 \times 2325 + 2.16 \times 2000 + 2.88 \times 1725 \\
&= 13{,}078 \text{ lb} \\
W_f &= 0.433G(LA_p - 0.294W_r) \\
&= 0.433 \times 0.89(6050 \times 3.142 - 0.294 \times 13{,}078) \\
&= 5844 \text{ lb} \\
\alpha &= \frac{SN^2}{70{,}500} = \frac{54 \times 18 \times 18}{70{,}500} = 0.248 \\
W_{\max} &= W_f + W_r(1 + \alpha) \\
&= 5844 + 13{,}078(1 + 0.248) = 22{,}165 \text{ lb} \\
\text{Stress} &= \frac{W_{\max}}{A_3} = \frac{22{,}165}{0.785} = 28{,}236 \text{ psi}
\end{aligned}
$$

The above stress greatly exceeds the recommended maximum of 22,000 psi, and emphasizes the problems resulting from the presence, in a well fluid, of hydrogen sulfide and other corrosive agents. In a noncorrosive fluid, the rods described could withstand working stresses of 30,000 psi without excessive failures. Use of a chemical inhibitor would be indicated in a well such as the one described in this example. Corrosive fluids effectively treated in this manner (i.e., with an inhibitor) may be considered noncorrosive, in which case the higher working stress is applicable.

5.10. Pump Displacement and Production Rate

Having developed expressions in Sec. 5.7 for effective plunger stroke, we can now define the theoretical pump displacement as

$$\begin{aligned} V &= A_p \text{ in.}^2 \times S_p \frac{\text{in.}}{\text{stroke}} \times N \frac{\text{strokes}}{\text{min}} \times \frac{1440 \text{ min/day}}{9702 \text{ in.}^3/\text{bbl}} \\ &= 0.1484 A_p S_p N \text{ bbl/day} \end{aligned} \tag{5.52}$$

For a given plunger diameter, the term $0.1484A_p$ in Eq. (5.52) is independent of surface operating conditions and is called the pump constant, K. Then Eq. (5.52) becomes

$$V = KS_pN \tag{5.53}$$

Table 5.2 shows plunger areas and pump constants for all API pump sizes.

Example 5.9. Calculate the pump constant for a 2-in. plunger.

SOLUTION:

$$K = 0.1484A_p = 0.1484(\pi/4)2^2 = 0.466 \text{ bbl/day/in./spm}$$

The ratio between the fluid actually handled and the pump displacement is the volumetric efficiency of the pump:

$$E_v = q/V \quad \text{or} \quad q = VE_v \quad \text{or} \quad V = q/E_v \tag{5.54}$$

where q is the rate of production from the well in barrels per day. Volumetric efficiency is an extremely important factor in design problems, yet it cannot be known to any degree of certainty until the production rate has been established, and even then the difficulties inherent in the determination of plunger stroke make calculation of volumetric efficiency something less than exact. Volumetric efficiencies, which vary over rather wide ranges, are affected by many factors, including nature of the fluid produced, type of pump, depth of pump, surface operating conditions, and gas interference. The last factor is particularly significant. For example, wells with gaseous or foamy liquid conditions will often have efficiencies as low as 25 to 50 per cent. Gaseous wells with fair separation of formation gas in the hole will have efficiencies of 50 to 70 per cent. For wells with good separation in the hole and ample pump submergence, efficiencies will be of the order 70 to 80 per cent, and for a high fluid level well with no gas interference, volumetric efficiency may approach 100 per cent.[11] When designing an installation in a particular area, it is best to rely on previous experience in *that* area in estimating volumetric efficiency. If this is not possible, and if little is known as to the nature of the well fluid, a good estimate would be 75 to 80 per cent.

Example 5.10. Calculate the volumetric efficiency for the pump of Example 5.4.

SOLUTION:

$K = 0.357$ (from Table 5.2)

$V = KS_pN = 0.357 \times 57.7 \times 20.5 = 422$ bbl/day

$q = 344$ bbl/day

$E_v = q/V = 344/422 = 0.82$

5.11. *Counterbalance Design*

It will be recalled from Sec. 5.4 that the primary function of the counterbalance system is to store energy on the downstroke, when polished rod load is low, and to release energy on the upstroke, when polished rod load is high, thus distributing more uniformly, throughout the pumping cycle, the loads and torques which must be sustained by the prime mover. Theoretically, the ideal counterbalance effect, C_i, would be such that the prime mover would carry the same average loads on the upstroke and on the downstroke, that is,

$$W_{\max} - C_i = C_i - W_{\min} \tag{5.55}$$

The ideal counterbalance effect is then

$$C_i = 0.50(W_{\max} + W_{\min}) \tag{5.56}$$

Substituting Eqs. (5.40) and (5.41) into (5.56),

$$C_i = 0.5W_f + W_r(1 - 0.127G) \tag{5.57}$$

From Eq. (5.57) it is seen that the ideal counterbalance effect balances one-half the fluid load plus the weight of rods in fluid.

Figure 5.4 shows that the counterbalance effect C_w, due to a counterweight of W_c pounds, depends on the geometry of the pumping unit and on the stroke length, as well as on the weight and position of the counterweight. In Fig. 5.4, β is the angle between the walking beam and the horizontal at any given time, γ is the angle between the pitman and the vertical, and θ is the angle between the crank and the vertical. The distance from crankshaft to center of gravity of the counterweight is d, and the distance from crank-

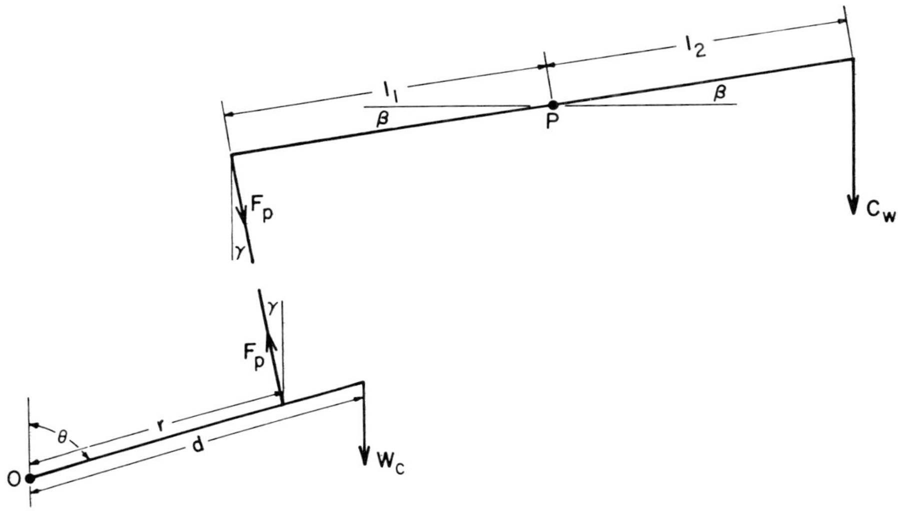

Fig. 5.4 Counterbalance effect of the counterweight.

shaft to pitman bearing is r. F_p is the force in the pitman, which for crank counterbalance, is always tension. Point O represents the crankshaft and point P represents the saddle bearing. Distances from saddle bearing to tail bearing and to bridle are l_1 and l_2, respectively. Taking moments about point P,

$$\begin{aligned} C_w(l_2 \cos \beta) &= (F_p \cos \gamma)(l_1 \cos \beta) \\ &\quad + (F_p \sin \gamma)(l_1 \sin \beta) \end{aligned}$$

$$F_p = \frac{C_w l_2 \cos \beta}{l_1[(\cos \gamma)(\cos \beta) + (\sin \gamma)(\sin \beta)]} \tag{5.58}$$

Taking moments about point O,

$$\begin{aligned} W_c(d \sin \theta) &= (F_p \cos \gamma)(r \sin \theta) \\ &\quad + (F_p \sin \gamma)(r \cos \theta) \end{aligned}$$

$$F_p = \frac{W_c d \sin \theta}{r[(\cos \gamma)(\sin \theta) + (\sin \gamma)(\cos \theta)]} \tag{5.59}$$

Equating (5.58) and (5.59) gives a relationship between the counterweight and its resulting counterbalance effect. The relationship can be simplified by assuming the angle γ to be very small at all times (which is a good assumption, provided the length of the pitman is very much greater than r).

In this case, sin γ is approximately zero and the cos γ is approximately unity. Equating (5.58) and (5.59) now gives

$$\frac{C_w l_2 \cos\beta}{l_1 \cos\beta} = \frac{W_c\, d \sin\theta}{r \sin\theta}$$

or

$$C_w = W_c(d/r)(l_1/l_2) \tag{5.60}$$

In addition to the effect of counterweights, some counterbalance effect may be provided by structural unbalance of the surface installation itself. If we call this effect C_s, the total counterbalance effect at the polished rod is

$$C = C_s + W_c(d/r)(l_1/l_2) \tag{5.61}$$

Example 5.11. For the well of Examples 5.4 and 5.5, the polished rod stroke length was 64 in., specific gravity of well fluid was 0.87, and rod and fluid loads were calculated to be 7433 lb and 3311 lb, respectively. The surface equipment includes a Lufkin C-160D-64-23* pumping unit assembly with 6460R** crank and 3CR counterweight. For this equipment the effect of structural unbalance (according to the manufacturer) is +671 lb, the saddle bearing is centered between bridle and tail bearing, and the length of the crank is 59.5 in. If the 5250-lb 3CR counterweight has its center of gravity 7.05 in. from its outer edge, how should it be positioned in order to obtain the theoretical ideal counterbalance effect?

SOLUTION:

$$\begin{aligned} C_i &= 0.5W_f + W_r(1 - 0.127G) \\ &= 0.5 \times 3311 + 7433(1 - 0.127 \times 0.87) = 8267 \text{ lb} \end{aligned}$$

Since $l_1 = l_2$, $r = S/2 = 64/2 = 32$ in., and thus

$$d = \frac{r(C - C_s)}{W_c} = \frac{32(8267 - 671)}{5250} = 46.3 \text{ in.}$$

The distance of outer edge of counterweight from end of crank is

$$59.5 - 46.3 - 7.05 = 6.2 \text{ in.}$$

5.12. *Calculation of Torque*

Closely related to the problem of counterbalance design is the matter of torque calculation. As has been stated, pumping units are rated in accordance with the maximum permissible torque on the gear reducer.

*For this pumping unit designation, "C" indicates crank counterbalance, "160" is the API rating, "D" indicates double reduction gear, "64" is the maximum available stroke length in inches, and "23" is the beam load capacity in thousands of pounds. Most manufacturers use similar designations.

**This designation indicates a maximum polished rod stroke of 64 in. and a crank radius of approximately 60 in.

If the geometry of the surface installation is not considered (i.e., $l_1 = l_2$, and structural unbalance is neglected) Eq. (5.61) reduces to

$$C = 2W_c d/S \tag{5.62}$$

since $r = S/2$. It should be understood that any calculations based on Eq. (5.62) are only approximations. Figure 5.5 shows forces acting on the crank at the instant the crank makes an angle θ with the vertical, as measured clockwise from crank position at the beginning of the upstroke. The polished rod load, W, is transmitted to the crank through the pitman, which is presumed to be vertical. The net torque about O (which is the torque on the gear reducer) is

$$T = Wr \sin\theta - W_c d \sin\theta \tag{5.63}$$

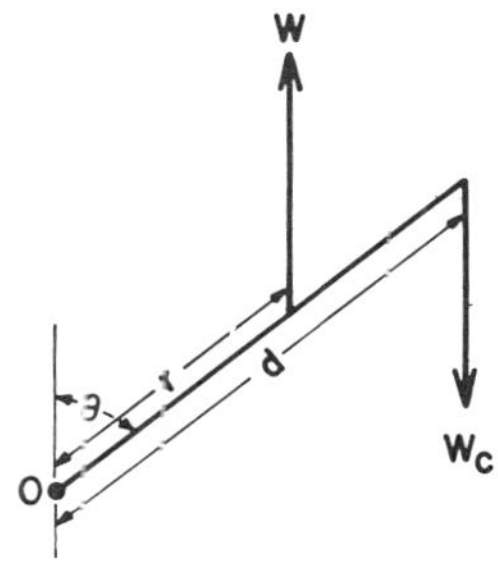

Fig. 5.5 Instantaneous torque on the gear box.

and from Eq. (5.62)

$$W_c d = CS/2 \tag{5.64}$$

Substituting Eq. (5.64) in (5.63),

$$T = W(S/2)\sin\theta - C(S/2)\sin\theta$$

$$T = (W - C)(S/2)\sin\theta \tag{5.65}$$

Equation (5.65) is an approximate expression for instantaneous torque on the gear box.

The highest possible values for variables W and $\sin\theta$ are, respectively, peak polished rod load, W_{max}, and sin 90° (= 1). Peak torque is, then

$$T_p = (W_{max} - C)(S/2) \tag{5.66}$$

In attempting to predict peak torque from calculations of ideal counterbalance, allowance must be made for the fact that the actual counterbalance of the unit may be somewhat different from the ideal. Usually, a unit can be counterbalanced to within 5 to 10 per cent of the ideal value. Assuming the counterbalance is 95 per cent of the ideal value, the expression for predicting peak torque is

$$T_p = (W_{max} - 0.95C_i)(S/2) \tag{5.67}$$

Example 5.12. Assuming that the unit of Example 5.11 will be no more than 5 per cent out of balance, calculate peak torque on the gear box. Can the unit described handle this torque satisfactorily?

SOLUTION: From Example 5.5,

$$W_{\max} = 13{,}600 \text{ lb}$$

From Example 5.11,

$$C_i = 8267 \text{ lb}$$

$$T_p = (13{,}600 - 0.95 \times 8267)(64/2) = 183{,}000 \text{ in.-lb}$$

The unit described is rated at 160,000 in.-lb, and the gear reducer would probably fail if it were subjected to a peak torque to 183,000 in.-lb.

5.13. Speed Reduction from Prime Mover to Crankshaft

Power is transmitted from the engine sheave, of diameter d_e, to the unit sheave, of diameter d_u, by means of V-belts. If the speed of the prime mover engine is N_e revolutions per minute, the belt velocity is

$$v_b = \pi d_e N_e \text{ in./min} \tag{5.68}$$

and the speed of the unit sheave is

$$N_u = v_b/\pi d_u = N_e(d_e/d_u) \text{ rpm} \tag{5.69}$$

If the gear ratio at the gear reducer is Z, pumping speed is

$$N = N_u/Z = N_e d_e/Z d_u \tag{5.70}$$

Example 5.13. A 1200 rpm electric motor is to be used as the prime mover for a unit operating at 20 spm. The double reduction gear reducer has a gear ratio of 29.32 to 1, and the unit sheave has a pitch diameter of 19.25 in. Calculate the diameter of the sheave to be used on the motor.

SOLUTION:

$$\begin{aligned} d_e &= Z d_u(N/N_e) = 29.32 \times 19.25 \times 20/1200 \\ &= 9.4 \text{ in.} \end{aligned}$$

5.14. Power Requirements of the Prime Mover

Two power loads must be considered in moving fluid from pump to surface. Hydraulic horsepower, H_h, is a statement of useful expenditure of energy. If q barrels per day of a fluid having specific gravity G are lifted from a depth L feet, the useful power involved is

$$\begin{aligned} H_h &= \frac{q \text{ bbl/day} \times 350G \text{ lb/bbl} \times L \text{ ft}}{1440 \text{ min/day} \times 33{,}000 \text{ ft-lb/min/hp}} \\ &= 7.36 \times 10^{-6} qGL \text{ hp} \end{aligned} \tag{5.71}$$

Equation (5.71) assumes the pump to be at the working fluid level, and neglects the effect of tubing pressure. As a more general statement,

$$H_h = 7.36 \times 10^{-6}\, qGL_N \tag{5.72}$$

The net lift, L_N, is the pressure differential through which the fluid passes (as it moves from the pump to the surface) expressed as feet of produced fluid. Net lift differs from pump setting depth because of effects of casing pressure (which represents a force tending to lift fluid) and tubing pressure (which represents a force tending to prevent fluid from being lifted). In terms of feet of fluid, the effect of casing pressure is the difference between pump depth and the working fluid level in the casing, i.e., $L - D$. The effect of tubing pressure, p_t, is to add to the actual lift a lift equivalent to

$$\frac{p_t \text{ lb/in.}^2 \times 144 \text{ in.}^2/\text{ft}^2}{62.4G \text{ lb/ft}^3} = \frac{p_t}{0.433G} \text{ ft} \tag{5.73}$$

The net lift is then given by

$$L_N = L - (L - D) + (p_t/0.433G) = D + (2.31p_t/G) \tag{5.74}$$

The second subsurface power load which must be considered is the frictional energy loss between pump and polished rod. It has been found empirically[12] that frictional energy loss per stroke can be estimated from

$$\tfrac{1}{8}W_r \times 2S = 0.25W_rS \text{ in.-lb} \tag{5.75}$$

For a pumping speed of N strokes per minute, friction horsepower is

$$H_f = \frac{0.25W_rSN \text{ in.-lb/min}}{12 \text{ in./ft} \times 33{,}000 \text{ ft-lb/min/hp}}$$

$$= 6.31 \times 10^{-7}\, W_rSN \text{ hp} \tag{5.76}$$

The total polished rod horsepower will, of course, be the sum of hydraulic and friction horsepowers. Brake horsepower for the prime mover must involve polished rod horsepower and a safety factor to account for unpredictable power losses within the surface equipment. A satisfactory safety factor is believed to be 1.5,*[,11] in which case the brake horsepower requirement for the prime mover would be

$$H_b = 1.5(H_h + H_f) \tag{5.77}$$

In selecting the prime mover, it must be remembered that the nameplate horsepower rating of the engine must in many cases be substantially reduced to arrive at the actual engine capacity. For example, a gas engine will have reduced capacity when used at higher altitudes. For an electric motor the nameplate rating must be reduced to account for horsepower

*Recent investigations reveal that a safety factor of 1.25 is being adopted.

capacity which is lost because of motor heating resulting from cyclic loading imposed by rod pumping.[11]

Example 5.14. For the well of Examples 5.4 and 5.5, the production was 355 bbl/day of a 0.87 gravity fluid at a polished rod stroke length of 64 in. and a pumping speed of 20.5 spm. Depth to the pump was 4560 ft, and weight of the sucker rod string was 7433 lb. Assuming the pump to be set at the working fluid level, and neglecting tubing pressure, calculate the minimum nameplate horsepower rating of the electric motor to be used as the prime mover if the manufacturer recommends a 25 per cent reduction in nameplate rating because of cyclic loading.

SOLUTION:

$$H_h = 7.36 \times 10^{-6} qGL = 7.36 \times 10^{-6} \times 355 \times 0.87 \times 4560$$

$$= 10.38 \text{ hp}$$

$$H_f = 6.31 \times 10^{-7} W_r SN = 6.31 \times 10^{-7} \times 7433 \times 64 \times 20.5$$

$$= 6.15 \text{ hp}$$

$$H_b = 1.5(H_h + H_f) = 1.5(10.38 + 6.15) = 24.8 \text{ hp}$$

$$\text{Motor rating} = \frac{24.8}{0.75} = 33 \text{ hp}$$

Example 5.15. What is the net lift for the well of the previous example if tubing pressure is 50 psig and pump submergence is 250 ft? Assume the casing to be at atmospheric pressure.

SOLUTION:

$$D = 4560 - 250 = 4310 \text{ ft}$$

$$L_N = D + 2.31 p_t / G = 4310 + 2.31 \times 50/0.87 = 4443 \text{ ft}$$

5.15. Summary of Design Equations

Summarized here are the equations derived in Secs. 5.6 through 5.14. Also given are a number of other expressions which have been developed for analysis of well problems. More detailed explanations of the latter group of equations can be found in references cited.

Listed below are symbols (with definitions and units) which are used in the *derived* equations.

$A_1, A_2 \ldots$	Cross-sectional areas of individual sections of a tapered sucker rod string, sq in.
A_p	Cross-sectional area of plunger, sq in.
A_t	Cross-sectional area of tubing wall, sq in.
C	Total counterbalance effect at polished rod, lb
C_i	Theoretical ideal counterbalance effect, lb
C_s	Counterbalance effect due to structural unbalance, lb

D	Depth to working fluid level, ft
E	Modulus of elasticity for steel, psi
E_v	Volumetric efficiency of pump
G	Specific gravity of well fluid
H_b	Brake horsepower of prime mover, hp
H_f	Subsurface frictional power loss, hp
H_h	Hydraulic horsepower to lift fluid, hp
K	Pump constant, barrels per day of production per inch of plunger stroke per spm pumping speed
L	Length of sucker rod string, ft
$L_1, L_2 \ldots$	Lengths of individual sections of a tapered sucker rod string, ft
L_N	Net lift of fluid, ft
$M_1, M_2 \ldots$	Unit weights of individual sections of a tapered sucker rod string, lb/ft
N	Pumping speed, spm
N_e	Prime mover engine speed, rpm
p_t	Tubinghead pressure, psi
q	Production rate, bbl/day
$R_1, R_2 \ldots$	Lengths of individual sections of a tapered sucker rod string expressed as fractions of total length of the rod string
S	Polished rod stroke length, in.
S_p	Effective plunger stroke length, in.
T_p	Peak torque on gear reducer, in.-lb
V	Pump displacement, bbl/day
W_c	Weight of crank counterweights, lb
W_f	Fluid load, lb
$W_{\max}$	Maximum (peak) polished rod load, lb
$W_{\min}$	Minimum polished rod load, lb
W_r	Weight of sucker rod string, lb
Z	Gear ratio of gear reducer
d	Distance along crank from center of crankshaft to center of gravity of counterweights, in.
d_e	Diameter of engine sheave, in.
d_u	Diameter of unit sheave, in.
l_1	Distance along walking beam from saddle bearing to tail bearing, in.
l_2	Distance along walking beam from saddle bearing to bridle, in.
r	Distance along crank from crankshaft to pitman bearing, in.
α	Acceleration factor for simple harmonic motion

In this list, and in the design equations which follow, when subscripts 1, 2, . . . , n are used, subscript 1 refers to the lowest section of the tapered sucker rod string, subscript 2 refers to the second lowest section, etc.

Equation for acceleration factor:

$$\alpha = SN^2/70{,}500$$

Equations used in designing tapered rod string of n sections:

$$\frac{0.433A_p + R_1M_1}{A_1} = \frac{0.433A_p + R_1M_1 + R_2M_2 + \ldots + R_nM_n}{A_n}$$

$$R_1 + R_2 + \ldots + R_n = 1$$

Equation for effective plunger stroke:

$$S_p = S + \frac{40.8L^2\alpha}{E} - \frac{5.20GDA_p}{E}\left[\frac{L}{A_t} + \frac{L_1}{A_1} + \frac{L_2}{A_2} + \ldots\right]$$

For anchored tubing, $L/A_t = 0$.

Equations for polished rod loads:

$$W_r = M_1L_1 + M_2L_2 + \ldots$$

$$W_f = 0.433G(LA_p - 0.294W_r)$$

$$W_{max} = W_f + W_r(1 + \alpha)$$

$$W_{min} = W_r(1 - \alpha - 0.127G)$$

Equations involving production rate:

$$K = 0.1484A_p$$

$$V = KS_pN$$

$$q = VE_v$$

Equations for counterbalance and torque:

$$C_i = 0.5W_f + W_r(1 - 0.127G)$$

$$C = C_s + W_c(d/r)(l_1/l_2)$$

$$T_p = (W_{max} - C)(S/2)$$

Equation relating prime mover and pumping speeds:

$$N = N_ed_e/Zd_u$$

Equations involving power requirements:

$$L_N = D + 2.31p_t/G$$

$$H_h = 7.36 \times 10^{-6}\, qGL_N$$

$$H_f = 6.31 \times 10^{-7}\, W_rSN$$

$$H_b = 1.5(H_h + H_f)$$

In attempting to arrive at a more empirically accurate expression for polished rod load to be used in calculating peak torque, Slonneger has proposed the expression[10]

$$W_{\max} = (W_r + W_f)(1 + SN/5400) \tag{5.78}$$

The net polished rod load is then taken to be the larger of the values given by Eqs. (5.79),

$$W_N = 0.4W_{\max} \quad \text{or} \quad W_N = W_{\max} - C \tag{5.79}$$

The peak torque on the gear reducer will then be

$$T_p = W_N(S/2) \tag{5.80}$$

The steps outlined in Eqs. (5.78) through (5.80) constitute the API method for calculating peak torque. In these three equations, all terms are as defined for the derived equations, with the exception of fluid load, which, for use in Eq. (5.78) only, is defined as

$$W_f = 0.433GLA_p \tag{5.81}$$

Example 5.16. In Example 5.12 the peak torque of a certain unit was calculated to be 183,000 in.-lb, based on peak polished rod load as given by the Mills equation. Calculate peak torque using the API method.

SOLUTION:

$W_f = 0.433 \times 0.87 \times 4560 \times 2.405 = 4131$ lb

$W_{\max} = (7433 + 4131)(1 + 64 \times 20.5/5400) = 14{,}382$ lb

$W_N = 0.4 \times 14{,}382 = 5752$ lb

$T_p = 5752 \times 64/2 = 184{,}000$ in.-lb

The investigations of Marsh and Coberly relative to effective plunger stroke took no account of the effect of rod vibrations. Rieniets,[13] in an attempt to apply the theory of vibrations to calculations of plunger travel developed an expression which can be written

$$S_p = S\cos\phi + W_tB/J - 2W_f/J - F/J \tag{5.82}$$

Effective plunger stroke S_p, polished rod stroke length S and fluid load W_f are used here as previously defined. ϕ is the phase angle between natural and forced vibrations of the rod string. It can be determined from

$$\phi = \arc\tan\left[\frac{\Gamma(\tau_1/\tau_2)}{1 - (\tau_1/\tau_2)^2}\right] \tag{5.83}$$

where Γ is the damping factor for the vibrations and τ_1 and τ_2 are periods for natural and forced vibrations, respectively. Γ depends on the nature of the fluid surrounding the vibrating rods, and varies from 0.4 for wells producing pipeline oil to 2 for well fluids of above 90 per cent water. If the velocity of stress transmission in the rod string is v feet per second and the length of the rod string is L feet, then the period of natural vibration is

$$\tau_1 = 4L/v \tag{5.84}$$

The validity of Eq. (5.84) will be established in Sec. 5.17. If the pumping speed is N strokes per minute, the period of forced vibration is

$$\tau_2 = 60/N \text{ sec} \tag{5.85}$$

In Eq. (5.82), W_t is the total disturbing force acting on the system, and B is the magnification factor for the disturbing force:

$$W_t = W_f(1 + \alpha) + 2W_r\alpha \tag{5.86}$$

$$B = \{[1 - (\tau_1/\tau_2)^2] + \Gamma^2(\tau_1/\tau_2)^2\}^{-0.5} \tag{5.87}$$

Again, the rod weight W_r and acceleration factor α have been introduced previously. In the last term of (5.82), F is the rod friction load. This can be estimated from dynamometer tests on completed installations, but is generally a completely unknown factor in design problems, and must often be neglected. J is the spring constant for the rod string, and can be calculated from

$$J = \frac{0.0833E}{L_1/A_1 + L_2/A_2 + \ldots} \tag{5.88}$$

E being the modulus of elasticity, and L and A being lengths and cross-sectional areas of the sections of a tapered string.

Other investigators, notably Langer and Lamberger, have applied vibration theory to solution of pumping well problems.[14] The equations below for effective plunger stroke and peak polished rod load are simplified versions of more complex expressions developed by Langer and Lamberger.

$$S_p = \frac{S}{\cos{(0.0004LN)^\circ}} - \frac{W_dL}{2 \times 10^6}\left[\frac{1}{A_r} + \frac{1}{A_t}\right] \tag{5.89}$$

$$W_{\max} = W_r - W_{rb} + W_d + 0.75SN[17A_r \tan{(0.0004LN)^\circ} + A_N] \tag{5.90}$$

The only terms in the above equations not previously defined in this section are: W_d, differential fluid load; A_r, cross-sectional area of sucker rods; W_{rb}, buoyant force on the rods; and A_N, net plunger area. The differential fluid load is the net fluid head on the full plunger area:

$$W_d = 0.433GDA_p \tag{5.91}$$

The buoyant force on the rods is the weight of fluid displaced by the rods:

$$W_{rb} = 62.4G(W_r/490) = 0.127GW_r \tag{5.92}$$

In terms of full plunger area and rod area,

$$A_N = A_p - A_r \tag{5.93}$$

The Langer and Lamberger equations can then be written

$$S_p = S \sec \omega - 50 \times 10^{-8} W_d L(1/A_r + 1/A_t) \tag{5.94}$$

$$W_{\max} = W_r(1 - 0.127G) + W_d + 0.75SN(17A_r \tan \omega + A_N) \tag{5.95}$$

where $\omega = (0.0004LN)°$.

Example 5.17. A pumping well in the Lorraine Field (Kansas) has a pump with a $2\frac{1}{2}$-in. plunger set in 3-in. tubing on a 2650-ft string of $\frac{7}{8}$-in. sucker rods. The production rate is 47.5 bbl of oil and 437 bbl of water per day at a stroke length of 54 in. and a pumping speed of 21 spm. Specific gravity of the produced fluid is 0.788, and working fluid level is 1181 feet. Compare the values for effective plunger stroke as calculated by the equations of Marsh and Coberly, the Rieniets equation, and the Langer and Lamberger equation. Also, compare values for peak polished rod load as calculated by the Mills equation, the Slonneger (API) equation, and the Langer and Lamberger equation. Assume the modulus of elasticity for the rods to be 30×10^6 psi and the velocity of stress transmission in the rods (i.e., the velocity of sound in steel) to be 15,800 fps. Neglect the effect of rod friction in the Rieniets equation.

Solution: We first calculate the individual terms of the six equations.

$A_p = 4.909$ sq in. ($2\frac{1}{2}$-in. plunger)

$A_t = 2.590$ sq in. (3-in. tubing)

$A_r = 0.601$ sq in., $M = 2.16$ lb/ft ($\frac{7}{8}$-in. rods)

Water cut $= 437/(437 + 47.5) = 0.902$, $\Gamma = 2$

$W_r = ML = 2.16 \times 2650 = 5724$ lb

$W_f = 0.433G(LA_p - 0.294W_r) = 0.3413(13{,}009 - 1683) = 3866$ lb

$\alpha = SN^2/70{,}500 = 54 \times 21 \times 21/70{,}500 = 0.338$

$W_t = W_f(1 + \alpha) + 2W_r\alpha = 5173 + 3869 = 9042$ lb

$W_d = 0.433GDA_p = 0.433 \times 0.7883 \times 1181 \times 4.909 = 1978$ lb

$\omega = 0.0004LN = 0.0004 \times 2650 \times 21 = 22.3°$

$A_N = A_p - A_r = 4.909 - 0.601 = 4.308$ sq in.

$$\frac{\tau_1}{\tau_2} = \frac{4L/v}{60/N} = \frac{NL}{15v} = \frac{21 \times 2650}{15 \times 15{,}800} = 0.235$$

$$J = \frac{0.0833E}{L/A_r} = \frac{0.0833 \times 30 \times 10^6 \times 0.601}{2650} = 567 \text{ lb/in.}$$

$$\phi = \arctan\left[\frac{\Gamma(\tau_1/\tau_2)}{1 - (\tau_1/\tau_2)^2}\right] = \arctan\left[\frac{0.470}{1 - 0.055}\right] = \arctan 0.497 = 26.4°$$

$$B = \{[1 - (\tau_1/\tau_2)^2]^2 + \Gamma^2(\tau_1/\tau_2)^2\}^{-0.5} = \{[1 - 0.055]^2 + 0.220\}^{-0.5} = 0.948$$

For S_p, the equation of Marsh and Coberly gives

$$S_p = S + \frac{40.8L^2\alpha}{E} - \frac{5.20GDA_p}{E}\left[\frac{L}{A_t} + \frac{L}{A_r}\right]$$

$$= 54 + 3.23 - 4.30 = 52.9 \text{ in.}$$

the Rieniets equation,

$$S_p = S\cos\phi + W_tB/J - 2W_f/J$$

$$= 48.35 + 15.12 - 13.64 = 49.8 \text{ in.}$$

and the Langer and Lamberger equation:

$$S_p = S\sec\omega - 50 \times 10^{-8}W_dL(1/A_r + 1/A_t)$$

$$= 58.38 - 5.37 = 53.0 \text{ in.}$$

For $W_{\max}$, the Mills equation gives

$$W_{\max} = W_f + W_r(1 + \alpha)$$

$$= 3866 + 7659 = 11{,}500 \text{ lb}$$

the Slonneger equation,

$$W_{\max} = (W_r + 0.433GLA_p)(1 + SN/5400)$$

$$= 10{,}162 \times 1.210 = 12{,}300 \text{ lb}$$

and the Langer and Lamberger equation,

$$W_{\max} = W_r(1 - 0.127G) + W_d + 0.75SN(17A_r\tan\omega + A_N)$$

$$= 5152 + 1978 + 850.5 \times 8.497 = 14{,}400 \text{ lb}$$

5.16. Steps in the Design of a Pumping Installation

The objective of a sucker rod pumping installation is the lifting of fluid from a formation to the surface at a desired rate. Therefore, the primary design factors to be considered are (1) mass production rate, and (2) net lift of the fluid. A knowledge of these factors makes possible the determination of optimum plunger size for the installation, i.e., the plunger size which will result in minimum load on the rods and the surface equipment, minimum torque on the gear box, and minimum power requirement for the prime mover. Determination of plunger size allows selection of tubing size, rod sizes and lengths, stroke length, pumping speed, torque rating of the unit, and power rating of the prime mover. The interdependency of these variables would make their selection extremely difficult if the design problem were approached entirely from the mathematical standpoint without benefit of previous experience.

If certain assumptions as to well conditions are made, it becomes possible to prepare charts which greatly reduce the number of trial calculations necessary in designing an installation. The assumptions generally made are (1) a specific gravity of 1.00 for the fluid being lifted, and (2) a net lift and a working fluid level equal to the pump setting depth. Examples of charts which simplify the design problem are presented here as Fig. 5.6 and

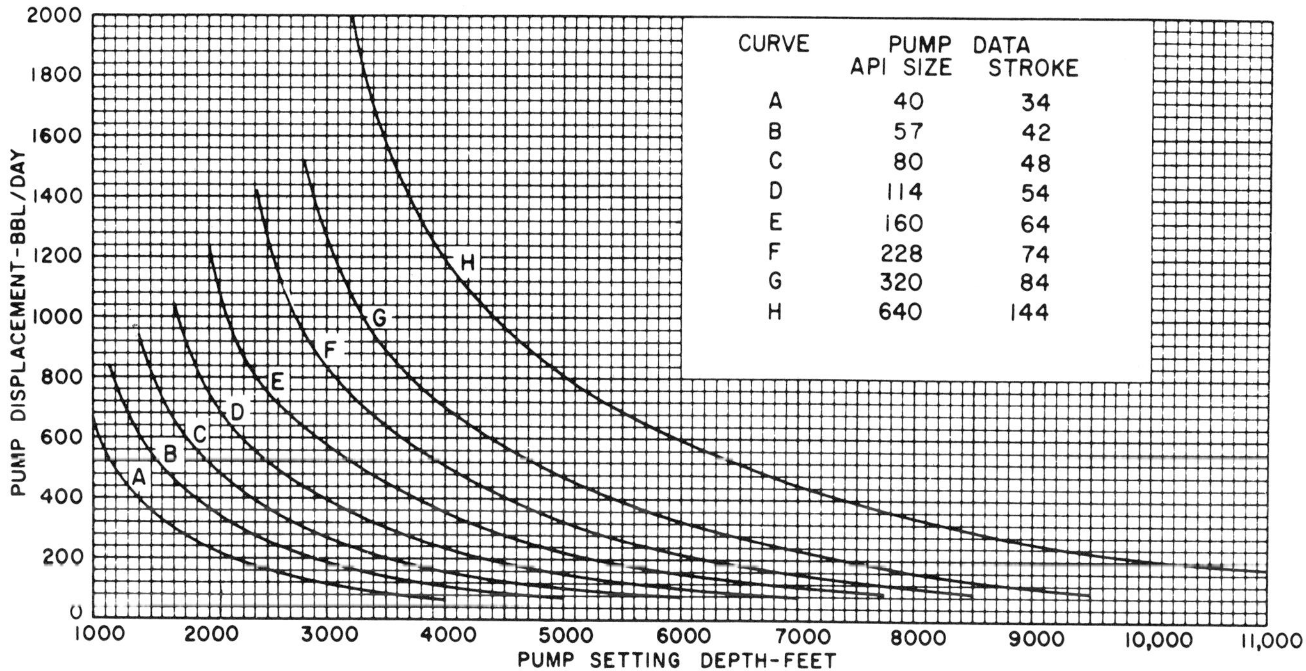

Fig. 5.6 Pump unit and stroke length selection chart. (After Kelley and Willis,[15,16] *The Petroleum Engineer.*)

Tables 5.5 through 5.12. These charts (which were prepared by the staff of the Atlantic Refining Company) were developed, essentially, from the conventional pumping equations previously derived, using the simplifying assumptions mentioned above.[15,16] They represent, graphically, the result of a large number of calculations concerned with determination of optimum pumping conditions to achieve given productions from given depths. After dimensions and operating conditions of the installation have been selected from the charts, it is necessary to check, by means of mathematical calculations, resulting loads, torques, production, and other pertinent factors. It must be realized that the charts are to be used as guides, and chart values may have to be adjusted to fit a particular situation. For example, presence of tubing already in a well to be put on pump dictates the maximum pump size, which might very well be smaller than the optimum size indicated by the selection chart.

Table 5.5. DESIGN DATA FOR API SIZE 40 UNIT WITH 34-INCH STROKE

(*After Kelley and Willis*[15,16])

Pump depth, ft	Plunger size, in.	Tubing size, in.	Rod sizes, in.	Pumping speed, strokes/min
1000–1100	$2\frac{3}{4}$	3	$\frac{7}{8}$	24–19
1100–1250	$2\frac{1}{2}$	3	$\frac{7}{8}$	24–19
1250–1650	$2\frac{1}{4}$	$2\frac{1}{2}$	$\frac{3}{4}$	24–19
1650–1900	2	$2\frac{1}{2}$	$\frac{3}{4}$	24–19
1900–2150	$1\frac{3}{4}$	$2\frac{1}{2}$	$\frac{3}{4}$	24–19
2150–3000	$1\frac{1}{2}$	2	$\frac{5}{8}$–$\frac{3}{4}$	24–19
3000–3700	$1\frac{1}{4}$	2	$\frac{5}{8}$–$\frac{3}{4}$	22–18
3700–4000	1	2	$\frac{5}{8}$–$\frac{3}{4}$	21–18

Table 5.6. DESIGN DATA FOR API SIZE 57 UNIT WITH 42-INCH STROKE

(*After Kelley and Willis*[15,16])

Pump depth, ft	Plunger size, in.	Tubing size, in.	Rod sizes, in.	Pumping speed, strokes/min
1150–1300	$2\frac{3}{4}$	3	$\frac{7}{8}$	24–19
1300–1450	$2\frac{1}{2}$	3	$\frac{7}{8}$	24–19
1450–1850	$2\frac{1}{4}$	$2\frac{1}{2}$	$\frac{3}{4}$	24–19
1850–2200	2	$2\frac{1}{2}$	$\frac{3}{4}$	24–19
2200–2500	$1\frac{3}{4}$	$2\frac{1}{2}$	$\frac{3}{4}$	24–19
2500–3400	$1\frac{1}{2}$	2	$\frac{5}{8}$–$\frac{3}{4}$	23–18
3400–4200	$1\frac{1}{4}$	2	$\frac{5}{8}$–$\frac{3}{4}$	22–17
4200–5000	1	2	$\frac{5}{8}$–$\frac{3}{4}$	21–17

Table 5.7. DESIGN DATA FOR API SIZE 80 UNIT WITH 48-INCH STROKE

(*After Kelley and Willis*[15,16])

Pump depth, ft	Plunger size, in.	Tubing size, in.	Rod sizes, in.	Pumping speed, strokes/min
1400–1550	$2\frac{3}{4}$	3	$\frac{7}{8}$	24–19
1550–1700	$2\frac{1}{2}$	3	$\frac{7}{8}$	24–19
1700–2200	$2\frac{1}{4}$	$2\frac{1}{2}$	$\frac{3}{4}$	24–19
2200–2600	2	$2\frac{1}{2}$	$\frac{3}{4}$	24–19
2600–3000	$1\frac{3}{4}$	$2\frac{1}{2}$	$\frac{3}{4}$	23–18
3000–4100	$1\frac{1}{2}$	2	$\frac{5}{8}$–$\frac{3}{4}$	23–18
4100–5000	$1\frac{1}{4}$	2	$\frac{5}{8}$–$\frac{3}{4}$	21–17
5000–6000	1	2	$\frac{5}{8}$–$\frac{3}{4}$	19–17

Table 5.8. DESIGN DATA FOR API SIZE 114 UNIT WITH 54-INCH STROKE

(*After Kelley and Willis*[15,16])

Pump depth, ft	Plunger size, in.	Tubing size, in.	Rod sizes, in.	Pumping speed, strokes/min
1700–1900	$2\frac{3}{4}$	3	$\frac{7}{8}$	24–19
1900–2100	$2\frac{1}{2}$	3	$\frac{7}{8}$	24–19
2100–2700	$2\frac{1}{4}$	$2\frac{1}{2}$	$\frac{3}{4}$	24–19
2700–3300	2	$2\frac{1}{2}$	$\frac{3}{4}$	23–18
3300–3900	$1\frac{3}{4}$	$2\frac{1}{2}$	$\frac{3}{4}$	22–17
3900–5100	$1\frac{1}{2}$	2	$\frac{5}{8}$–$\frac{3}{4}$	21–17
5100–6300	$1\frac{1}{4}$	2	$\frac{5}{8}$–$\frac{3}{4}$	19–16
6300–7000	1	2	$\frac{5}{8}$–$\frac{3}{4}$	17–16

Table 5.9. DESIGN DATA FOR API SIZE 160 UNIT WITH 64-INCH STROKE

(*After Kelley and Willis*[15,16])

Pump depth, ft	Plunger size, in.	Tubing size, in.	Rod sizes, in.	Pumping speed, strokes/min
2000–2200	$2\frac{3}{4}$	3	$\frac{7}{8}$	24–19
2200–2400	$2\frac{1}{2}$	3	$\frac{7}{8}$	23–19
2400–3000	$2\frac{1}{4}$	$2\frac{1}{2}$	$\frac{3}{4}$–$\frac{7}{8}$	23–19
3000–3600	2	$2\frac{1}{2}$	$\frac{3}{4}$–$\frac{7}{8}$	23–18
3600–4200	$1\frac{3}{4}$	$2\frac{1}{2}$	$\frac{3}{4}$–$\frac{7}{8}$	22–17
4200–5400	$1\frac{1}{2}$	2	$\frac{5}{8}$–$\frac{3}{4}$–$\frac{7}{8}$	21–17
5400–6700	$1\frac{1}{4}$	2	$\frac{5}{8}$–$\frac{3}{4}$–$\frac{7}{8}$	19–15
6700–7750	1	2	$\frac{5}{8}$–$\frac{3}{4}$–$\frac{7}{8}$	17–15

Table 5.10. Design data for API size 228 unit with 74-inch stroke

(After Kelley and Willis[15,16])

Pump depth, ft	Plunger size, in.	Tubing size, in.	Rod sizes, in.	Pumping speed, strokes/min
2400–2600	$2\frac{3}{4}$	3	$\frac{7}{8}$	24–20
2600–3000	$2\frac{1}{2}$	3	$\frac{7}{8}$	23–18
3000–3700	$2\frac{1}{4}$	$2\frac{1}{2}$	$\frac{3}{4}-\frac{7}{8}$	22–17
3700–4500	2	$2\frac{1}{2}$	$\frac{3}{4}-\frac{7}{8}$	21–16
4500–5200	$1\frac{3}{4}$	$2\frac{1}{2}$	$\frac{3}{4}-\frac{7}{8}$	19–15
5200–6800	$1\frac{1}{2}$	2	$\frac{5}{8}-\frac{3}{4}-\frac{7}{8}$	18–14
6800–8000	$1\frac{1}{4}$	2	$\frac{5}{8}-\frac{3}{4}-\frac{7}{8}$	16–13
8000–8500	$1\frac{1}{16}$	2	$\frac{5}{8}-\frac{3}{4}-\frac{7}{8}$	14–13

Table 5.11. Design data for API size 320 unit with 84-inch stroke

(After Kelley and Willis[15,16])

Pump depth, ft	Plunger size, in.	Tubing size, in.	Rod sizes, in.	Pumping speed, strokes/min
2800–3200	$2\frac{3}{4}$	3	$\frac{7}{8}$	23–18
3200–3600	$2\frac{1}{2}$	3	$\frac{7}{8}$	21–17
3600–4100	$2\frac{1}{4}$	$2\frac{1}{2}$	$\frac{3}{4}-\frac{7}{8}-1$	21–17
4100–4800	2	$2\frac{1}{2}$	$\frac{3}{4}-\frac{7}{8}-1$	20–16
4800–5600	$1\frac{3}{4}$	$2\frac{1}{2}$	$\frac{3}{4}-\frac{7}{8}-1$	19–16
5600–6700	$1\frac{1}{2}$	$2\frac{1}{2}$	$\frac{3}{4}-\frac{7}{8}-1$	18–15
6700–8000	$1\frac{1}{4}$	$2\frac{1}{2}$	$\frac{3}{4}-\frac{7}{8}-1$	17–13
8000–9500	$1\frac{1}{16}$	$2\frac{1}{2}$	$\frac{3}{4}-\frac{7}{8}-1$	14–11

Table 5.12. Design data for API size 640 unit with 144-inch stroke

(After Kelley and Willis[15,16])

Pump depth, ft	Plunger size, in.	Tubing size, in.	Rod sizes, in.	Pumping speed, strokes/min
3200–3500	$2\frac{3}{4}$	3	$\frac{7}{8}-1$	18–14
3500–4000	$2\frac{1}{2}$	3	$\frac{7}{8}-1$	17–13
4000–4700	$2\frac{1}{4}$	$2\frac{1}{2}$	$\frac{3}{4}-\frac{7}{8}-1$	16–13
4700–5700	2	$2\frac{1}{2}$	$\frac{3}{4}-\frac{7}{8}-1$	15–12
5700–6600	$1\frac{3}{4}$	$2\frac{1}{2}$	$\frac{3}{4}-\frac{7}{8}-1$	14–12
6600–8000	$1\frac{1}{2}$	$2\frac{1}{2}$	$\frac{3}{4}-\frac{7}{8}-1$	14–11
8000–9600	$1\frac{1}{4}$	$2\frac{1}{2}$	$\frac{3}{4}-\frac{7}{8}-1$	13–10
9600–11,000	$1\frac{1}{16}$	$2\frac{1}{2}$	$\frac{3}{4}-\frac{7}{8}-1$	12–10

Considering the manner in which the selection charts are constructed, the first factors encountered in the design problem are pump displacement and pump setting depth. Pump displacement is calculated from maximum anticipated fluid production and estimated volumetric efficiency of the pump. To check the selections (as to loads, torque, etc.) it is necessary to know the specific gravity of the well fluid and the working fluid level. Therefore the *minimum* information consists of (1) fluid production and (2) depth to pump. Additional information which should be known or estimated consists of (1) volumetric efficiency of the pump, (2) specific gravity of the fluid, and (3) working fluid level. Although these five factors are the most important, it would be an oversimplification of the problem to fail to point out that certain other information would be desirable. For example, a knowledge of the *reservoir* characteristics would be helpful, since the presence of a water drive would most certainly affect the total fluid production necessary to achieve a given oil production. In this connection, it cannot be overemphasized that a pumping installation must be designed to handle not current production but anticipated future production. We will assume, in the examples which follow, that a value for maximum desired fluid production has been established, without consideration of the source of this information. When the volumetric efficiency of the pump cannot be better estimated it will be taken to be 0.80. When the specific gravity of the fluid cannot be determined it will be assumed to be 1.00. When the working level of the fluid cannot be estimated it will be assumed to be at the pump depth.

Given below are the steps to be taken in designing a pumping installation. It is suggested that all calculations be made with equations derived in Secs. 5.6 through 5.14, as summarized in Sec. 5.15.

1. From maximum anticipated fluid production and estimated volumetric efficiency, calculate the pump displacement.
2. From Fig. 5.6 determine the stroke length and API rating of the pumping unit to be used. From manufacturers' literature, select a unit which has the desired stroke length and API rating.
3. From the appropriate chart of Tables 5.5 through 5.12, select tubing size, plunger size, rod sizes, and pumping speed corresponding to the pump setting depth.
4. Calculate the fractional length of each section of the rod string, using the data of Tables 5.2 and 5.4.
5. Calculate the length of each section of the rod string to the nearest 25 ft.
6. Calculate the acceleration factor.
7. Determine the effective plunger stroke length.
8. Using the estimated volumetric efficiency, determine the probable production rate and check it against the desired production rate.
9. Calculate the dead weight of the rod string.
10. Calculate the fluid load.

11. Determine peak polished rod load and check it against the maximum beam load for the unit selected.

12. Calculate the maximum stress at the top of the rod string and check it against the maximum permissible working stress for the rods to be used.

13. Calculate the ideal counterbalance effect and check it against the counterbalance available for the unit selected.

14. From the manufacturer's literature, determine the position of the counterweight to obtain the ideal counterbalance effect.

15. On the assumption that the unit will be no more than five per cent out of counterbalance, calculate the peak torque on the gear reducer and check it against the API rating of the unit selected.

16. Calculate hydraulic horsepower, friction horsepower, and brake horsepower of the prime mover. Select the prime mover.

17. From the manufacturer's literature obtain the gear reduction ratio and unit sheave size for the unit selected, and the speed of the prime mover. From this determine the engine sheave size to obtain the desired pumping speed.

If, at any step, the unit or any component of the unit is found to be either undersized or greatly oversized as to load, torque, or production capacity, the design should be changed accordingly.

Example 5.18. A well in the Seminole Field (Oklahoma) is to be put on pump. The proposed setting depth of the pump is 4200 ft, and considering the water cut, anticipated production is 500 bbl per day of a fluid having a specific gravity of 0.83 and containing no hydrogen sulfide. It is believed that the working fluid level will be low. The prime mover is to be a two-cylinder gas engine. Sucker rods are to be 1040 carbon steel, which have a working stress of 30,000 psi in noncorrosive fluid. Select the subsurface and surface equipment for the unit and calculate the brake horsepower which must be available at the prime mover.

SOLUTION: The steps shown below correspond to the suggested steps outlined previously.

1. Estimated volumetric efficiency = 80 per cent

$$V = q/E_v = 500/0.80 = 625 \text{ bbl/day}$$

2. L = 4200 ft. Figure 5.6 indicates use of an API size 320 unit with 84-in. stroke. An example of a unit which would meet these specifications is the Lufkin C-320D-84-30 pumping unit assembly with 8482R crank and OR counterweight. This unit has a beam load capacity of 30,000 lb and, according to manufacturer's literature, a maximum counterbalance effect (at the 84-in. stroke) of 21,645 lb (without auxiliary weights).

3. From Table 5.11, the plunger size is 2 in., the tubing size $2\frac{1}{2}$-in., the rod sizes $\frac{3}{4}$, $\frac{7}{8}$, and 1 in., and the pumping speed 19.5 spm (by interpolation). From Table 5.1,

$$A_1 = 0.442 \text{ sq in.} \qquad M_1 = 1.63 \text{ lb per ft}$$
$$A_2 = 0.601 \text{ sq in.} \qquad M_2 = 2.16 \text{ lb per ft}$$
$$A_3 = 0.785 \text{ sq in.} \qquad M_3 = 2.88 \text{ lb per ft}$$

From Table 5.2, $A_p = 3.142$ sq in. and $K = 0.466$. From Table 5.3, $A_t = 1.812$ sq in.

4. For a $\frac{3}{4}$–$\frac{7}{8}$–1 combination,

$$R_1 = 0.664 - 0.0894A_p = 0.664 - 0.281 = 0.383$$
$$R_2 = 0.181 + 0.0478A_p = 0.181 + 0.150 = 0.331$$
$$R_3 = 0.155 + 0.0416A_p = 0.155 + 0.131 = 0.286$$

5. $L_1 = LR_1 = 4200 \times 0.383 = 1609$ ft (1600 ft)

$L_2 = LR_2 = 4200 \times 0.331 = 1390$ ft (1400 ft)

$L_3 = LR_3 = 4200 \times 0.286 = 1201$ ft (1200 ft)

6. $\alpha = SN^2/70{,}500 = 84 \times 19.5 \times 19.5/70{,}500 = 0.453$

7. Since the working fluid level is low, it will be assumed that $D = L$. Taking $E = 30 \times 10^6$,

$$S_p = S + \frac{40.8L^2\alpha}{E} - \frac{5.20GDA_p}{E}\left[\frac{L}{A_t} + \frac{L_1}{A_1} + \frac{L_2}{A_2} + \frac{L_3}{A_3}\right]$$
$$= 84 + \frac{40.8 \times 4200 \times 4200 \times 0.453}{30 \times 10^6}$$
$$- \frac{5.20 \times 0.83 \times 4200 \times 3.142}{30 \times 10^6}\left[\frac{4200}{1.812} + \frac{1600}{0.442} + \frac{1400}{0.601} + \frac{1200}{0.785}\right]$$
$$= 84 \times 10.87 - 18.60 = 76.3 \text{ in.}$$

8. $q = KA_pNE_v = 0.466 \times 76.3 \times 19.5 \times 0.80 = 555$ bbl per day. The desired production rate of 500 bbl per day can be obtained.

9. $W_r = M_1L_1 + M_2L_2 + M_3L_3$

$= 2608 + 3024 + 3456 = 9090$ lb

10. $W_f = 0.433G(LA_p - 0.294W_r)$

$= 0.359(13{,}196 - 2668) = 3780$ lb

11. $W_{\text{max}} = W_f + W_r(1 + \alpha) = 3780 + 13{,}205 = 17{,}000$ lb

The beam load is within the allowable limit of 30,000 lb.

12. The maximum rod stress is

$$\frac{W_p}{A_3} = \frac{17{,}000 \text{ lb}}{0.785 \text{ sq in.}} = 21{,}600 \text{ psi}$$

which is within the allowable limit of 30,000 psi.

13. $C_i = 0.5W_f + W_r(1 - 0.127G) = 1890 + 8134$

$= 10{,}000$ lb

The ideal counterbalance effect can be obtained since a maximum effect of 21,645 lb is available.

14. The weight of the OR counterweight is 13,590 lb and the center of gravity is 18.27 in. from the outer edge. The crank length of the 8482R crank is 82 in. For the

C-320D-84-30 unit assembly, the saddle bearing is centered between tail bearing and bridle, and the counterbalance effect of structural unbalance is +1194 lb. In the equation

$$C = C_s + W_c(d/r)(l_1/l_2)$$

$l_1 = l_2$ and $r = S/2$, so that

$$C = C_s + 2W_c d/S$$

and

$$d = \frac{S(C - C_s)}{2W_c} = \frac{84(10{,}000 - 1194)}{2 \times 13{,}590} = 27.3 \text{ in.}$$

The distance from outer end of crank to outer edge of counterweight is 82 − 27.3 − 18.27 = 36.4 in. (Note: In most instances, pumping unit manufacturers supply charts from which the counterweight position for a given counterbalance effect can be determined directly without calculations of the type shown here. In this connection, see Fig. 5.49.)

15. $T_p = (W_{\text{max}} - 0.95C_i)S/2$

$$= 7462 \times 42 = 313{,}000 \text{ in.-lb}$$

An API size 320 unit has a rating of 320,000 in.-lb, and so the peak torque is within allowable limits.

16. Assuming net lift and pump setting depth to be identical,

$$H_h = 7.36 \times 10^{-6}\, qGL_N$$
$$= 7.36 \times 10^{-6} \times 555 \times 0.83 \times 4200 = 14.2 \text{ hp}$$
$$H_f = 6.31 \times 10^{-7}\, W_r SN$$
$$= 6.31 \times 10^{-7} \times 9090 \times 84 \times 19.5 = 9.39 \text{ hp}$$
$$H_b = 1.5(H_h + H_f)$$
$$= 1.5 \times 23.6 = 35.4 \text{ hp}$$

17. For the C-320D-84 unit, the gear ratio is 30.12 and the standard sheave has a pitch diameter of 25 in. The gas engine is rated at 35 hp at a continuous speed of 750 rpm. The engine sheave size is

$$d_e = Zd_u \frac{N}{N_e} = \frac{30.12 \times 25 \times 19.5}{750} = 19.5 \text{ in.}$$

5.17. Factors in Sucker Rod and Tubing Failure

In many wells, frequent replacement of sucker rods and tubing (necessitated by excessive failures of these members) constitutes serious economic problems. As a matter of fact, sucker rod performance is generally considered to be a direct indication of the efficiency with which a well is being pumped. For this reason it is advisable to mention several of the more probable causes of rod (and tubing) failure.

Assuming that the pumping installation has been properly designed, the first and perhaps the most important factor in prevention of unnecessary rod failure is the proper handling of rods during shipping, storage, running, and pulling. While many investigations have been made and much has been published relative to corrosion and the other factors to be discussed, it has been demonstrated that the effects of these factors can be substantially reduced by careful application of certain practices in handling sucker rods. For example, the presence, on the surface of a rod, of pits or cracks (even of microscopic size) would allow corrosive agents to gain access to the metal more readily, thereby accelerating the rate at which failure is approached. The rules (recommended by the American Petroleum Institute) covering the care and handling of sucker rods are given in API RP 11BR.[17]

For purposes of this discussion it will be assumed that, for an elastic member, failure has occurred when Hooke's law no longer applies, i.e., when there is no longer a constant ratio between applied stress and the resulting deformation (elongation) of the member. The stress at which this failure occurs is called the proportional limit of the material. The proportional limit cannot, however, be used as a criterion for establishing maximum permissible working stress for sucker rods, since repeated cyclic loads imposed on rods result in fatigue failures, which generally occur at stresses well below the proportional limit. Thus we must consider the endurance limit, i.e., the maximum stress (without failure) which can be applied for an infinite number of loadings or stress reversals. However, in determining experimentally the endurance limit, a material obviously cannot be made to undergo an infinite number of loadings. For steel, it has been found that the endurance limit can be taken as the maximum stress which permits 10,000,000 reversals without failure.[18]

The endurance limit of sucker rods depends on (1) the components (such as carbon, manganese, silicon, nickel, chromium, and molybdenum) present in the steel, (2) the corrosive agents (such as hydrogen sulfide, carbon dioxide, and oxygen) present in the well fluid, and (3) the range of stresses to which the rods are subjected. Table 5.13 summarizes the grades of steel generally used in sucker rod manufacture, giving compositions and endurance limits.[19]

In any instance, data obtained from laboratory studies must be applied with extreme caution to field problems. This is especially true in the case of experimentally determined endurance limits, and experience may dictate values much lower than those given in Table 5.13. If laboratory tests are to be of any use, the stress reversals must, of course, take place more rapidly than they would occur under field conditions. (At a pumping speed of 20 spm, 347 days would be required to reach 10,000,000 stress reversals.) However, where corrosion is involved, time is an important factor, and corrosion fatigue stresses for a rod string which has been in contact with a corrosive fluid for 347 days would be much lower than those for a laboratory

Table 5.13. RECOMMENDED ENDURANCE LIMITS OF SUCKER RODS
(*from Sucker Rod Handbook*[19])

Rod type	AISI-SAE steel grade	Endurance limit in corrosive fluid[a], psi With H_2S	Without H_2S	Endurance limit in non-corrosive fluid for stress range[b], psi $< 60\%$	$> 60\%$	$< 30\%$
Carbon Steel	C1033			30,000		
	C1035			30,000		
	C1036			30,000		
	C1038			30,000		
	C1039			30,000		
	C1040			30,000		
	C1042			30,000		
	C1043			30,000		
Alloy Steel	Mn 1335				30,000	40,000
	Ni-Cr 3310					40,000
	Ni-Cr (Mayari)	22,000	30,000		30,000	40,000
	Ni-Mo 4620	22,000	30,000		30,000	40,000
	Ni-Mo 4621	22,000	30,000		30,000	40,000
	Ni-Mo 4800 series	22,000				

[a]Corrosive fluids effectively treated with a chemical inhibitor may be considered noncorrosive.

[b]Stress range is given as per cent of maximum operating stress.

sample which has been tested at, say, 1750 reversals per minute, and therefore exposed to corrosive agents for only four days.[19]

When corrosive fluids are present in a well, two approaches to the problem are possible. The sucker rod string can be constructed of the higher strength alloys which are more resistant to corrosion, or the fluid may be treated chemically in the well in such a manner as to reduce substantially the harmful effects of the corrosive agents. The latter method, i.e., the use of chemical inhibitors, is now generally recognized as the more desirable solution, and use of the more expensive alloys is largely confined to the deeper wells, where large rod loads make high strength rods necessary.

A possible factor in premature rod failure is synchronous pumping speed. The idea that synchronous speeds might adversely affect rod performance was first introduced by Slonneger,[20] and the idea has since received the attention of many authorities in the field of oil well pumping. The agreement is by no means unanimous as to the effects of synchronous pumping speeds, and many experienced field men discount it entirely as a factor in rod performance. However, the subject merits some discussion, and in any case where frequent or otherwise unexplainable rod failures occur, it is desirable to compare the pumping speed with the natural frequency of the rod string for possible synchronous conditions.

Consider a rod of length L anchored at point p as shown in Fig. 5.7. This corresponds to a sucker rod string of length L, point b being the position of the plunger and p being the point at which the polished rod passes through

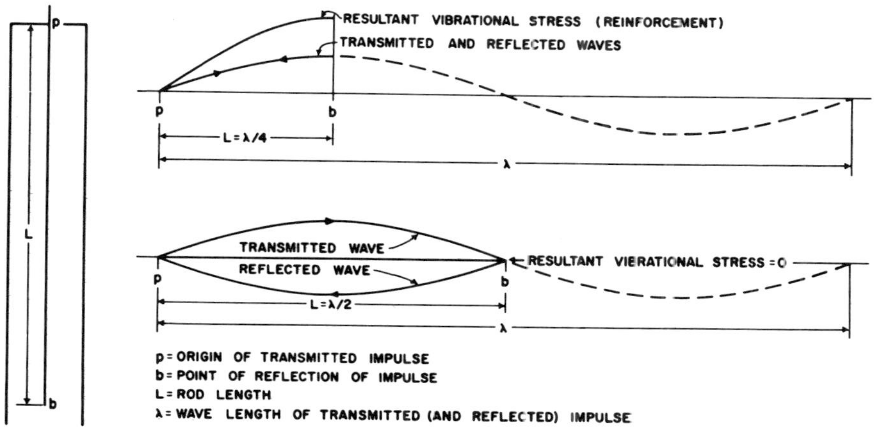

Fig. 5.7 Relationships between length of rod string and wave length of impulse for synchronous pumping speed (top) and non-synchronous pumping speed (bottom).

the stuffing box. If an impulse is supplied at point p, it will travel to point b (at the speed of sound in the metal) and be reflected. If point b is at an antinode, the reflected wave will be in phase with the transmitted wave, and will reinforce it. If point b is at a node, the reflected wave will be 180° out of phase with the transmitted wave, and the vector sum of the amplitudes at any point will be zero. In the first case (reinforcement), the rod is to be vibrating at its natural frequency, and any vibrational stresses present would be accentuated and made more severe.

The simplest case of reinforcement, that is, the first appearance of an antinode, would be for

$$L = \lambda/4 \quad \text{or} \quad \lambda = 4L \tag{5.96}$$

where λ is the wavelength of the transmitted (and reflected) impulse. Wavelength, velocity of stress transmission v, and frequently f for wave motion are related by

$$v = f\lambda \quad \text{or} \quad \lambda = v/f \tag{5.97}$$

For the case described by Eq. (5.96),

$$v/f = 4L \quad \text{or} \quad f = v/4L \tag{5.98}$$

Assuming the velocity of stress transmission in the rod string (i.e., the speed of sound in steel) to be 15,800 fps,

$$f = \frac{15{,}800 \text{ ft/sec} \times 60 \text{ sec/min}}{4 \times L \text{ ft}}$$

$$= 237{,}000/L \text{ vibrations/min} \tag{5.99}$$

Equation (5.99) gives the fundamental frequency of the rod. Thus, if impulses were to be transmitted to the rod at a frequency f as defined by Eq. (5.99), i.e., if the pumping speed for the well were f strokes per minute, the rod would be set into vibration at its natural frequency, and resonance would result.

Let t_1 be the time required for an impulse to travel distance L at velocity v:

$$L = vt_1 \tag{5.100}$$

Let the pumping speed, N_1, be such that time between impulses is t_1. Then, if the rod is vibrating at its natural frequency,

$$N_1 = 237{,}000/L = 237{,}000/vt_1 \tag{5.101}$$

If the pumping speed N_2 is such that time between impulses is $t_2 = 2t_1$, an impulse will be imparted to the rods when the previous impulse has traveled a distance $2L$, and the rod is still in resonance. For this,

$$N_2 = 237{,}000/vt_2 = 237{,}000/2vt_1 = 237{,}000/2L \tag{5.102}$$

In general, N represents a synchronous pumping speed if

$$N = 237{,}000/nL \tag{5.103}$$

where n is any integer.

The discussion to this point has been confined to the case where the rod is vibrating at its fundamental frequency, or first harmonic. The next case of reinforcement occurs when point b is at the second antinode. In this case,

$$L = 3\lambda/4 \quad \text{or} \quad \lambda = 4L/3 \tag{5.104}$$

The implication of Eq. (5.103) is that resonance will occur if the relationship between the pumping speed and the natural frequency of the rods is such that

$$f = nN \tag{5.105}$$

and resonance will occur if

$$\lambda = v/nN \tag{5.106}$$

Combining Eqs. (5.104) and (5.106),

$$4L/3 = v/nN \tag{5.107}$$

$$N = \frac{3v}{4nL} = \frac{3 \times 15{,}800 \times 60}{4nL} = \frac{237{,}000}{nL} \times 3 \tag{5.108}$$

The next case of resonance will occur when point b falls at the third antinode, in which case

$$L = 5\lambda/4 \quad \text{or} \quad \lambda = 4L/5 \tag{5.109}$$

By the line of reasoning followed in obtaining Eqs. (5.105) through (5.108),

$$N = \frac{237{,}000}{nL} \times 5 \tag{5.110}$$

Resonance therefore exists if

$$N = 237{,}000m/nL \tag{5.111}$$

where m is any odd integer. However, damping effects cause the amplitude of the reflected wave to be insignificant except in the case of the rod vibrating at its fundamental frequency. In this instance $m = 1$, and Eq. (5.103) is sufficient to define a synchronous pumping speed.

If synchronous performance is considered to be a factor in rod failure, undesirable pumping speeds are those which make n an integer in the expression

$$n = 237{,}000/NL \tag{5.112}$$

Any pumping speed which would not make n an integer would be nonsynchronous. However, to keep as far as possible from synchronous speeds, pumping speeds should be selected so as to make $n = 1\frac{1}{2}, 2\frac{1}{2}, 3\frac{1}{2}$, etc.

Example 5.19. Frequent sucker rod failures have occurred in a well which is being pumped at 20 spm with the pump set at 6000 ft. Comment on synchronous pumping speed as a possible factor.

SOLUTION:

$$n = 237{,}000/NL = 237{,}000/20 \times 6000 = 1.98$$

A pumping speed of 20 spm is, for all practical purposes, synchronous. If rod loads are within recommended limits, and if well fluid is noncorrosive (or, being corrosive, has been treated with an inhibitor), synchronous speed may be given serious consideration as a factor in rod failure. Even if other factors are present, synchronous speed may be a contributing factor, and a change in stroke length, which would permit use of a different pumping speed, is advisable.

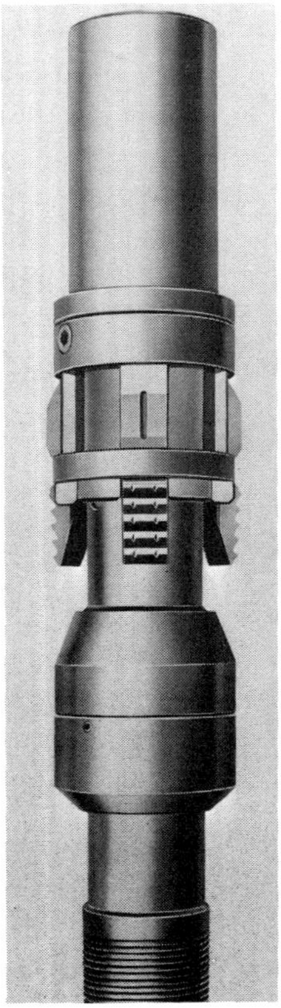

Fig. 5.8 A tension tubing anchor. (Courtesy Baker Oil Tools, Inc., Los Angeles, California.)

In some wells, excessive rod failures and serious tubing leaks have resulted from wear caused by contact between rod and tubing. Cases have been cited[21] in which tubing pulling jobs were necessary every other week. In the past, failures of this nature have been attributed to deviation of the hole from the vertical. Recent mathematical and experimental investigations by Lubinski and Blenkarn[21] have demonstrated the important role played by tubing buckling. Methods suggested by these investigators for minimizing the detrimental effects of buckling include use of tension anchors, tail pipe, compression anchors, sucker rod guides, and corrosion inhibitors. In addition to prevention of buckling, the use of tubing anchors eliminates the effect of tubing stretch described in Sec. 5.7, thus increasing the effective plunger stroke.

A tension anchor is a tubing anchor which permits elongation but prevents shortening of the tubing (see Fig. 5.8). This device therefore prevents tubing buckling, which takes place only on the upstroke when, and if, stretch is allowed to come out of the tubing. One method of installing a tension anchor is to set the anchor on the tubing, and then pick up the tubing at the surface until it is in tension. Charts for determining proper pickup (for 2-in. and $2\frac{1}{2}$-in. tubing) have been developed by Lubinski and Blenkarn, and are presented here as Figs. 5.9 and 5.10. Reference to these figures will show that some knowledge of well fluid level is needed to use the charts. If it is necessary to estimate fluid level in the well at the time the anchor is set, it is better to estimate it too high rather than too low. If it is necessary to estimate working fluid level, a future estimated level rather than a present estimated level (the future level being lower than the present level) should be used. Adherence to these rules will prevent the calculated pickup from being too small. The pickup given by the chart is effective pickup, to which should be added pickup needed for expansion of the anchor, a value which depends on the type of anchor. Actual pickup should generally be measured in inches rather than pounds, since chart values for pickup in pounds do not consider frictional forces.

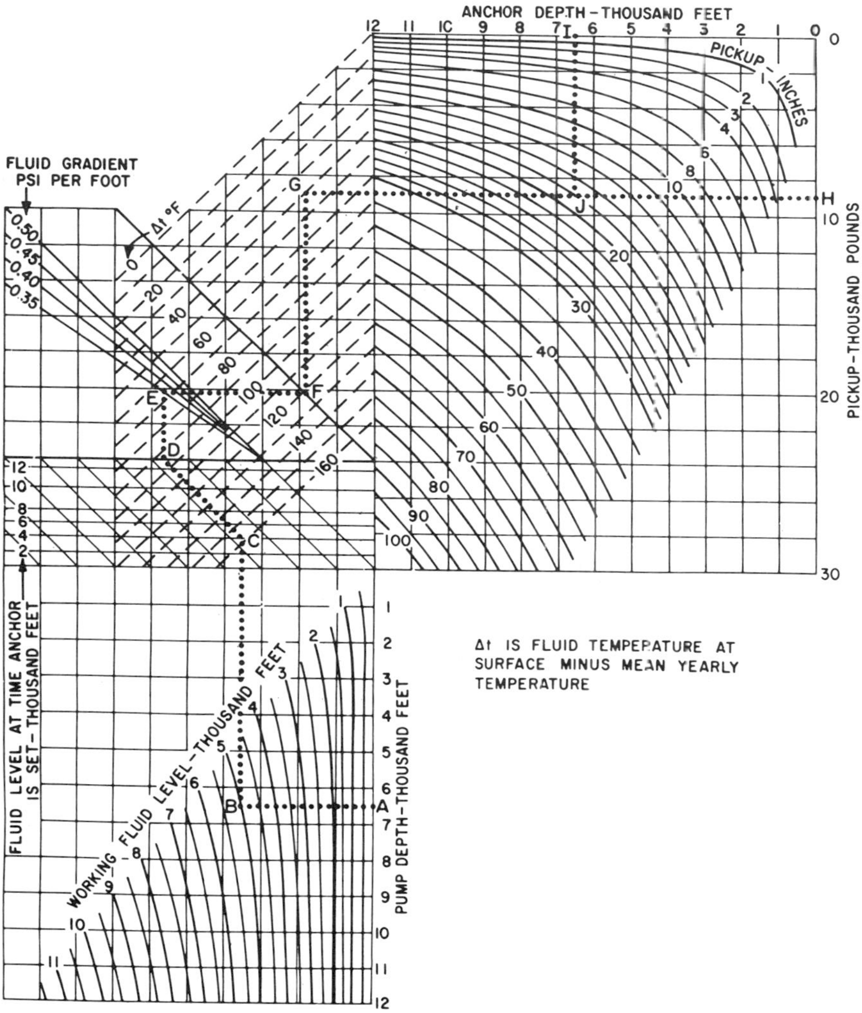

Fig. 5.9 Tubing pickup for tension anchor (2-in. tubing). (Lubinski and Blenkarn,[21] *Trans.* AIME.)

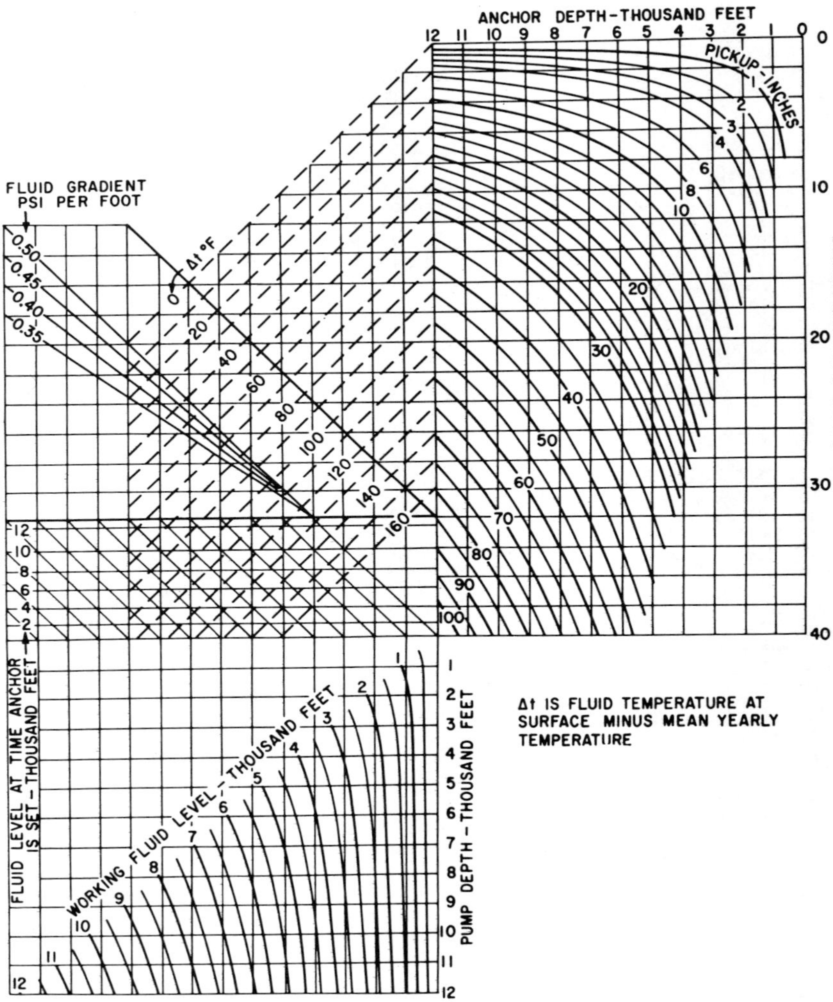

Fig. 5.10 Tubing pickup for tension anchor ($2\frac{1}{2}$-in. tubing). (Lubinski and Blenkarn,[21] *Trans.* AIME.)

Example 5.20. Determine the pickup necessary for a tension anchor being set at 6500 ft in 2-in. tubing if the following conditions obtain:

Pump depth = 6500 ft

Estimated working fluid level = 5000 ft

Estimated fluid level when anchor is set = 4000 ft

Specific gravity of fluid = 0.81

Fluid temperature at surface = 100°F

Mean yearly surface temperature = 60°F

SOLUTION: To determine pickup for 2-in. tubing, use Fig. 5 9.

Fluid gradient = 0.81 × 0.433 = 0.35 psi/ft

$\Delta t = 100 - 60 = 40°F$

Enter chart at point *A* (pump depth = 6500 ft).

Move horizontally to point *B* (working fluid level = 5000 ft).

Move vertically to point *C* (fluid level at time anchor is set = 4000 ft).

Move diagonally to point *D* (intersection with the horizontal reference line).

Move vertically to point *E* (fluid gradient = 0.35 psi/ft).

Move horizontally to point *F* (intersection with the diagonal reference line).

Move vertically to point *G* (Δt = 40°F).

Move horizontally to point *H* (intersection with pickup axis). Read pickup = 9000 lb.

Enter chart at point *I* (anchor depth = 6500 ft).

Move vertically to point *J* (intersection with line *GH*).

Read pickup = 18 in.

Tail pipe is simply pipe suspended below the pump in nonanchored tubing, which is of sufficient weight to have a straightening effect and thus prevent buckling above the pump. The weight may be obtained by an extension of the tubing itself, or, if the depth of the hole will not permit this, by using relatively short lengths of heavy-walled pipe (for example, drill collars).

Lubinski and Blenkarn, in Reference 21 cited previously, have shown that the force which causes buckling in tubing can be given by

$$F_b = A_p \Delta p \tag{5.113}$$

where Δp is the pressure differential across the plunger, as defined by Eq. (5.17). If M_p is unit weight (pounds per foot) and L_p is length (feet) of the tail pipe, the weight in air of the tail pipe is

$$W_a = M_p L_p \tag{5.114}$$

In Sec. 5.8, it was shown that the buoyant force on steel when submerged in a fluid of specific gravity G is

$$\text{Buoyant force} = 0.127W_aG \tag{5.115}$$

To prevent buckling, the effective weight of tail pipe in well fluid must be at least equal to the buckling force:

$$A_p\Delta p = W_a - 0.127W_aG = W_a(1 - 0.127G) \tag{5.116}$$

Substituting Eqs. (5.17) and (5.114) into (5.116),

$$0.433GDA_p = M_pL_p(1 - 0.127G) \tag{5.117}$$

and the length of tail pipe to prevent buckling is

$$L_p = \frac{0.433GDA_p}{M_p(1 - 0.127G)} \tag{5.118}$$

Example 5.21. Suppose, in Example 5.20, it is desired to prevent buckling by use of tail pipe rather than by a tension anchor. If the pump has a $1\frac{1}{2}$-in. plunger, calculate the length of tail pipe consisting of an extension of the tubing.

SOLUTION:

$A_p = 1.767$ sq in. (Table 5.2)

$M_p = 4.70$ lb/ft (Table 5.3)

$$L_p = \frac{0.433 \times 0.81 \times 5000 \times 1.767}{4.70(1 - 0.127 \times 0.81)} = 735 \text{ ft}$$

A tubing anchor which permits shortening but prevents elongation of tubing is called a compression or hookwall anchor. Compression anchors are used primarily to increase the effective plunger stroke by eliminating the effect of tubing stretch. A compression anchor will not, by itself, eliminate buckling (since it permits shortening), and for prevention of buckling it must be used in conjunction with tail pipe.

With a compression anchor at the pump, a greater weight of tail pipe is required than for freely suspended tubing. This results from the need to subject tubing to more tension at anchor setting time if future thermal buckling (when the tubing is heated by hot formation crude) is to be avoided.[21] In obtaining an expression for the amount of tail pipe needed in this instance, it is assumed that at the time the anchor is set, tubing temperature varies linearly with depth as shown by line AB in Fig. 5.11, A being the mean yearly surface temperature and B the temperature at anchor depth L_a. This is equivalent to an assumption that the tubing is in thermal equilibrium with surrounding formations and that the geothermal gradient is linear. (Seasonal surface changes AA_1 and AA_2 affect only a negligible length of tubing.) It is further assumed that, during pumping, tubing temperature varies linearly with depth as shown by the line CB. Then AC, designated Δt, is the temperature change of tubing near the

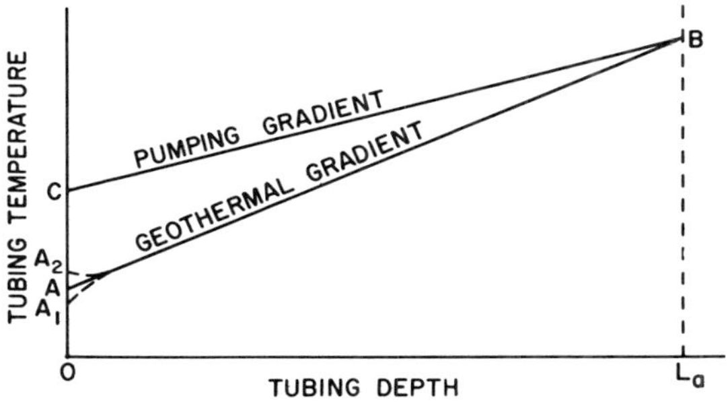

Fig. 5.11 Effect of pumping on tubing temperature. (After Lubinski and Blenkarn,[21] *Trans.* AIME.)

surface between the time the anchor is set and a later time when the well is being pumped. This means that $\Delta t/2$ is the average temperature change of the tubing between these times. If the average temperature of tubing of length L_a is increased by $\Delta t/2$ and the tubing is permitted to elongate freely, then its elongation ϵ is

$$\epsilon = L_a \delta \Delta t/2 \tag{5.119}$$

where δ is the thermal coefficient of expansion of the tubing. The force F_c required to restore the tubing to its original length is related to the elongation ϵ, the length L_a, and the cross-sectional area of tubing wall A_t by

$$E = \frac{F_c/A_t}{\epsilon/L_a} = \frac{F_c L_a}{A_t L_a \delta \Delta t/2} = \frac{2F_c}{A_t \delta \Delta t} \tag{5.120}$$

For steel, $E = 30 \times 10^6$ and $\delta = 6.9 \times 10^{-6}$. Solving for F_c,

$$F_c = \frac{E A_t \delta \Delta t}{2} = \frac{30 \times 10^6 \times 6.9 \times 10^{-6} A_t \Delta t}{2} = 104 A_t \Delta t \tag{5.121}$$

The force defined by Eq. (5.121) must be added to the force to prevent buckling, in order to obtain the minimum effective weight of tailpipe in fluid when a compression anchor is used:

$$M_p L_p (1 - 0.127G) = 0.433GDA_p + 104A_t\Delta t \tag{5.122}$$

$$L_p = \frac{0.433GDA_p + 104A_t\Delta t}{M_p(1 - 0.127G)} \tag{5.123}$$

Example 5.22. With reference to Examples 5.20 and 5.21, determine the length of tail pipe (2-in. tubing) necessary to prevent buckling if a compression anchor is to be set at the pump.

Solution:

$A_t = 1.304$ sq in. (Table 5.3)

$$L_p = \frac{0.433 \times 0.81 \times 5000 \times 1.767 + 104 \times 1.304 \times 40}{4.70(1 - 0.127 \times 0.81)}$$

$$= 2020 \text{ ft}$$

Sucker rod guides, which were devised primarily for use in crooked holes, have some usefulness also in minimizing the effects of buckling, but much of their effectiveness can be lost by improper spacing. Figure 5.12[21] gives

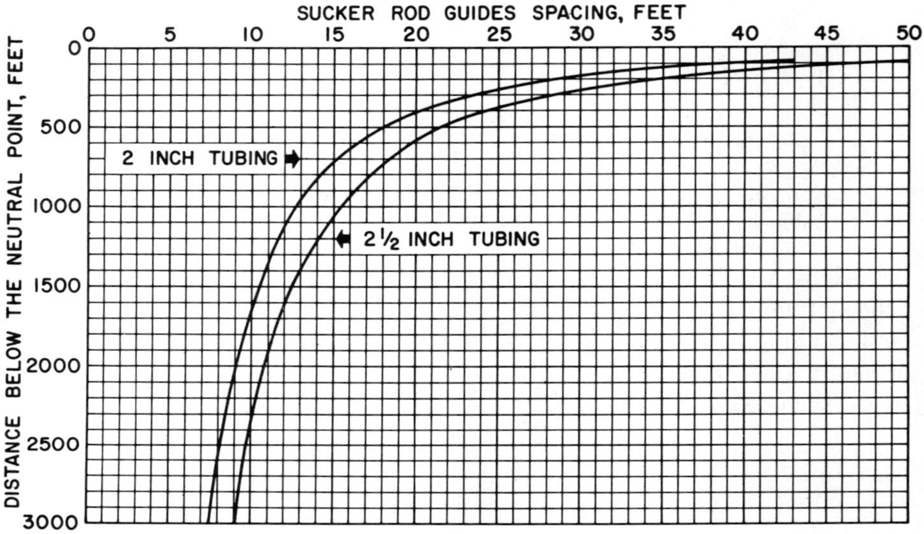

Fig. 5.12 Spacing of sucker rod guides. (Lubinski and Blenkarn,[21] *Trans.* AIME.)

the distance between two successive guides in terms of the distance of the lower of the two guides below the neutral point of the tubing. For our purposes, the neutral point may be taken to be the point in the tubing below which buckling can occur. Lubinski and Blenkarn[21] have shown that the distance of the neutral point above the pump can be given by

$$L_x = F_b/Q \tag{5.124}$$

where Q is the weight (in pounds per foot) of tubing in well fluid. If the weight of tubing in air is Q_a pounds per foot, then

$$Q = Q_a(1 - 0.127G) \tag{5.125}$$

The buckling force has previouly been defined as

$$F_b = A_p\Delta p = 0.433GDA_p \tag{5.126}$$

If L_o is depth to the neutral point, then

$$L = L_x + L_o \tag{5.127}$$

Combining Eqs. (5.124) through (5.127),

$$L_o = L - \frac{0.433GDA_p}{Q_a(1 - 0.127G)} \tag{5.128}$$

Once the location of the neutral point is determined in this manner, Fig. 5.12 can be used to determine the spacing of sucker rod guides. Since no buckling occurs above the neutral point, no guides are needed above that point to control buckling, although they may be necessary for other reasons.

Example 5.23. For the well of Examples 5.20 and 5.21, a sucker rod guide is set 140 ft above the pump. What should be the spacing between this guide and the next guide up the rod string?

SOLUTION:

$$L_o = 6500 - 735 = 5765 \text{ ft}$$

The depth to the lower guide is

$$6500 - 140 = 6360 \text{ ft}$$

The distance of the lower guide below neutral point is

$$6360 - 5765 = 595 \text{ ft}$$

From Fig. 5.12, the spacing is 17 ft.

5.18. Use of the Recording Dynamometer in Well Analysis

This section will be concerned with a brief description of a typical recording dynamometer and with its use in determination of well loads. Figure 5.13 shows a mechanical type of dynamometer which is characterized by simplicity of installation and operation, portability, and accuracy. This instrument makes use of the principle of the proving ring, as developed by the Bureau of Standards for calibrating tensile testing machines.

As shown in Fig. 5.13, the proving ring (behind the instrument case) is held in compression between carrier bar and polished rod clamp. The ring is thus subjected to total polished rod load, and its resulting deflection is mechanically amplified and transmitted to the main (load) stylus. (In Fig. 5.13, the tip of the main stylus is visible at the left front of the instrument case.) An increasing load on the ring causes this stylus to move from left to right across a waxed chart on a drum. The drum, because of its linkage with the large pulley to the left of the instrument case, can be made to rotate by pulling the cord which is wrapped around the pulley. The object at the end of the cord is a magnetic clamp. When a dynamometer card is being recorded, the magnet is clamped to a stationary object below the carrier bar, for example, the stuffing box. As the carrier bar, follows the up-and-down motion of the polished rod, the drum rotates alternately clockwise and counterclockwise. As a result, the figure inscribed on the waxed chart by the main stylus is a plot of polished rod load

Fig. 5.13 A recording dynamometer. (Courtesy of Johnson-Fagg Engineering Company, Tulsa, Oklahoma.)

versus polished rod position. This plot is referred to as a dynamometer card. The timing bar (outside the instrument case near its right front edge) can be caused to vibrate at set frequencies. It actuates the timer stylus (at the right of the chart), which traces on the chart a wavy line used in determining instantaneous polished rod velocities.

The dynamometer is placed in service in the following manner. A second (lower) clamp is placed on the polished rod just above the stuffing box and the carrier bar is then lowered until there is enough space between carrier bar and upper (original) polished rod clamp to accommodate the ring of the dynamometer. The dynamometer is placed on the carrier bar (with no force on the ring) and the cord is pulled to turn the chart drum. This causes the main stylus to inscribe the zero load line on the chart, and all subsequent loads are determined by deflections from this reference line. The carrier bar is then raised until the ring supports the upper clamp, at which time the lower clamp is removed. The load on the ring is now the polished rod load, and the position of the main stylus with reference to the zero load line can be used to calculate polished rod load.

The instrument can also be installed by placing the ring above the original polished rod clamp, rather than between polished rod clamp and carrier bar. If this is done, the second clamp is placed above the ring. After the upper clamp is tightened, the lower clamp is loosened, causing the top clamp to support the entire polished rod load. When the instrument is used in this manner, pump spacing is not changed. Installation above the clamp is recommended when change in pump spacing would affect pump operation, for example, when gas interference at the pump might change the load characteristics of the well.

The counterbalance effect at the polished rod can be determined in the following manner. The pumping unit is stopped at the position where the counterbalance effect is greatest, i.e., the point at which the crank is horizontal on the upstroke. A polished rod clamp is then placed a few inches above the stuffing box. A chain is placed above the clamp just installed, wrapped around the polished rod, and attached to the flow line. If the brake is then released, the load on the ring results only from the counterbalance, and if the cord is pulled, the main stylus will inscribe on the chart a line representing counterbalance effect. If the polished rod load is greater than the counterbalance effect, the rods will fall when the brake is released. In this case no chain is needed and the lower clamp should be put directly above the stuffing box (with crank horizontal) before the counterbalance effect is recorded.[22]

After the zero load line and the line representing counterbalance effect have been obtained, the chain and the lower polished rod clamp are removed, and the well is then pumped at its normal pumping speed. This gives a dynamometer card which permits analysis of load distributions throughout the pumping cycle. The main stylus (and also the timer stylus) can be locked away from the chart or released to engage the chart as desired. This makes it possible to record loads over a single pumping cycle, thus eliminating the confusion which would result from several dynamometer cards superimposed on each other.

The appearance of a typical dynamometer card and the analysis of such cards will be treated in subsequent sections. However, it should be emphasized here that no two dynamometer cards, even from the same well, are likely to be identical. In order to be reasonably certain that a *representative* card is obtained for the well being tested, it is desirable to make at least three or four cards, preferably at intervals of 15 to 20 min.

5.19. Interpretation of a Dynamometer Card

Consider an idealized pumping system wherein the following conditions obtain: (1) the well is pumped very slowly, i.e., there are no acceleration forces; (2) there are no vibrational forces within the system; (3) there are no frictional forces; (4) the standing valve opens and the traveling valve closes instantaneously at the beginning of the upstroke; (5) the standing valve closes and the traveling valve opens instantaneously at the beginning of the downstroke; and (6) there are no changes in rod length due to fluid load transfer. If such a set of conditions were possible, the dynamometer card for the well would be a rectangle such as that shown in Fig. 5.14. Line AB represents the upstroke, where the polished rod load is the fluid load plus weight of the rods in fluid. At the end of the upstroke, the entire fluid load is immediately transferred to the standing valve, and line CD, which represents the downstroke, indicates that the polished rod load is simply the weight of the rods in fluid. Thus, the shape of the ideal card is dictated by only two load factors, namely, fluid load and weight of the rods in fluid.

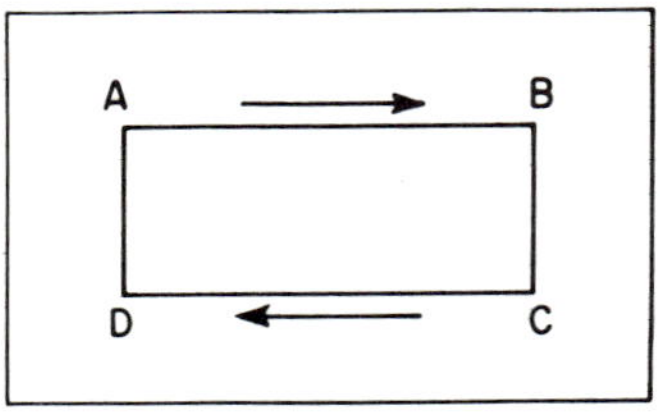

Fig. 5.14 Dynamometer card for an ideal pumping system.

For any actual pumping system it is, of course, impossible for all six of the ideal conditions to obtain. For most actual pumping systems it is

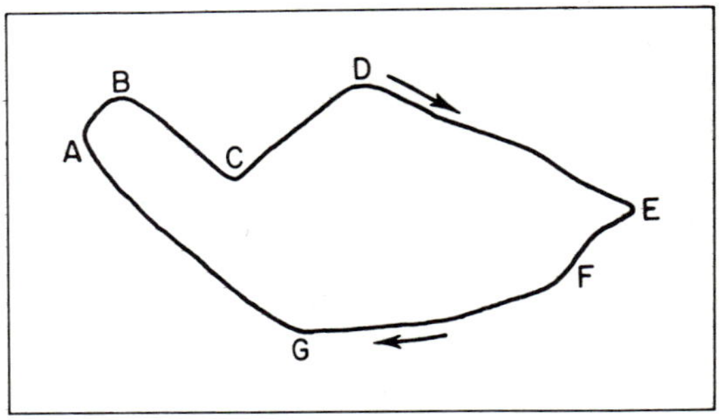

Fig. 5.15 Dynamometer card for an actual pumping system. (*Sucker Rod Handbook.*[9])

impossible to obtain *any* of the six conditions. Therefore, eight load factors contribute to the shape of actual cards, and most of these cards bear little, if any, resemblance to the ideal card shown in Fig. 5.14. Figure 5.15 shows a card similar to those obtained from a normal pumping well. While it is not possible here to point out the influence of all eight load factors, this card does illustrate several features which are characteristic of dynamometer cards in general. The discussion which follows is based on an article by Russell.[23]

Point *A* (Fig. 5.15) represents the end of the downstroke and the beginning of the upstroke for the polished rod. As the traveling valve closes, the polished rod begins to pick up the fluid load. This accounts for the increase in load from *A* to *B*. The momentary decrease in polished rod load from *B* to *C* is a result of rod stretch which occurs when the rods take over the fluid load. As the rods move upward (in approximate simple harmonic motion), the acceleration load increases until it reaches a maximum at point *D*, theoretically near the middle of the upstroke. From point *D* to point *E*, the acceleration load decreases as rod velocity decreases to zero. Point *E* represents the end of the upstroke and the beginning of the downstroke. As the rods fall, the traveling valve opens and the standing valve closes. At point *F* the standing valve has taken over the fluid load, and there is a marked decrease in polished rod load. The acceleration load (which, on the downstroke, decreases polished rod load) increases to point *G*, near the middle of the downstroke, where minimum polished rod load occurs. From *G* to *A* the (negative) acceleration load decreases, causing an increased polished rod load.

No account has yet been taken of the influence of vibration and friction on the shape of the dynamometer card. These factors, although difficult to isolate and analyze on the card, are always present, and contribute significantly to total polished rod load. As pointed out in Sec. 5.17, vibration loads can be particularly severe in cases where pumping speed is synchronous with natural frequency of rod vibration.

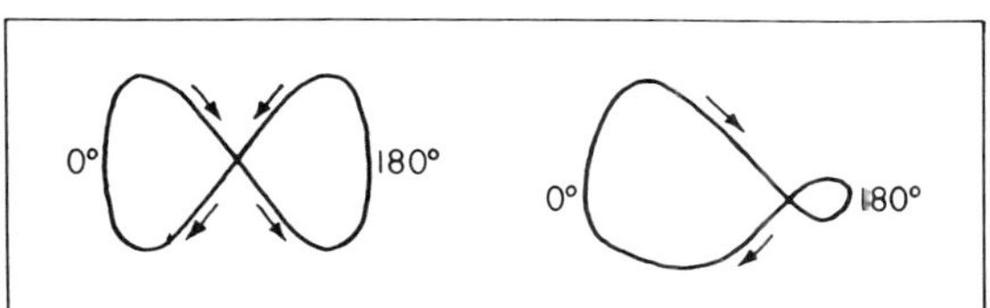

Fig. 5.16 Development of a theoretical dynamometer card for a well being pumped at synchronous speed (second order). Vibrational load throughout one pumping cycle (left) is superimposed on ideal card to obtain card at right. (Slonneger,[20] *Drilling and Production Practice,* API.)

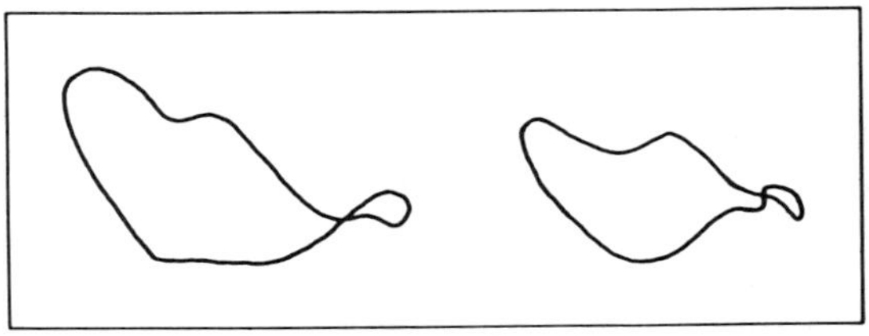

Fig. 5.17 Actual dynamometer cards for wells being pumped at synchronous speed (second order). (*Sucker Rod Handbook.*[9])

Slonneger,[20] by a theoretical approach, has been able to predict, to a high degree of accuracy, the shapes of dynamometer cards for wells being pumped at synchronous speeds. Figures 5.16 and 5.17 show theoretical and actual second-order cards, that is, cards for which $n = 2$ in Eq. (5.112). Figure 5.18 shows theoretical and actual third-order cards, or cards for which $n = 3$. Likewise, Fig. 5.19 shows fourth-order cards. It can be

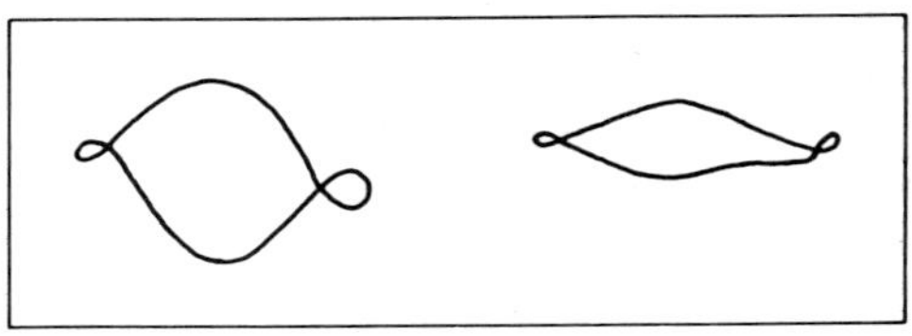

Fig. 5.18 Theoretical (left) and actual (right) third order cards. (Slonneger,[20] *Drilling and Production Practice,* API and *Sucker Rod Handbook.*[9])

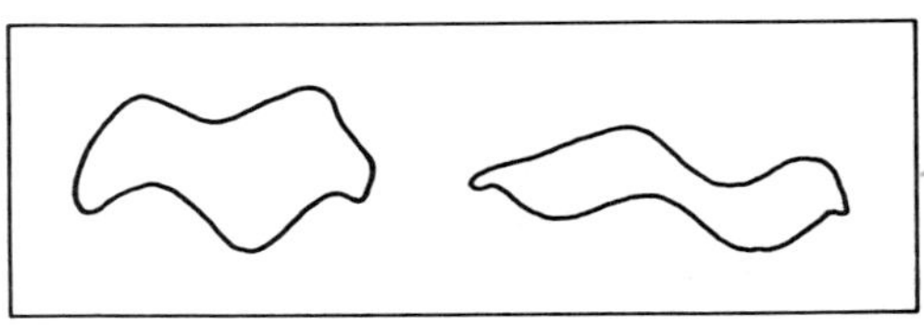

Fig. 5.19 Theoretical (left) and actual (right) fourth order cards. (Slonneger,[20] *Drilling and Production Practice,* API and *Sucker Rod Handbook.*[9])

seen that cards of each order have definitely recognizable characteristics, and that synchronous pumping speeds are easily identified by means of dynamometer cards.

Of course, in any system involving contact between moving parts energy must be expended to overcome friction. In a pumping system, however, *excessive* friction may occur for any one of several reasons.

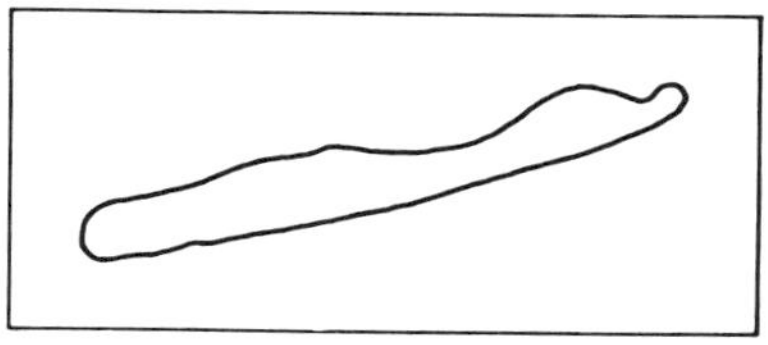

Fig. 5.20 Card indicating restriction in well.

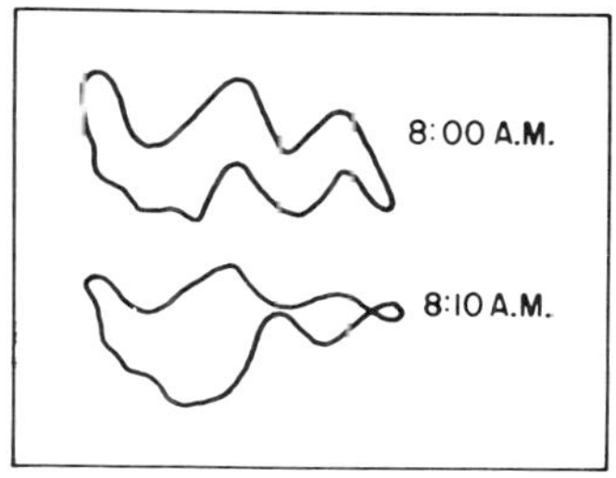

Fig. 5.21 Cards indicating sticking plunger.

Figure 5.20[24] shows a card for which there is a continuous load increase on the upstroke. This suggests a restriction of some sort in the well. Figure 5.21 shows two cards for the same well, taken ten minutes apart. The unpredictable manner in which the standing valve takes over the load indicates a sticking plunger. For shallow wells, plunger sticking can also be detected by feeling the polished rod and noting the jerky motion of the rod string.[23] Figure 5.22 shows other cases of excessive friction. In any instance where the dynamometer card shows a large area (indicating a large amount of work per cyle) with little to show for the work by way of fluid production, and in any instance where the dynamometer card is jerky and erratic from cycle to cycle, friction should be suspected.

Several other well problems produce cards with characteristic features. A notable example is fluid pound, which results from failure of the pump

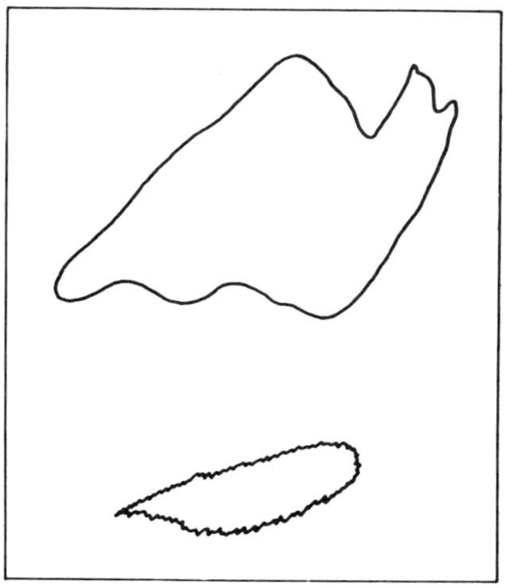

Fig. 5.22 Cards indicating excessive friction in pumping system.

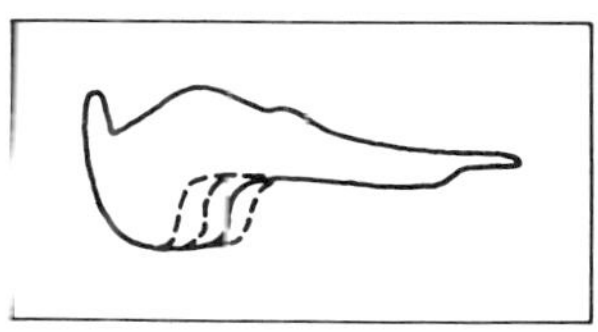

Fig. 5.23 Card indicating fluid pound.

barrel to fill completely on the upstroke. Fluid pound causes unnecessary buckling and flexing of rods, and thereby considerably shortens sucker rod life. Figure 5.23 shows a card typical for this condition, characterized by a sudden decrease in load near the end of the downstroke, the decrease occurring at different polished rod positions from cycle to cycle. Assuming sufficient submergence of the pump below the working fluid level, fluid pound is caused by a withdrawal rate which is in excess of the producing capacity of the formation. The remedy, therefore, is to reduce stroke length, pumping speed, or plunger size, thereby decreasing pump displacement. In this connection, it might be noted that many pump manufacturers recommend that average plunger speeds be kept below an upper limit of approximately 4.7 fps.[25]

Example 5.24. Estimate the average plunger speed (over one complete pumping cycle) if the polished rod stroke is 84 in. and the pumping speed is 15 spm.

Solution: Assuming plunger and polished rod strokes to be equal, the plunger travels $2 \times 84 = 168$ in. in one pumping cycle, i.e., in 1/15 min. Therefore the average speed is

$$\frac{168 \times 15 \text{ in./min}}{12 \text{ in./ft} \times 60 \text{ sec/min}} = 3.5 \text{ ft/sec}$$

Gas pound, resulting from partial filling of the barrel with gas, is indicated by the card of Fig. 5.24. The card for gas pound is similar to that for fluid pound, the essential difference being that the load decrease is not as sudden in the case of gas pound. It is possible for the pump to

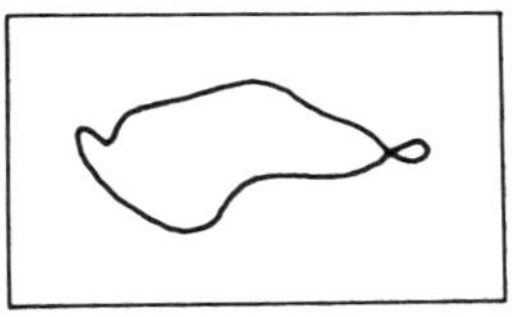

Fig. 5.24 Card Indicating gas pound.

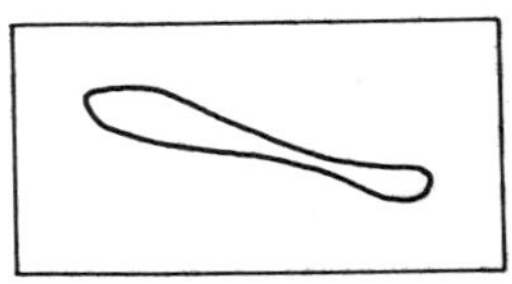

Fig. 5.25 Card Indicating gas lock.

become gas-locked, as a result of failure of valves to open properly. This situation would result in a card similar to the one shown in Fig. 5.25. The small area indicates practically no work being done by the pump, since pump motion merely reflects compression of entrapped gas on the downstroke and expansion of the gas on the upstroke. Schmoe[26] has given two reasons for the loss in efficiency caused by gas interference. First, the traveling valve does not open at the start of the plunger downstroke. Instead, its action is delayed until the plunger has compressed the gas in

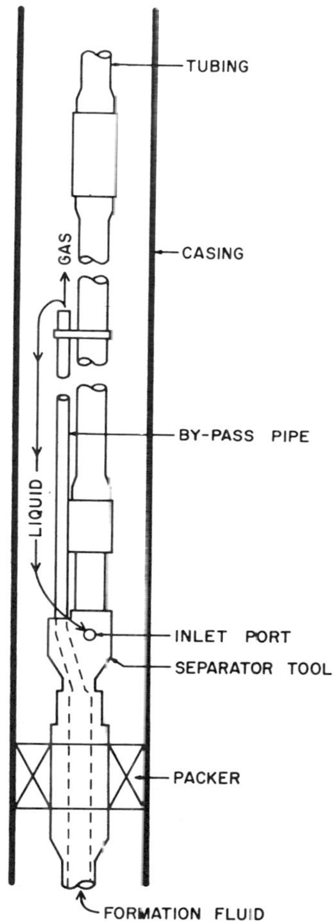

Fig. 5.26 A typical bottom-hole separator. Formation fluid passes up through the packer and into the separator tool, where it is diverted into a by-pass pipe. Length of this pipe is usually one to four joints, depending on fluid characteristics. As fluid emerges from the pipe, pressure drop, turbulence, and gravity segregation cause gas to continue up the annulus while liquid drops back to the annular region immediately above the packer. Liquid then enters the seating nipple through inlet ports of the separator tool, where it is picked up by the pump and produced through tubing to the surface. (After Schmoe,[26] *World Oil.*)

the pump to a pressure equal to the hydrostatic fluid head in the tubing. Second, the standing valve does not open at the start of the upstroke. Its opening is delayed until the pressure inside the pump is reduced to the producing bottom-hole pressure.

There are two approaches to problems involving gas at the pump. First, the distance between standing valve and plunger (at the end of the downstroke) should be reduced to a minimum. Second, gas problems are minimized by the use of devices which restrict admission of gas to the working barrel, i.e., gas anchors or bottom-hole separators.

Uren[27] describes the gas anchor as follows. "In its simplest form, a gas anchor consists of a joint or two of tubing suspended from the shoe of the working barrel, capped at its lower end and provided with one or more holes near its upper end through which oil may enter. The holes in this tube fix the lowermost level to which fluid may be depressed by action of the pump. Oil entering this tubing is permitted to enter the pump only through a smaller tubing, concentric with the outer tubing, the upper end of which is screwed to the standing valve foot piece, whereas its lower end is well submerged beneath the surface of the fluid in the outer tubing. Separation of gas from oil is accomplished by causing the oil to flow down through the annular space between the two tubes, one telescoped within the other. During this downward movement of the oil, the gas escapes and moves upward by reason of its lower density, escaping, through apertures above the oil inlet, to the space between the casing and the tubing from which it may be permitted to escape at the side outlets of the casing head." The simplest type of gas anchor is shown in Fig. 5.1.

Schmoe[26] describes three types of bottom-hole separator whose use in the Rocky Mountain area has been responsible for increased daily oil production through reduction of gas-oil ratio and, consequently, of gas problems at the pump. One type of bottom-hole separator is shown in Fig. 5.26.

Undertravel or overtravel of the pump plunger, if present, is reflected distinctly in the dynamometer card for the well. Tilting of the card upward from left to right (Fig. 5.27) indicates undertravel. Tilting of the

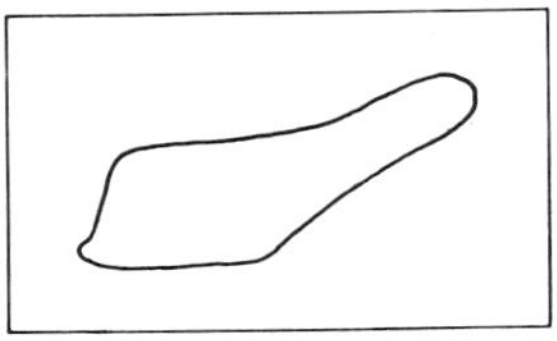

Fig. 5.27 Card indicating plunger undertravel.

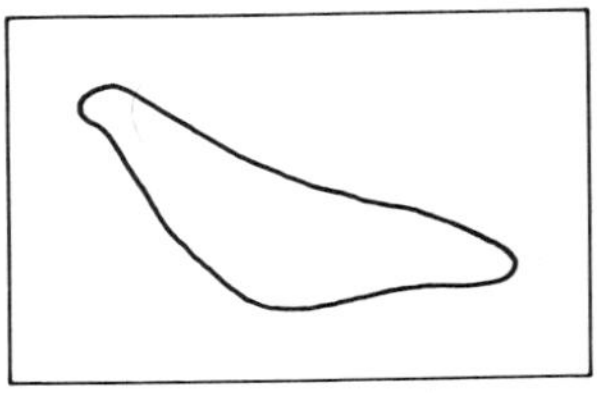

Fig. 5.28 Card indicating plunger overtravel.

card downward from left to right (Fig. 5.28) indicates overtravel. Undertravel results from excessive loading. Overtravel (due to acceleration) usually occurs at pumping speeds in excess of 20 spm.[23]

Dynamometer card interpretation is a skill which must be acquired by study and experience. The cards shown here point out only the most obvious problems; in most actual cases, interpretation is much more difficult. Also, it should be noted that it is virtually impossible to analyze a well merely by study of a single dynamometer card. Several cards should be taken. Additionally, as much as possible should be known about the characteristics and the production history of the well.

5.20. Applications of Dynamometer Cards

In addition to information which can be gained by simply recording the polished rod load over a complete pumping cycle, the recording dynamometer can be applied to the solution of specific problems involving valve conditions, counterbalance, and certain other pumping phenomena.

Fagg[22] gives the following methods for checking valve conditions. To check the traveling valve, the unit should be stopped at about the center of the upstroke. If the cord (mentioned in Sec. 5.18) is pulled at once, the line recorded on the chart by the main stylus represents the weight of sucker rods plus the weight of fluid on the plunger. With the unit remaining at rest, the cord should be pulled quickly at about one-second intervals. Any reduction in polished rod load indicates leakage of

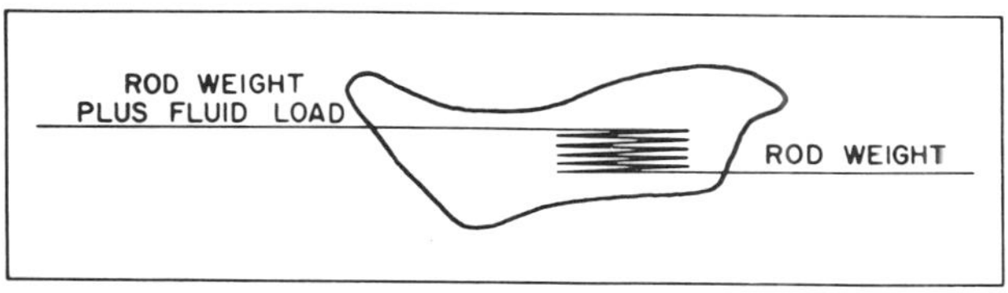

Fig. 5.29 Card indicating fluid leakage past traveling valve. (After Russell,[23] *World Oil.*)

fluid past the traveling valve or the plunger. Figure 5.29 indicates a plunger with leaking traveling valve. Rate of leakage can be determined from the time required for the load to reach the dead weight of the rods.

The standing valve can be checked by stopping the unit near the end of the downstroke, and immediately pulling the cord. Since at this point in the pumping cycle the standing valve should be closed and the traveling valve open, the line thus recorded represents the dead weight of the rods. If load remains constant with time, there is no leakage past the standing valve. If load increases as in Fig. 5.30, the indication is that leakage is occurring through the standing valve and that some of the fluid load is being transferred to the plunger and the rod string.

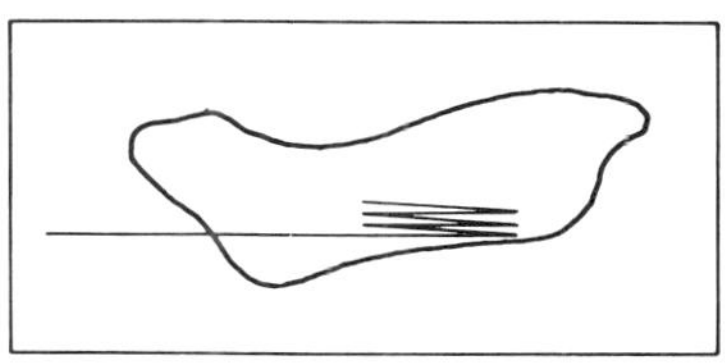

Fig. 5.30 Card indicating fluid leakage past standing valve. (After Russell,[23] *World Oil.*)

The dynamometer can also be used to determine instantaneous torques throughout the pumping cycle (with the objective of evaluating counterbalance effect) and instantaneous horsepower. Mills,[28] for example, has shown how counterweight and well-load torque can be calculated from rig geometry and dynamometer load data. Kemler[29] has shown how instantaneous polished rod velocities and dynamometer loads can be used to obtain instantaneous horsepower. Both procedures involve detailed graphical analysis of the pumping unit. Johnson[30] describes a method of calculating instantaneous torque which involves use of the timing wave (recorded on the dynamometer card) to obtain polished rod velocity, and use of a recording tachometer to obtain instantaneous prime mover speed. This is probably the most accurate method available, short of those involving complex analyses of the geometry of the pumping unit.

Fagg[22] has outlined an approximate method for calculating instantaneous torque, requiring only information available from a dynamometer card. This method assumes simple harmonic motion for the rods (i.e., uniform angular velocity for the crank). It also makes use of Eq. (5.65), which assumes the pitman to be vertical at all times and ignores the geometry of the surface installation. Even with these limitations, the method is sufficiently accurate for most purposes, and can be employed quickly and easily by the field engineer.

The method can best be understood by reference to Fig. 5.31. The essential information source is the dynamometer card, including lines

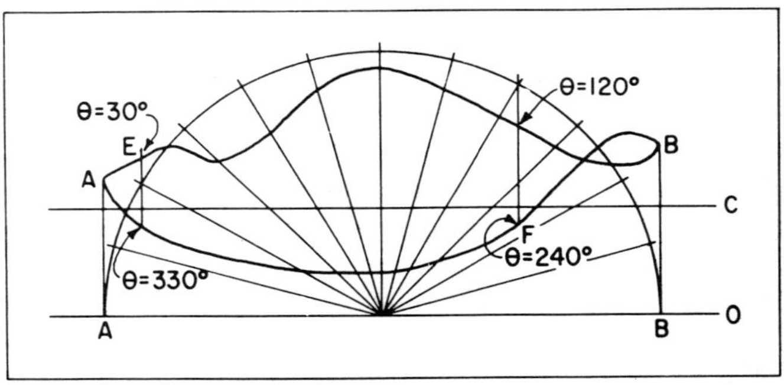

Fig. 5.31 Approximate method for calculating instantaneous torque.

representing counterbalance effect and zero load. Let points A and B represent the beginning of the upstroke and the beginning of the downstroke, respectively. Also, let θ be the angle between the crank and the vertical, as measured clockwise from the crank position at the beginning of the upstroke. If A and B are projected vertically onto the zero load line and a semicircle is drawn on this projection as a diameter, points on the dynamometer card corresponding to given crank angles can be determined. This is done by constructing radii so as to divide the semicircle into equal segments of, say, 15° each, and then projecting vertically the intersections of the radii with the semicircle. Intersections of these projections with the load curve represent instantaneous values of polished rod load W at various crank angles θ. Instantaneous torque is then calculated from

$$T = (W - C)(S/2) \sin \theta \tag{5.65}$$

In determining loads corresponding to crank angles, account should be taken of the fact that on the upstroke (point A to point B) crank angle varies from 0° to 180°, and on the downstroke (point B to point A) crank angle varies from 180° to 360°.

Example 5.25. In Fig. 5.31, the distance of point E above the zero load line is $1\frac{3}{16}$ in., the distance of point F above the zero load line is $\frac{11}{16}$ in., and the distance of line C (counterbalance effect) above the zero load line is $\frac{13}{16}$ in. The instrument constant for the dynamometer is 4000 lb per in. Calculate the instantaneous torque on the gear reducer (a) when the crank angle is 30° and (b) when the crank angle is 240°. The polished rod stroke length is 72 in.

SOLUTION:

$$C = 4000 \text{ lb/in.} \times \tfrac{13}{16} \text{ in.} = 3250 \text{ lb}$$
$$S/2 = 72/2 = 36 \text{ in.}$$

(a) $\sin \theta = \sin 30° = 0.500$

$$W = 4000 \text{ lb/in.} \times 1\tfrac{3}{16} \text{ in.} = 4750 \text{ lb}$$
$$T = (4750 - 3250) \times 36 \times 0.500 = 27{,}000 \text{ in.-lb}$$

(b) $\sin \theta = \sin 240° = -0.866$

$$W = 4000 \text{ lb/in.} \times \tfrac{11}{16} = 2750 \text{ lb}$$
$$T = (2750 - 3250) \times 36 \times (-0.866) = 15{,}600 \text{ in.-lb}$$

The effectiveness of counterbalance can be graphically demonstrated by plotting instantaneous torque against crank angle over the full pumping cycle. Figure 5.32 shows the torque curve for a well which is under-

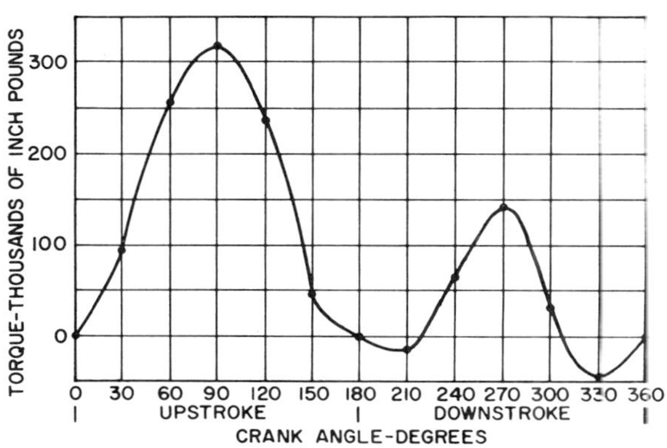

Fig. 5.32 Torque curve for undercounterbalanced well.

counterbalanced. It should be noticed that the peak torque on the upstroke is very much greater than the peak torque on the downstroke. Figure 5.33 shows the results of overcounterbalancing. Here the peak

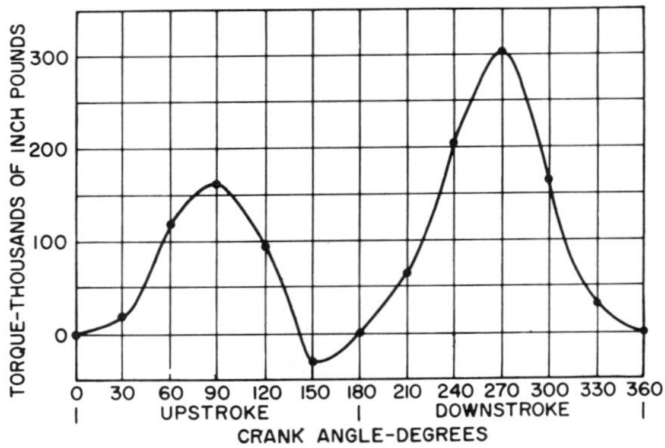

Fig. 5.33 Torque curve for overcounterbalanced well.

torque is excessively high on the downstroke. For a well which is properly counterbalanced, the torque curve will have approximately equal peak torques on upstroke and downstroke, as shown in Fig. 5.34. This allows the minimum possible torque load to be placed on the gear reducer.

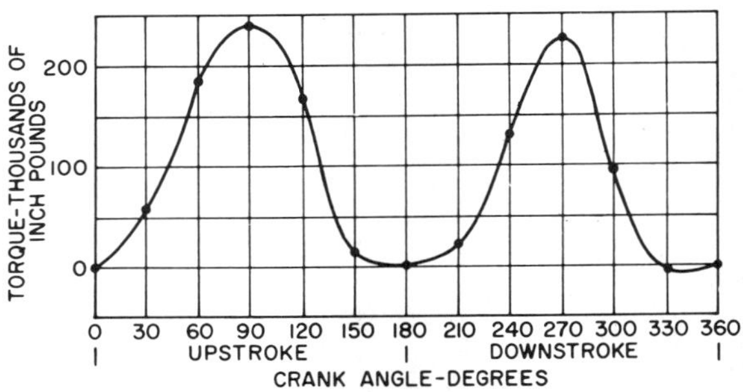

Fig. 5.34 Torque curve for correctly counterbalanced well.

If a dynamometer card is available for a pumping well, the correct counterbalance is generally considered to be the mean between maximum and minimum loads, as shown by the card. The existing counterbalance can then be adjusted accordingly. To obtain confirmation of correct counterbalance, another card should be run after counterbalance has been adjusted, and a torque curve should be plotted from the latter card.

Example 5.26. For the card of Example 5.25, the point representing maximum polished rod load is $1\frac{1}{2}$ in. above the zero load line, and the point representing minimum polished rod load is $\frac{5}{16}$ in. above the zero load line. The outer edge of the counterweight is 32 in. from the end of the 82-in. long crank. The distance from the center of gravity to the outer edge of the counterweight is 7 in. Neglecting effects of structural unbalance and rig geometry, calculate the position to which the counterweight should be moved in order to counterbalance the unit properly.

SOLUTION:

$W_{\max} = 4000 \text{ lb/in.} \times 1\frac{1}{2} \text{ in.} = 6000 \text{ lb}$

$W_{\min} = 4000 \text{ lb/in.} \times \frac{5}{16} \text{ in.} = 1250 \text{ lb}$

Desired counterbalance $= 0.5(6000 + 1250) = 3625$ lb

Existing counterbalance $= 3250$ lb (Example 5.25)

From Eq. (5.62), $C = 2W_c d/S$, and

$$C/d = 2W_c/S = \text{const.}$$

if neither counterweight nor stroke length is changed. Letting the subscript 1 refer to existing conditions and subscript 2 refer to desired conditions.

$$C_1/d_1 = C_2/d_2$$

Let x be the distance from center of gravity to outer edge of counterweight, and let y be the distance from outer edge of counterweight to end of crank. Since the crank is 82 in. long,

$$82 = d_1 + x + y_1 = d_2 + x + y_2$$

Using this and the preceding equation,

$$d_2 = 82 - x - y_2 = d_1 \frac{C_2}{C_1} = (82 - x - y_1) \frac{C_2}{C_1}$$

$$y_2 = 82 - x - (82 - x - y_1) \frac{C_2}{C_1} = 82 - 7 - (82 - 7 - 32) \frac{3625}{3250}$$

$$= 27 \text{ in.}$$

Agnew[31] has suggested the use of certain dynamometer card data to determine pumping bottom-hole pressure and net lift for a well. Data needed can be obtained from standing valve and traveling valve tests as previously described. The weight of rods plus fluid, as measured by a traveling valve check, will be designated here as W_{rfm}, and the weight of rods, as measured by a standing valve check, will be designated as W_{rm}. The measured fluid load is then

$$W_{fm} = W_{rfm} - W_{rm} \tag{5.129}$$

The calculated weight of the rods (in air) and the calculated fluid load (neglecting the effect of bottom-hole pressure) will be called W_{rc} and W_{fc}, respectively.

Since the tubing may be filled with a mixture of salt water, oil, and gas, the average specific gravity of tubing fluid is not easily determined from surface data. It can, however, be calculated from the standing valve test, since the difference between calculated and measured weights of the rod string is the weight of displaced fluid of effective or average specific gravity G. Section 5.8 and Eq. (5.115) show that

$$W_{rm} = W_{rc}(1 - 0.127G) \tag{5.130}$$

Once the effective gravity is determined from Eq. (5.130), fluid load is calculated from

$$W_{fc} = 0.433GLA_p \tag{5.131}$$

The difference between the calculated and measured fluid loads is the force resulting from the bottom-hole pressure p_w acting over the plunger area A_p:

$$p_w = \frac{W_{fc} - W_{fm}}{A_p} \tag{5.132}$$

The net lift L_N is the actual lift from the pump, L, less the lift supplied by the bottom-hole pressure:

$$L_N = L - \frac{p_w}{0.433G} \tag{5.133}$$

Equation (5.133) can be written

$$L_N = L\left(1 - \frac{p_w}{0.433GL}\right) \tag{5.134}$$

Substituting the value of p_w given in Eq. (5.132),

$$L_N = L\left(1 - \frac{W_{fc} - W_{fm}}{0.433GLA_p}\right) \tag{5.135}$$

Substituting W_{fc} as given by Eq. (5.131),

$$L_N = L\left(1 - \frac{W_{fc} - W_{fm}}{W_{fc}}\right) = L\frac{W_{fm}}{W_{fc}} \tag{5.136}$$

Example 5.27. A well in the Ventura Field (California) has a $1\frac{3}{4}$-in. plunger set on a sucker rod string consisting of 3500 ft of $\frac{3}{4}$-in. rods and 2000 ft of $\frac{7}{8}$-in. rods. Standing and traveling valve tests on the well give values of 8910 lb for rod load and 12,370 lb for rod plus fluid load. Using the method suggested by Agnew, calculate the pumping bottom-hole pressure and the net lift of the well.

Solution:

$A_p = 2.405$ sq in., $M_1 = 1.63$ lb/ft, $M_2 = 2.16$ lb/ft

$W_{fm} = W_{rfm} - W_{rm} = 12{,}370 - 8910 = 3460$ lb

$W_{rc} = M_1L_1 + M_2L_2 = 3500 \times 1.63 + 2000 \times 2.16 = 10{,}025$ lb

$W_{rm} = W_{rc}(1 - 0.127G)$

$G = 7.87(1 - W_{rm}/W_{rc}) = 7.87(1 - 8910/10{,}025) = 0.874$

$L = L_1 + L_2 = 3500 + 2000 = 5500$ ft

$W_{fc} = 0.433GLA_p = 0.433 \times 0.874 \times 5500 \times 2.405 = 5000$ lb

$$p_w = \frac{W_{fc} - W_{fm}}{A_p} = \frac{5000 - 3460}{2.405} = 640 \text{ psi}$$

$$L_N = L \times \frac{W_{fm}}{W_{fc}} = \frac{5500 \times 3460}{5000} = 3810 \text{ ft}$$

5.21. Problem Well Testing

Merryman and Lawrence[32] have devised an excellent system of classifying problem wells and isolating the cause of an unsatisfactory performance. Data necessary for well classification are (1) a dynamometer card, showing polished rod loads over a full pumping cyle, (2) calculated static rod load W_{rc}, (3) calculated static rod plus fluid load W_{rfc}, (4) measured static rod load W_{rm} as determined by a standing valve test, and (5) measured static rod plus fluid load W_{rfm} as determined by a traveling valve test. The classification is based essentially on the condition of the valves (as indicated by the tests), and on the agreement (or lack of agreement) between calculated and measured loads. For purposes of this discussion, problem wells will be designated as Types A, B, C, D, and E.

Type A wells are those for which (1) calculated and measured fluid loads are in good agreement and (2) no valve leaks are indicated. A card from a Type A well is shown in Fig. 5.35.

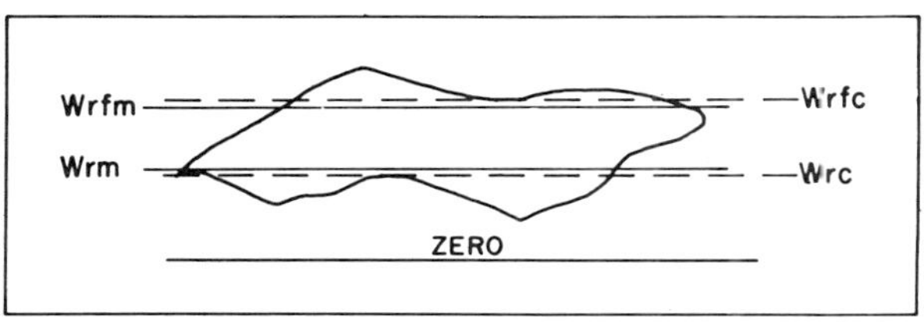

Fig. 5.35 Card from Type A well. (After Merryman and Lawrence,[32] *Pumping Well Problem Analysis.*)

Type B wells are those for which (1) calculated and measured fluid loads are not in good agreement and (2) no valve leaks are indicated. A card from a Type B well is shown in Fig. 5.36. Since no load changes appear to occur during standing and traveling valve tests, satisfactory valve conditions are indicated. And, since calculated and measured rod loads are not significantly different, the difference between loads represented by the lines W_{rfc} and W_{rfm} must be attributed to different calculated and measured fluid loads.

Type C wells are those for which a leaking valve is indicated, either by a

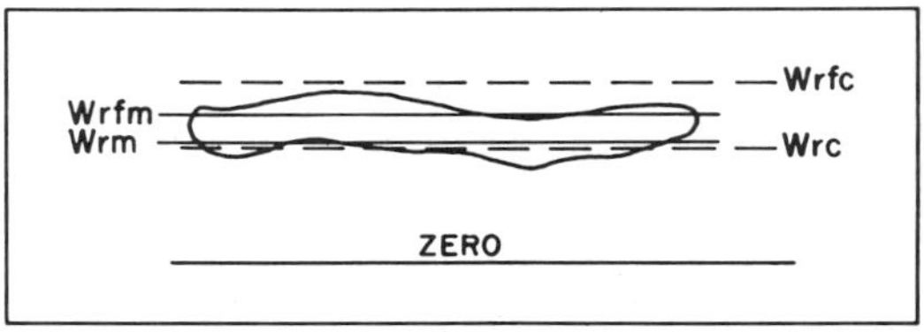

Fig. 5.36 Card from Type B well. (After Merryman and Lawrence,[32] *Pumping Well Problem Analysis.*)

traveling valve test or by a standing valve test. Cards from Type C wells are shown in Fig. 5.37. In both cases, a traveling valve leak is indicated. In the case of the lower card, the leak is exceptionally severe, as shown by the line representing W_{rfm}.

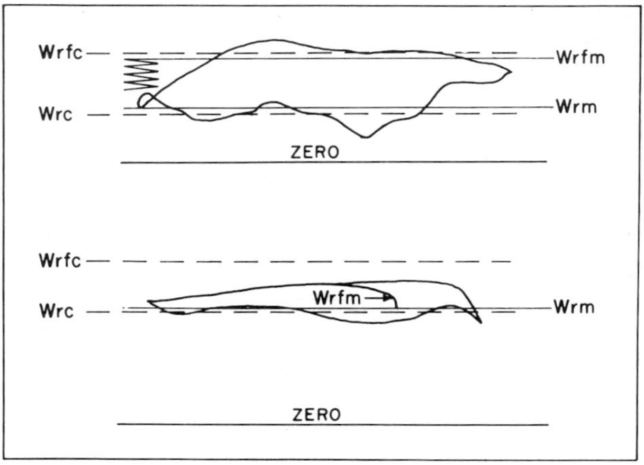

Fig. 5.37 Cards from Type C wells. (After Merryman and Lawrence,[32] *Pumping Well Problem Analysis.*)

Type D wells are those for which traveling and standing valve tests show the same load line. Comparison with calculated values shows which of the loads (rod or rod plus fluid) is represented by the line. Thus, for the upper card of Fig. 5.38, the recorded line represents rod plus fluid load,

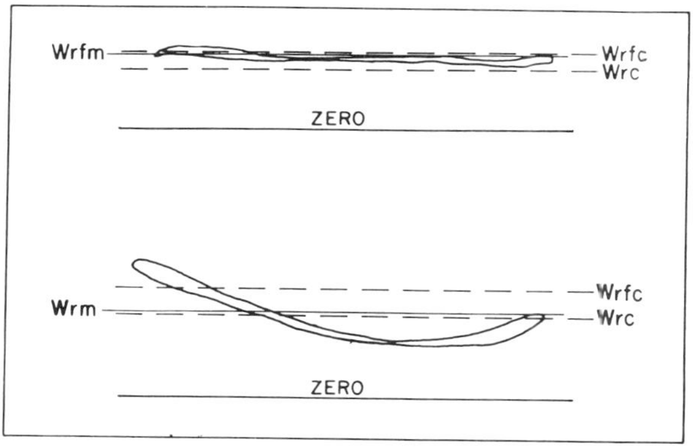

Fig. 5.38 Cards from Type D wells. (After Merryman and Lawrence,[32] *Pumping Well Problem Analysis.*)

whereas for the lower card, the recorded line represents only rod load, since it more nearly coincides with the calculated rod load.

Type E wells are those for which abnormal loads are indicated, either

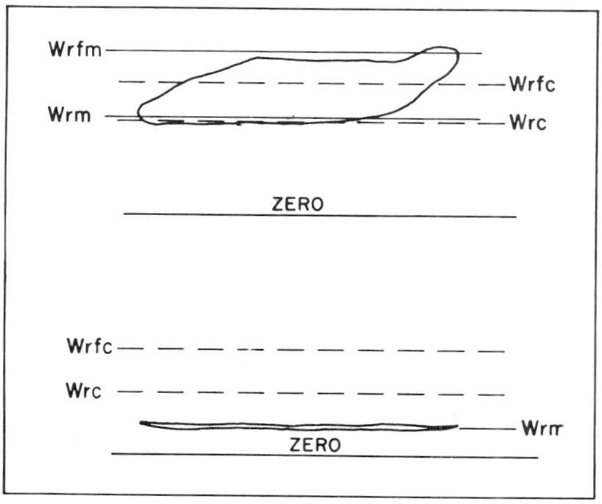

Fig. 5.39 Cards from Type E wells with abnormal loads indicated by valve measurements. (After Merryman and Lawrence,[32] *Pumping Well Problem Analysis.*)

by valve tests or by card shape. In the case of valve tests two possibilities exist, as shown by Fig. 5.39. For the upper card, a traveling valve test shows the measured rod plus fluid load to be greatly in excess of the calculated value. For the lower card, a standing valve test shows the measured rod load to be very much less than the calculated value. Figure 5.40 shows a card from a Type E well for which calculated and measured static loads are in reasonable agreement, but for which abnormal peak loads are indicated by the irregular shape of the dynamometer card.

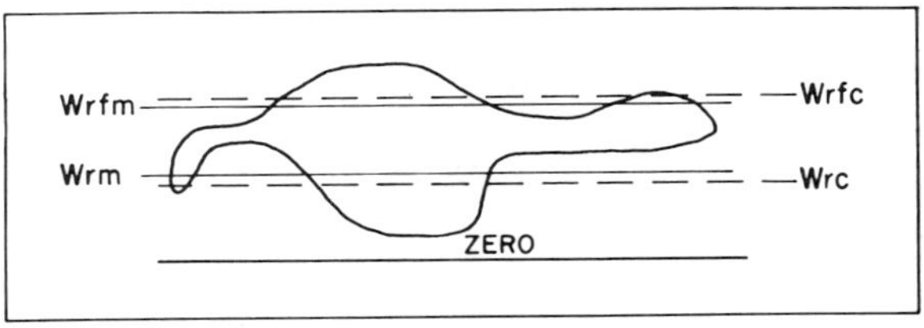

Fig. 5.40 Card from Type E well with abnormal load indicated by card shape. (After Merryman and Lawrence,[32] *Pumping Well Problem Analysis.*)

After a well has been classified according to the above system, it is subjected to certain test and evaluation procedures which have as their purpose the isolation of pumping problems with a minimal expenditure of time and money. Recommended procedures for problem well analysis are shown in Figs. 5.41 through 5.45.

5.22. Determination of Pumping Efficiencies

Certain computations of pumping equipment performance can be made by detailed analysis of the dimensions of the dynamometer card.[33] The measurements required are those illustrated in Fig. 5.46, i.e., area within the load loop, area between load loop and zero load line, length of the load loop, and distances above the zero load line of points on the loop representing maximum and minimum loads. If the instrument constant for the dynamometer is known, load range, average loads, and correct counterbalance can be determined.

In addition to this, knowledge of the polished rod stroke length and the pumping speed will permit calculation of polished rod horsepower. The below-ground efficiency is calculated by comparing the polished rod horsepower with the hydraulic horsepower as computed from net lift, production rate, and specific gravity of the fluid. The above-ground efficiency is calculated by comparing the polished rod horsepower with the power input to the prime mover.

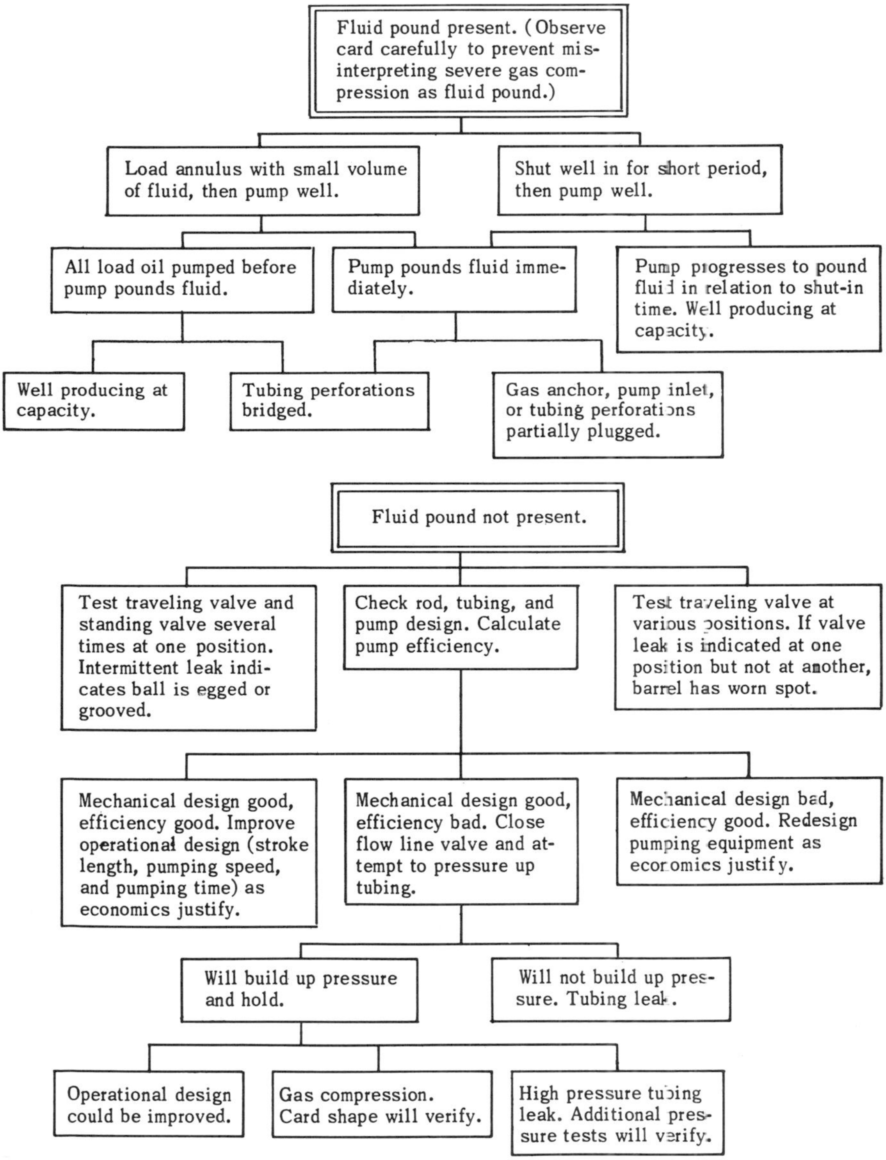

Fig. 5.41 Analytical procedures for Type A wells. (After Merryman and Lawrence,[32] *Pumping Well Problem Analysis*.)

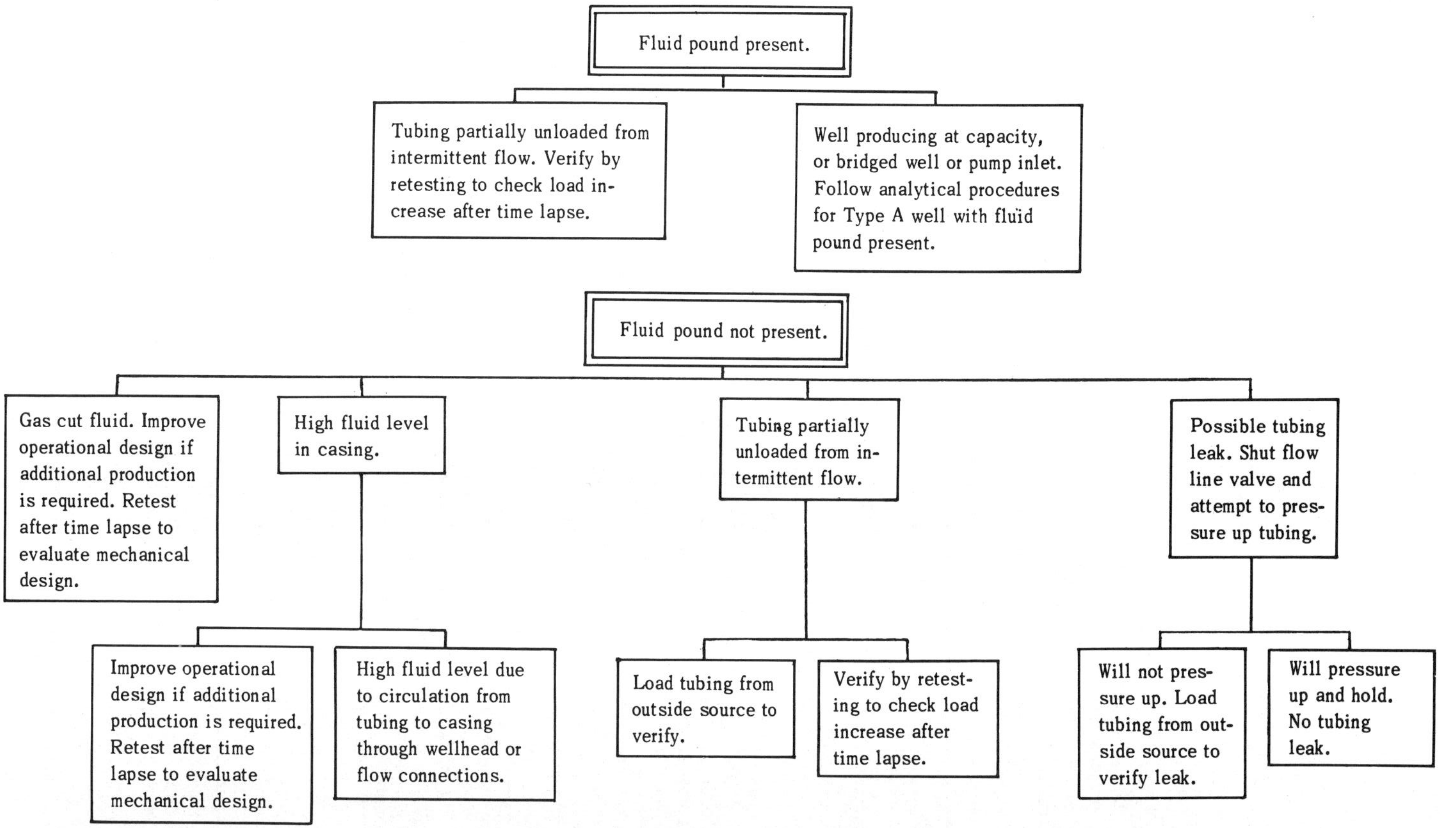

Fig. 5.42 Analytical procedures for Type B wells. (After Merryman and Lawrence,[32] *Pumping Well Problem Analysis.*)

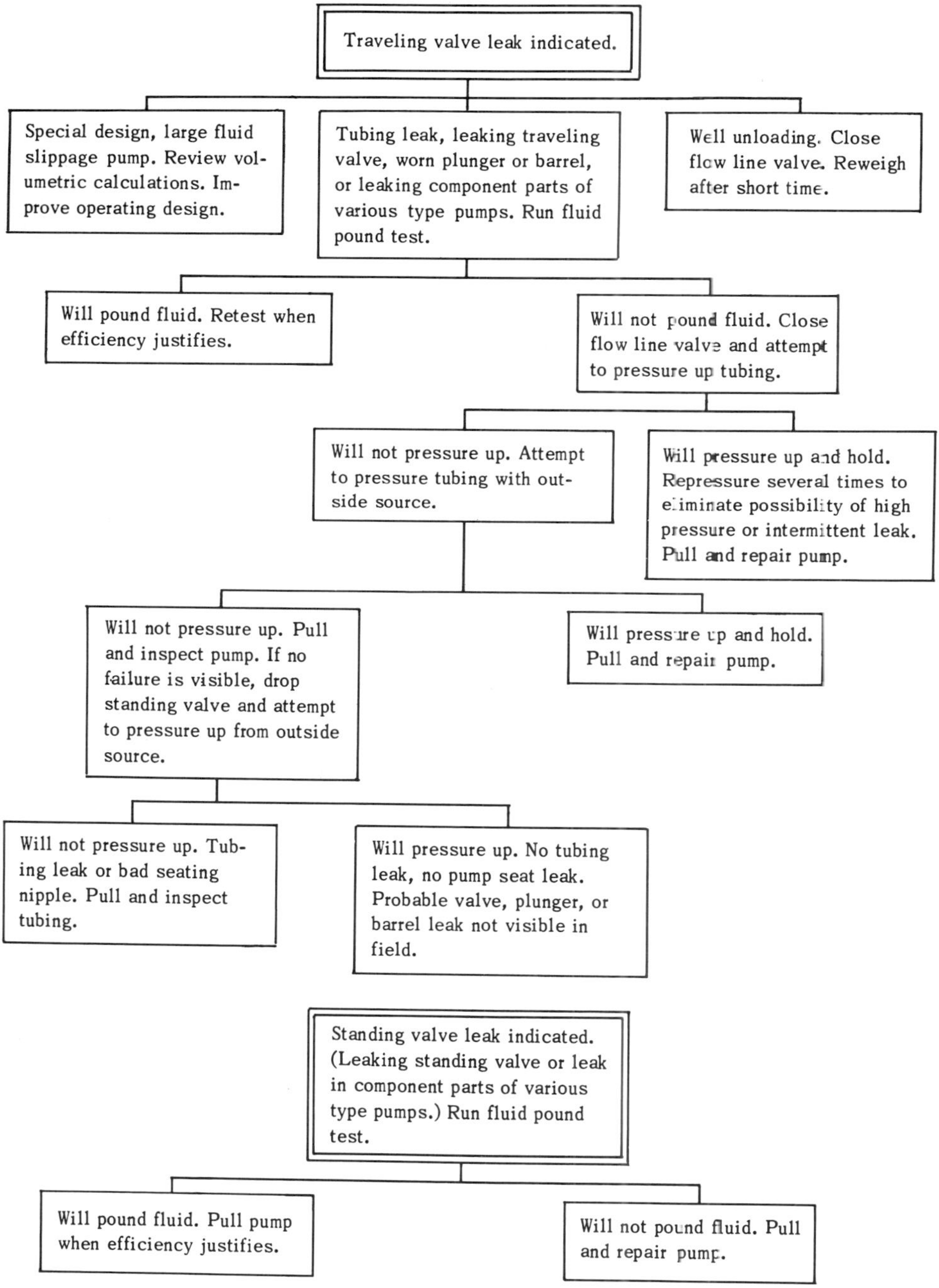

Fig. 5.43 Analytical procedures for Type C wells. (After Merryman and Lawrence,[32] *Pumping Well Problem Analysis.*)

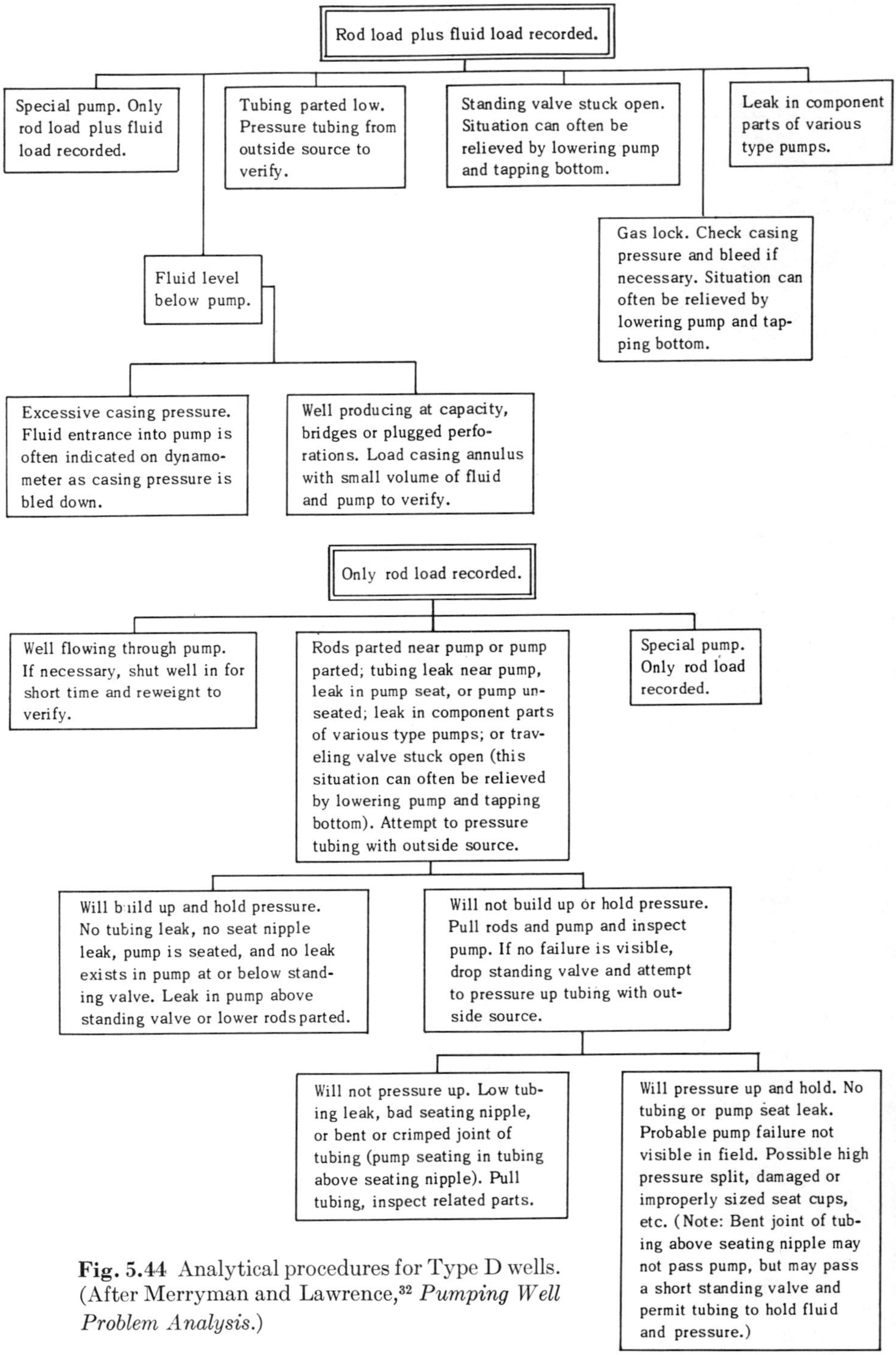

Fig. 5.44 Analytical procedures for Type D wells. (After Merryman and Lawrence,[32] *Pumping Well Problem Analysis*.)

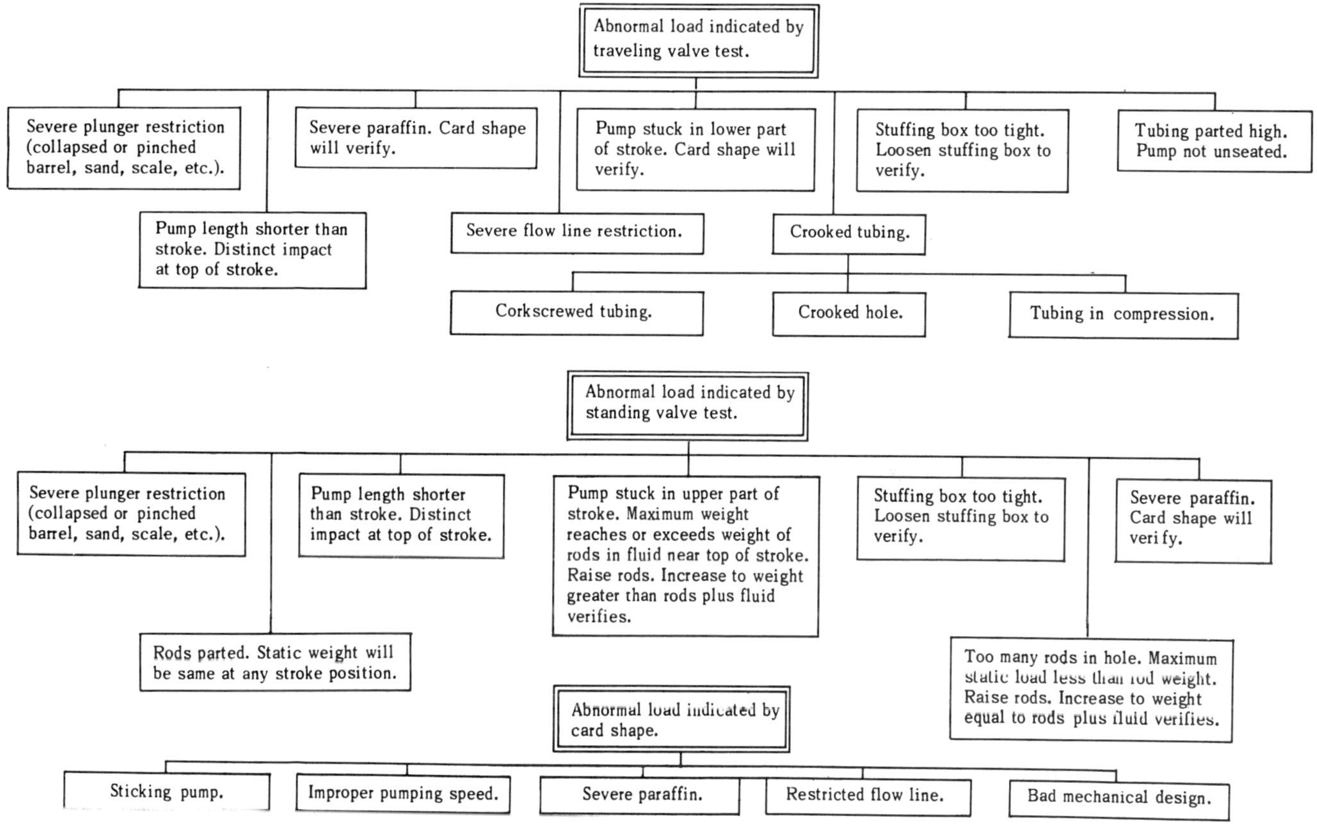

Fig. 5.45 Analytical procedures for Type E wells. (After Merryman and Lawrence,[32] *Pumping Well Problem Analysis*.)

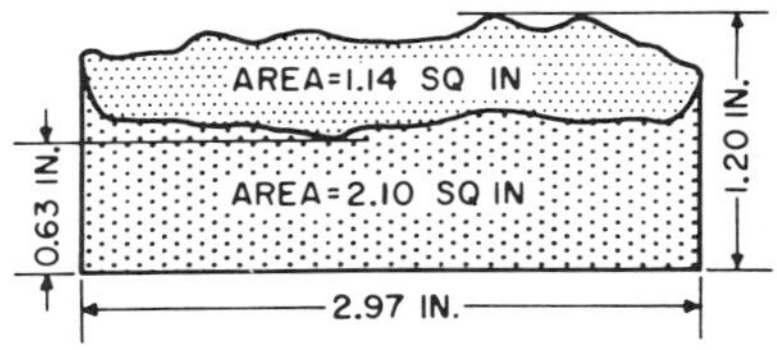

Fig. 5.46 Required measurements for detailed analysis of a dynamometer card. (Marsh and Watts,[33] *Drilling and Production Practice*, API.)

Example 5.28. The dynamometer which recorded the card of Fig. 5.46 has an instrument constant of 6000 lb per in. The well for which the card was taken is being pumped at 20 spm with a 54-in. stroke, and the resulting production is 200 bbl per day of a fluid of specific gravity 0.85. The net lift of the fluid is 3500 ft, and the wattmeter on the line to the electric motor being used as the prime mover shows an average power input of 8 kw. With this information, compute (1) the load range (expressed as per cent of maximum polished rod load), (2) the correct counterbalance effect for the unit, and (3) the overall efficiency of the installation.

Solution: The card dimensions are shown in Fig. 5.46.

(1) Maximum load = 1.20 in. × 6000 lb/in. = 7200 lb

Minimum load = 0.63 in. × 6000 lb/in. = 3780 lb

$$\text{Load range} = \frac{7200 - 3780}{7200} = 47.5 \text{ per cent}$$

(2) On the upstroke, the area between the load curve and the zero load line is 1.14 + 2.10 = 3.24 sq in., and the average height of the load curve is

$$\frac{\text{Area}}{\text{Length}} = \frac{3.24}{2.97} = 1.09 \text{ in.}$$

Therefore the average upstroke load is

$$1.09 \text{ in.} \times 6000 \text{ lb/in.} = 6540 \text{ lb}$$

On the downstroke, the area between load curve and zero load line is 2.10 sq in., and the average height of the load curve is

$$\frac{\text{Area}}{\text{Length}} = \frac{2.10}{2.97} = 0.71 \text{ in.}$$

Therefore the average downstroke load is

$$0.71 \text{ in.} \times 6000 \text{ lb/in.} = 4260 \text{ lb}$$

The correct counterbalance is the mean between the average upstroke and downstroke loads,

$$0.5(6540 + 4260) = 5400 \text{ lb}$$

Note: Designating the upper area (within the loop) as A_u, the lower area (below the loop) as A_l, the length of the loop as L_l, and the instrument constant as Y, we can write the ideal counterbalance effect as

$$C_i = 0.5\left[Y\frac{A_u + A_l}{L_l} + Y\frac{A_l}{L_l}\right] = \frac{Y}{L_l}\left[A_l + \frac{A_u}{2}\right] \tag{5.137}$$

(3) $\text{Power input} = \dfrac{8 \text{ kw}}{0.746 \text{ kw/hp}} = 10.72 \text{ hp}$

$$\begin{aligned}\text{Power output} &= \text{Hydraulic horsepower} = 7.36 \times 10^{-6}\, \gamma G L_N \\ &= 7.36 \times 10^{-6} \times 200 \times 0.85 \times 3500 \\ &= 4.38 \text{ hp}\end{aligned}$$

$$\text{Efficiency} = \frac{\text{Power output}}{\text{Power input}} = \frac{4.38}{10.72} = 40.9 \text{ per cent}$$

Note: The mean effective height of the loop is A_u/L_l inches. The mean effective force at the polished rod is therefore YA_u/L_l pounds, and the work done at the polished rod is SYA_u/L_l inch-pounds per cycle. The polished rod horsepower is

$$\begin{aligned}H_p &= \frac{SYA_u}{L_l}\frac{\text{in.-lb}}{\text{cycle}} \times \frac{N \text{ cycles/min}}{12 \times 33{,}000 \text{ in.-lb/min/hp}} \\ &= \frac{SYA_uN}{396{,}000L_l} \text{ hp}\end{aligned} \qquad (5.138)$$

For this example,

$$H_p = \frac{54 \times 6000 \times 1.14 \times 20}{396{,}000 \times 2.97} = 6.28 \text{ hp}$$

$$\text{Above-ground efficiency} = \frac{6.28}{10.72} = 58.6 \text{ per cent}$$

$$\text{Below-ground efficiency} = \frac{4.38}{6.28} = 69.7 \text{ per cent}$$

$$\text{Overall efficiency} = 0.586 \times 0.697 = 40.9 \text{ per cent}$$

PROBLEMS

Note: Except where otherwise indicated, it can be assumed in the following problems, that pump setting depth, pumping fluid level, and net lift are identical, and that tubing pressure is negligible.

5.1. For a well in the Jennings Field (Louisiana), polished rod stroke length is 68 in., pumping speed is 16 spm, and length of the sucker rod string is 7325 ft. What would be the increase in effective plunger stroke if pumping speed were increased to 20 spm? *Ans.* 10.1 in.

5.2. A well in the Oklahoma City Field (Oklahoma) has a sucker rod string consisting of 4600 ft of $\frac{3}{4}$-in. rods and 1950 ft of $\frac{7}{8}$-in. rods. The plunger diameter is $1\frac{1}{2}$ in., polished rod stroke length is 64 in., and average specific gravity of the well fluid is 0.825. At what pumping speed will polished rod stroke length and effective plunger stroke length be equal if the tubing is anchored? *Ans.* 20.6 spm.

5.3. Ni-Mo 4620 alloy steel sucker rods are being used in a well containing a corrosive fluid with hydrogen sulfide. What are the smallest-sized rods which should be used in the top section of the tapered string if peak polished rod load is expected to be 15,000 lb? *Ans.* 1 in.

5.4. For a well in the Selman City Field (Texas), the plunger diameter is 2 in., pumping speed is 23 spm, the calculated effective plunger stroke is 30.8 in., and production is 20 bbl of oil and 180 bbl of water per day. Calculate the pump constant and the volumetric efficiency of the pump in this well. *Ans.* 0.466; 60 per cent.

5.5. For a well in the Atlanta Field (Arkansas), the $1\frac{3}{4}$-in. plunger is set in 2-in. tubing on 4275 ft of $\frac{3}{4}$-in. sucker rods. There is no tubing anchor, and the working fluid level is known to be low. When the well is pumped at 18 spm with a 64-in. stroke, the production is 283 bbl per day of a fluid of specific gravity 0.825. From these data calculate the volumetric efficiency of the pump, peak polished rod load, ideal counterbalance effect, and polished rod horsepower. *Ans.* 84 per cent; 12,000 lb; 7710 lb; 12.4 hp.

5.6. Assuming the well of Prob. 5.5 to be ideally counterbalanced, calculate peak torque on the gear reducer, using the derived formulas. Compare with peak torque as calculated by the API method. *Ans.* 137,000; 154,000 in.-lb.

5.7. A dynamometer test was run on the well of Prob. 5.5. An analysis of the resulting dynamometer card shows a peak load of 11,500 lb, a minimum polished rod load of 3800 lb, and a polished rod horsepower of 11.02. From results of the dynamometer test, calculate the ideal counterbalance effect, maximum sucker rod stress, and peak torque if the unit is counterbalanced to within 5 per cent of the ideal value. *Ans.* 7650 lb; 26,000 psi; 135,000 in.-lb.

5.8. From dynamometer data for the well of Prob. 5.5, determine the stress range (at the top of the sucker rod string) expressed as a percentage of maximum stress. *Ans.* 67 per cent.

5.9. Calculate synchronous pumping speeds for a 5000-ft string of sucker rods for 1st, 2nd, 3rd, 4th, and 5th orders of synchronization. *Ans.* 47.4; 23.7; 15.8; 11.8; 9.5 spm.

5.10. For a certain pumping installation production is 200 bbl per day of a fluid having a specific gravity of 0.845 from a depth of 2500 ft. What is the below-ground efficiency for this installation if a dynamometer test shows a polished rod horsepower of 7.4 hp? *Ans.* 42 per cent.

5.11. For a well in the Bryan Pool Field (Texas) a $1\frac{3}{4}$-in. plunger is to be set with 3200 feet of $\frac{5}{8}$-in. sucker rods. It is estimated that a pumping speed of 17 spm and a polished rod stroke length of 42 in. will give the desired production of 157 bbl per day. It is further estimated that the working fluid level will be about 2000 ft, and that the specific gravity of the produced fluid will be 0.825. Assuming a 30 per cent reduction due to cyclic loading, calculate the minimum nameplate rating for the electric motor to be used as the prime mover. *Ans.* 7.67 hp.

5.12. For a certain well in the Chase Field (Kansas) a 2-in. plunger is set on a 3300-ft string of $\frac{7}{8}$-in. sucker rods. The tubing size is 3 in., and there is no tubing anchor or gas anchor. Production is 120 bbl per day of oil and 267 bbl per day of water, the average specific gravity of the tubing fluid being 0.83.

(For a fluid of this nature, the damping factor for use in the Rieniets Equation is approximately 0.80.) An Echometer shows the pumping fluid level to be at 2066 ft. The pumping speed is 18 spm at a polished rod stroke length of 64 in. Calculate the effective plunger stroke using the equations of Marsh and Coberly, the Rieniets equation, and the Langer and Lamberger equation. *Ans.* 62; 67; 62 in.

5.13. For the well of Prob. 5.12, calculate peak polished rod load using the Mills equation, the API method, and the Langer and Lamberger equation. *Ans.* 12,200; 12,300; 14,800 lb.

5.14. The area within the loop of a certain dynamometer card is 2.97 sq in., and the area between the loop and the zero load line is 3.82 sq in. The length of the loop is 4.56 in., and the instrument constant for the dynamometer is 5550 lb per in. Calculate the correct counterbalance effect for the unit *Ans.* 6450 lb.

5.15. For the well of Prob. 5.14, the pumping speed is 20 spm and stroke length is 48 in. Calculate the above-ground efficiency if the wattmeter (on the line to the electric motor being used as the prime mover) shows an average power input of 14.5 kw. *Ans.* 45 per cent.

5.16. In the design of a certain well, it is anticipated that production will be 375 bbl per day with a pump setting depth of 6500 ft. Assuming a volumetric efficiency of 75 per cent for the pump, calculate the rod size and the length of each section of the tapered string if sucker rods are available in lengths of 25 ft. *Ans.* 2925 ft of $\frac{3}{4}$-in. rods; 1925 ft of $\frac{7}{8}$-in. rods; 1650 ft of 1-in. rods.

5.17. For a well in the Pilot Butte Field (Wyoming) the sucker rod string consists of 3575 ft of $\frac{3}{4}$-in. rods and 2625 ft of $\frac{7}{8}$-in. rods. A standing valve test on the well gave a static rod load of 10,190 lb. What is the effective specific gravity of the fluid in the tubing? *Ans.* 0.90.

5.18. For the well of Prob. 5.17, plunger size is $1\frac{3}{4}$-in. If a traveling valve test shows a load of 14,725 lb, what pumping bottom-hole pressure is indicated? *Ans.* 530 psi.

5.19. What is the net lift for the well of Probs. 5.17 and 5.18? *Ans.* 4840 ft.

5.20. A dynamometer card for a pumping well in the Lake St. John Field (Louisiana) is shown in Fig. 5.47. Stroke length for the unit is 44 in., and pumping speed is 19.5 spm. The dynamometer used for the measurements has an instrument constant of 8000 lb per in. Calculate the torque on the gear reducer when the crank has turned through an angle of 330° from the beginning of the upstroke. *Ans.* 38,500 in.-lb.

5.21. For the well of Prob. 5.20, the depth to the plunger is 5000 ft, and production is 225 bbl per day at the pumping speed of 19.5 spm. Because of an increase in water cut, it is desired to obtain more production by increasing pumping speed to a value greater than 19.5 spm. At what *approximate* production rate would a synchronous pumping speed be encountered? *Ans.* 273 bbl per day.

5.22. Construct the torque curve over a complete pumping cycle for the well of Prob. 5.20.

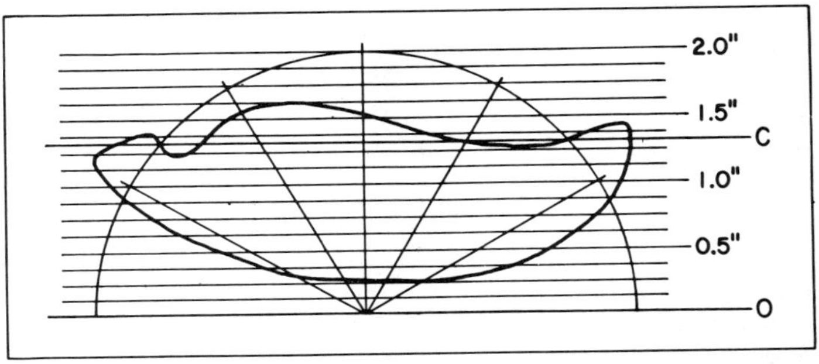

Fig. 5.47 Dynamometer card for Problem 5.20.

5.23. Construct the approximate torque curve which would result if the unit of Prob. 5.20 were counterbalanced ideally on the basis of the dynamometer card. What approximate reduction in peak torque would result from the change in counterbalance? *Ans.* 66,000 in.-lb.

5.24. A certain crank counterbalance has its center of gravity 33 in. from the center of the crankshaft when stroke length is 44 in. In order to reduce pumping speed, and thereby decrease peak polished rod load, it was decided to change the stroke length to 64 in. What will be the new distance from the center of gravity of the counterbalance to the center of the crankshaft if it is desired to keep the counterbalance effect the same? *Ans.* 48 in.

5.25. The dynamometer cards of Fig. 5.48 were obtained during tests on five problem wells. State the most obvious cause of poor performance in each case. *Ans.* (A) fluid pound; (B) friction; (C) overtravel; (D) gas lock; (E) synchronous pumping speed.

5.26. The overall efficiency of a certain pumping unit (utilizing an electric motor as the prime mover) is 39 per cent. If electric power costs $0.03 per kilowatt-hour, calculate the cost (for power only) of lifting a barrel of fluid of specific gravity 0.855 from a depth of 5400 ft. *Ans.* $0.047.

5.27. A well in the Steamboat Butte Field (Wyoming) is producing 584 bbl per day of a fluid having a specific gravity of 0.90. Pump setting depth is 5125 ft, pumping fluid level is 3200 ft, and tubing pressure is 50 psi. What is the net lift of the fluid? *Ans.* 3330 ft.

5.28. For the well of Prob. 5.27, the fuel requirement for the prime mover is 3.03 MCF/day of a gas having a net heating value of 1025 Btu/SCF. What is the overall efficiency of the installation? *Ans.* 25 per cent.

5.29. A pumping unit in the Pickens Field (Mississippi) has a single reduction gear with a gear ratio of 9:6. The unit sheave has a pitch diameter of 34 in., and the sheave on the multicylinder gas engine has a pitch diameter of 8 in. What is the engine speed when the pumping speed is 23 spm? *Ans.* 938 rpm.

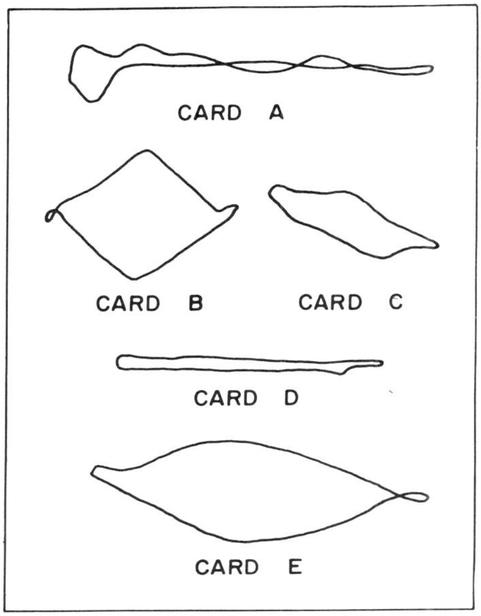

Fig. 5.48 Dynamometer cards for Problem 5.25.

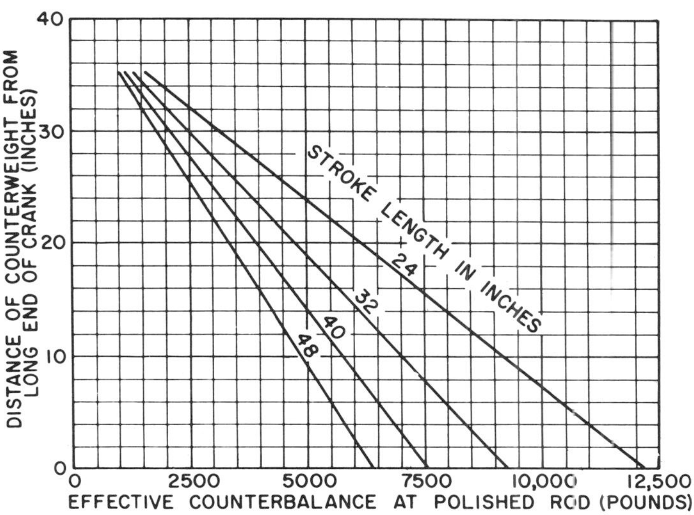

Fig. 5.49 Counterbalance chart for Problem 5.30. (Courtesy of Lufkin Foundry and Machine Company, Lufkin, Texas.)

5.30. The chart of Fig. 5.49 (for a Lufkin C-114DA-48-14 pumping unit assembly with 4846R crank and 5AR counterweight) gives the counterbalance effect as a function of the distance from outer edge of counterweight to end of crank, for stroke lengths of 24, 32, 40, and 48 in. For the 4846R crank, the distance from crankshaft to end of crank is 46 in. For the pumping unit assembly, the effect of structural unbalance is 565 lb, the distance from tail bearing to saddle bearing is 6 ft, and the distance from bridle to saddle bearing is 6 ft. Calculate, for the 5AR counterweight, the weight and the distance from the center of gravity to the outer edge. *Ans.* 3370 lb; 4.91 in.

5.31. A pumping well in the Livingston Field (Texas) is producing, through $2\frac{1}{2}$-in. tubing, a fluid of specific gravity 0.92. The temperature of the produced fluid at the surface is 95°F and the approximate mean yearly surface temperature is 70°F. Pump depth is 6550 ft, plunger diameter is $1\frac{3}{4}$ in., and the working fluid level is estimated to be 6530 ft. It is decided to set a tension anchor to reduce sucker rod failures believed to result from tubing buckling. How much pickup is necessary if the anchor is set at 6600 ft when the static fluid level is 3500 ft? *Ans.* 15,000 lb; 22 in.

5.32. Concerning Prob. 5.31, what extension of tubing below the pump (as tail pipe) would be necessary to prevent buckling without use of an anchor? What extension would be necessary if a compression anchor were to be set? *Ans.* 1090; 1910 ft.

5.33. For the well of Prob. 5.31, the total depth is 6800 ft, and it would therefore be impossible to use a tailpipe of length as great as that calculated in Prob. 5.32. Suppose that worn drill collars, no longer satisfactory for drilling, were available for use as tailpipe. What length of $5\frac{1}{2}$-in. OD by $2\frac{1}{4}$-in. ID collars (weight = 60 lb per ft) would be necessary with the compression anchor? *Ans.* 207 ft.

5.34. Calculate the depth to the neutral point of tubing under the following conditions: specific gravity of well fluid = 0.825; tubing size = $2\frac{1}{2}$ in.; plunger size = $1\frac{1}{4}$ in.; pump depth = 5000 ft; working fluid level = 3200 ft. *Ans.* 4760 ft.

5.35. At what points should sucker rod guides be placed to minimize effects of tubing buckling in the well of Prob. 5.34? *Ans.* 4760; 4809; 4857; 4899; 4935; 4968; 5000 ft.

5.36. It is desired to put on pump a Gulf Coast well having a static fluid level of 1000 ft. The well fluid has a 40 per cent water cut, with oil of gravity 40° API and water of specific gravity 1.015. The installation is to be designed for an oil production of 125 bbl per day. To prevent tubing buckling, it is decided to set a tension anchor at the pump depth of 4500 ft. Reservoir studies and well fluid analyses indicate a probable pump volumetric efficiency of at least 80 per cent. The mean yearly surface temperature at the well is 70°F.
(a) Determine the API rating of the pumping unit assembly to be purchased for this installation. *Ans.* 160,000 in.-lb.
(b) Determine, for the proposed installation, polished rod stroke length, optimum plunger size, and resulting tubing size and pumping speed. *Ans.* 64 in.; $1\frac{1}{2}$ in.; 2 in.; 20 spm.

(c) Design the sucker rod string, assuming sucker rods to be available in lengths of 25 ft. *Ans.* 1725 ft of $\frac{5}{8}$-in. rods; 1475 ft of $\frac{3}{4}$-in. rods; 1300 ft of $\frac{7}{8}$-in. rods.
(d) Calculate the actual *oil* production for the assumed volumetric efficiency and the operating conditions of part (b). *Ans.* 151 bbl per day.
(e) An example of a pumping unit assembly which meets the requirements of parts (a) and (b) is the Bethlehem 160D-20S-64. This unit has a beam load rating of 20,700 lb, an available counterbalance effect (without auxiliary counterweights) of 11,770 lb, and a peak torque rating of 160,000 in.-lb. Compare these values with calculated values of peak polished rod load, ideal counterbalance effect, and probable peak torque under the selected operating conditions. *Ans.* 12,100 lb; 7530 lb; 158,000 in.-lb.
(f) What is the maximum sucker rod stress under the selected operating conditions; what type of rods could be used if the well fluid contains hydrogen sulfide? *Ans.* 20,100 psi; Ni-Cr (Mayari) or Ni-Mo alloy steel rods.
(g) Determine the minimum brake horsepower rating for the electric motor to be used as the prime mover. *Ans.* 20 hp.
(h) The standard unit sheave for the Bethlehem 160D-20S-64 pumping assembly has a pitch diameter of 24 in., and the double reduction gear has a reduction ratio of 26.4. If motor speed is 1200 rpm, what should be the pitch diameter of the motor sheave to obtain the desired pumping speed? *Ans.* 10.6 in.
(i) After setting the anchor on the tubing it is necessary to provide sufficient tension for prevention of buckling. Determine the effective pickup for surface fluid temperatures of 110°, 100°, and 90°F. *Ans.* 15, 13, 11 in.

5.37. For a certain pumping installation, the pump setting depth is 5125 ft and the working fluid level is 3200 ft. The well is making 580 bbl per day of a fluid of specific gravity 0.90. Assuming negligible tubing pressure, calculate the overall efficiency of the installation if the wattmeter on the line to the electric motor being used as the prime mover shows an average power input of 23 kilowatts. *Ans.* 40 per cent.

5.38. To obtain a desired production, a well is to be pumped at 20 spm with a polished rod stroke length of 64 in. Calculations based on existing subsurface equipment show rod load to be 7430 lb, fluid load to be 3310 lb, and the ideal counterbalance effect to be 8270 lb. Considering the pumping units available (according to the pumping unit selection chart) and assuming that the unit can be counterbalanced to within 5 per cent of the ideal value, determine the smallest API size unit with a torque rating sufficient to handle this job. *Ans.* API size 228.

5.39. What is the average plunger speed for a well being pumped at 20 spm if the effective plunger stroke is 58 in.? *Ans.* 3.22 fps.

5.40. It is desired to design a pumping installation to obtain a production of 360 bbl per day at a pump setting depth of 3600 ft and an assumed volumetric efficiency of 80 per cent. The appropriate design data (Table 5.9) shows two possible plunger sizes for these conditions. Considering the possibility of excessive sucker rod fatigue resulting from synchronous pumping speed, which plunger size should be selected? *Ans.* 2 in.

5.41. For a certain pumping well the following data are available: pump setting depth = 3500 ft; rod size = $\frac{3}{4}$ in.; tubing size = $2\frac{1}{2}$ in. (tubing not anchored); plunger size = $1\frac{3}{4}$ in.; polished rod stroke length = 54 in.; specific gravity of produced fluid = 0.85; pumping speed = 18 spm; working fluid level = 3150 ft. Calculate the volumetric efficiency of the pump if the well is producing 180 bbl per day. *Ans.* 60 per cent.

5.42. A well is producing 240 bbl per day of a fluid having a gravity of 0.80. The pump is set at 3500 ft on $\frac{3}{4}$-in. rods, and the plunger size is $1\frac{3}{4}$ in. Calculate the ideal counterbalance effect for the unit. *Ans.* 6290 lb.

REFERENCES

[1]*World Oil* (Feb. 15, 1959), p. 144.

[2]*API Specification for Oil Well Pumps* (API Standard 11A), 7th Ed. (Oct., 1944).

[3]*API Specification for Sucker Rods* (API Standard 11B), 12th Ed. (May, 1957).

[4]*API Specification for Miscellaneous Pumping Equipment* (API Standard 11D), 5th Ed. (Jan., 1952).

[5]J. Zaba, "Oil Well Pumping Practices No. 14 — Counterbalancing, Purpose and Principles," *Oil and Gas Journal* (Sept. 30, 1943), pp. 51–53.

[6]*API Specification for Pumping Units* (API Standard 11E), 7th Ed. (May, 1956).

[7]Hallan N. Marsh, "High Volumetric Efficiency in Oil Well Pumping and Its Practical Results," *Production Bulletin,* No. 207 (API, June 30, 1931), pp. 47–57.

[8]C. J. Coberly, Discussion of Reference 7, *Production Bulletin,* No. 207 (API, June 30, 1931), pp. 57–61.

[9]Example and problem data for wells (where fields are mentioned by name) were taken, by permission, from *Sucker Rod Handbook* (Bethlehem: Bethlehem Steel Company, 1953).

[10]"Standardization Committee Report on Pumping Equipment and Engines (Exhibit A, Well-Load Formulas)," *Production Bulletin,* No. 230 (API, Nov., 1943), pp. 442–462.

[11]Howard L. Kelley and Roy M. Willis, "Manual for Selection of Beam Pumping Equipment — Part 3," *The Petroleum Engineer* (Sept., 1954), pp. B77–B97.

[12]J. Zaba and W. T. Doherty, *Practical Petroleum Engineers Handbook,* 4th Ed. (Houston: Gulf Publishing Company, 1956), p. 532.

[13]R. W. Rieniets, "Plunger Travel of Oil-Well Pumps," *Drilling and Production Practice* (API, 1937), pp. 159–178.

[14]B. F. Langer and E. H. Lamberger, "Calculation of Load and Stroke in Oil-Well Pump Rods," *Oil and Gas Journal* (July 2, 1942), pp. 27–32.

[15]Howard L. Kelley and Roy M. Willis, "Manual for Selection of Beam Pumping Equipment — Part 1," *The Petroleum Engineer* (July, 1954), pp. B37–B40.

[16]Howard L. Kelley and Roy M. Willis, "Manual for Selection of Beam Pumping Equipment — Part 2," *The Petroleum Engineer* (Aug., 1954), pp. B40–B54.

[17]*API Recommended Practice for Care and Handling of Sucker Rods* (API RP 11BR), 2nd Ed. (Jan., 1955).

[18]J. Zaba, "Oil-Well Pumping Practices No. 5 — Fatigue of Sucker Rods," *Oil and Gas Journal* (July 29, 1943), pp. 229–233.

[19]*Sucker Rod Handbook* (Bethlehem: Bethlehem Steel Company, 1953), pp. 20–25.

[20]J. C. Slonneger, "Vibration Problems in Oil Wells," *Drilling and Production Practice* (API, 1937), pp. 179–188.

[21]Arthur Lubinski and K. A. Blenkarn, "Buckling of Tubing in Pumping Wells, Its Effects and Means for Controlling It," *Trans. AIME*, **210**, 73–88 (1957).

[22]L. W. Fagg, "Dynamometer Cards and Well Weighing," *Trans. AIME*, **189**, 165–174 (1950).

[23]J. H. Russell, Jr., "Interpretation of Dynamometer Cards," *World Oil* (July, 1953), pp. 175–188.

[24]Dynamometer cards shown in Figs. 5.20 through 5.28 are taken from *Sucker Rod Handbook* and from the article by Russell cited as Reference 23.

[25]Marvin L. Thompson, "In Deep Well Pumping, Know Your Well Loads," *Oil and Gas Journal* (April 27, 1959), pp. 156–163.

[26]W. P. Schmoe, "Bottom Hole Separators Increase Production," *World Oil* (April, 1959), pp. 152–156.

[27]L. C. Uren, *Petroleum Production Engineering* (*Oil Field Exploitation*), 3rd Ed. (New York: McGraw-Hill Book Company, Inc., 1953), pp. 281–282.

[28]Kenneth N. Mills, "Counterbalancing and Determining Peak Loads on Pumping Units," *World Oil* (Nov., 1948), pp. 176–182.

[29]Emory N. Kemler, "Counterbalancing of Oil-Well Pumping Machines," *Drilling and Production Practice* (API, 1943), pp. 87–107.

[30]D. O. Johnson, "Counterbalancing of Beam Pumping Units," *Drilling and Production Practice* (API, 1951), pp. 234–241.

[31]Bob G. Agnew, "The Dynamometer as a Production Tool," *Drilling and Production Practice* (API, 1957), pp. 161–168.

[32]J. M. Eubanks, B. L. Franks, D. K. Lawrence, T. E. Maxwell, and C. J. Merryman (Engineering Department, Sun Oil Company), *Pumping Well Problem Analysis*, (Midland: Joe Chastain, 1958), pp. 155–181.

[33]H. N. Marsh and E. V. Watts, "Practical Dynamometer Tests," *Drilling and Production Practice* (API, 1938), pp. 162–182.

CHAPTER SIX

Gas Lift

6.1. Introduction

The development of modern gas lift should be ranked high among the progressive steps taken by the oil industry over the last 20 years. The first mention of the application of compressed air to lifting oil was made by Eaton,[*,1] who described the use of air lift for raising oil in Venango County, Pennsylvania, in 1864. The principle was that of the two-pipe system: air was introduced into a blast pipe, the lower end being turned up so as to enter the cuplike opening of the discharge pipe. The early applications of gas lift have been described by Stirling,[2] Schneider,[3] Shaw,[4,5] Nicklin,[6] and Rayne, Craft, and Hawkins.[7] Shaw[8] and Cromer and Huntington,[9] in a study of gas-lift efficiencies, found four gas injection rates that were significant in the operation of gas-lift installations, as follows:

1. Injection rates of no flow, where the amount of gas flowing up the tubing is insufficient to lift the fluid. If gas enters the tubing at an extremely low rate, it will rise to the surface in small semispheres which are flat on the bottom.

2. Injection rate of maximum efficiency, where a minimum volume of gas is required to lift a barrel of fluid. Shaw[10] observed that this rate is determined by drawing a line from the origin tangent to the curve of production versus gas-oil ratio, as shown in Fig. 6.1.

*References throughout the text are given at the end of each chapter.

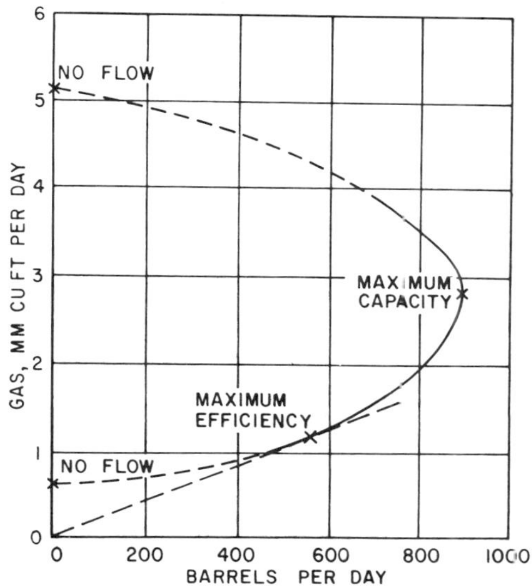

Fig. 6.1 Significant points in continuous-flow gas lift. (Shaw,[10] *The Texas Engineering Experiment Station.*)

3. Injection rate of maximum flow, which will yield maximum production.

4. Injection rate of no flow due to excess gas, which is reached when the friction produced by the gas prevents fluid from entering the tubing.

Early application of gas lift utilized the simple "U-tube" principle and was used in wells producing large quantities of fluid from shallow depths. High pressure gas introduced into either the annulus or the tubing depressed the fluid to the bottom of the tubing, and the well was produced by means of aeration. The high pressure required to displace the fluid in the annulus (kick-off pressure) and the necessary lowering of pressure after kick-off made this type of lift very inefficient. It was impractical for high fluid level, low capacity wells where small volumes of fluid were to be lifted. Such wells could be unloaded only by gradually lowering the tubing in stages and allowing the well to flow at each step until the tubing was hung at the desired depth. It was this type of well which resulted in the development of the kick-off valve (flow valve or unloading valve) shown in Fig. 6.2, which admitted gas into the tubing at selected points above bottom.[11] The valves operated automatically in response to the differential pressure between casing and tubing. When this differential decreased to a predetermined pressure, the valve opened, and when it exceeded the differential of the valve, the valve closed. The valves were spaced in the tubing below the static fluid level, 250 ft apart for a 100 psi

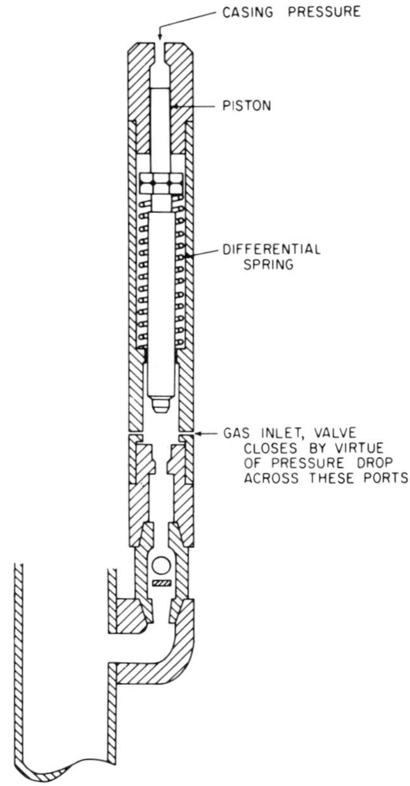

Fig. 6.2 Spring-loaded differential valve. (After Kirkpatrick,[11] *Division of Production,* API.)

differential valve and an 0.4 psi per ft fluid. The gas injected in the annulus uncovered the first valve which then unloaded the column of fluid in the tubing and remained open until the differential pressure exceeded the operating differential of the valve. During this production interval the second valve was uncovered and the unloading continued until the well was flowing from the lowest valve at which it was desired to admit gas. Gas-lift valves rapidly became essential in the design of gas-lift installations to accomplish one or more of the following aims:

1. Reduce unloading (kick-off) pressures.
2. Permit deeper injection depths.
3. Lower producing gas-oil ratios.
4. Eliminate swabbing of high fluid level wells which loaded up with water.

Another problem was encountered in wells of high capacity and low bottom-hole pressure. High gas pressure in the casing annulus, acting directly against the reservoir, reduced well productivity. This operating problem was resolved by placing a packer between the tubing and casing which held the annulus gas off the formation between injection cycles. The operation of low-capacity, low bottom-hole pressure wells was still unsatisfactory until a standing valve was placed at the bottom of the tubing as a supplement to the annulus packer. The installation was then a *closed* system and could be economically used to deplete low-capacity wells. Until the introduction of this type of installation, gas lift was intermediate between flowing and pumping as a means of producing oil.

One more significant improvement in gas-lift performance was obtained

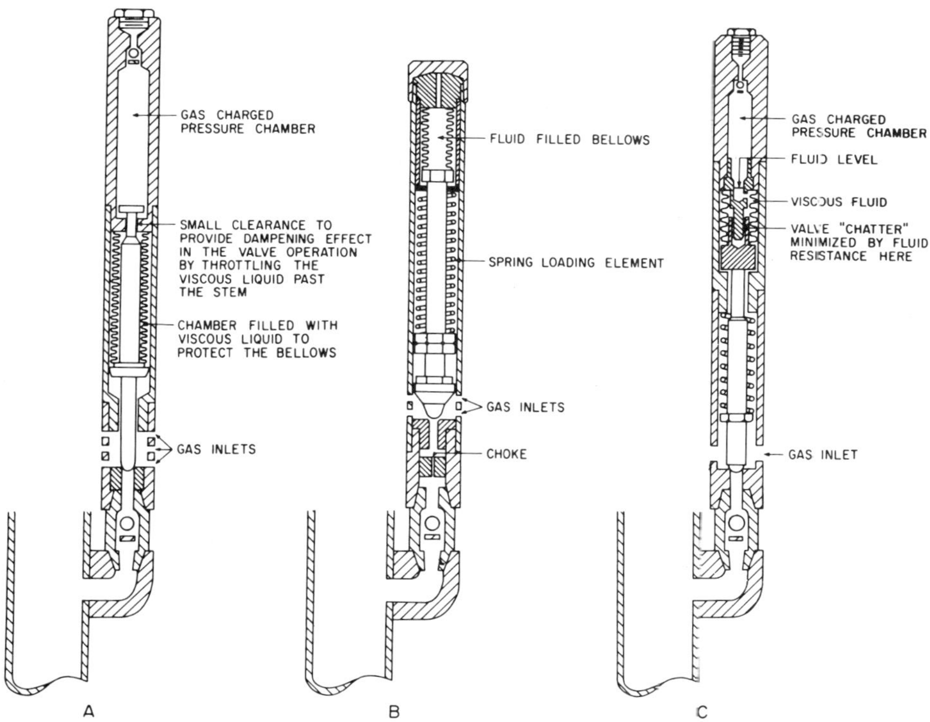

Fig. 6.3 (A) Gas-charged bellows intermitting valve (B) Spring-loaded continuous-flow valve. (C) Combination gas-charged bellows and spring-loaded intermitting valve. (After Kirkpatrick,[11] *Division of Production*, API.)

with the use of bellows and a spring-loaded type of valve[11] (Fig. 6.3). The bellows flexibility is achieved by convolutions which are pressed in three-ply, thin-walled seamless tubing. Advantages of this valve type are:

1. The operating pressure is constant.
2. A longer service life is provided by tungsten carbide stem tip and Monel seat.
3. The valve operating pressures can be changed in the field.
4. Fluid flow can be prevented by a reverse-flow check valve.
5. The valves are adaptable to either intermittent or continuous flow.

6.2. Application

Gas lift is a mechanical process of lifting fluid from a well where gas at relatively high pressure ($\geq$ 250 psi) is used as the lifting medium. Several factors must be investigated when a method of artificial lift is considered for a given well or reservoir. Gas lift is installed, either during completion or in later workover operations. These installations are very versatile in that they may be applied to wells of any depth, any reservoir pressure, and any rate of flow up to tens of thousands of barrels of fluid per day. Frequently, when future need for gas lift is anticipated, valves are installed at the time of original completion and later placed in operation using wire-line services. The factors that limit or impair the efficiency of gas lift are the presence of sour gas or crudes of high viscosity.

In a comprehensive study of gas-lift methods, Babson[12] noted that the behavior of a producing well is the resultant of both *well* and *equipment* characteristics. The most important well characteristics are (1) bottom-hole pressure, (2) productivity index, (3) formation gas-oil ratio, (4) per cent water, (5) depth, (6) size of tubing and casing, and (7) type of reservoir mechanism. The important equipment characteristics of any gas-lift method are (1) back pressure held against the formation and (2) gas consumption.

There are four categories of wells to be considered in gas-lift application: (1) high productivity index, high bottom-hole pressure wells; (2) high productivity index, low bottom-hole pressure wells; (3) low productivity index, high bottom-hole pressure wells; (4) low productivity index, low bottom-hole pressure wells. Wells having a productivity index of 0.50 or less are classified as low productivity wells, and those with a productivity index greater than 0.50 are classified as high productivity wells. High bottom-hole pressures will support a fluid column equal to 70 per cent of the well depth; and low bottom-hole pressures will support a fluid column of less than 40 per cent of the well depth.

Gas lift is accomplished in one of two ways: (1) by continuous gas injection into the tubing or casing at some predetermined depth to reduce the pressure opposite the producing formation (this is called *continuous-*

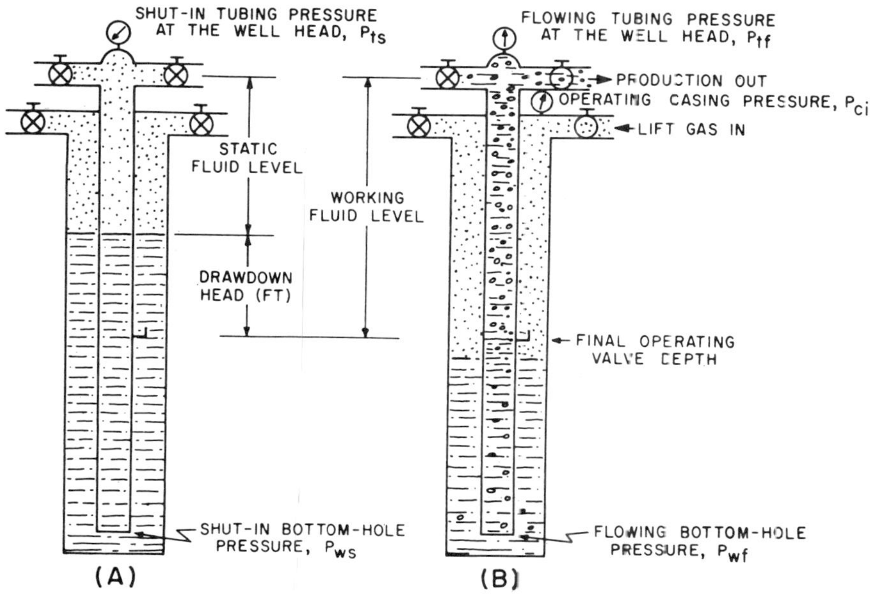

Fig. 6.4 (A) Well shut-in and (B) flowing continuously. (After Kirkpatrick,[13] *The Power of Gas.*)

flow gas lift; see Fig. 6.4), or (2) by injection of gas at a high instantaneous rate for a short period, with ejection of a slug of fluid up the tubing at controlled time intervals (this is referred to as *intermittent-flow* gas lift; see Fig. 6.5). Figure 6.6 shows the three basic types of gas-lift installations, commonly referred to as open, semiclosed, and closed.[13] Wells with a high bottom-hole pressure and a high productivity index (≥ 0.5) are normally designed for semiclosed, continuous flow. Wells with a high bottom-hole pressure and low productivity index are designed for semiclosed, intermittent flow. Wells with a low bottom-hole pressure and either a high or a low productivity index are usually designed for closed, intermittent flow.

By the continuous flow method (Fig. 6.7), a small volume of high pressure gas is introduced into the tubing to aerate or lighten the fluid column, so that the flowing bottom-hole pressure, with the aid of the expanding injection gas, will deliver the fluid to the surface.[11] To accomplish this efficiently, it is desirable to design a system that will permit injection through a single valve at the greatest depth possible with the available injection pressure.

Intermittent flow (slug lifting) consists of injecting high pressure gas into the tubing at sufficient volume and pressure to lift the fluid head accumulated above the valve to the surface with maximum velocity, in

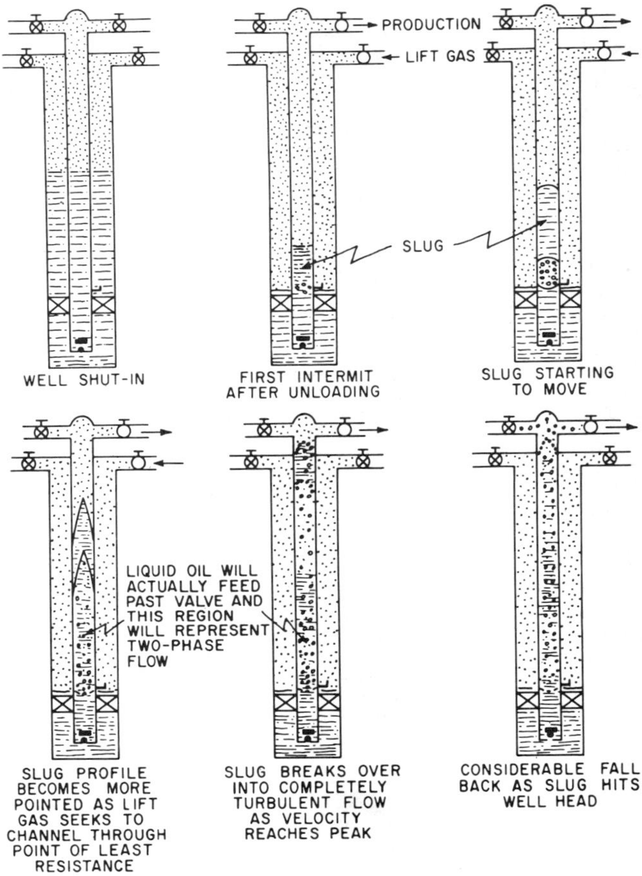

Fig. 6.5 Intermittent or ballistic-type flow. (After Kirkpatrick,[13] *The Power of Gas.*)

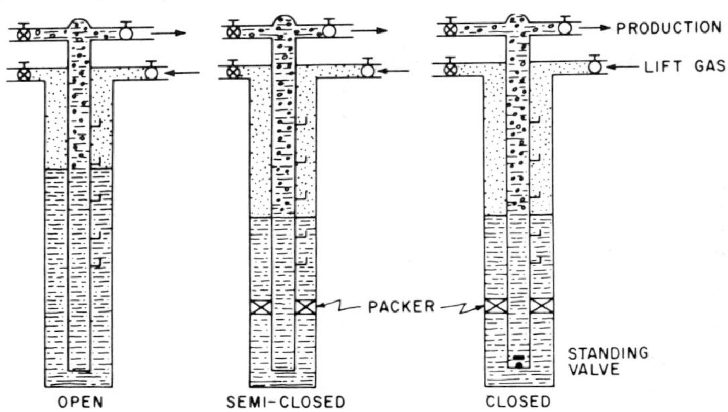

Fig. 6.6 Basic types of gas-lift installations. (Kirkpatrick,[13] *The Power of Gas.*)

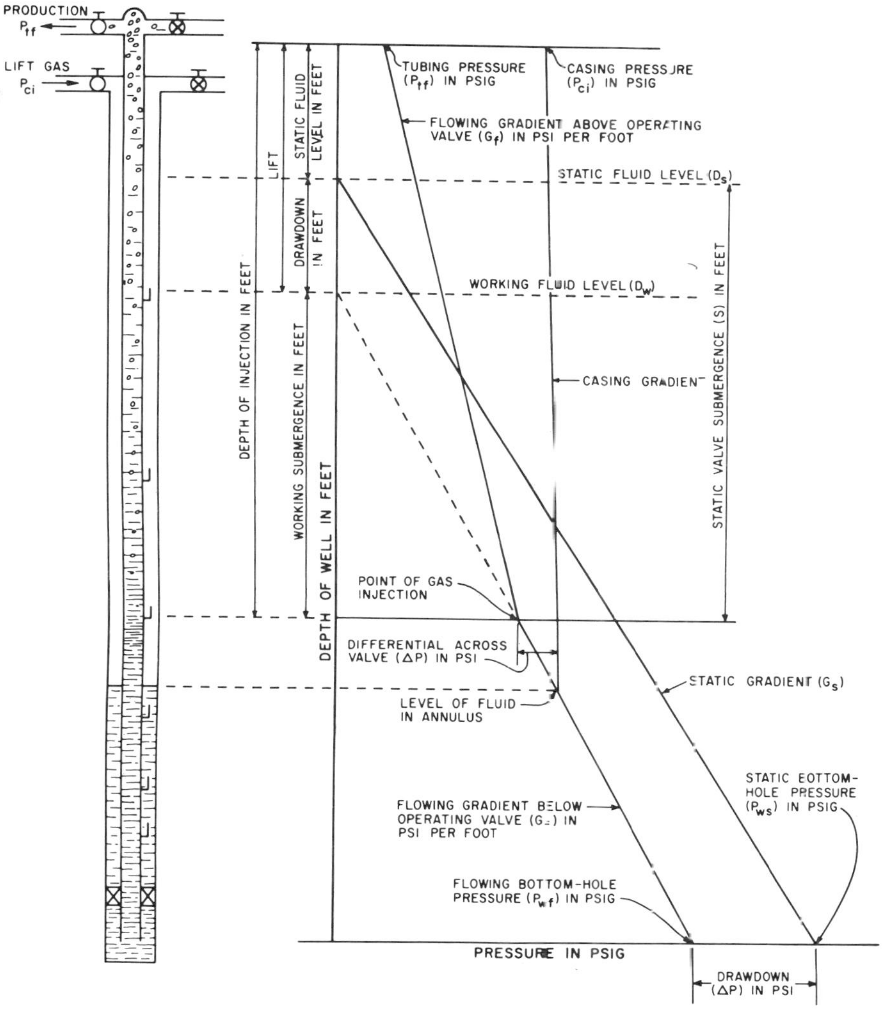

Fig. 6.7 Basic gas lift terms, continuous-flow gas lift. (After Kirkpatrick,[11] *Division of Production*, API.)

order to minimize aeration and fluid fall-back. This type of flow is accomplished by the action of a mechanical valve with a large port and by control of the gas pressure and volume in the annulus with a surface-controlled intermitter. The point of maximum efficiency and maximum capacity is determined by experimenting with different cycle[14] and injection periods as shown in Table 6.1. The maximum producing rate with intermittent lift is governed primarily by the gas injection cycle frequency, injection pressure, and tubing size.

Table 6.1. SELECTION OF INJECTION CYCLE AND PERIOD

Injection period, hr	Injection time, sec	Fluid produced, bbl/day	Gas-fluid ratio, cu ft/bbl
0.5	62	24.39	4720
0.5	46	20.31	4280
0.5	33	20.31	3830
0.5	21	16.23	1928
1	64	28.41	2042
1	44	28.53	1533
1	34	28.44	1110
1	26	12.18	2035
2	60	20.31	1400
2	46	20.40	1082
2	37	16.23	1084
2	31	16.23	938

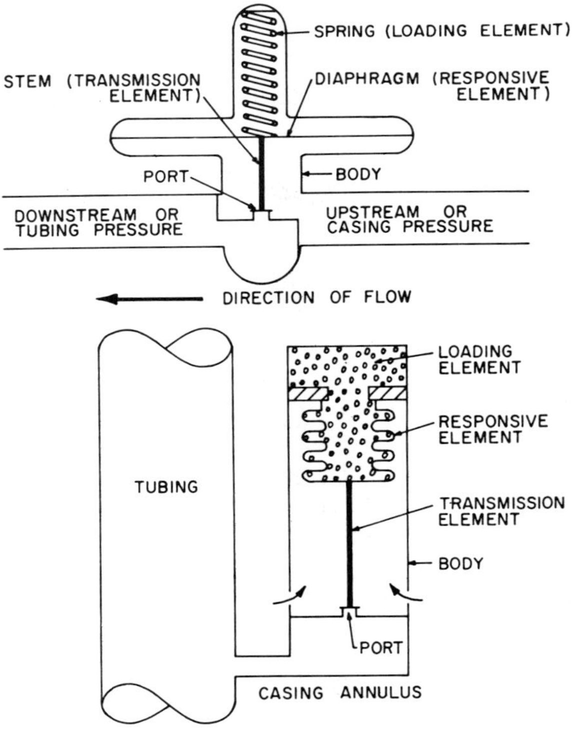

Fig. 6.8 Back-pressure regulator (top) and pressure-charged gas-lift valve (bottom). (Kirkpatrick,[11] *Division of Production*, API.)

6.3. Valve Mechanics

Gas-lift valves operate in accordance with certain basic principles, and are most simply compared with pressure regulators. The component parts are: (1) body, (2) loading element (spring, gas, or combination of both), (3) responsive element (metal bellows, piston, or rubber diaphragm), (4) transmission element (metal rod or rubber diaphragm), and (5) metering element (orifice or port). The loading pressures (spring for the regulator and gas for the gas-lift valve) for the regulators and gas-lift valves, shown in Figs. 6.8 through 6.10, act over a large area of responsive element

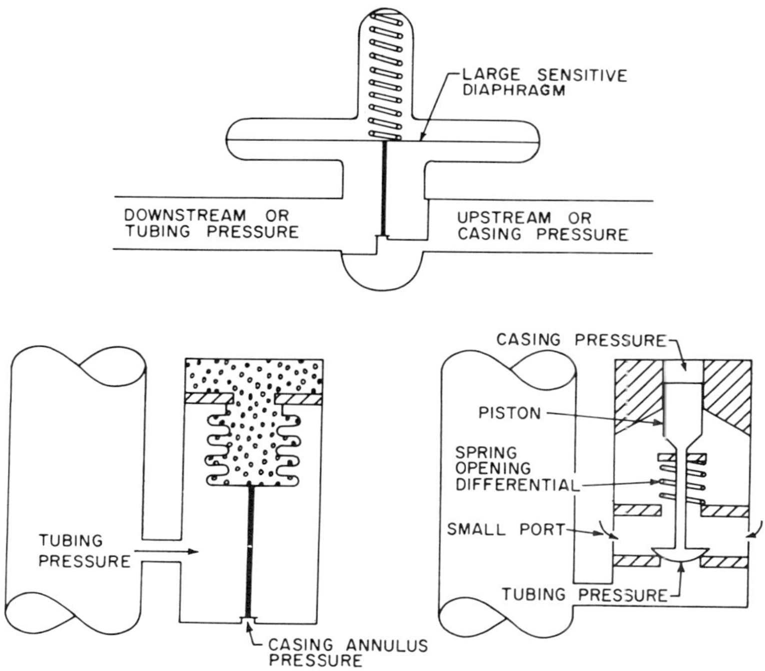

Fig. 6.9 Pressure regulator sensitive to downstream pressure (top), fluid-operated valve (bottom left), and spring-loaded differential valve (bottom right). (Kirkpatrick,[11] *Division of Production*, API.)

(diaphragm for regulator and bellows for the valve).[11] When the area of the responsive element is large compared with the valve port, the valve is relatively insensitive to the tubing pressure, i.e., the effect of liquid column build-up in the tubing, helping to open the valve, is small.

A force balance is used to analyze the forces acting to open and close the pressure-operated flow valve. A combination pressure-operated

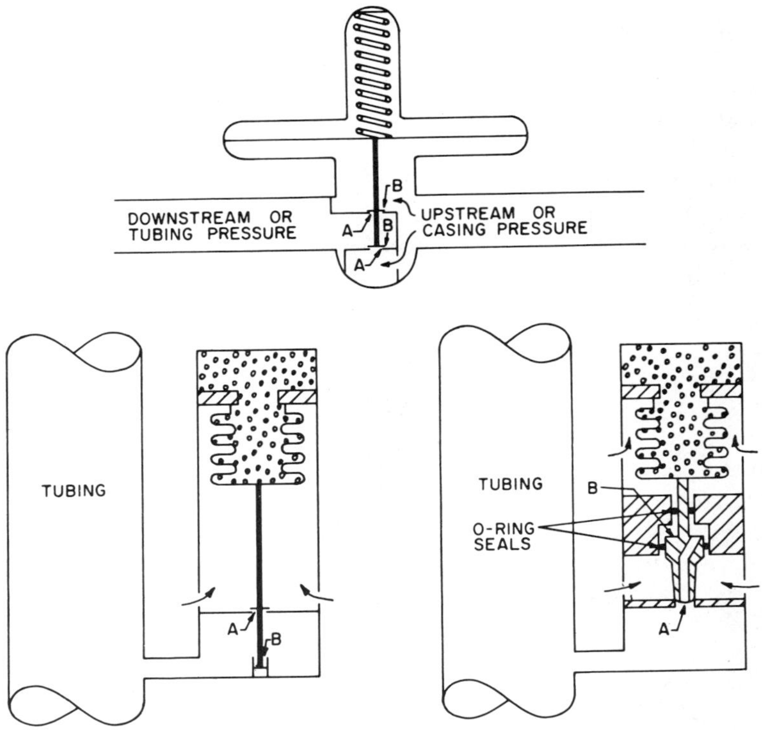

Fig. 6.10 Back-pressure balanced regulator (top), gas-lift valves with tubing pressure balanced (bottom). (Kirkpatrick,[11] *Division of Production*, API.)

(bellows) and spring-loaded valve is used in deriving expressions for the valve operating characteristics. A spring is introduced into this type of valve (Fig. 6.11) to supply a portion of the closing force of the valve. Normally the effect of a spring is expressed in force per unit length of compression; however, for valve mechanics it is convenient to express this spring effect in terms of equivalent pressure that the spring exerts in opening and closing the valve.

Opening pressure of valve under operating conditions. The force balance for the valve of Fig. 6.11b,[11] which is closed and ready to open, is

$$F_c = F_o$$

where F_c is the force in pounds holding the valve closed and F_o the force in pounds trying to open the valve. Substituting terms for F_c and F_o,

$$F_c = p_{bT}A_b + S_t(A_b - A_v) \tag{6.1}$$

$$F_o = p_o(A_b - A_v) + p_tA_v \tag{6.2}$$

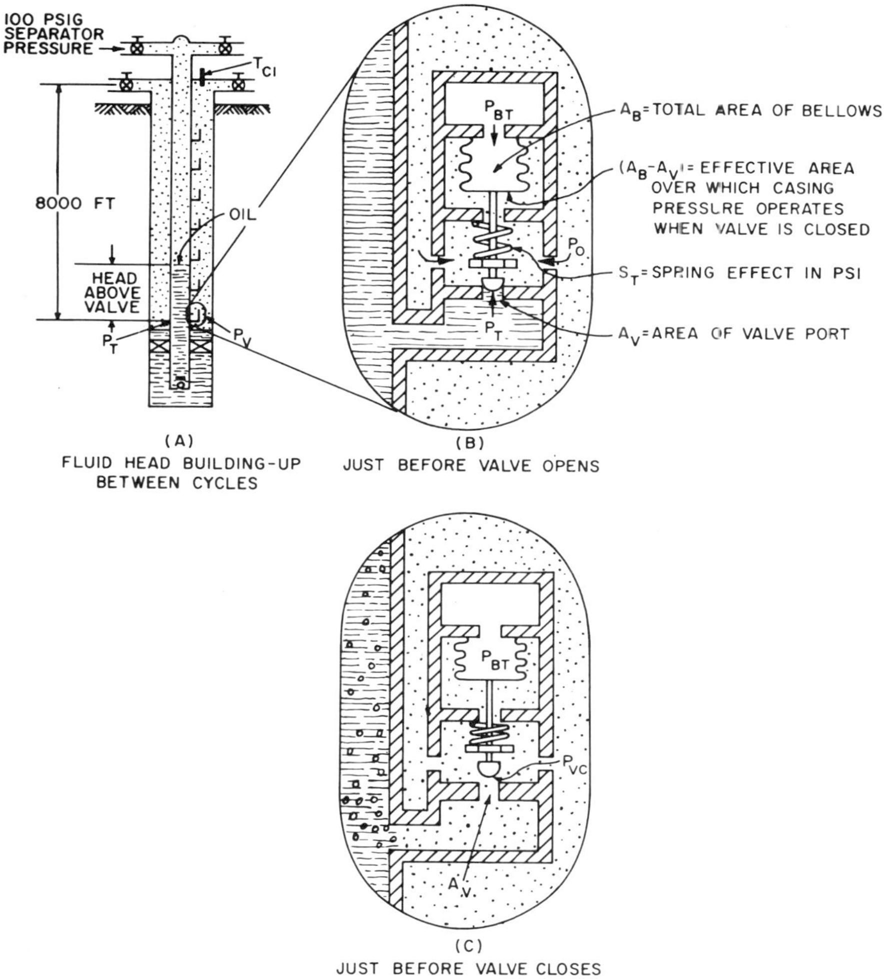

Fig. 6.11 Closed and open position of pressure-charged valves. (After Kirkpatrick,[11] *Division of Production*, API.)

where p_b = pressure inside the bellows at 60°F, psig

p_{bT} = pressure inside the bellows at operating temperature, psig

A_b = total effective bellows area, sq in.

p_o = opening pressure of valve with tubing effect, psig

p_{vo} = opening pressure of valve in a tester without tubing effect ($p_t = 0$), psig

p_t = pressure in the tubing (opposite the valve) just before opening, psig

A_v = area of the valve port, sq in.

S_t = effective spring tension, psig

Equating Eqs. (6.1) and (6.2), dividing by A_b, and solving for p_o,

$$p_o = \frac{p_{bT}}{1 - (A_v/A_b)} + S_t - p_t\left[\frac{A_v/A_b}{1 - (A_v/A_b)}\right] \tag{6.3}$$

The term $p_{bT}/(1 - A_v/A_b)$ is the pressure necessary to overcome the bellows charge at operating temperature. The full effect of the spring, S_t, must be overcome. The tubing effect on the opening pressure of the flow valve when the tubing pressure opposite the valve is known is

$$\text{T.E.} = p_t\left[\frac{A_v/A_b}{1 - (A_v/A_b)}\right] \tag{6.4}$$

For example, if there is a pressure in the tubing opposite the flow valve of 300 psig and the ratio A_v/A_b is 0.11, the tubing effect is

$$\text{T.E.} = 300\left[\frac{0.11}{1 - 0.11}\right] = 37 \text{ psig}$$

or the opening pressure is reduced by 37 psig.

In a valve tester, the tubing pressure opposite the valve is zero. With a force balance, opening and closing forces are equated,

$$p_{vo}(A_b - A_v) = p_bA_b + S_t(A_b - A_v)$$

or

$$p_{vo} = \frac{p_b}{(1 - A_v/A_b)} + S_t \tag{6.5}$$

or, in terms of p_b,

$$p_b = (p_{vo} - S_t)(1 - A_v/A_b) \tag{6.6}$$

Closing pressure of valve under operating conditions. The valve is now open and ready to close, and the casing pressure is operating over the total effective bellows area (Fig. 6.11c). The force balance is

$$F_c = F_o$$

Substituting terms for F_c and F_o,

$$F_c = p_{bT}A_b + S_t(A_b - A_v) \tag{6.7}$$

$$F_o = p_{vc}A_b \tag{6.8}$$

where p_{vc} is the closing pressure in the casing opposite the valve and the other terms are as previously defined. Equating (6.7) and (6.8) and solving for p_{vc},

$$p_{vc} = p_{bT} + S_t(1 - A_v/A_b) \tag{6.9}$$

or

$$p_{vc} = (1 - A_v/A_b)\left[\frac{p_{bT}}{(1 - A_v/A_b)} + S_t\right]$$

On substitution of (6.6) into (6.9), the valve closing pressure in a tester (where the tubing pressure is zero) is

$$p_{vc} = p_{vo}(1 - A_v/A_t) \tag{6.10}$$

Equation (6.3) can be stated in conceptual form as:

[pressure to open] = [bellows effect] + [spring effect] − [tubing effect]

Since at charging conditions the tubing effect is zero, the pressure to open the valve consists of the bellows and spring effects. The bellows of most valves are charged at 60°F and the volume is assumed constant. The bellows pressure at operating conditions is found by substituting in the nonideal gas law,

$$\frac{p_b}{z_{60}(460 + 60)} = \frac{p_{bT}}{z_T(460 + T_v)}$$

whence

$$p_{bT} = p_b \times \frac{z_T}{z_{60}} \times \frac{(460 + T_v)}{(460 + 60)} \tag{6.11}$$

where p_{bT} is the pressure in the bellows at operating temperature, p_b is the pressure in the bellows at 60°F, z_T and z_{60} are the deviation factors, and T_v is the temperature at the valve. The opening pressure of the valve at operating conditions is obtained by substituting p_{bT} for p_b in Eq. (6.3). Deviation factors for nitrogen as the bellows gas have been tabulated by Sage and Lacy[15] from earlier measurements.

Valve spread. The difference between the opening and closing pressures of a bellows-type valve, Δp, is called the *spread* and is a measure of the difference between the effective area of the bellows and area of the port. In connection with the effect of spread on the injection gas–oil ratio, consider an intermittent-flow well when the injection-cycle controller has stopped the flow of gas into the annulus. The valve will remain open until the pressure in the annulus is reduced to the closing pressure. If the annular volume is large, a large volume of gas will flow through the gas-lift valve after the intermitter has closed. A low spread where the annular volume is large will considerably reduce the injection gas–oil ratio.

To calculate the spread, consider the valve of Fig. 6.11b which is about to open. The spread Δp is equal to the opening pressure minus the closing pressure, or Eq. (6.3) minus Eq. (6.9). Solving for Δp,

$$\Delta p = \frac{p_{bT}}{1 - A_v/A_b} + S_t - p_t\left[\frac{A_v/A_b}{(1 - A_v/A_b)}\right] - (1 - A_v/A_b)\left[\frac{p_{bT}}{(1 - A_v/A_b)} + S_t\right] \tag{6.12}$$

$$= \left[\frac{A_v/A_b}{1 - A_v/A_b}\right][p_{bT} + S_t(1 - A_v/A_b) - p_t] \tag{6.13}$$

For example, a valve has a bellows pressure of 888 psi at well conditions, a spring effect of 100 psi, a ratio A_v/A_b of 0.11, and a pressure in the tubing opposite the valve (p_t) of 300 psi. Substituting in Eq. (6.13),

$$\Delta p = \left[\frac{0.11}{1 - 0.11}\right][888 + 100(1 - 0.11) - 300] = 84 \text{ psi}$$

Static gas column pressures. Accurate calculation of the gas injection pressure at any depth is essential for designing and analyzing most gas-lift installations. The gas pressure at any depth in terms of the operating pressure at the casing wellhead[16] is

$$p_v = p_{wh} \exp\left(\frac{0.01875\gamma D}{z_{\text{avg}}T_{\text{avg}}}\right) \tag{6.14}$$

where p_v = casing pressure at the valve at depth D, psia
p_{wh} = gas pressure at the wellhead, psia
γ = gas specific gravity (air = 1.0)
D = depth to valve, feet
z_{avg} = compressibility factor for average pressure and average temperature of gas column
T_{avg} = average temperature of the gas column, °R

For gas-lift calculations, Eq. (6.14) can be expressed in an alternate form in terms of common logarithms as

$$p_v = p_{wh}10^{\gamma D/122.7z_{\text{avg}}T_{\text{avg}}} \tag{6.15}$$

Because the pressure and temperature increase with depth, the compressibility factor in Eq. (6.15) is for average temperature and average pressure of the gas column; therefore, this equation must be solved by trial and error. If the temperature at the total depth is known, the temperature at any depth is calculated from the temperature gradient. For the Gulf Coast area, if the temperature is unknown it can be estimated from

$$T_v = 0.015D + 80°\text{F} \tag{6.16}$$

or

$$T_v = 0.016D + 70°\text{F} \tag{6.17}$$

where T_v is the temperature at the valve and D is the depth to the valve in feet. Before the average pressure can be calculated, the pressure opposite the point of gas injection must be estimated from

$$\Delta p = 0.25(p_{wh}/100)(D/100) \tag{6.18}$$

If the calculated gas pressure at the point of injection results in a compressibility factor which differs from the factor based on Eq. (6.18), the calculations are repeated. The compressibility factors given in Fig. 6.12 were obtained by Winkler[17] from the work of Standing and Katz[18] and of the California Natural Gasoline Association.[19] Example 6.1 illustrates the use of Eqs. (6.15), (6.16), (6.18), and Fig. 6.12.

Example 6.1. Calculate the pressure opposite a gas-lift valve, given:

Depth to valve = 6000 ft

Injection pressure at surface = 600 psig

Gas gravity = 0.70

Temperature at the surface = 100°F

SOLUTION: From Eq. (6.16) the temperature at 6000 ft is

$$T_v = 0.015 \times 6000 + 80°\text{F} = 170°\text{F}$$

and

$$T_{\text{avg}} = (100 + 170)/2 = 135°\text{F}$$

From Eq. (6.18) the increase in pressure is

$$\Delta p \cong 0.25 \times 6.15 \times 60 = 92 \text{ psi}$$

and the pressure at the valve is 692 psig, which results in an average pressure of

$$p_{\text{avg}} = p_{wh} + \frac{\Delta p}{2} = 600 + \frac{92}{2} = 646 \text{ psig}$$

From Fig. 6.12, $z = 0.912$, and from Eq. (6.15) the pressure at 6000 ft is

$$p_v = 614.7 \times 10^{0.70\times6000/122.7\times0.912\times595}$$
$$= 710.6 \text{ psia or } 695.9 \text{ psig}$$

For practical purposes this answer is close enough to the assumed value of 692 psig, but to illustrate the computation another value will be assumed. The average pressure for a p_v of 696 psig is $(600 + 696)/2$ or 348 psig, and the average compressibility from Fig. 6.12 is 0.910. The pressure at 6000 ft is

$$p_v = 614.7 \times 10^{0.70\times6000/122.7\times0.910\times595} = 711.2 \text{ psia or } 696.5 \text{ psig}$$

The gas pressure at depth will change only slightly from the first approximation if the estimated pressure opposite the valve is reasonably accurate.

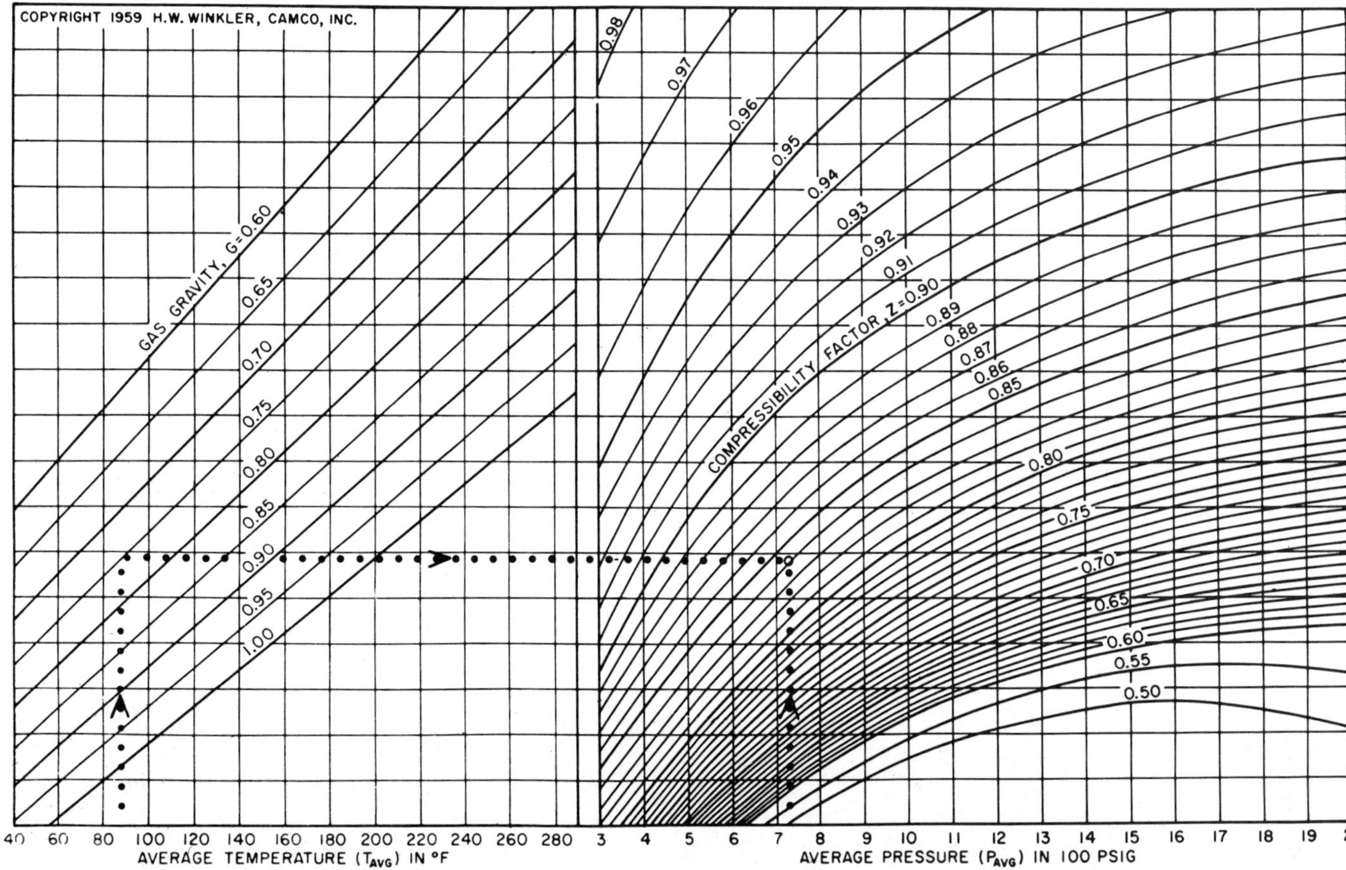

Fig. 6.12 Compressibility factors for natural gases, Copyright 1959. (Winkler,[17] *World Oil.*)

Since Eq. (6.11), for bellows pressure at operating temperature, and Eq. (6.14), for casing pressure at depth, require trial-and-error solutions, they are presented in graphical form[13] in Fig. 6.13 (for nitrogen-charged bellows valves). To reduce the large number of variables entering into these equations, several base conditions are assumed. The *approximate* correction for other than base conditions is given by

$$\Delta p_{\text{cor}} = \Delta p_{\text{chart}} \times \frac{\gamma_g}{0.6} \times \frac{T_{\text{chart}}}{T_{\text{act}}} \tag{6.19}$$

where Δp_{cor} = corrected pressure due to gas column weight, psi

Δp_{chart} = chart value of pressure due to gas column weight, psi

γ_g = gas specific gravity

T_{chart} = average temperature for chart basis, °R

T_{act} = average temperature for actual conditions, °R

The use of Fig. 6.13 is shown in Example 6.2.

Example 6.2. Calculate the surface casing pressure required to open a nitrogen-charged bellows and spring-loaded valve, given:

Depth to valve = 6000 ft

Temperature at valve = 160°F

Surface opening pressure at 60°F = 600 psig

Spring tension = 75 psi

Tubing pressure at the valve = 100 psig

Specific gravity of the gas = 0.70

Temperature at wellhead = 70°F

$A_v/A_b = 0.11$

SOLUTION: From Eq. (6.6) the bellows pressure is

$$\begin{aligned} p_b &= (p_{vo} - S_t)(1 - A_v/A_b) \\ &= (600 - 75)(1 - 0.11) = 467 \text{ psig at } 60°\text{F} \end{aligned}$$

To obtain a bellows pressure at 6000 ft, using Fig. 6.13, begin at 467 psig on the abscissa, go straight up to the 60°F line, follow diagonally upward and to the right, parallel to the 450 psig and 500 psig nitrogen-charge pressure lines on either side, as far as the intersection of the 160°F line, and then read straight down to the abscissa a bellows pressure of 560 psig. The valve opening pressure at 6000 ft from Eq. (6.3) is

$$p_o = \frac{560}{(1 - 0.11)} + 75 - 100\left(\frac{0.11}{1 - 0.11}\right) = 692 \text{ psig}$$

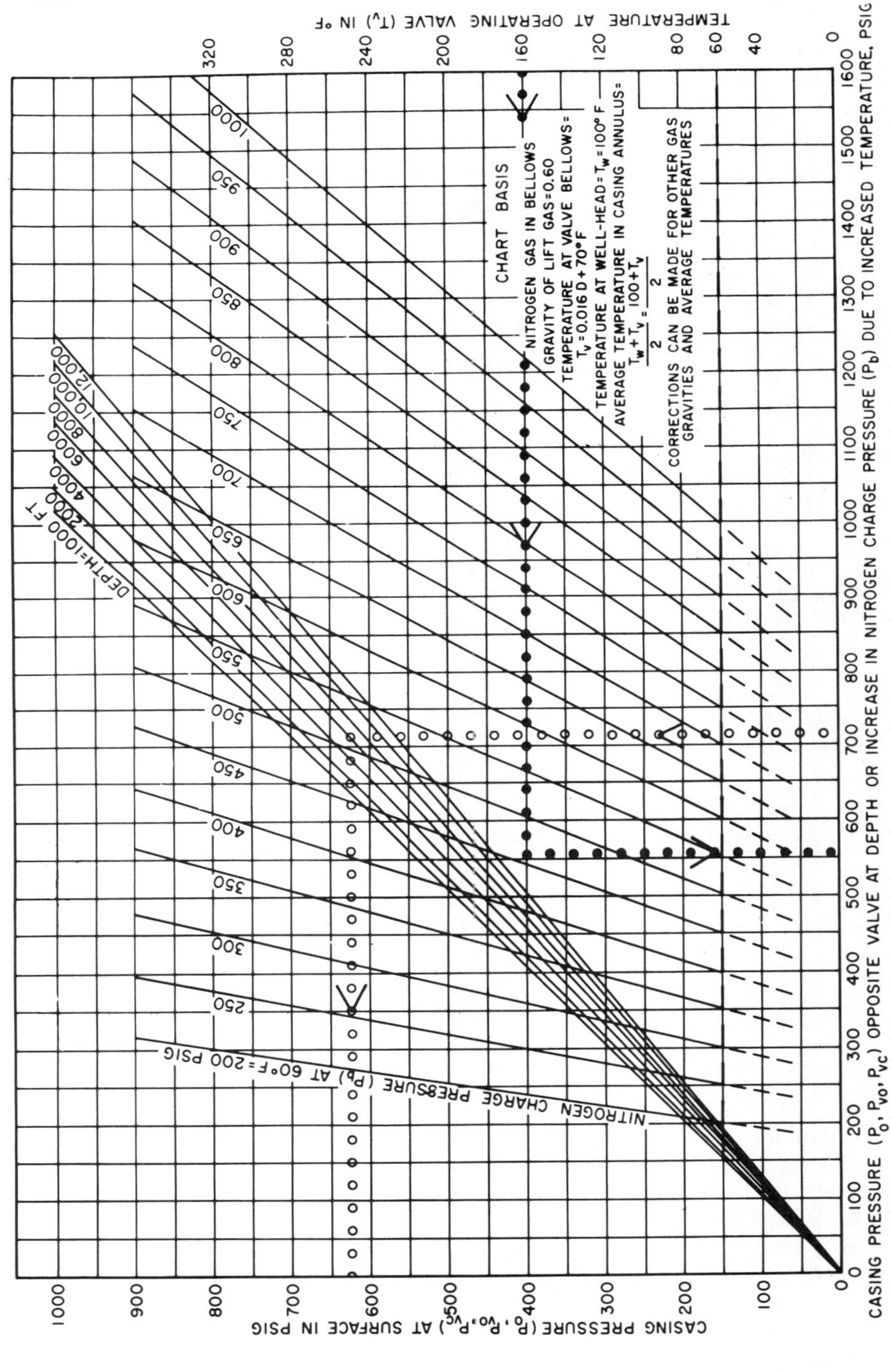

Fig. 6.13 Composite chart for determining valve string design and operation of nitrogen-charged bellows valves.

From the casing pressure opposite the valve of 692 psig on the abscissa of Fig. 6.13, proceed upward to the 6000-ft depth line and read on the ordinate a *surface* casing opening pressure of 605 psig. If the gas gravity had been 0.60 and the wellhead temperature 100°F (chart conditions), the valve opening pressure would have been 605 psig; however, it is necessary to correct for well conditions. Δp_{chart} is (692 − 605) psig or 87 psi. T_{chart} is equal to

$$\frac{100 + 166}{2} = 133°\text{F or } 593°\text{R}$$

and T_{act} is

$$\frac{70 + 160}{2} = 115°\text{F or } 575°\text{R}$$

Then from Eq. (6.19)

$$\Delta p_{\text{cor}} = 87 \times \frac{0.7}{0.6} \times \frac{593}{575} = 105 \text{ psi}$$

and the actual surface opening pressure is

$$p_c = 692 - 105 = 587 \text{ psig}$$

Flowing gas column pressures. For small conduit wells ($\frac{3}{4}$-, 1-, and 2-in. combinations), the pressure drop of the lift gas is critical. Buthod and Whitely[20] have developed an equation for calculating the pressure traverse of downward flowing gas,

$$p_v^2 = B(p_{ci}^2 - A) + A \tag{6.20}$$

where p_v = subsurface pressure at the valve in psia

p_{ci} = surface injection pressure in psia

$B = \exp\left(\dfrac{0.0376\gamma D}{z_{\text{avg}} T_{\text{avg}}}\right)$ (gas column weight factor)

$A = 1.51 \times 10^{-5} \times \dfrac{(qT_{\text{avg}} z_{\text{avg}})^2}{d^{5.23}}$ (friction factor)

d = equivalent flow channel diameter, in.

q = gas flow rate, MSCF/day

γ = gas specific gravity

D = depth, ft

Example 6.3 shows the use of Eq. (6.20) when the equivalent diameter is 50 per cent of the actual annular area. An equivalent diameter based on percentage reduction in the annular area is more practical for field use because of the irregularity of the annular area. It is of interest to note the possibility of either a positive or a negative gas-pressure gradient for the

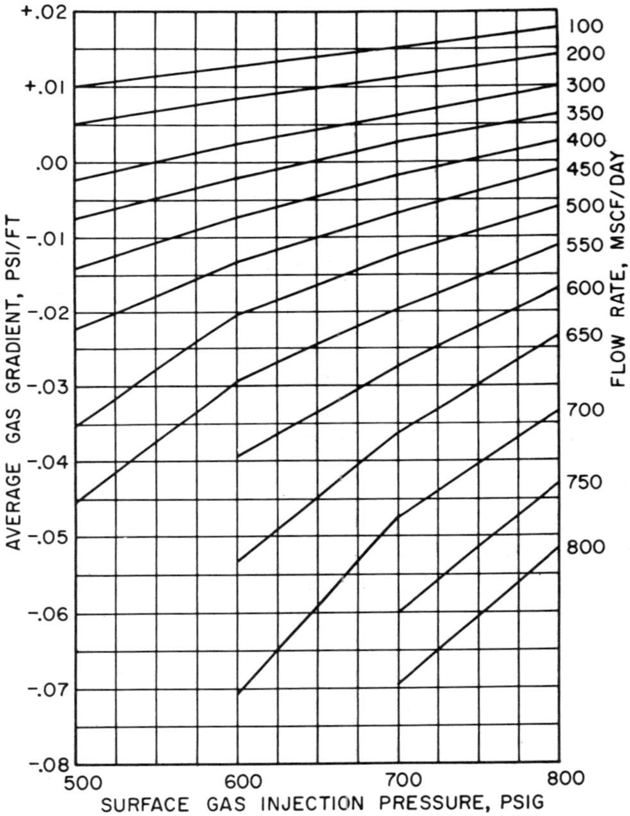

Fig. 6.14 Gas gradients for 1-in. and 1-in. by 2-in. annulus. Well data: depth = 5000 ft, average temperature = 100° F, average gas gravity = 0.60, and average gas compressibility factor = 0.97. (Kirkpatrick,[21] *Petroleum Division,* ASME.)

downward flow of gas (Fig. 6.14).[21] At low flow rates the gradient is positive, whereas at high flow rates it is negative, i.e., as the flow increases, the energy loss increases.

Example 6.3. Calculate the subsurface pressure from surface injection pressure, given:

Depth to operating valve = 5000 ft

Well completed with $\frac{3}{4}$-in. tubing inside 2-in. tubing

Annular area = 2.406 sq in.

Average temperature in annulus = 100°F

Gas gravity = 0.60

$z_{avg} = 0.97$

Available surface injection pressure = 600 psia

Injection rate = 800,000 SCF/day

SOLUTION: The equivalent flow channel area is 2.406/2 or 1.203 sq in. and the equivalent diameter is

$$d = \sqrt{\frac{1.203}{0.7854}} = 1.237 \text{ in.}$$

The gas column weight factor is

$$B = \exp\left(\frac{0.0376 \times 0.6 \times 5000}{0.97 \times 560}\right) = 1.23$$

and the friction factor is

$$A = 1.51 \times 10^{-5}\left[\frac{(800 \times 560 \times 0.97)^2}{1.237^{5.23}}\right]$$

$$= 9.38 \times 10^5 \text{ psia}^2$$

Substituting in Eq. (6.20),

$$p_v^2 = 1.23(600^2 - 9.38 \times 10^5) + 9.38 \times 10^5$$

$$= 2.271 \times 10^5 \text{ psi}^2$$

and

$$p_v = 475 \text{ psia}$$

Unloading. As the injection pressure is applied to the annulus of the well shown in Fig. 6.15, all valves in the tubing string are opened and the fluid is displaced through the open valves into the tubing and, in open-type installations, is U-tubed around bottom. When the first valve (valve 1) is uncovered, gas is injected into the unaerated column of fluid above the valve in the tubing. As the annulus pressure is greater than the tubing pressure, the liquid level in the annulus will drop until valve 2 is uncovered. If the unloading is continuous, when valve 2 is uncovered there is a two-phase flow gradient above valve 1 and a dead liquid (static) gradient between valves 1 and 2. The pressure in the *tubing* opposite valve 2 just before it is uncovered is composed of the tubing pressure, an aerated flowing gradient above valve 1, and an unaerated gradient between valves 1 and 2. When valve 2 is uncovered, both valves will be open. With most continuous-flow installations, a gas injection choke is placed in the input casing line to regulate the gas volume. If the well is unloaded with a choke of the proper size, with both valves 1 and 2 open, gas flows at a higher rate into the tubing than into the casing, and the surface pressure on the casing will drop. As the top valve has a higher closing pressure than the second valve, it will close first, and the second valve will pass gas until the third valve is un-

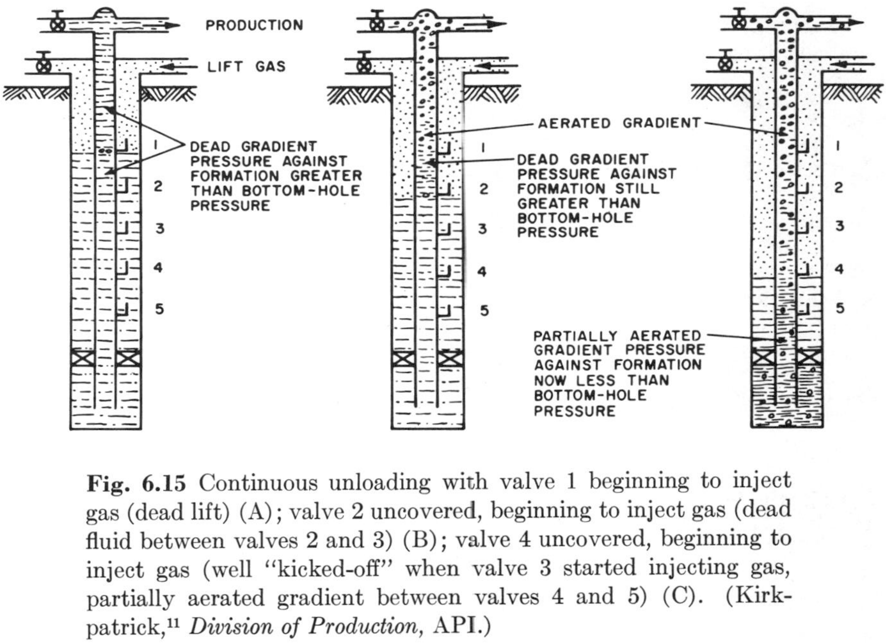

Fig. 6.15 Continuous unloading with valve 1 beginning to inject gas (dead lift) (A); valve 2 uncovered, beginning to inject gas (dead fluid between valves 2 and 3) (B); valve 4 uncovered, beginning to inject gas (well "kicked-off" when valve 3 started injecting gas, partially aerated gradient between valves 4 and 5) (C). (Kirkpatrick,[11] *Division of Production,* API.)

covered and begins operating. This cycle will repeat itself. In working down to the operating valve where gas will be injected continuously, owing to the reservoir pressure drawdown, fluids will start entering the well bore. At this point the flowing gradient below the valve passing gas will become less than that for the dead fluid and the gradient above the point of gas injection will increase owing to the increased volume of fluid entering the tubing.

Valve spacing. The proper spacing of gas-lift valves is determined by writing a force or pressure balance for the valve. Only the spacing of pressure-loaded valves will be considered. The depth to the first valve depends upon the injection pressure available, the static fluid gradient, the static fluid level of the well, and the wellhead back pressure, i.e., whether the well is unloaded into a separator or into a pit. In conventional workover practice the well fluid is displaced with salt water or mud, and upon completion the mud is displaced with (salt) water, so that in either case the well is left full of water. The method of spacing pressure-operated valves illustrated here is basically to set the valves at successively lower opening pressures as they are installed deeper in a well. The spacing formulas may be used for either continuous- or intermittent-flow installations. The pressure balance[22] on a combination valve (Fig. 6.16) if it unloads into a pit is

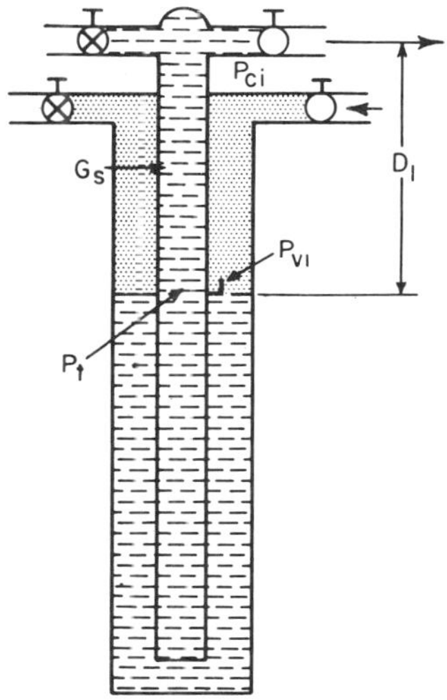

Fig. 6.16 Static fluid level at or near the surface. (Kirkpatrick,[22] *The Oil and Gas Journal.*)

$$(D_1G_s) = p_{v1} \tag{6.21}$$

Solving for D_1,

$$D_1 = \frac{p_{v1}}{G_s} \tag{6.22}$$

or, if the well is unloaded into a separator,

$$D_1 = \frac{p_{v1} - p_{sp}}{G_s} \tag{6.23}$$

where D_1 = depth to the first valve (regardless of type), ft

p_{v1} = casing pressure opposite the valve, usually taken as the available casing pressure at the surface (p_{ci}), psi

G_s = static (dead-liquid) unloading gradient, usually 0.5 psi per ft to assure the unloading of the well

p_{sp} = separator pressure, psi

When the static fluid level is below the depth calculated by use of Eqs. (6.22) or (6.23), then the first valve is placed at the approximate fluid

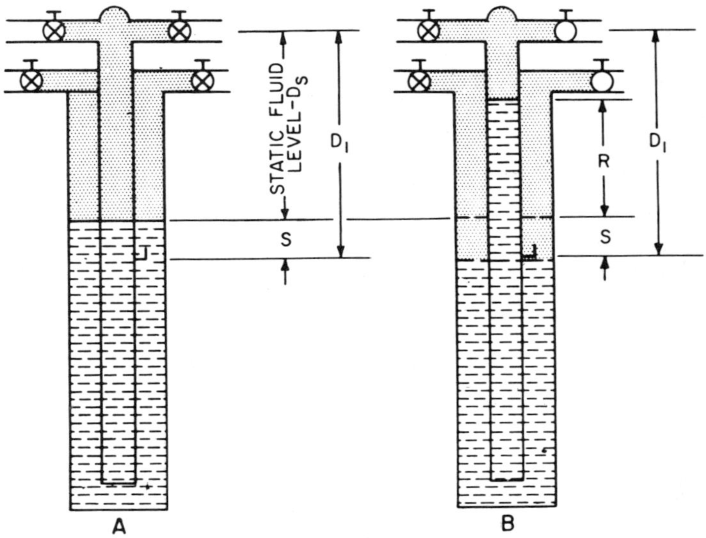

Fig. 6.17 Static fluid level below the surface. (Kirkpatrick,[22] *The Oil and Gas Journal.*)

level, unless the depth to the fluid level with tubing in the hole is accurately known. From Fig. 6.17, it is seen that

$$D_1 = D_s + S \tag{6.24}$$

where D_s is the depth to the static fluid level in feet and S is the submergence of the valve below the static fluid level, also in feet. Also from Fig. 6.17a,

$$SA_a = RA_t \quad \text{or} \quad R = SA_a/A_t \tag{6.25}$$

where A_a = cross-sectional area of the annulus

A_t = cross-sectional area of the tubing

R = liner distance in the tubing corresponding to the volume measured off by the linear distance (S) in the annulus

Referring to Fig. 6.17b, the pressure balance on the first valve is

$$(S + R)G_s = p_{v1} \tag{6.26}$$

Substituting Eq. (6.25) into Eq. (6.26),

$$S(1 + A_a/A_t)G_s = p_{v1} \tag{6.27}$$

From Eq. (6.24), substituting $(D_1 - D_s)$ for S into Eq. (6.27),

$$(D_1 - D_s)(1 + A_a/A_t)G_s = p_{v1}$$

Solving for D_1,

$$D_1 = D_s + \frac{p_{v1}}{(1 + A_a/A_t)G_s} \tag{6.28}$$

The values of A_a/A_t for various sizes of tubing and casing[14] are given in Tables 6.2 and 6.3.

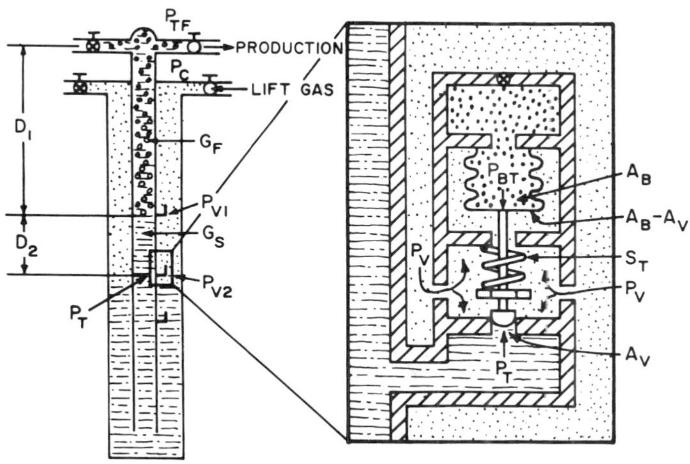

Fig. 6.18 Spacing of pressure-loaded gas-lift valves. (After Kirkpatrick,[22] *The Oil and Gas Journal.*)

To calculate the depth to the second valve, refer to Fig. 6.18. The pressure balance is

$$p_{v2} = G_f D_1 + G_s D_2 + p_{sp} \tag{6.29}$$

Solving for D_2,

$$D_2 = \frac{p_{v2} - G_f D_1 - p_{sp}}{G_s} \tag{6.30}$$

where D_2 = distance from first valve to second valve, ft

p_{v2} = casing pressure at valve 2, which can be taken as the opening pressure of valve 2 at 60°F or the closing pressure of the valve above at depth, psi

G_f = unloading gradient, psi per foot

Table 6.2. ANNULUS — TUBING AREA RATIOS AND ANNULAR VOLUMES[a]

(Courtesy Merla Tool Company)

Dimensions of outer casing string			Dimensions of inner tubing string							
			Nom $1\frac{1}{4}$ in. OD 1.660 in.		Nom $1\frac{1}{2}$ in. OD 1.900 in.		Nom 2 in. OD 2.375 in.		Nom $2\frac{1}{2}$ in. OD 2.875 in.	
OD, in.	Weight, lb/ft	ID, in.	ratio	bbl/100 ft	ratio	bbl/100 ft	ratio	bbl/100 ft	ratio	bbl/100 ft
$4\frac{1}{2}$	9.50	4.090	7.34	1.36	5.05	1.27	2.78	1.08	1.42	0.82
	11.60	4.000	6.95	1.29	4.77	1.20	2.60	1.01	1.30	0.75
	13.50	3.920	6.62	1.23	4.53	1.14	2.44	0.94	1.19	0.69
5	11.50	4.560	9.47	1.75	6.62	1.67	3.80	1.47	2.10	1.22
	13.00	4.494	9.16	1.69	6.39	1.61	3.65	1.41	2.00	1.16
	15.00	4.408	8.75	1.62	6.09	1.54	3.46	1.34	1.87	1.09
	18.00	4.276	8.15	1.51	5.65	1.43	3.17	1.23	1.68	0.97
$5\frac{1}{2}$	13.00	5.044	11.91	2.21	8.41	2.12	4.97	1.92	2.88	1.67
	14.00	5.012	11.74	2.17	8.28	2.08	4.89	1.89	2.83	1.64
	15.50	4.950	11.42	2.12	8.04	2.03	4.73	1.83	2.72	1.58
	17.00	4.892	11.12	2.05	7.82	1.98	4.59	1.78	2.63	1.52
	20.00	4.778	10.54	1.95	7.40	1.87	4.31	1.67	2.44	1.41
	23.00	4.670	10.00	1.85	7.01	1.77	4.06	1.57	2.27	1.32
7	17.00	6.538	20.99	3.88	15.07	3.79	9.31	3.60	5.79	3.35
	20.00	6.456	20.44	3.78	14.66	3.70	9.04	3.51	5.61	3.24
	23.00	6.366	19.83	3.67	14.21	3.58	8.75	3.38	5.41	3.14
	26.00	6.276	19.23	3.56	13.77	3.47	8.47	3.28	5.22	3.03
	29.00	6.184	18.63	3.46	13.33	3.37	8.18	3.17	5.03	2.90
	32.00	6.094	18.05	3.35	12.91	3.26	7.90	3.06	4.85	2.80
	35.00	6.004	17.48	3.24	12.49	3.15	7.63	2.96	4.66	2.71
	38.00	5.920	16.95	3.14	12.10	3.05	7.38	2.85	4.49	2.60

[a]Couplings and upsets disregarded, maximum error 2 per cent.

Table 6.3. ANNULUS–INNER STRING AREA RATIOS AND ANNULAR VOLUMES[a]

(*Courtesy Merla Tool Company*)

Dimensions of outer string		Dimensions of inner string							
		Nom $\frac{1}{2}$ in. OD 0.840 in.		Nom $\frac{3}{4}$ in. OD 1.050 in.		Nom 1 in. OD 1.315 in.		Nom $1\frac{1}{4}$ in. OD 1.660 in.	
Nom size in.	ID, in.	ratio	bbl/100 ft	ratio	bbl/100 ft	ratio	bbl/100 ft	ratio	bbl/100 ft
1[b]	1.049	1.02	0.038[c]	—	—	—	—	—	—
$1\frac{1}{4}$	1.380	3.10	0.117	1.18	0.078	0.16	0.017[c]	—	—
$1\frac{1}{2}$	1.610	4.88	0.183	2.20	0.145	0.78	0.084	—	—
2	1.995	8.46	0.318	4.24	0.280	2.05	0.219	0.64	0.119
$2\frac{1}{2}$	2.441	13.47	0.510	7.16	0.472	3.84	0.411	1.68	0.311
3	2.992	21.30	0.801	11.57	0.763	6.56	0.702	3.25	0.602
4	3.958	38.65	1.44	21.46	1.415	12.67	1.354	6.78	1.25

[a]Couplings and upsets disregarded, maximum error 2 per cent.
[b]Standard string; all other strings are tubing.
[c]Body of inner string will pass through outer string but coupling or upset will not pass.

The unloading gradients[23] for total rate of fluid production and size of tubing are given in Fig. 6.19; however, a minimum unloading gradient of 0.04 psi per ft is recommended for intermittent unloading. When Fig. 6.19

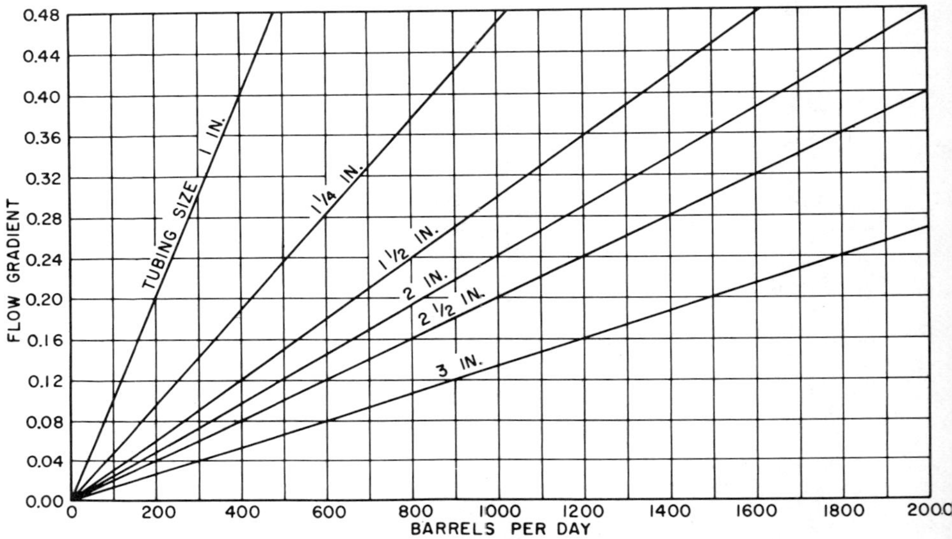

Fig. 6.19 Unloading gradients. (Courtesy of *Otis Engineering Corporation.*)

is used, the unloading rate for continuous flow is assumed equal to the desired producing rate. The depth separating all lower valves is derived in a similar manner, i.e., the distance between the second and third valve, D_3, is

$$D_3 = \frac{p_{v3} - (D_1 + D_2)G_f - p_{sp}}{G_s} \tag{6.31}$$

where p_{v3} is the opening pressure of valve 3 at 60°F or the closing pressure of the valve in the well. It is advisable to set the pressure of the first valve at 50 psi less than the available gas pressure; however, the full line pressure can be used.

The above calculations for low productivity index wells are discontinued when the spacing between two valves, in feet, becomes less than the opening pressure of the upper valve. Additional valves are so spaced that the distance in feet between the valves is not less than the opening pressure of the valve above. For example the distance between a 500 psig opening pressure valve and a 475 psig valve below it need not be less than 500 ft.

After calculation of the setting depths, the closing pressures of the valves under operating conditions are checked to be sure that they decrease with depth. If this is not the case, it will be difficult if not impossible to work down to the final operating valve, and a greater pressure drop between valves will have to be used. Example 6.4 shows conventional valve spacing based on the opening pressures of the valves at 60°F and also a closing pressure analysis. It is also possible to calculate the valve spacing on the basis of the closing pressure of the upper valve. When this method is used, it is necessary to make an opening pressure analysis to determine the *set* (opening) pressure at 60°F. Example 6.5 shows the valve spacing and opening pressure analysis based on the closing pressure of the upper valve.

Example 6.4. Calculate the valve spacing for a well to be placed on intermittent flow, on the basis of opening pressure at 60°F, given:

Available pressure = 600 psig
Fluid production = 100 bbl/day
Productivity index = 0.3 bbl/day/psi
Tubing size = 2 in.
Well unloaded into pit
Well full of water, gradient = 0.5 psi/ft
Bottom valve to be set at 4250 ft
Intermittent unloading gradient = 0.04 psi/ft
Gas gravity = 0.60
Spring tension = 100 psi
$A_v/A_b = 0.11$
Valve opening pressures = 550, 525, 500, 475, and 450 psig

Solution: From Eq. (6.22), the depth to the first valve is

$$D_1 = \frac{600}{0.5} = 1200 \text{ ft}$$

From Eq. (6.30), the distance from valve 1 to valve 2 is

$$D_2 = \frac{525 - (1200 \times 0.04)}{0.5} = 954 \text{ ft}$$

and the setting depth is 2154 ft. From Eq. (6.31), the distance from valve 2 to valve 3 is

$$D_3 = \frac{500 - (2154 \times 0.04)}{0.5} = 828 \text{ ft}$$

and the setting depth of valve 3 is 2982 ft. The distance between valve 3 and valve 4 is

$$D_4 = \frac{475 - (2982 \times 0.04)}{0.5} = 712 \text{ ft}$$

and the setting depth is 3694 ft. The distance between valve 4 and valve 5 is

$$D_5 = \frac{450 - (3694 \times 0.04)}{0.5} = 604 \text{ ft}$$

and valve 5 is set at the desired depth of 4250 ft.

The closing pressure analysis of the valve spacing calculated above is shown in Table 6.4. Columns (1) and (2) are taken from the spacing analysis. The temperature at valve depth, Col. (3), is calculated from Eq. (6.17). The pressure in the

Table 6.4. SURFACE CLOSING PRESSURE ANALYSIS TO DETERMINE OPERATING VALVE SPACING

(1) p_{vo} Valve opening pressure at 60°F, psig	(2) D Depth of valve, ft	(3) T_v Temp at valve, °F	(4) p_b Bellows pressure at 60°F, psig	(5) p_{bT} Bellows pressure at T_v, psig	(6) p_{vc} Valve closing pressure at depth, psig	(7) p_{vc} Surface closing pressure, psig
550	1200	89	401	422	511	489
525	2154	104	378	412	501	475
500	2982	118	356	397	486	452
475	3694	129	334	385	474	434
450	4250	138	312	360	449	410

bellows at 60°F, Col. (4), is calculated from Eq. (6.6). The pressures in Col. (5) are read from Fig. 6.13. The following approximate temperature correction factor[43] can be used to calculate the nitrogen charge pressure inside the bellows at well temperature:

$$p_{bT}\ (\text{at}\ T_v) = p_b\ (\text{at}\ 60°\text{F})\ F_{\text{temp}}$$

where

$$F_{\text{temp}} = 1 + 0.00215(T_v - 60)$$

The valve closing pressure at valve depth, Col. (6), is obtained by substituting Col. (5) into Eq. (6.9), and Col. (7) is Col. (6) less gas column weight and is read from Fig. 6.13 by starting on the abscissa, going straight up to the diagonal depth line, and then reading the surface pressure on the left-hand ordinate.

Example 6.5. Calculate the valve spacing for a well to be placed on intermittent flow, on the basis of closing pressure of the valve above, given:

Available pressure = 800 psig

Fluid production = 40 STB/day

Separator pressure = 75 psig

Depth to operating valve = 7450 ft

Well loaded with oil, gradient = 0.40 psi/ft

Injection gas gravity and temperatures same as chart basis for Fig. 6.13.

Spring tension = 75 psi

$A_v/A_b = 0.112$

Desired surface closing pressures of valves = 750, 740, 730, 720, and 690 psig

SOLUTION: The depth to the first valve, calculated from Eq. (6.23), is

$$D_1 = \frac{800 - 75}{0.40} = 1813 \text{ ft}$$

The closing pressure of valve 1 at 1813 ft, 782 psig, is found by adding the gas column weight from Fig. 6.13 to the 750 psig surface closing pressure. This is the pressure used in the spacing equation for the next valve. The spacing of the second valve, from Eq. (6.30), is

$$D_2 = \frac{782 - 75 - (1813 \times 0.04)}{0.40} = 1585 \text{ ft}$$

and the depth to valve 2 is 3398 ft. The closing pressure of valve 2 at 3398 ft, from Fig. 6.13, is 799 psig. The spacing of valve 3, from Eq. (6.31), is

$$D_3 = \frac{799 - 75 - (3398 \times 0.04)}{0.40} = 1470 \text{ ft}$$

and the depth to valve 3 is 4868 ft. The closing pressure of valve 3 at 4868 ft, from Fig. 6.13, is 810 psig. In a similar manner, the depth to valve 4 is found to be 6218 ft and the depth to valve 5 is 7458 ft.

The opening pressure analysis is given in Table 6.5. Columns (1), (2), and (4) are taken from the first part of the problem. Column (3) is from a temperature survey of the well. Column (5) is calculated by solving for p_{bT} in Eq. (6.9). The

Table 6.5. VALVE OPENING PRESSURE ANALYSIS FOR DESIRED SURFACE CLOSING PRESSURES

(1) p_{vc} Surface closing pressure of valve, psig	(2) D Depth of valve, ft	(3) T_v Temp at valve, °F	(4) p_{vc} Valve closing pressure at depth, psig	(5) p_{bT} Bellows pressure at T_v, psig	(6) p_b Bellows pressure at 60°F, psig	(7) p_{vo} Valve opening pressure at 60°F, psig
750	1813	122	782	715	630	784
740	3398	141	799	731	622	775
730	4868	159	810	742	612	764
720	6218	176	820	753	603	754
690	7458	191	803	736	575	723

pressures in Col. (6) are read from Fig. 6.13, by starting at well temperature on the right-hand ordinate, going horizontally to the left to the intersection of the bellows pressure at well temperature, reading vertically up from the abscissa, then reading downward diagonally to the left paralleling the bellows charge pressure lines to the intersection of the 60°F line, and then reading straight down to the abscissa, the bellows pressure at 60°F. Column (7) is obtained from Eq. (6.5). The pressures in Col. (7) are the set pressures ordered from the manufacturer.

6.4. Design of a Continuous-Flow Gas-Lift System

The optimum depth to the point of gas injection is located either graphically or by the volumetric balance method.

Graphical method. The well data[24] used in the calculations are given in Table 6.6. By use of a suitable scale, the bottom-hole pressure is plotted on the abscissa against the depth on the ordinate with zero depth at the top

Table 6.6. WELL DATA USED IN GRAPHICAL SOLUTION OF CONTINUOUS-FLOW GAS-LIFT SYSTEM

Tubing size = $2\frac{1}{2}$ in.
Static bottom-hole pressure = 3200 psig
Productivity index = 2.5 bbl fluid/day/psi
Gravity of oil = 40° API
Specific gravity of salt water = 1.07
Water-oil ratio = 6:1
Desired production = 100 STB/day (700 bbl of fluid per day)
Surface kick-off and operating pressure = 850 psig
Wellhead temperature = 100°F
Bottom-hole temperature = 220°F
Flowing tubing pressure = 50 psig
Flowing tubing temperature = 120°F
Well depth = 8200 ft
Retrievable valves without springs.
Valve ratio A_v/A_b = 0.11
Injection gas gravity and average temperature of gas column in casing same as chart basis for Fig. 6.13.

(Fig. 6.20). The specific gravity of the oil is

$$\gamma_o = \frac{141.5}{131.5 + °\text{API}} = \frac{141.5}{131.5 + 40} = 0.825$$

Then the oil gradient is 0.825 × 0.433 or 0.357 psi per ft and the salt water gradient is 1.07 × 0.433 or 0.463 psi per ft. The composite static gradient is

$$\begin{aligned} G_s &= f_o \times G_o + f_w \times G_w \\ &= \tfrac{1}{7} \times 0.357 + \tfrac{6}{7} \times 0.463 = 0.45 \text{ psi/ft} \end{aligned}$$

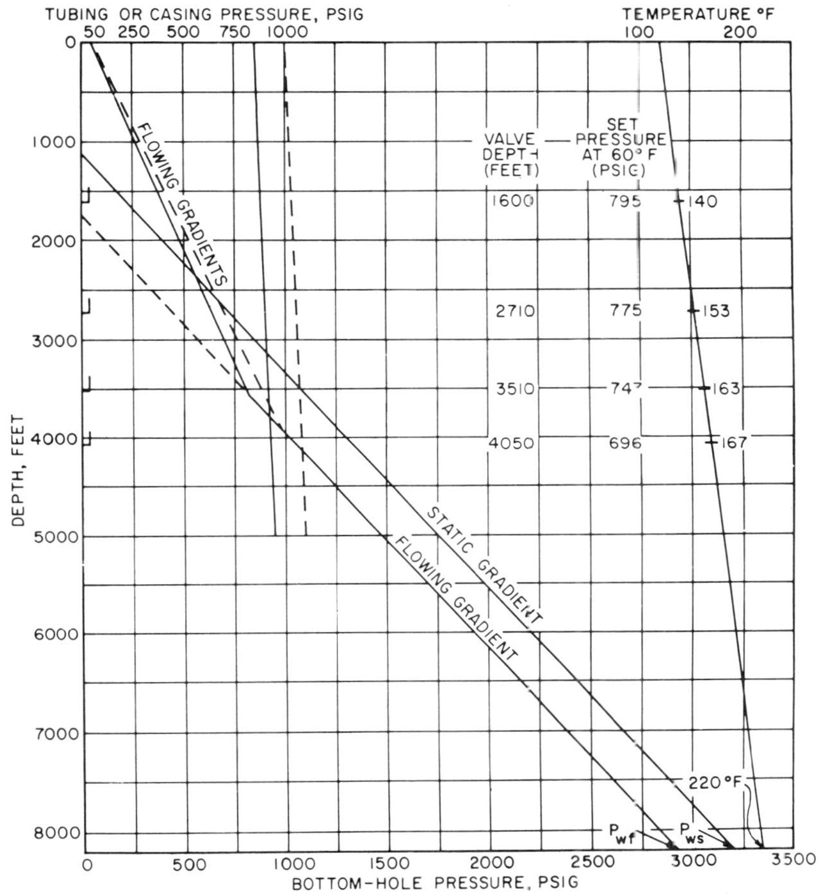

Fig. 6.20 Determination of the optimum injection point. (Courtesy of Otis Engineering Corporation.[23])

The static bottom-hole pressure is 3200 psig at 8200 ft and the pressure p at 2000 ft is found from

$$G_s = \frac{\Delta p}{\Delta D} = 0.45 = \frac{3200 - p}{8200 - 2000}$$

or

$$p = 410 \text{ psig}$$

The static bottom-hole pressure and the pressure at 2000 ft are now plotted at their respective depths and the static gradient line is drawn through the plotted points and extrapolated to the ordinate. The drawdown required to obtain the desired production is calculated from the productivity index,

$$\Delta p = \frac{q_o + q_w}{J}$$

where J = productivity index in barrels of fluid produced per day per pound per square inch drop in reservoir pressure

q_o = barrels of oil produced per day

q_w = barrels of water produced per day

Since the well is to produce 100 bbl of oil per day with 600 bbl of salt water and the productivity index is 2.5, the drawdown is

$$\Delta p = \frac{700}{2.5} = 280 \text{ psi}$$

and the flowing bottom-hole pressure is 3200 − 280 or 2920 psig. Assuming the flowing gradient below the point of gas injection to be equal to the static gradient, the flowing gradient is drawn from the flowing bottom-hole pressure (at 8200 ft) parallel to the static gradient line.

To establish the casing gradient, it is necessary to calculate the pressure in the casing at approximately half the well depth. To simplify the procedure for calculating the gas lift valve depths and opening pressures, the specific gravity of the injection gas and the average temperature of the gas column in the casing are assumed to be the same as the chart basis for Fig. 6.13. If actual conditions vary significantly from the chart basis, curves of gas pressure versus depth similar to those in Fig. 6.13 should be plotted for the actual data and then the installation can be designed as outlined in this example.

The casing gradient is plotted on Fig. 6.20 by drawing a line between the surface casing pressure (850 psig) and the pressure at 4000 ft (930 psig). The intersection of the casing gradient line and the flowing gradient line is the point at which the pressure in the tubing is equal to the pressure in the casing. In order to establish the approximate depth of the operating gas-lift valve, assume a differential of 100 psi across the valve. The depth at which there is a 100 psi differential across the valve (between tubing and casing) is approximately 3540 ft. The 100 psi differential is a desirable allowance for variation in injection gas line pressure and decreases in upper valve opening pressures to assist the unloading process.

The flowing gradient above the point of injection is obtained by connecting the point of gas injection with the wellhead pressure (50 psig), and is found to be 0.22 psi per ft. It is easily seen that if the injection pressure is 1000 psig, the point of injection is 3900 ft and the flowing gradient above the point of injection is then 0.24 psi per ft. This shows that as the gradient becomes steeper, less gas is injected and the gas-liquid ratio is considerably less, i.e., the greater the available gas pressure, the smaller is the volume required to produce the well at a given rate of flow. It should also be noted that the tubing pressure at the wellhead works directly against the injection pressure, with the result that the lower the wellhead pressure, the more efficiently the well is produced.

A method of locating graphically the depth of the operating gas lift valve is given on Fig. 6.20. The flowing-temperature gradient has been drawn on the same figure by connecting the bottom-hole temperature of 220°F with the estimated flowing wellhead temperature of 120°F. The flowing temperature at the valve depth is needed in order to calculate the valve opening pressure at 60°F in a tester.

The depths of all gas lift valves except the top one are calculated on the basis of the closing pressure of the valve above at valve depth. The depth of the top valve is calculated in a conventional manner using Eq. (6.23). Depths of the remaining valves are calculated using Eqs. (6.30) and (6.31). From Fig. 6.19, the unloading gradient for 700 bbl per day through $2\frac{1}{2}$-in. tubing is 0.14 psi per ft. The valve depth calculations are given in the table on the next page.

The third valve, at 3510 ft, is the operating valve (this depth is close enough to the approximate desired depth of gas injection 3540 ft, to require no adjustment). The bottom valve, at 4050 ft, is used in case the productivity index is in error or the bottom-hole pressure declines.

The surface opening pressures p_{vo} of the valves for this installation are calculated from data obtained in a tester at 60°F in Table 6.7, by use of Fig. 6.13 and Eq. (6.5) without the spring term (S_t).

Volumetric balance method. In the preceding graphical method of determining the pressure traverse, it was assumed that all fluid in the well was at some average density, and therefore straight lines were obtained for the flowing traverse. The mass flow rate m in pounds of production of gas, oil, and water per barrel of stock-tank oil, is the same at any point in the tubing, but the respective volumes of gas and liquid, expressed as composite volume factors B_t, are variable. Therefore, the density of production m/B_t is different for each pressure point in the tubing and the pressure traverse is a curve rather than a straight line. From a study of the laws governing the multiphase flow of oil and gas mixtures in vertical tubing, various methods for predicting the actual pressure traverse have been proposed.[25–32] Poettmann and Carpenter[31] made a significant contribution by developing a generalized correlation, based on field data from a large number of flowing and gas-lift wells, that would predict the pressure traverse in vertical flow strings for known conditions of flow. In the case of flowing wells, this provided a means of evaluating the effects of tubing size, flow rate, and bottom-hole pressure on one another. In the case of gas-lift

VALVE DEPTH CALCULATION

Valve no.	p_{vc} Surface closing pressure of valve, psig	D Depth of valve, ft	p_{vc} Valve closing pressure at depth, psig	Calculation of valve depth
1	800	1600	830	$D_1 = \frac{850 - 50}{0.50} = 1600 \text{ ft}$
2	780	2710	830	$D_2 = \frac{830 - 50 - 0.14(1600)}{0.50} = 1110 \text{ ft} + 1600 \text{ ft} = 2710 \text{ ft}$
3	750	3510	810	$D_3 = \frac{830 - 50 - 0.14(2710)}{0.50} = 800 \text{ ft} + 2710 \text{ ft} = 3510 \text{ ft}$
4	700	4050		$D_4 = \frac{810 - 50 - 0.14(3510)}{0.50} = 540 \text{ ft} + 3510 \text{ ft} = 4050 \text{ ft}$

Table 6.7. VALVE OPENING PRESSURE ANALYSIS FOR DESIRED SURFACE CLOSING PRESSURES

(1) p_{vc} Surface closing pressure of valve, psig	(2) D Depth of valve, ft	(3) T_v Temp at valve, °F	(4) p_{vc} Valve closing pressure at depth, psig	(5) p_{bT} Bellows pressure at T_v, psig	(6) p_t Bellows pressure at 60°F, psig	(7) p_{vo} Valve opening pressure at 60°F, psig
800	1600	140	830	830	708	795
780	2710	153	830	830	690	775
750	3510	163	810	810	665	747
700	4050	169	765	765	620	696

installations, it provided such information as the optimum depth, pressure, and rate for gas injection, horsepower requirements for lifting the oil, and the effect of production rate and tubing size on these quantities.

The following derivations were developed by Poettmann and Carpenter in obtaining their correlation. The basis for any fluid-flow calculation consists of a total energy balance in the fluid flowing between any two points in the tubing. According to the law of conservation of energy, the total energy entering a system by virtue of the fluid flow must equal the energy leaving the system plus the energy interchanged between the fluid and its surroundings. The energy balance based on a 1-lb weight of flowing fluid is expressed in differential form as

$$V\,dp + dh + d\left(\frac{v^2}{2g}\right) + dW_f + dW_s = 0$$

or, in terms of finite differences,

$$\int_{p_1}^{p_2} V\,dp + \Delta h + \Delta\left(\frac{v^2}{2g}\right) + W_f + W_s = 0 \tag{6.32}$$

where V = specific volume of flowing fluid, cubic feet per pound

p = pressure, pounds per square foot absolute

Δh = difference in height above datum plane, in feet (well or injection depth)

$\Delta\left(\frac{v^2}{2g}\right)$ = change in kinetic energy of one pound of flowing fluid between initial and final state of flow (v equals velocity in feet per second, g = 32.17 ft per sec^2)

W_f = energy losses from irreversibilities of the fluid in flow (slippage or liquid hangup), from frictional effects caused by the surface of the tubing, from change in surface areas between the mixture of liquid and gas, and from combinations of any of these effects

W_s = work done by the fluid while in flow (similar to shaft work in driving a turbine)

Assuming homogeneous steady-state flow of the mixture of gas and liquid between any two points in the tubing, the external work W_s done by the fluid is zero. The value of the kinetic energy function (increase in velocity head) $\Delta(v^2/2g)$ is small compared with other terms in the equation and can be neglected. Equation (6.32) reduces to

$$\int_{p_1}^{p_2} V\,dp + \Delta h + W_f = 0 \tag{6.33}$$

The absolute value of the term

$$\int_{p_1}^{p_2} V\,dp$$

must always be greater than, or equal to, Δh for upward flow. This means that W_f must always be zero or positive. If the reverse is found to be true, then the field or *PVT* data do not apply.

Assuming the energy losses W_f can be correlated by the Fanning equation as developed for incompressible flow, then

$$W_f = \frac{4f\bar{v}^2\Delta h}{2gd} \tag{6.34}$$

or

$$f = \frac{2gW_f d}{4\bar{v}^2\Delta h} \tag{6.35}$$

where f is the friction factor and v is the *integrated* average value of velocity between pressure limits p_1 and p_2. By substituting and rearranging Eqs. (6.33) and (6.34),

$$\Delta h = \frac{-\int_{p_1}^{p_2} V\,dp}{1 + \dfrac{4\bar{v}^2 f}{2gd}} = \frac{\int_{p_2}^{p_1} V\,dp}{1 + \dfrac{4\bar{v}^2 f}{2gd}} \tag{6.36}$$

In terms of field data:

$$\bar{v} = \frac{q_t}{A} = \frac{q_o\bar{V}_m}{A}\ \text{ft/day}$$

and

$$\bar{v}^2 = \frac{q_o^2\bar{V}_m^2}{(86{,}400A)^2} = \frac{q_o^2\bar{V}_m^2}{(86{,}400)^2(0.7854)^2d^4}\ \text{ft}^2/\text{sec}^2 \tag{6.37}$$

where q_t = integrated average of the total fluid flowing, in cubic feet per day

q_o = stock-tank barrels of oil produced per day

V_m = cubic feet of mixed gas, oil, and water at pressure p per stock-tank barrel of oil at any point in the flow string

d = inside diameter of flow string, in feet

$\bar{V}_m$ = integrated average value of V_m between the pressure limits of p_1 and p_2, or

$$\bar{V}_m = \frac{\int_{p_2}^{p_1} V_m \, dp}{p_1 - p_2} \tag{6.38}$$

Equation (6.36) can now be written in the form

$$\Delta h = \frac{\frac{1}{m}\int_{p_2}^{p_1} V_m \, dp}{1 + \frac{4\bar{v}^2 f}{2gd}} \tag{6.39}$$

Then substituting Eq. (6.37) into Eq. (6.39),

$$\Delta h = \frac{\frac{1}{m}\int_{p_2}^{p_1} V_m \, dp}{1 + \frac{f q_o^2 \bar{V}_m^2}{7.41 \times 10^{10} \, d^5}} \tag{6.40}$$

where m is the total weight of gas, oil, and water associated with one barrel of stock-tank oil flowing into and out of the flow string, in pounds per stock-tank barrel. The units are

$$\Delta h = \frac{(\text{STB/lb}) \times (\text{cu ft/STB}) \times (\text{lb/ft}^2)}{1 + \frac{(\text{STB})^2/\text{day}^2 \times \text{ft}^6/(\text{STB})^2}{2 \times 32.17 \text{ ft/sec}^2 \times (86{,}400)^2 \text{ sec}^2/\text{day}^2 \times (0.7854)^2 \text{ ft}^5}}$$

The volume V_m is calculated from the composite (two-phase) formation volume factor B_t and the producing water-oil ratio q_w/q_o; then

$$V_m = 5.615\left(B_t + \frac{q_w}{q_o}\right) \text{ cu ft/STB} \tag{6.41}$$

or the total volume of formation fluid produced per day is

$$V_t = 5.615(q_o B_t + q_w B_w) \text{ cu ft/day} \tag{6.42}$$

The formation volume factor of the water, B_w, is usually taken as unity.

The bubble-point pressure p_b is estimated from the producing (solution) gas-oil ratio, the gravity of the separator gas, the API gravity of the tank oil, and the reservoir temperature, by using the correlation chart prepared by Standing[33] and reproduced in Fig. 6.21. For example, if the gas-oil ratio is 350 SCF per STB, the gas gravity 0.75, and the tank-oil gravity 30°API, the bubble-point pressure is 1930 psia. If the static bottom-hole pressure is greater than 1930 psia the oil is undersaturated. For gas-lift purposes Fig. 6.21 is used in reverse order to check the measured gas-oil ratio.

The composite (gas plus liquid phase) volume factor B_t likewise is estimated from the solution gas-oil ratio, the separator gas gravity, the API gravity of the tank oil, the reservoir temperature, and the reservoir pressure, by using the correlation prepared by Standing[33] and given in Fig. 6.22. If the solution gas-oil ratio is 1500 SCF per STB, the separator gas gravity 0.80, the tank-oil gravity 40° API, the temperature 200°F, and the pressure 1000 psia, the composite volume factor is 5 bbl per STB. The total weight m of fluid flowing per barrel of stock-tank oil is calculated as follows:

$$m = 350.17\left(\gamma_o + \frac{q_w}{q_o}\gamma_w\right) + R\rho_a\gamma_g \tag{6.43}$$

where B_t = composite volume factor, in barrels of space occupied by both the free gas phase and the liquid oil phase at flowing pressure and temperature per stock-tank barrel of oil

q_o = oil production, barrels per day

q_w = water production, barrels per day

γ_o = specific gravity of oil at 60°F

γ_w = specific gravity of water of 60°F

γ_g = specific gravity of separator gas

R = gas-oil ratio, standard cubic feet per stock-tank barrel

ρ_a = density of air at 60°F = 0.0764 lb per cu ft

350.17 = weight of water in lb per bbl at 60°F.

The value of the term $\int_{p_2}^{p_1} V_m\,dp$ is obtained by plotting V_m against absolute pressure and determining the area under the curve between the pressure limits p_1 and p_2 by graphical integration; in other words, the curve is divided into a series of trapezoids and the total area is determined by multiplying the width of each trapezoid by the arithmetical average of the length of its two sides and then adding up all the areas. Upon dividing the difference between the flowing bottom-hole pressure and the tubing

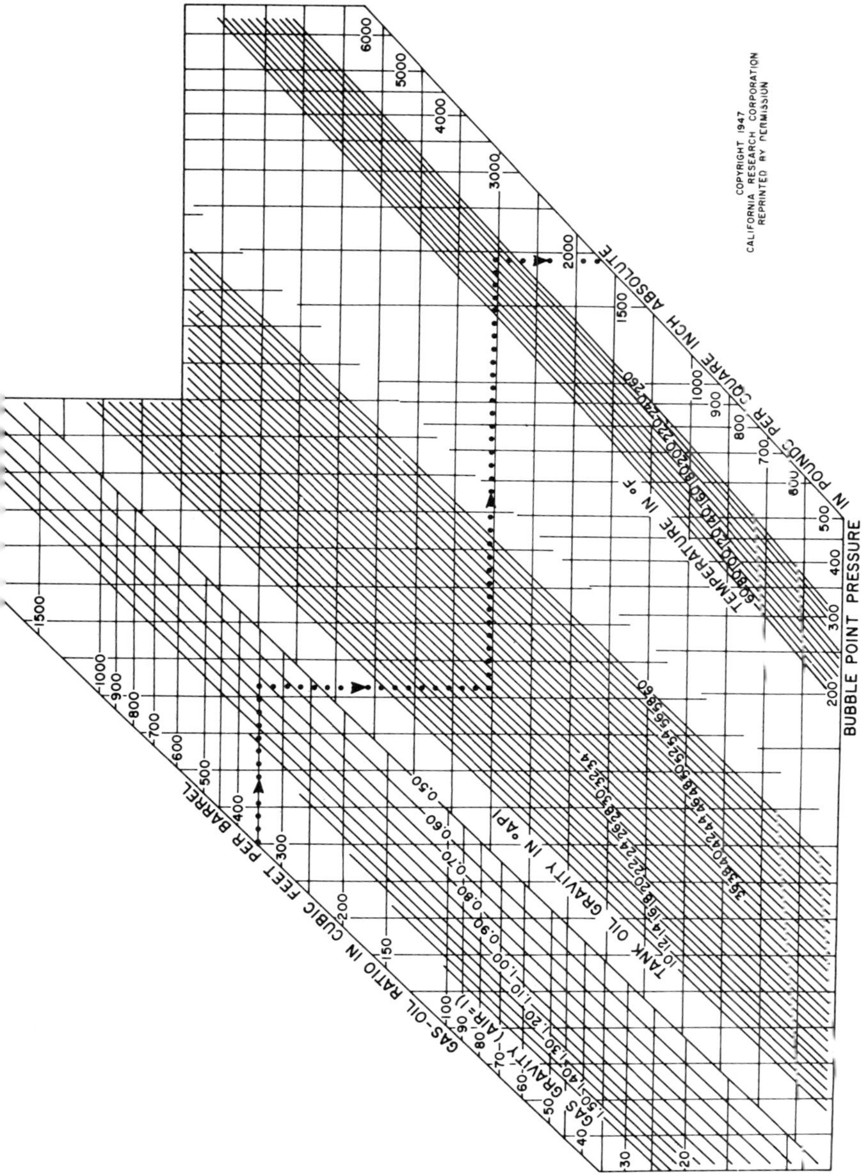

Fig. 6.21 Standing's solution gas-oil ratio chart. (Kirkpatrick,[11] *Division of Production*, API.)

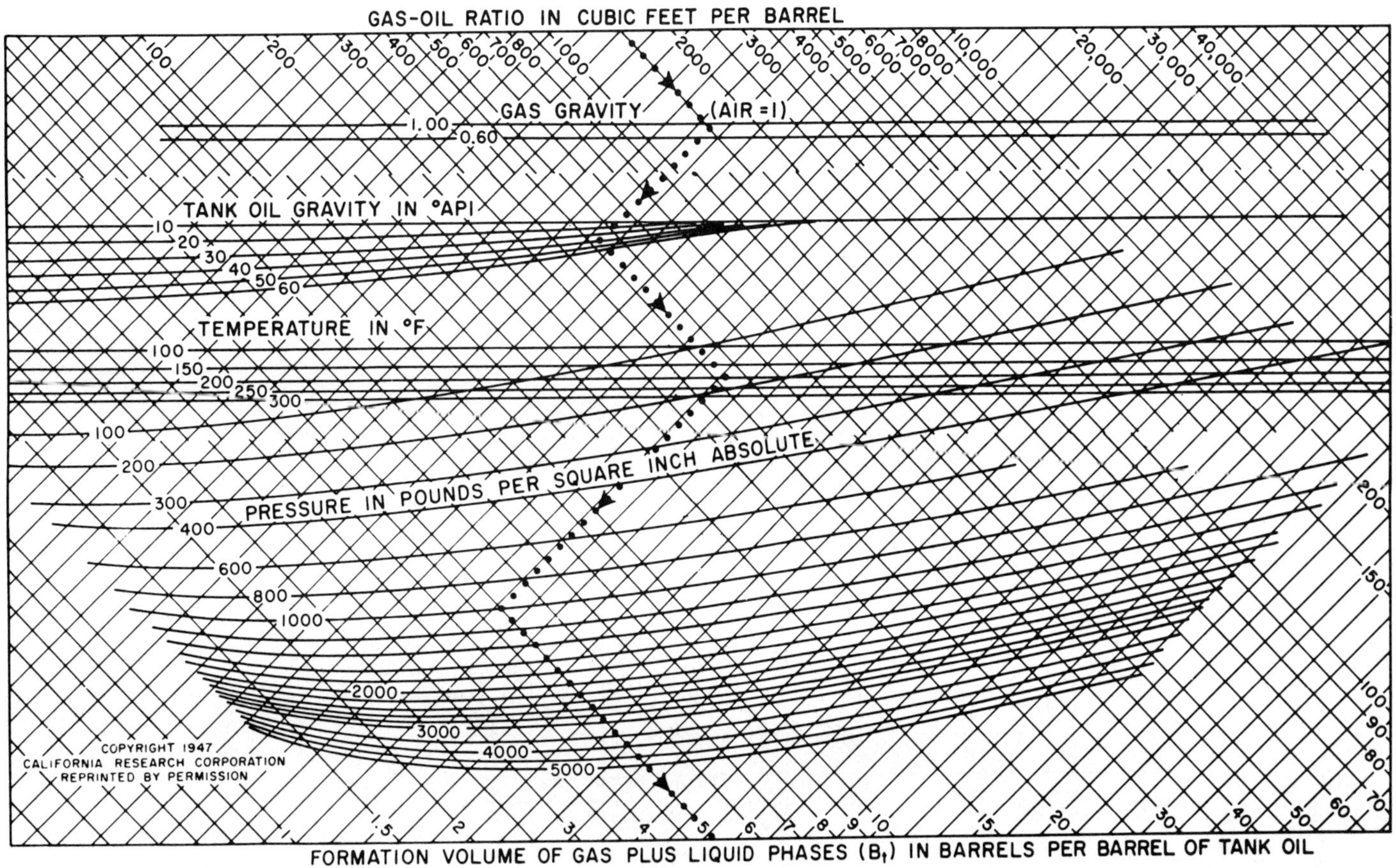

Fig. 6.22 Standing's composite volume-factor chart. (Kirkpatrick,[11] *Division of Production*, API.)

pressure, Δp, by Eq. (6.40), the following expression for pressure gradient results:

$$\frac{\Delta p}{\Delta h} = \frac{\left(1 + \dfrac{f q_o^2 \bar{V}_m^2}{7.41 \times 10^{10} d^5}\right) \Delta p}{\dfrac{1}{m} \displaystyle\int_{p_2}^{p_1} V_m \, dp} \tag{6.44}$$

From the production data, PVT data, and pressure and temperature traverses of a large number of flowing and gas-lift wells, Poettmann and Carpenter[31] correlated f with the product of the inside diameter of the

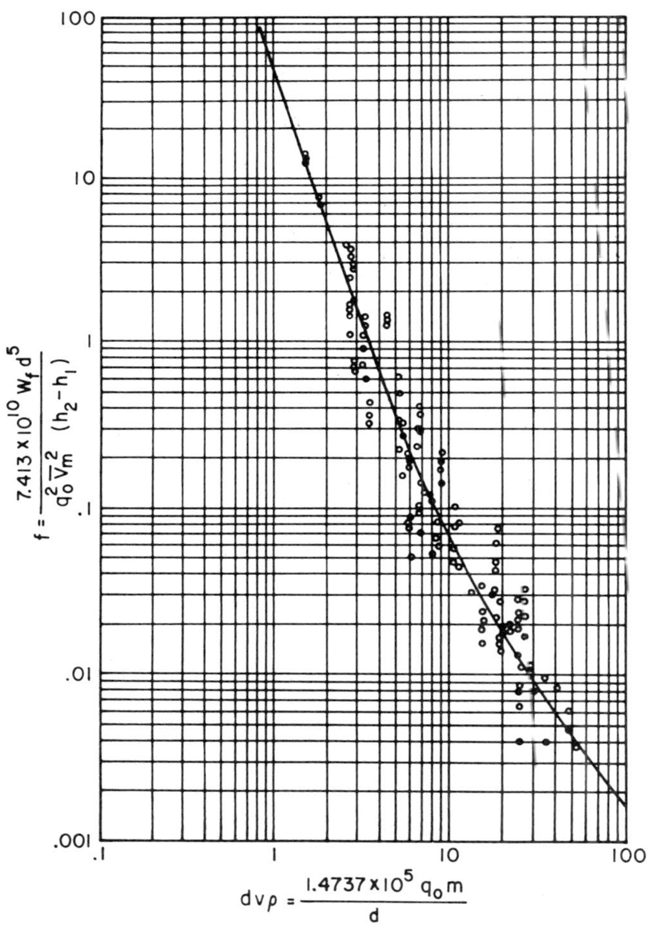

Fig. 6.23 Correlation of field data on flowing and gaslift wells. (Poettmann and Carpenter,[31] *Drilling and Production Practice*, API.)

tubing in feet, and the mass velocity $v\rho$ of the fluid flowing in the tubing (Fig. 6.23). In terms of field units obtained from well data:

$$dv\rho = \frac{dq_oV_m}{86{,}400A} \times \frac{m}{V_m}$$

where the velocity term is

$$\frac{q_oV_m}{86{,}400A} = \frac{q_oV_m}{86{,}400 \times 0.785d^2}$$

and the density term is m/V_m. With these substitutions, the expression reduces to

$$dv\rho = 1.4737 \times 10^{-5}\frac{q_om}{d} \tag{6.45}$$

$dv\rho$ corresponds to the numerator of the formula for the Reynolds number. It should be noted that viscosity is not one of the variables in multiphase flow because the degree of turbulence is of such magnitude that, of the total energy loss W_f, the portion resulting from viscous shear is negligible. The value of the friction factor f, from Eq. (6.35), is

$$f = \frac{7.41 \times 10^{10}\,W_fd^5}{q_o^2\bar{V}_m^2(h_2 - h_1)} \tag{6.46}$$

Figure 6.23 and the following equations are used to construct pressure gradient curves. The integrated average flowing density $\bar{\rho}$ is given by

$$\bar{\rho} = \frac{m}{\bar{V}_m} = \frac{m}{\int_{p_2}^{p_1} V_m dp \Big/ (p_1 - p_2)} \tag{6.47}$$

Letting $\Delta p = p_1 - p_2$, the expression for the pressure gradient is obtained by combining Eq. (6.44) with Eq. (6.47):

$$\frac{\Delta p}{\Delta h} = \bar{\rho}\left(1 + \frac{fq_o^2\bar{V}_m^2}{7.41 \times 10^{10}d^5} \times \frac{m^2}{m^2}\right)$$

$$= \bar{\rho}\left(1 + \frac{fq_o^2m^2}{7.41 \times 10^{10}\bar{\rho}^2d^5}\right) \tag{6.48}$$

or

$$\frac{\Delta p}{\Delta h} = \bar{\rho} + \bar{K}\bar{\rho} \tag{6.49}$$

where

$$\bar{K} = \frac{fq_o^2m^2}{7.41 \times 10^{10}\bar{\rho}^2d^5} \tag{6.50}$$

For a given tubing size, $\bar{K}$ is a function only of q_om and $\bar{\rho}$, and f is a function only of q_om. The empirical gradient curves of Figs. 6.24 and 6.25 were constructed for several tubing sizes by combining the friction factor data of Fig. 6.23 with Eq. (6.49). The product of q_o and m is plotted on the vertical scale in units of pounds of total fluid (oil, water, and gas) flowing per day, and the gradient ($\Delta p/\Delta h$) is plotted on the abscissa in units of pounds per square inch per foot of depth. The curves plotted on Figs. 6.24 and 6.25 are for constant integrated average density $\bar{\rho}$ between pressures p_1 and p_2, i.e., the gradient corresponding to values of $\bar{\rho}$. The pressure traverse is established by calculating the integrated average density $\bar{\rho}$ between the limits

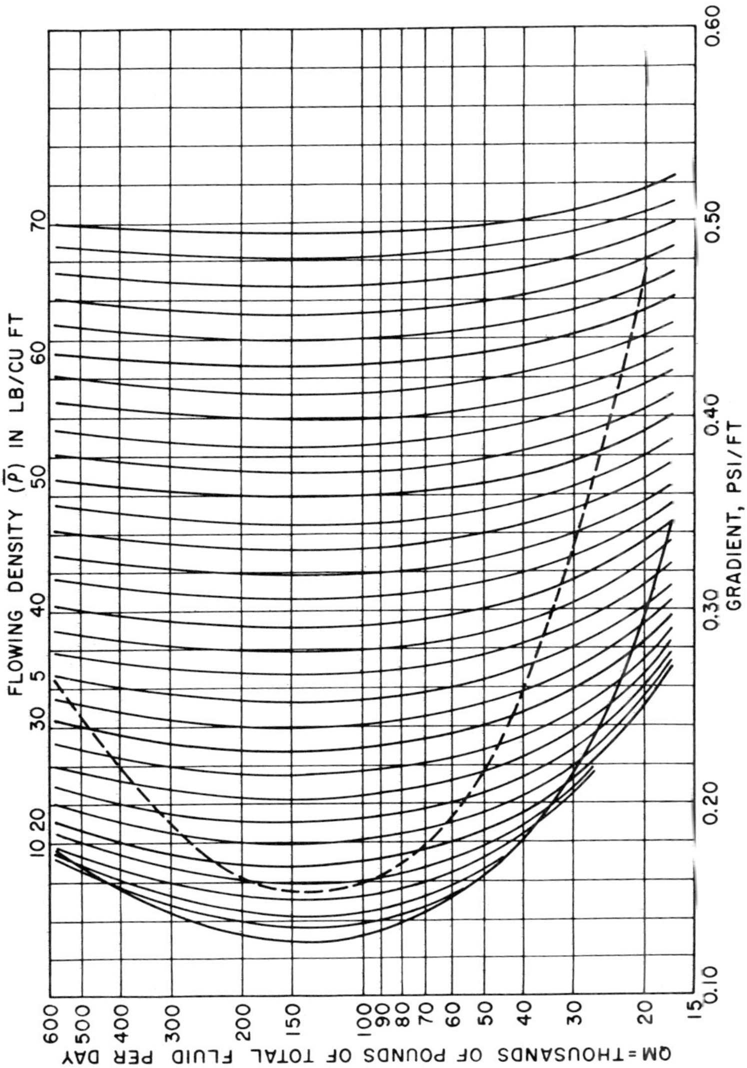

Fig. 6.24 Pressure gradients for 2.0-in. nominal tubing, 1.995-in. ID. (After Poettmann and Carpenter,[31] *Drilling and Production Practice,* API.)

of p_1 and p_2 by the use of PVT data. The curvature of the gradient lines is significant. At high rates of flow no slippage or liquid hangup occurs; the energy loss in this region is the result of frictional effects. At low rates of flow considerable slippage or liquid hangup exists in the flow strings. Example 6.6 shows how field data were used to obtain a point on Fig. 6.23.

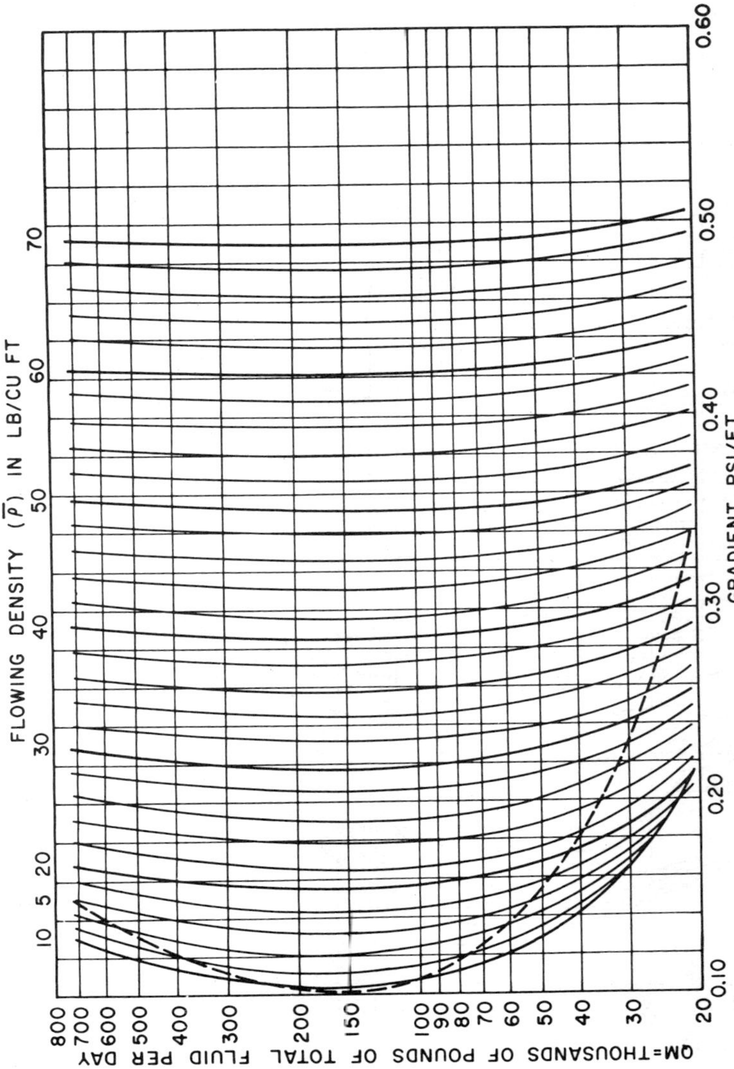

Fig. 6.25 Pressure traverses for $2\frac{1}{2}$-in. nominal tubing, 2.441-in. ID. (After Poettmann and Carpenter,[31] *Drilling and Production Practice*, API.)

Example 6.6. Correlate the field data of a flowing well and calculate pressure gradient, given:

Oil production = 302 STB/day

Gravity of oil = 42.3° API

Gas-oil ratio = 936 SCF/STB

Gravity of gas = 0.816

Tubing size = $2\frac{1}{2}$ in., 6.5 lb/ft, ID 2.441 in.

Tubing pressure = 860 psia

Flowing bottom-hole pressure = 2530 psia

Depth to middle of perforations = 6850 ft

Average flowing temperature = 100°F

No water production

SOLUTION: The specific gravity of the oil at 60°F is

$$\gamma_o = \frac{141.5}{131.5 + 42.3} = 0.814$$

Substituting in Eq. (6.43), the weight rate of flow is

$$\begin{aligned} m &= 350.17 \times 0.814 + 0.0764 \times 0.816 \times 936 \\ &= 343.39 \text{ lb/STB} \end{aligned}$$

where 350.17 is the weight of water in lb/bbl and 0.0764 is the density of air in lb/cu ft at 60°F and 14.7 psia.
From Eq. (6.45),

$$dv\rho = \frac{1.4737 \times 10^{-5} \times 302 \times 343.39}{2.441/12} = 7.51$$

The value of $\int_{p_2}^{p_1} V_m \, dp$ in Eq. (6.40) is

$$\begin{aligned} \int_{p_2}^{p_1} V_m \, dp &= \int_{p_2}^{p_1} 5.615 \times 144 \left(B_t + \frac{q_w}{q_o} \right) dp \\ &= 808.56 \int_{p_2}^{p_1} B_t \, dp \end{aligned}$$

This is evaluated by graphical integration over the pressure interval $p_1 - p_2$ or $(2530 - 860) = 1670$ psia.

The values shown in the table were read from Fig. 6.22 for a gas-oil ratio of 936 SCF per STB and a temperature of 100°F.

p, psia	B_t, bbl/STB	p, psia	B_t, bbl/STB
800	3.05	1800	1.65
1000	2.50	2000	1.57
1200	2.15	2200	1.50
1400	1.92	2400	1.44
1600	1.77	2600	1.40

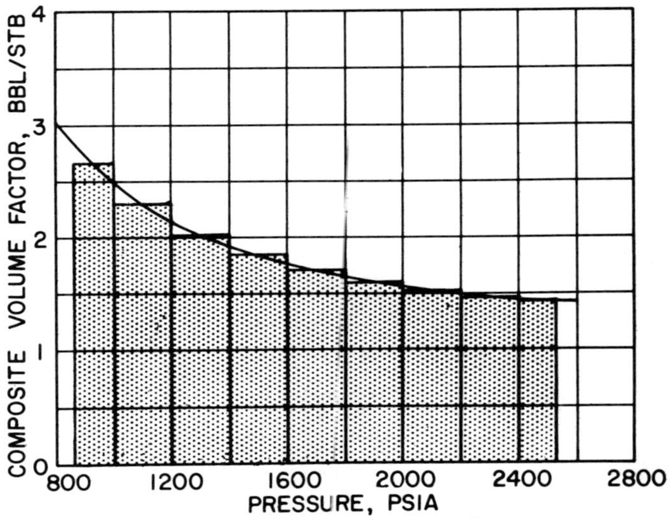

Fig. 6.26 Graphical integration of $\int_{p_2}^{p_1} B_t\, dp$.

These data are plotted in Fig. 6.26 and the area is 3072 psi bbl/STB. Then the value of the integral is

$$\int_{p_2}^{p_1} V_m\, dp = 808.56 \times 3072 = 2{,}483{,}896 \text{ ft-lb/STB}$$

where dp is in psi. Substituting the numerator of Eq. (6.39) for the first term of Eq. (6.33),

$$-\frac{1}{m}\int_{p_2}^{p_1} V_m\, dp + \Delta h + W_f = 0$$

The energy losses are

$$W_f = \frac{1}{m}\int_{p_2}^{p_1} V_m\, dp - \Delta h$$

$$= \frac{1}{343.39} \times 2{,}483{,}896 - 6850 = 383 \text{ ft}$$

From Eq. (6.38) the value of $\bar{V}_m$ is

$$\bar{V}_m = \frac{\int_{p_2}^{p_1} V_m\, dp}{p_1 - p_2} = \frac{2{,}483{,}896}{144(2530 - 860)} = 10.33 \text{ cu ft/STB}$$

The value of the friction factor f is

$$f = \frac{7.41 \times 10^{10}\, W_f d^5}{q_o^2 \bar{V}_m^2 \Delta h}$$

Substituting,

$$f = \frac{7.41 \times 10^{10} \times 383(2.441/12)^5}{(302)^2 \times (10.33)^2 \times 6850} = 0.148$$

From the data for this well, the plotting point on Fig. 6.23 is $dv\rho = 7.51$ and $f = 0.148$.

To locate a point on a pressure traverse, consider that the rate of flow of oil, gas, and water, $q_o m$, is 100,000 lb per day in $2\frac{1}{2}$-in. tubing. Substituting in Eq. (6.45),

$$dv\rho = \frac{1.4737 \times 10^{-5} \times 100{,}000}{2.441/12} = 7.24$$

From Fig. 6.23, for a value of $dv\rho = 7.24$, f is equal to 0.125. For an average flowing density $\bar{\rho}$ of 40 lb per cu ft, $\overline{K}$ from Eq. (6.50) is

$$\overline{K} = \frac{f q_o^2 m^2}{7.41 \times 10^{10} \bar{\rho}^2 d^5} = \frac{0.125(100{,}000)^2}{7.41 \times 10^{10} \times 40^2 \times (2.441/12)^5} = 0.03029$$

Substituting in Eq. (6.49),

$$\frac{\Delta p}{\Delta h} = 40 + 40 \times 0.03029 = 41.2 \text{ lb/cu ft}$$

$$= 41.2\,\frac{\text{lb/ft}^2}{\text{ft}} \times \frac{\text{ft}^2}{144\text{ in.}^2} = 0.286 \text{ psi/ft}$$

Kirkpatrick[21] has calculated similar pressure traverses for small tubing. Using the data of Baxendell,[34] Benham and Poettmann[35] have described a similar technique based on equivalent diameter which will permit the logical design of gas-lift installations where the flow is through the *annulus*.

The application of the gradient curves (Figs. 6.24 and 6.25) in determining pressure traverses for gas-lift wells can now be considered. The following well data are needed for the volumetric balance calculation:

1. Metered producing liquid and gas rates.
2. Operating tubing wellhead pressure.
3. Specific gravities of the oil, water, and gas.
4. Static bottom-hole pressure.
5. Productivity index.
6. Wellhead and bottom-hole temperature.
7. Size of tubing and depth set.
8. Depth to midpoint of perforated interval.

The procedure for optimum design of continuous-flow gas-lift systems includes the use of equations previously derived and will be illustrated in Example 6.7. For the producing rate desired the first step is to calculate, from the productivity index, solution gas-oil ratio, and shut-in bottom-hole pressure, points on the flowing gradient *below* the point of gas injection. The pressure drawdown is given by

$$p_{wf} = p_{ws} - \frac{q_t}{J} \tag{6.51}$$

where p_{wf} = flowing bottom-hole pressure, psia

p_{ws} = static bottom-hole pressure, psia

q_t = total fluid desired, barrels per day

J = productivity index, barrels of fluid produced per day per psi drop in pressure

The weight in pounds, m, of 1 bbl of stock-tank oil plus the associated gas and water is calculated from Eq. (6.43).

Assumed pressure points are established at regular intervals of 200 psi

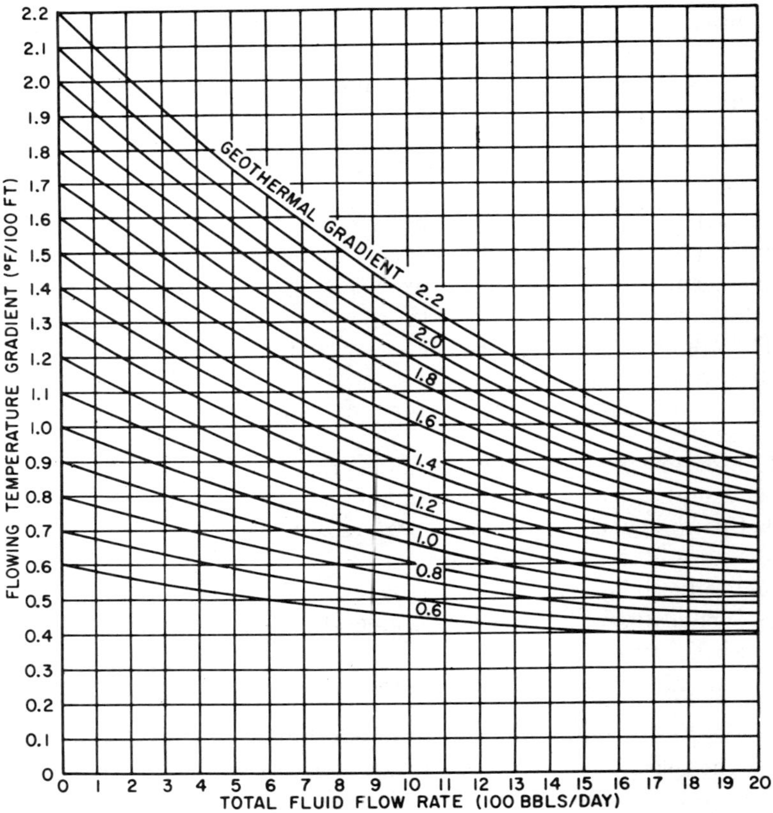

Fig. 6.27 Flowing temperature gradients for different flow rates, geothermal gradients, and tubing sizes. (Kirkpatrick,[11] *Division of Production*, API.) Chart to be used directly for $2\frac{1}{2}$-in. tubing; for 2-in. tubing, multiply actual flow rate by 2; for 3-in. tubing, divide the actual flow rate by 1.5.

from the flowing bottom-hole pressure (see Table 6.8), and the composite volume factor corresponding to each pressure and other well data are read from Fig. 6.22. The temperature used with Fig. 6.22 is the average temperature in the tubing under flowing conditions. The flowing temperature gradient may be estimated from the total fluid flow rate and the geothermal gradient for the area by using the correlation chart prepared by Kirkpatrick[11] which is presented in Fig. 6.27. For example, a well is producing 1400 bbl per day through $2\frac{1}{2}$-in. tubing from 5500 ft; the geothermal gradient for the area is 1.9°F per 100 ft; and the bottom-hole temperature is 175°F. At the intersection of 1400 bbl per day and the 1.9 geothermal gradient line, the gradient on the ordinate is 1°F/100 ft. The surface flowing temperature is 175 − (1.0)(55) or 120°F.

The composite volume factor for the oil, water, and gas is obtained by adding the barrels of water produced per barrel of stock-tank oil (WOR) to the value of B_t, at each pressure point. The density of production is obtained by dividing the mass rate of production (pounds per stock-tank barrel) by the composite volume factor for the oil, water, and gas and converting to pounds per cubic foot by dividing by 5.615 cu ft per bbl.

The gradient at each assumed pressure point is obtained from the pressure gradient curves (Fig. 6.24 or 6.25) for the value of $q_o m$ and the density $\bar{\rho}$ at each pressure point. The *average* gradient is the average value of the gradient between pressure increments, and the distance between pressure points is obtained by dividing the pressure difference between two respective points by the average gradient. For gradient points *above* the point of gas injection, pressure points established at regular intervals are added to the wellhead pressure and the composite volume factors are read for these pressures. If the surface tubing pressure is high enough to give a density readable on the gradient charts (Figs. 6.24 and 6.25), the distances between gradient points are added cumulatively to determine the depth of each pressure point, starting with the tubing pressure at zero depth. If the surface tubing pressure is not high enough to give a surface density that can be read on the gradient charts, an extrapolation technique is used. For example, on the 2-in. gradient curves of Fig. 6.24 the 5 lb per cu ft density curve departs considerably from the 10 lb per cu ft curve, which occurs in the case of systems with high gas-oil ratios and low tubing head pressures. This means that the kinetic energy is appreciable and that neglecting it in the derivation accounts for some of the abnormal calculated results. The correlations are unreliable in the low density range; therefore, the gradient curve must be plotted against an arbitrary depth and extrapolated to top-hole conditions as shown in Fig. 6.31 accompanying Example 6.7. The flowing gradients are calculated for different assumed injected gas-fluid ratios for the operating wellhead pressure. These gradients are calculated on the basis of the total gas–fluid ratio (Fig. 6.28). The intersection of each of the flowing gradients above the point of injection and the flowing gradient below the point of

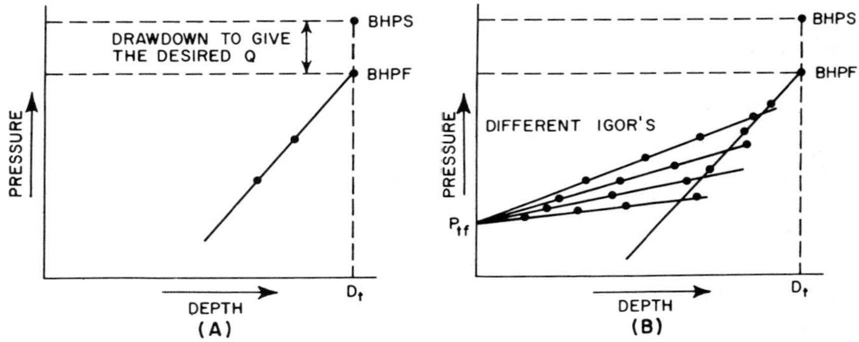

Fig. 6.28 Illustrative plot of flowing gradient below the point of injection (A) and flowing gradients both above and below the point of injection (B). (After Kirkpatrick,[11] *Division of Production,* API.)

injection provide the necessary data for calculation of the ideal adiabatic horsepower[13] H_a, or

$$H_a = \frac{144 q_o p_2 V_2}{1440 \times 33{,}000(1 - K)} \left[\left(\frac{p_2}{p_1} \right)^{(K-1)/K} - 1 \right] \tag{6.52}$$

where q_o = stock-tank barrels of oil per day

p_2 = injection pressure in psia

V_2 = cubic feet of gas injected (at pressure p_2 and temperature T_2) per stock-tank barrel of oil

K = ratio of specific heats for natural gas $\cong 1.3$ [see Eq. (6.53)]

p_1 = tubing pressure in psia

Plots are made of ideal adiabatic horsepower versus injection pressure as shown in Fig. 6.29 and also of injection gas-oil ratio versus injection pressure.[11] The optimum injection pressure is that corresponding to minimum horsepower, and this optimum pressure when applied to the graph of injection gas–fluid ratio versus injection pressure gives the optimum injection gas–fluid ratio. Example 6.7 shows the volumetric balance calculation for a continuous-flow gas-lift installation.

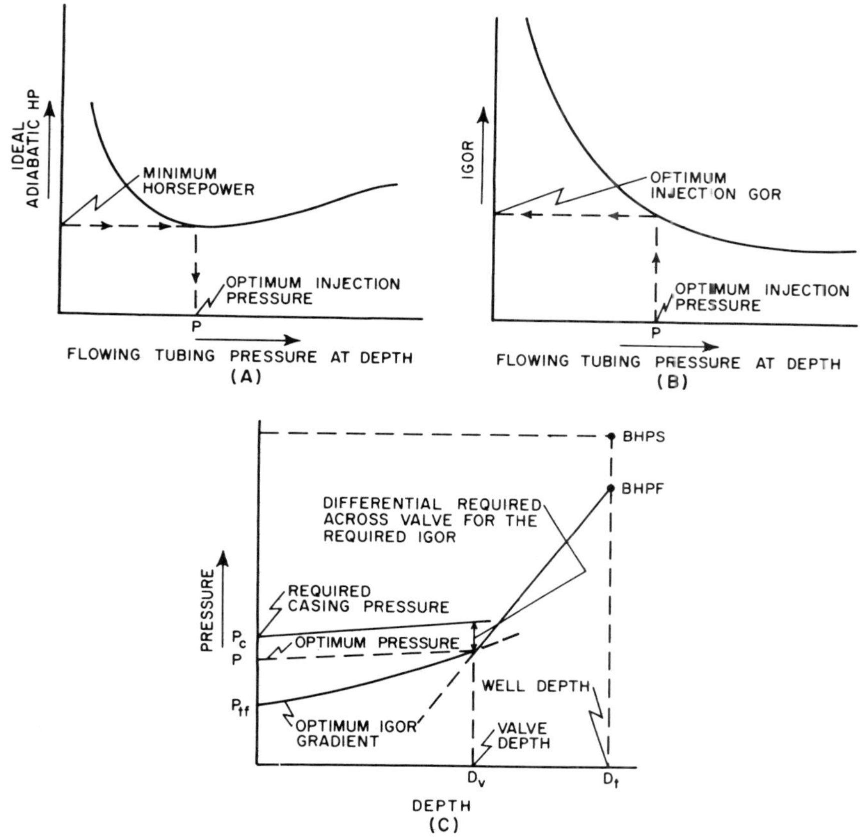

Fig. 6.29 Illustrative plot of ideal adiabatic horsepower and injection gas-oil ratio as a function of tubing pressure. (Kirkpatrick,[11] *Division of Production*, API.)

Example 6.7. Calculate the point of gas injection and the gas-oil ratio for a well on continuous-flow gas lift, given:

Depth of well = 8000 ft

Productivity index $J = 1.0$ bbl/day/psi

Static bottom-hole pressure $p_{ws} = 3000$ psia

Oil production desired $q_o = 40$ bbl/day

Water production $q_w = 400$ bbl/day

Total production $q_t = 440$ bbl/day

Gravity of oil = 40° API at 60°F ($\gamma_o = 0.825$)

Specific gravity of water = 1.15

Specific gravity of lift gas and separator gas = 0.60

Solution gas–oil ratio R_s = 680 SCF/STB (from Fig. 6.21 or volume metering)

Bottom-hole temperature = 200°F

Back pressure at tubing = 100 psia

Size of tubing = $2\frac{1}{2}$ in., 6.5 lb/ft

Surface temperature = 70°F

Solution: For a production rate of 440 bbl of fluid per day, the flowing bottom-hole pressure is

$$p_{wf} = p_{ws} - \frac{q_t}{J} = 3000 - \frac{440}{1.0} = 2560 \text{ psia}$$

The mass rate of production in pounds per stock-tank barrel, which is a constant below the point of gas injection, is found from Eq. (6.43):

$$\begin{aligned} m &= 350.17 \times 0.825 + 10 \times 350.17 \times 1.15 \\ &\quad + 680 \times 0.0764 \times 0.60 \\ &= 288.89 + 4027.0 + 31.17 = 4347.1 \text{ lb/STB} \end{aligned}$$

and

$$q_o m = 40 \times 4347.1 = 173{,}884 \text{ lb/day}$$

Since the surface temperature is 70°F, the static temperature gradient is

$$G_t = \frac{200 - 70}{8000} = 1.6°\text{F}/100 \text{ ft}$$

and from Fig. 6.27 the flowing temperature gradient is 1.31°F per 100 ft. The flowing surface temperature is

$$T_{tf} = 200 - 1.31 \times 80 = 95°\text{F}$$

and the average tubing temperature is

$$T_{\text{avg}} = \frac{95 + 200}{2} = 148°\text{F}$$

Taking into account the low rate of production and the low injection gas-liquid ratio, an average temperature of 148°F was used to calculate the pressure traverse above and below the point of gas injection. In high volume wells, however, the average temperature of the flowing traverse *below* the point of gas injection would be near reservoir temperature, and a lower average flowing temperature would be used to calculate the pressure traverse *above* the point of gas injection, on the basis of average flowing tubing temperature and bottom-hole temperature. At the pressure given in Col. (1) of Table 6.8, and a temperature of 148°F, and using the other well data, the two-phase volume factors, Col. (2), are read from Fig. 6.22 for each pressure increment. The water-oil ratio (10:1) is added to the two-phase volume factor of Col. (2) and placed in Col. (3). The density of production $\bar{\rho}$ in pounds per cubic foot, Col. (4), is obtained by dividing the weight rate of production

Table 6.8. CONTINUOUS-FLOW GAS-LIFT ANALYSIS[a]

(1) Tubing pressure, psia	(2) Two-phase volume factor B_t, bbl/STB	(3) Total composite vol. factor, $B_t + q_w/q_o$	(4) Density of production $\bar{\rho}$, $(m/5.615)(3)$, lb/cu ft	(5) Gradient G_f, psi/ft	(6) Average gradient, psi/ft	(7) Distance between pressure points, ft	(8) Depth to pressure point, ft
m = 4347.1 lb/STB $q_o m$ = 173,884 lb/day R_s = 680 SCF/STB							
2560	1.38	11.38	68.03	0.478			8000
2360	1.43	11.43	67.73	0.474	0.476	420	7580
2160	1.49	11.49	67.38	0.472	0.473	423	7157
1960	1.54	11.54	67.09	0.470	0.471	425	6732
m = 4517.7 lb/STB $q_o m$ = 180,708 lb/day R = 4400 SCF/STB R_i = 3720 SCF/STB							
100	128.0	138.0	5.83				
300	49.5	59.5	13.52	0.118		1772	1772
500	28.5	38.5	20.90	0.148	0.133	1504	3276
700	20.0	30.0	26.82	0.199	0.174	1149	4425
m = 4462.8 lb/STB $q_o m$ = 178,511 lb/day R = 3200 SCF/STB R_i = 2520 SCF/STB							
100	92.0	102.0	7.79				
300	37.5	47.5	16.73	0.136		1512	1512
500	22.0	32.0	24.84	0.185	0.161	1242	2574
700	15.5	25.5	31.17	0.227	0.206	971	3725
900	11.6	21.6	36.80	0.264	0.246	813	4538

[a] T_{avg} for all data = 148°F

m by the total composite volume factor at each pressure point, and is converted to pounds per cubic foot by dividing by 5.615 cu ft per bbl.

The gradient in Col. (5) corresponding to the values of $q_o m$ and $\bar{\rho}$ is read from the pressure gradient curves for $2\frac{1}{2}$-in. tubing (Fig. 6.25). The average gradient is the arithmetical average of the gradients of Col. (6), and the distance between pressure points, Col. (7), is obtained by dividing the pressure increment (200 psi in this case) by the average gradient between the pressure points. The depth to the assumed pressure points, Col. (8), is obtained by cumulative subtraction of the distance between the pressure points from the total depth. These points are plotted on Fig. 6.30.

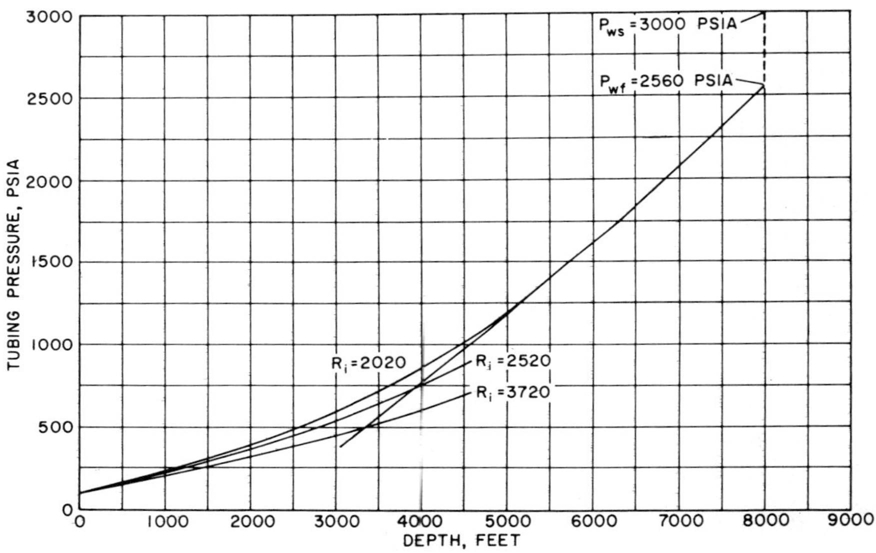

Fig. 6.30 Detailed analysis of continuous flow.

The data for the flowing gradients *above* the point of gas injection at different assumed injection gas–oil ratios are calculated in a similar manner. These gradients are calculated on the basis of total gas–oil ratio, which includes both solution and injection gas. At a total gas–oil ratio of 4400 SCF per stock-tank barrel, the pounds of fluid per stock-tank barrel are

$$m = 350.17 \times 0.825 + 10 \times 350.17 \times 1.15 + \frac{4400 \times 29 \times 0.60}{379.4}$$

$$= 288.89 + 4027.0 + 201.8 = 4517.7 \text{ lb/STB}$$

and

$$q_o m = 40 \times 4517.7 = 180{,}708 \text{ lb/day}$$

The starting pressure point is the surface tubing pressure, and the assumed pressure increments (200 psi in this case) are added to the tubing pressure of Col. (1). Using the average temperature in the tubing under flowing conditions, the ratio of solution plus injection gas to oil, and other well data, values of the two-phase volume factor B_t are read from Fig. 6.22 for each pressure point. The water-oil ratio is added to the two-phase volume factor to give the total composite volume factor, Col. (3). The density of production in pounds per cubic foot, Col. (4), is calculated as previously described, and the average gradients are read from the pressure traverse (Fig. 6.25). Since the 5 lb per cu ft density curve departs considerably from the 10 lb per cu ft curve, the gradient curve is plotted against an arbitrary depth datum and extrapolated into the low-pressure range. The 700, 500, and 300 psi pressure points are plotted against an arbitrary depth scale as shown in Fig. 6.31 and, from these

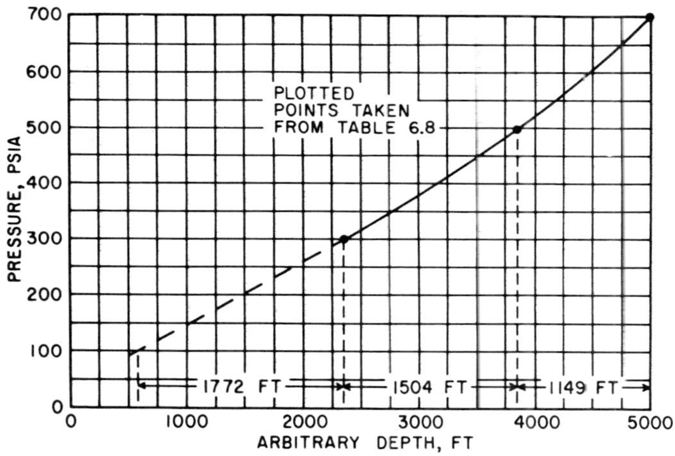

Fig. 6.31 Extrapolation of gradient curve to arbitrary depth datum.

points, the curve is extrapolated until it intersects the 100 psi wellhead pressure line. The difference between the depth to this intersection (575 ft) and the depth to the 300 psi pressure point (2347 ft) is the distance between the 100 and 300 psi pressure points, 1772 ft.

The depth to each pressure point is given in Col. (8). The values are plotted on Fig. 6.30. The intersection of each of the flowing gradient points *above* the point of injection and the flowing gradient *below* the point of injection furnish the necessary data, given in Table 6.9, for calculating the ideal adiabatic horsepower.

Table 6.9. DATA FROM FIGURE 6.30

Injection gas–oil ratio, SCF/STB	Tubing pressure at intersection, psia	Depth to intersection, ft
4220	500	3350
3720	520	3410
2520	740	3950
2020	1200	5000

The ideal adiabatic horsepower supplied by the gas in lifting the oil from each of the injection depths against a tubing pressure of 100 psia is calculated from Eq. (6.52). The injection gas–oil ratio at standard conditions of 14.7 psia and 60°F is corrected to injection conditions by use of the ideal gas law. For example, at an injection depth of 3410 ft, the temperature from Eq. (6.17) is 125°F. The volume of 3720 SCF of gas at injection conditions is

$$V_v = 3720 \times \frac{14.7}{520} \times \frac{585}{520} = 118.3 \text{ cu ft/STB}$$

and the adiabatic horsepower from Eq. (6.52) is

$$H_a = \frac{144 \times 40 \times 520 \times 118.3}{1440 \times 33{,}000(1 - 1.3)} \left[\left(\frac{520}{100} \right)^{(1.3-1)/1.3} - 1 \right]$$

$$= -24.9 \times 0.463 = -11.5 \text{ hp}$$

The data from Table 6.9 is used in the computation of the theoretical adiabatic horsepower required for the various injection pressures and gas–oil ratios. Figure

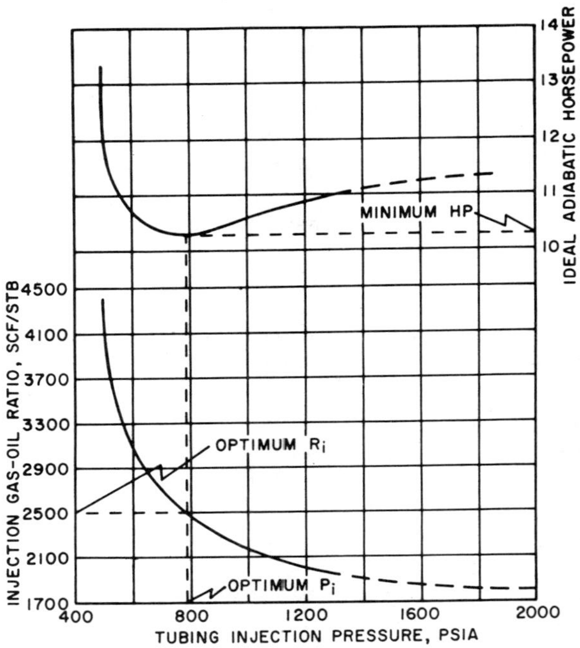

Fig. 6.32 Detailed analysis of continuous-flow gas lift.

6.32 is plotted from the data obtained from this computation. The optimum operating condition is the point of minimum horsepower, which for this well is an injection pressure of 780 psia and an injection gas–oil ratio of 2500 SCF per STB.

Choke size. The choke size for the valve or surface injection line is calculated from the equation[36]

$$q = \frac{155.5 C_d A p_1 \sqrt{2g \dfrac{K}{K-1} \left(r^{2/K} - r^{(K+1)/K}\right)}}{\sqrt{\gamma T}} \tag{6.53}$$

where q = gas flow in MCF per 24 hr at 14.65 psia and 60°F

C_d = discharge coefficient (determined experimentally)

A = area of choke in square inches

p_1 = upstream pressure in psia

g = 32.17 ft/sec²

p_2 = downstream pressure in psia

$K = c_p/c_v = \dfrac{\text{specific heat at constant pressure}}{\text{specific heat at constant volume}}$

$r = p_2/p_1 \geq r_o$

$r_o = \left(\dfrac{2}{K+1}\right)^{K/(K-1)}$ = critical flow pressure ratio

γ = specific gravity (1.00 for air)

T = inlet temperature, °R

The average values of the discharge coefficients obtained by Cook and Dotterweich[36] are given in Table 6.10. Example 6.8 shows the calculation of choke size by use of Eq. (6.53).

Table 6.10. FLOW AND DISCHARGE COEFFICIENTS FOR 6-IN. LONG CHOKES

Nominal diameter, in.	Area A, sq in.	Discharge coefficient C_d	AC_d
2/64	0.000767	0.613	0.0004702
3/64	0.001726	0.650	0.001122
4/64	0.003067	0.677	0.002076
5/64	0.00497	0.700	0.003479
6/64	0.00690	0.721	0.004975
7/64	0.00939	0.730	0.006855
8/64	0.01230	0.757	0.009311
12/64	0.02761	0.805	0.02223
16/64	0.0491	0.832	0.04085
20/64	0.0767	0.832	0.06381
24/64	0.1104	0.832	0.09185
28/64	0.1506	0.828	0.1247
32/64	0.1963	0.828	0.1625

Example 6.8. Calculate the choke diameter for a gas-lift well, given:

Injection gas–oil ratio = 2000 SCF/STB

Production = 40 STB/day

Casing injection pressure at the choke = 700 psia

Downstream pressure = 600 psia

Gas gravity = 0.60

Temperature at the choke = 150°F = 610°R

$K = 1.31$

SOLUTION:

$$r = \frac{600}{700} = 0.857$$

$$r_o = \left(\frac{2}{K+1}\right)^{K/(K-1)} = \left(\frac{2}{1.31+1}\right)^{1.31/0.31} = 0.545$$

Since r is greater than r_o, the value of r is used in Eq. (6.53). Equation (6.53) in terms of the discharge coefficient and choke area becomes

$$AC_d = \frac{q\sqrt{\gamma T}}{155.5p_1\sqrt{2g\,\frac{K}{K-1}\,(r^{2/K} - r^{(K+1)/K})}}$$

and substituting,

$$AC_d = \frac{40 \times 2\sqrt{0.60 \times 610}}{155.5 \times 700\sqrt{2 \times 32.17\left(\frac{1.31}{0.31}\right)[(0.857)^{2/1.31} - (0.857)^{2.31/1.31}]}}$$

$$= 0.005097 \text{ sq in.}$$

From Table 6.10 the choke diameter is $\frac{6}{64}$ in.

Flowing gradient curves. In view of the time required to calculate the flowing gradient curves of Fig. 6.30, a set of master curves has been prepared, which permits the solution of gas-lift problems without tedious calculations. One of these curves[37] is given in Fig. 6.33. Since many of the wells to be placed on gas lift have a high water cut, these curves are usually prepared for a fluid with a specific gravity of 1.074 (salt water). Each family of curves is prepared for a constant flow rate in a particular size of tubing and various gas-liquid ratios. With this master set of curves, it is possible to solve most continuous-flow gas-lift problems by superimposing coordinate paper over the curves to fit the particular well conditions. Because of the number of curves involved in this master set, it is impossible to include them in this book.

6.5. *Intermittent Flow*

Gas-lift design for intermittent flow is different from that for continuous flow. In continuous flow, the fluid column is aerated to reduce the bottom-hole pressure to the point that the differential pressure across the sand

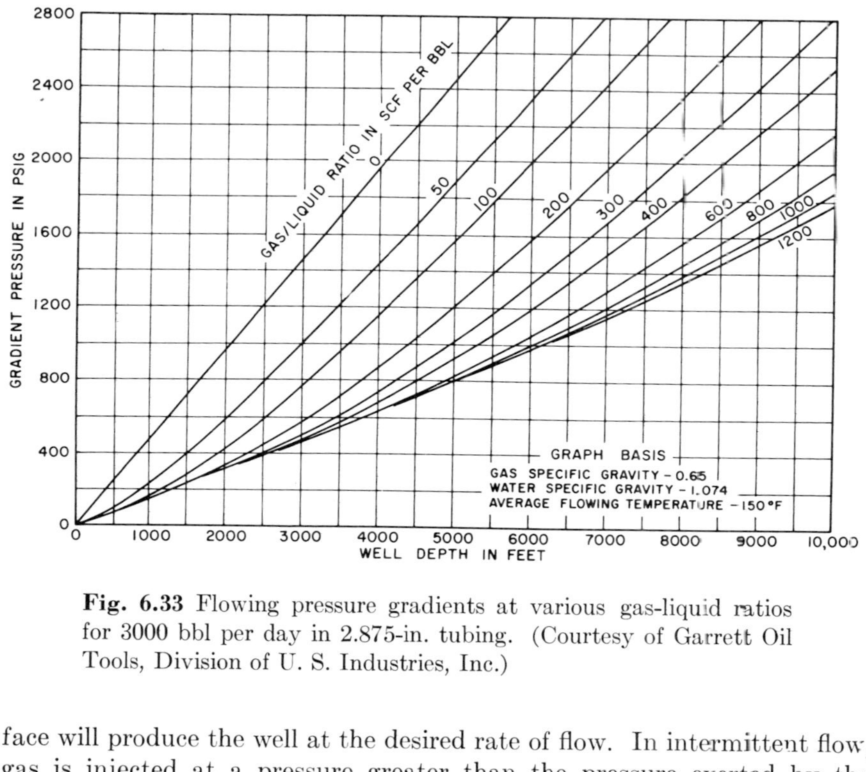

Fig. 6.33 Flowing pressure gradients at various gas-liquid ratios for 3000 bbl per day in 2.875-in. tubing. (Courtesy of Garrett Oil Tools, Division of U. S. Industries, Inc.)

face will produce the well at the desired rate of flow. In intermittent flow, gas is injected at a pressure greater than the pressure exerted by the column of fluid in the tubing above the operating valve, lifting the fluid by expansion and displacement — in effect, by ballistic lift.

In some wells the productivity index will remain constant over a wide variation in flow rate so that the flow rate is directly proportional to the bottom-hole pressure drawdown. In other wells, at higher flow rates the linearity fails and the productivity declines so that little benefit is gained by drawing down the static bottom-hole pressure more than 75 per cent.

Figure 6.34 shows the bottom-hole pressure of a well being produced by intermittent flow. The bottom-hole pressure at the instant the valve opens is indicated by point *A*. The pressure impulse results in an instantaneous pressure build-up at point *B*, which reaches a maximum at *C* after initial acceleration of the oil column. Figure 6.35 shows the intermittent flowing gradient which is the summation of the gradient of the gas above the slug, the gradient of the slug, and the gradient of the lift-gas and entrained liquids below the slug. Figure 6.36 shows the casing and tubing

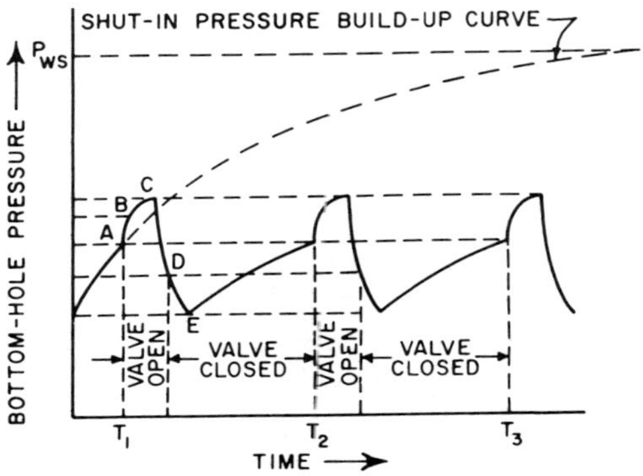

Fig. 6.34 Illustrative plot of the bottom-hole pressure of a well produced by intermittent flow. (After Kirkpatrick,[13] *The Power of Gas.*)

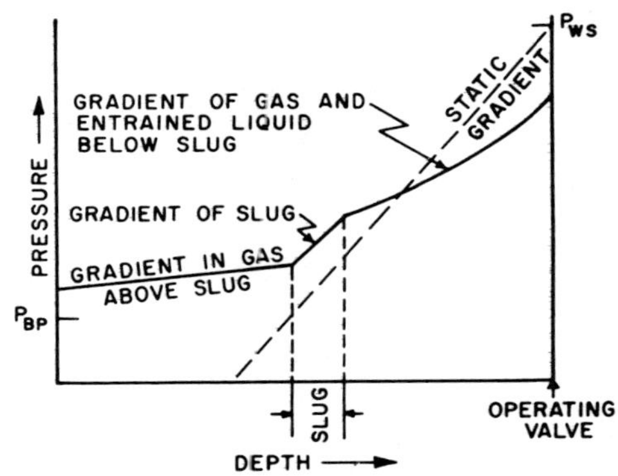

Fig. 6.35 Intermittent-flow gradient as liquid slug approaches midpoint of its tubing travel. (After Kirkpatrick,[13] *The Power of Gas.*)

pressures of wells on intermittent flow, with an analysis of the operating problems. The minimum cycle time is approximately three minutes per 1000 ft of depth.

Kirkpatrick[11] has illustrated the design technique for intermittent-flow gas lift given in Example 6.9.

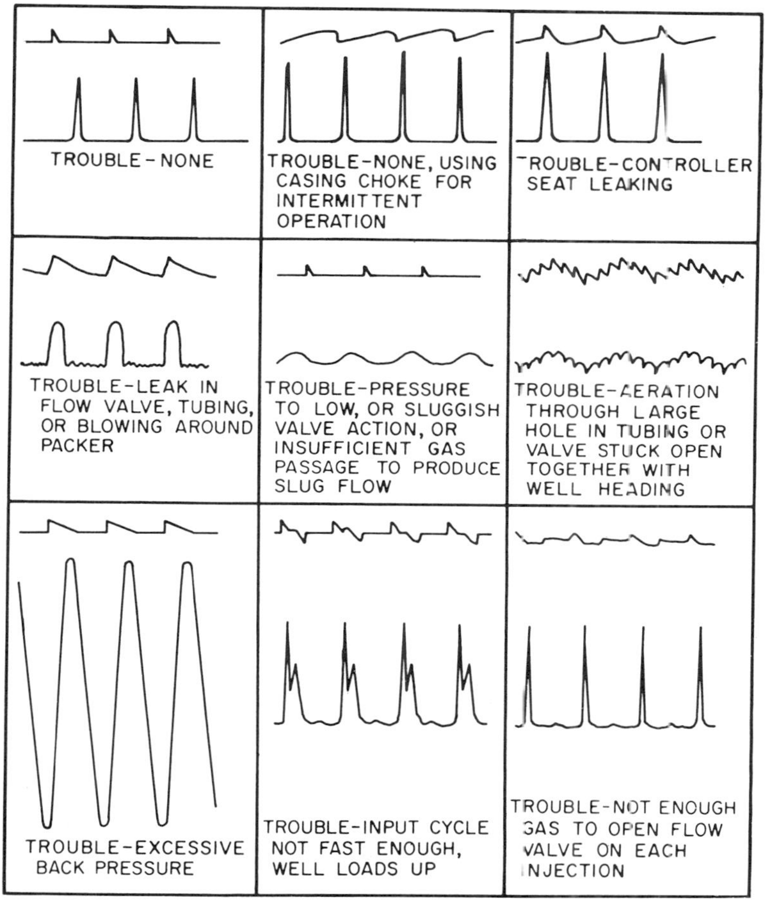

Fig. 6.36 Tubing and casing pressures of wells on intermittent flow (upper trace is the casing pressure, lower trace the tubing pressure). (Courtesy of Otis Engineering Corporation.)

Example 6.9. Calculate the depth to the operating valve and theoretical minimum injection gas–oil ratio, given:

Depth = 8000 ft

Available surface operating pressure = 800 psig

Tubing size = 2 in.

Casing size = $5\frac{1}{2}$ in., 20 lb/ft

No water production (reservoir under volumetric control)

API gravity of oil = 30°, sp. gr. = 0.8762

Bottom-hole pressure = 2000 psig

Well productivity index = 0.10

Tubing pressure = 50 psig

Oil production desired = 100 bbl/day

Average tubing temperature = 127°F

Cycle time = 45 min (estimated)

SOLUTION: The static gradient is

$$G_s = 0.8762 \times 0.433 = 0.379 \text{ psi/ft}$$

For a production of 100 bbl per day, the bottom-hole pressure drawdown is

$$\Delta p = \frac{100 \text{ bbl/day}}{0.10 \text{ bbl/day/psi}} = 1000 \text{ psi}$$

and the average flowing bottom-hole pressure is

$$p_{wf} = 2000 - 1000 = 1000 \text{ psig}$$

The depth to the *static* fluid level with 50 psig surface tubing pressure is

$$D_s = 8000 - \left(\frac{2000 - 50}{0.379}\right) = 2855 \text{ ft}$$

and 1000 psi drawdown is equivalent to a hydrostatic head of 1000/0.379 or 2639 ft. Therefore, the depth to the working fluid level is 2855 + 2639 or 5494 ft, which is the midpoint of the slug build-up (Fig. 6.37). The number of cycles required per day with 45 min per cycle is

$$\frac{1440 \text{ min/day}}{45 \text{ min/cycle}} = 32 \text{ cycles per day}$$

and for 100 bbls per day, the number of barrels per cycle is 100/32 or approximately 3 bbl/cycle. Operating experience has shown that, depending on the depth, from 30 to 60 per cent of the total liquid slug is lost due to slippage or fall-back. Assuming that 60 per cent of the starting slug is produced, the volume of the starting slug is 3/0.60 or 5.0 bbl and the capacity of the 2-in. tubing is 0.00387 bbl/ft. The slug length is 5.0/0.00387 or 1292 ft, and the operating valve should be located 1292/2 or 646 ft below the working fluid level. Since the depth to the working fluid level is 5494 ft, the *depth* to the operating valve is 6140 ft (Fig. 6.37). Consideration should be given to future well conditions, as a decrease in bottom-hole pressure or a change in productivity index would change the working fluid level. Normally, wells producing from volumetric reservoirs are equipped with valves to bottom.

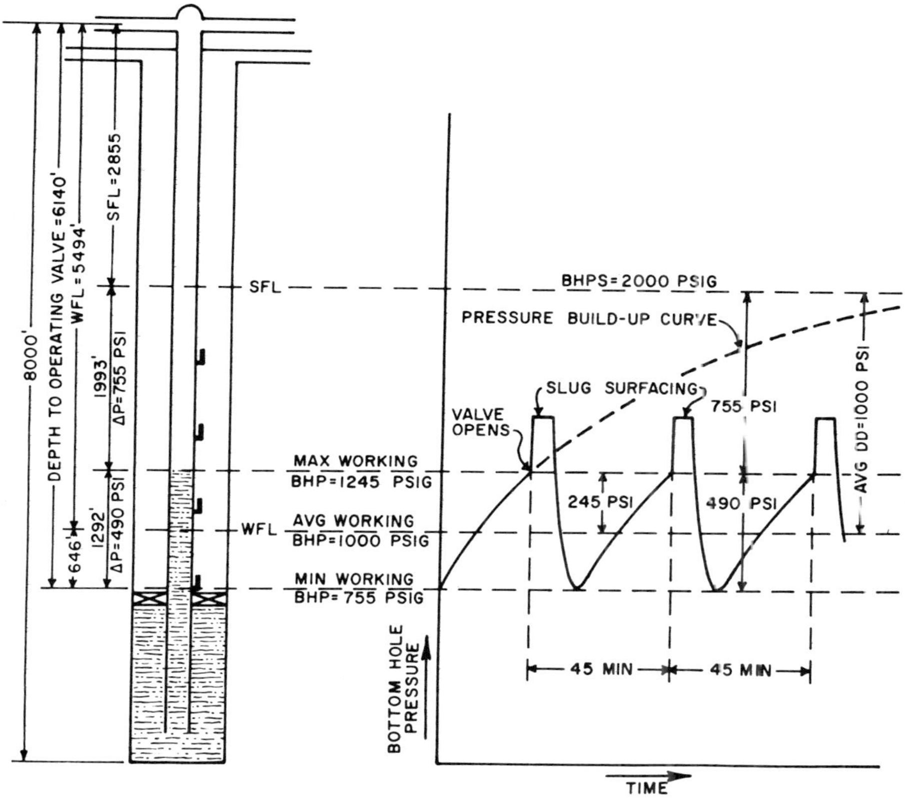

Fig. 6.37 Bottom-hole pressure for slug build-up and surfacing for intermittent flow. (After Kirkpatrick,[11] *Division of Production*, API.)

The pressure in the tubing opposite the valve with 50 psig surface pressure, neglecting the weight of the gas column, is

$$p_t = 50 + (1292 \times 0.379) = 540 \text{ psig}$$

To attain an average slug surfacing velocity of 1000 ft per min for minimum slippage and fall-back, the pressure in the casing opposite the operating valve at the instant the valve opens should be at least 50 per cent greater than the tubing pressure at valve depth with a minimum differential of 200 psi. Therefore, with a tubing pressure at valve depth of 540 psig, the instant the valve opens the minimum casing pressure at 6140 ft is

$$p_{v\,\min} = 540 + \frac{540}{2} = 810 \text{ psig}$$

The surface operating pressure from Fig. 6.13 is 707 psig.

The minimum volume of gas required to lift the slug to the surface will be that required to fill the tubing from injection depth to the surface less the volume occupied by the liquid slug. The 2-in. tubing has a capacity of 0.00387 bbl/ft, and with the injection depth at 6140 ft the volume of the tubing is 23.8 bbl. The length of the slug is 1292 ft (5.0 bbl); therefore, the space occupied by the gas just as the starting slug reaches the surface is 23.8 − 5.0 or 18.8 bbl (105.5 cu ft). The approximate pressure in the tubing immediately under a liquid slug at the instant the slug surfaces is equal to the pressure due to the slug length plus the wellhead tubing pressure. The approximate pressure under the slug the instant the slug surfaces is

$$p_t = 50 + \left[\frac{3.0 \text{ bbl}}{0.00387}\right] \times 0.379 = 344 \text{ psig}$$

The average pressure is

$$p_{\text{avg}} = \frac{810 + 344}{2} = 577 \text{ psig, or } 591.7 \text{ psia}$$

and the average temperature of the tubing is 127°F or 587°R. The injection gas gravity is usually close to 0.60; however, after the slug reaches the surface, the gas below the slug has a higher gravity due to slippage. The gas gravity is assumed to be 0.80, and the compressibility factor at an average pressure of 577 psig and an average temperature of 127°F, from Fig. 6.12, is 0.886. The minimum gas required is

$$V_{sc} = 105.5 \times \frac{591.7}{14.7} \times \frac{520}{587} \times \frac{1}{0.886}$$

$$= 4246 \text{ SCF per cycle}$$

Since there are 3 barrels of oil per cycle, the minimum injection gas–oil ratio is 4246/3 or 1415 SCF per STB.

6.6. Accumulation Chambers

The use of oil accumulation chambers is an adaptation of the closed intermittent installation. Their use provides an inexpensive type of lift which will produce stripper-type wells to depletion. There are two basic types of chamber installations: the conventional (two-packer) and the insert type,[11] shown in Fig. 6.38.

Through the proper use of chambers it is possible to:

1. Produce to depletion high productivity index, low bottom-hole pressure and low productivity index, low bottom-hole pressure wells.
2. Produce more fluid (slug volume) per cycle.
3. Increase daily production by permitting an overall lower average producing bottom-hole pressure.
4. Inject gas near total depth in wells with long perforated sections.
5. Obtain lower injection gas–fluid ratios.
6. Improve the efficiency of present intermittent gas-lift installations by minimizing fall-back, as the gas does not enter the tubing until the fluid is moving rapidly upward.

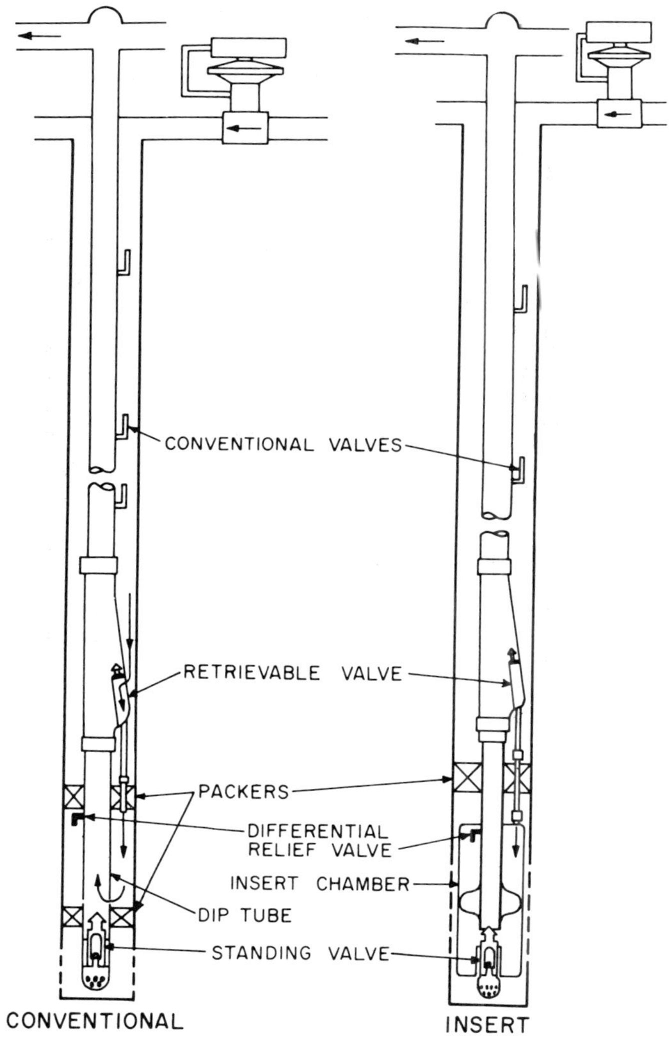

Fig. 6.38 Conventional and insert chamber installation. (Kirkpatrick,[11] *Division of Production*, API.)

The foregoing may be accomplished by various designs. For example, to demonstrate how the chamber provides increased slug volume per linear foot of *head* above the formation, consider a well with 7-in. OD, 26-lb casing and $2\frac{1}{2}$-in. tubing. The ratio R of the annular volume of the chamber to the volume of the tubing above the chamber is 3.03/0 579 or 5.2 to 1, that is, 1 ft of fluid in the chamber annulus is equal in volume to

5.2 ft of fluid in the tubing. A 100-ft fluid head build-up in the dip tube and chamber annulus represents a 520 + 100 or 620 ft fluid height in the dip tube and tubing when the chamber annulus is displaced by gas.

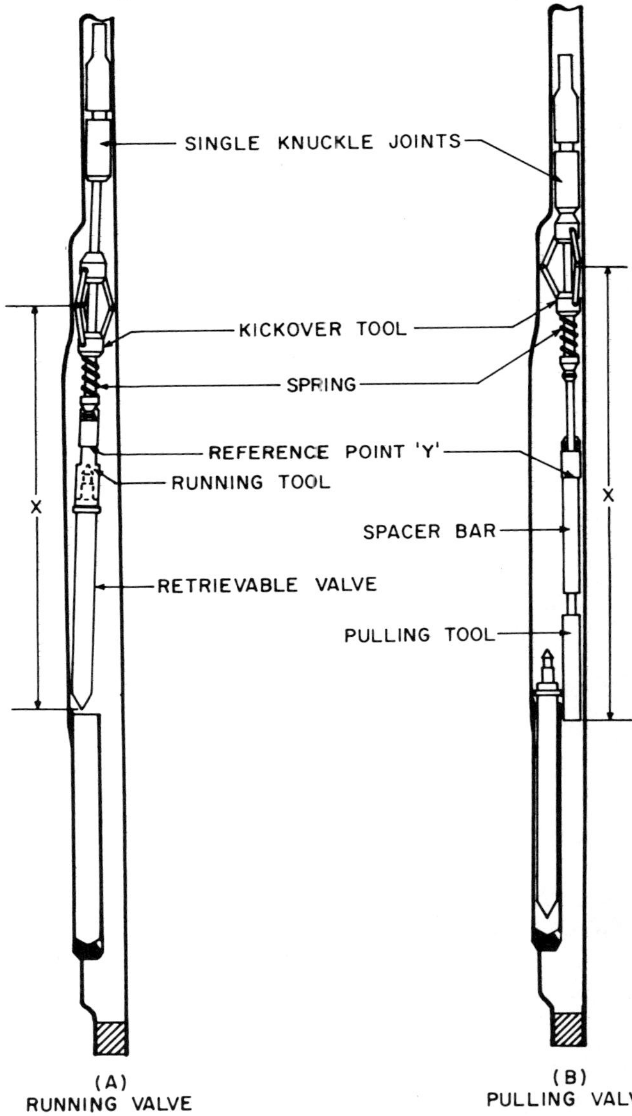

Fig. 6.39 Wire-line retrievable gas-lift valve with eccentric mandrel. When running and retrieving valve, the pulling tool must be run past the valve and then pulled back up as shown in (B). (After Kirkpatrick,[11] *Division of Production,* API.)

The chamber is designed to permit gas injection only through the operating valve, and the top of the chamber should not be located above the maximum working fluid level of the well. An excessive chamber length will result in high injection gas–oil ratios (if the chamber does not fill) and the long bleed-down time will reduce the number of cycles per day. The dip tube is spaced near the bottom of the chamber and, if possible, should be of the same diameter as the tubing to provide maximum slug volume per foot of head and to permit the use of a retrievable standing valve.

Retrievable equipment[11] (Fig. 6.39) provides maximum efficiency and eliminates costly tubing jobs. Retrievable valves can be *reset* to produce the well efficiently during changing reservoir pressures.

With a two-packer or insert-chamber installation, the chamber and dip tube are assumed to fill only to the top of the chamber, i.e., the tubing above the chamber is assumed to remain empty. The bottom packer is set as near the *top* of the perforations as possible, and the dip tube and chamber lengths are the same. Chambers are designed on the basis of injection gas pressures.[38] The maximum height of fluid the injection pressure will support is

$$H_i = \frac{p_i - p_{tf}}{G_s} \tag{6.54}$$

and, in terms of chamber length,

$$H_i = L_c + L_c R \tag{6.55}$$

Equating (6.54) and (6.55),

$$\frac{p_i - p_{tf}}{G_s} = L_c + L_c R$$

and

$$L_c = \frac{p_i - p_{tf}}{G_s(R + 1)} \tag{6.56}$$

where p_i = injection gas pressure inside the chamber, psig

p_{tf} = tubing pressure, psig

G_s = static gradient of produced fluid, psi per foot

L_c = length of the chamber, feet

R = ratio of annular volume of the chamber to the volume of the tubing above the chamber

Equation (6.56) is the conventional chamber formula. Example 6.10 shows the design of a conventional packer installation.

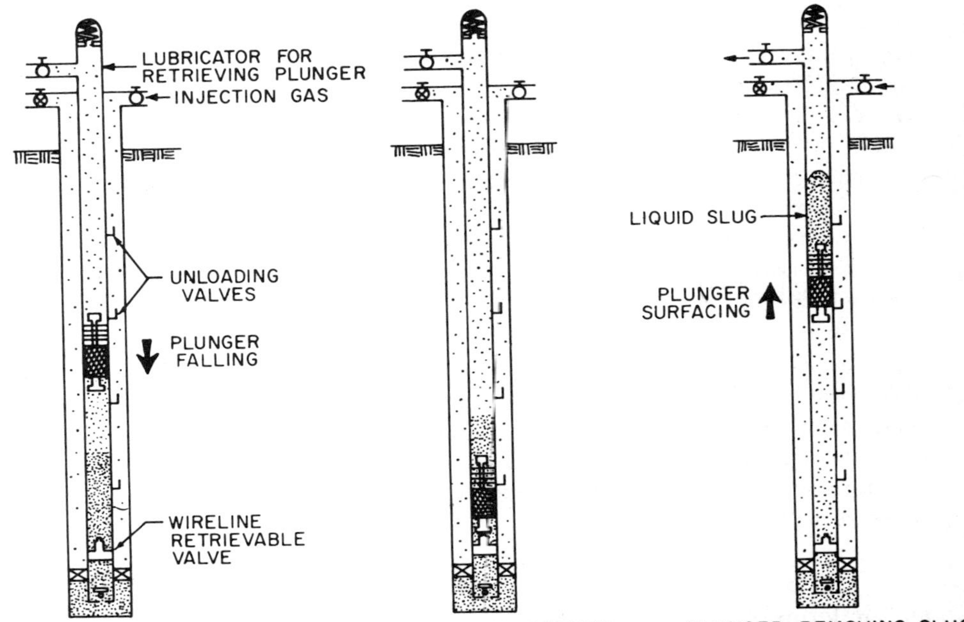

Fig. 6.40 Gas-lift plunger operation. (Kirkpatrick,[11] *Division of Production*, API.)

Example 6.10. Calculate the chamber length where the bottom packer is set just above the perforations, given:

Well depth = 6600 ft
Casing size = 7 in., 26 lb/ft
Tubing and dip tube size = $2\frac{1}{2}$ in.
Perforated interval = 6300 to 6500 ft
Static fluid gradient = 0.4 psi/ft
Tubing pressure = 50 psig
Injection gas pressure = 600 psig
Annular volume = 0.0303 bbl/ft (Table 6.2)
Tubing volume = 0.00579 bbl/ft (Table A.4)

SOLUTION:

$$R = \frac{0.0303}{0.00579} = 5.2$$

Substituting in Eq. (6.56),

$$L_c = \frac{600 - 50}{0.4(5.2 + 1)} = 222 \text{ ft}$$

6.7. Free Piston System of Artificial Lift

One of the most efficient methods of producing oil involves the use of the cycle-controlled *gas-lift plunger* (free piston). The basic idea of a gas-lift plunger is the introduction of a solid interface between the gas and fluid phases of the system (Fig. 6.40). The expanding segments[11] make continuous contact with the walls of the tubing, thereby forming an effective seal. At the surface, the fluid enters the flow line through an outlet in the flow tee followed by the plunger which enters the lubricator. A valve in the plunger is opened mechanically, permitting the plunger to fall through the gas and oil to the footpiece[11,39] (Fig. 6.41).

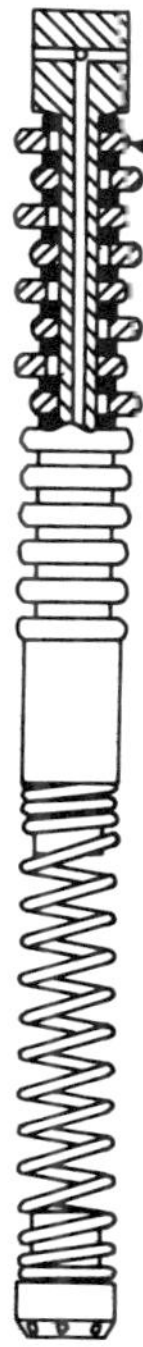

Fig. 6.41 Gas-lift plunger with self-adjusting washers as the sealing element. (Kirkpatrick,[11] *Division of Production*, API.)

In flowing wells, the free piston offers these advantages:

1. The flowing life of the well is prolonged.
2. Gas–oil ratios are usually considerably reduced.
3. The free piston serves as an automatic paraffin scraper.

The addition of the free piston to the gas-lift system results in these advantages:

1. Channeling of gas through the fluid and fall-back during lifting is eliminated.
2. Injection ratios per barrel of fluid are considerably lower because of a solid interface between gas and fluid phases of the system.
3. Lower injection pressures can be used, since a known and controlled volume with a known hydrostatic head is lifted each cycle.
4. Continuous swabbing action of the plunger prevents paraffin deposition.

The free piston can be applied to all wells that are within its capacity where sufficient gas and pressure are available, and to gas wells for removing liquid accumulations which hold back gas production. The maximum gross production[40] rates of wells ordinarily considered suitable for free piston operation are given in Table 6.11.

Lebeaux and Sudduth[41] have described the operation of the free piston, and have given practical limits of the device in prolonging the flowing

Table 6.11. Maximum production with free pistons
(*After Beeson, Knox, and Stoddard*[40])

Depth, ft	Maximum production, bbl/day	
	2-in. plunger	$2\frac{1}{2}$-in. plunger
4000	130	190
6000	115	175
8000	105	160
10,000	95	150
12,000	85	140

life of wells producing from volumetric reservoirs in West Texas. For the wells investigated, the theoretical lift ratio was taken as 225 SCF per STB per 1000 ft of depth and the actual gas requirements ranged from 400 to 450 SCF per STB per 1000 ft of lift. The total cost of an installation, including time cycle control, surface lubricator, and bottom stop, was approximately $1000 as compared with $15,000 for a pumping unit. The gas requirements of the free piston may be determined from the nomograph presented as Fig. 6.42.

PROBLEMS

6.1. (a) What are the four gas injection rates that are significant in continuous-flow gas lift?
(b) Make an illustrative plot of fluid production versus gas injection and indicate the point of maximum efficiency.

6.2. (a) What are the four well categories to be considered in gas-lift application?
(b) Using Fig. 6.7 as a reference, define the following terms: static fluid level, static gradient, static bottom-hole pressure, drawdown, working submergence, and productivity index.

6.3. (a) What are the three main types of gas lift installations?
(b) In what type of well is a standing valve necessary?
(c) In a well producing on continuous-flow gas lift without a packer and having a fluid seal, is the casing pressure putting additional back pressure on the formation?

6.4. (a) A well 8000 ft deep is standing full of salt water having a normal gradient of 0.465 psi per ft. What is the bottom-hole pressure? *Ans.* 3720 psi.
(b) A well having a bottom-hole pressure of 3200 psig at 12,000 ft was producing 32° API pipe-line oil at the time it went dead. What was the depth to the static fluid level? *Ans.* 3467 ft.
(c) If the well of part (b) had been producing 50 per cent salt water of specific gravity 1.10, what would have been the depth to the static fluid level? *Ans.* 4488 ft.

6.5. (a) If a well had a productivity index of 0.60 bbl fluid per day per psi and a static bottom-hole pressure of 2600 psig, what would be the flowing bottom-

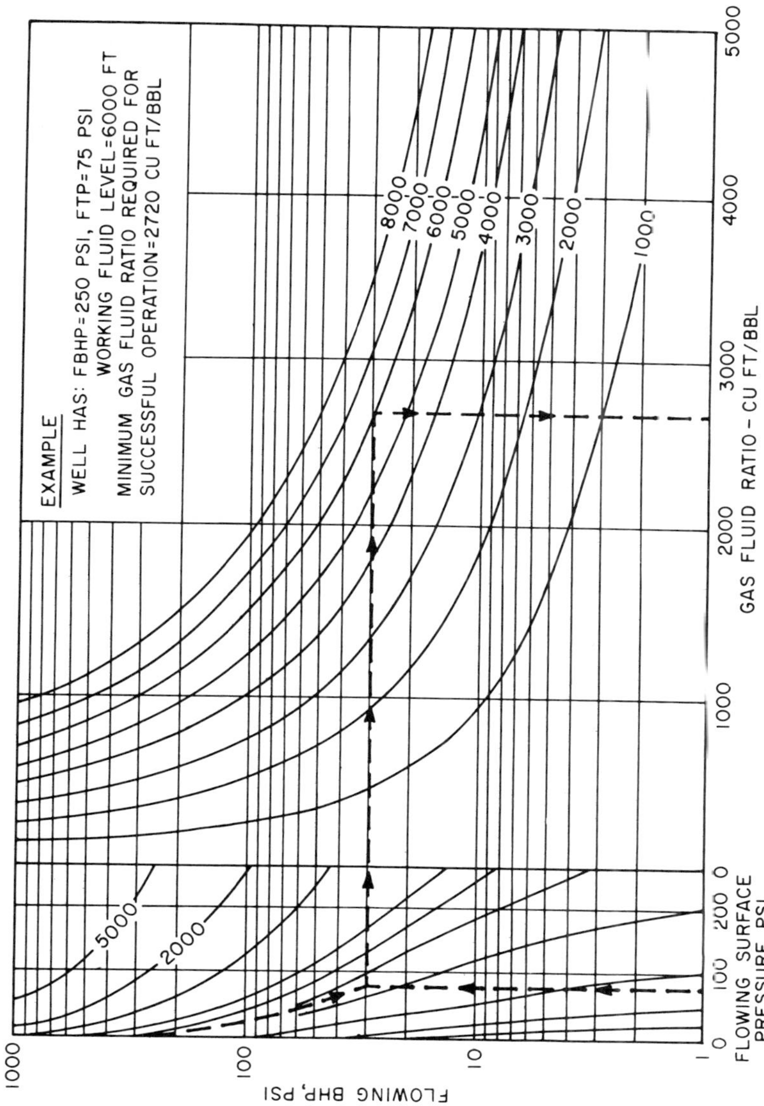

Fig. 6.42 Nomograph for estimating limits of free piston application. (Lebeaux and Sudduth,[41] *Jour. of Petr. Tech.* AIME.)

hole pressure at a production rate of 300 bbl per day? *Ans.* 1600 psig.
(b) A well having a productivity index of 2.20 will produce how many barrels per day if the pressure drop is 300 psi? *Ans.* 660 bbl/day.
(c) The productivity index would be expected to be most reliable for what type of reservoir mechanism?
(d) Explain how you measure the productivity index of a well.

6.6. From the well data of Table 6.1, what are the optimum operating conditions?

6.7. What is the force balance under operating conditions for the conventional fluid-operated valve[42] of Fig. 6.43? The notation is as follows: A_b = effective bellows area; A_s = area of stem; A_v = area of port; p_b = bellows charge pressure; p_c = injection pressure; p_t = tubing pressure.

6.8. Using the symbols given in Prob. 6.7, write the force balance under operating conditions for the partially balanced, fluid-operated valve of Fig. 6.43.

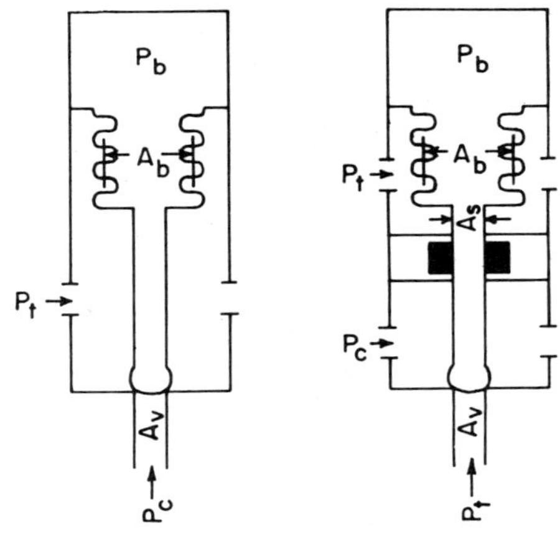

Fig. 6.43 Conventional fluid-operated valve (left) and partially balanced fluid-operated valve (right). (Winkler, *World Oil.*)

6.9. Calculate the casing pressure opposite a valve at 5000 ft using Eq. (6.15), and check with Fig. 6.13 for the following well conditions:

Geothermal gradient = 1.6°/100 ft	Wellhead pressure = 500 psia
Mean surface temperature = 70°F	Gas gravity = 0.60

Ans. 562 psia.

6.10. A pressure-operated valve without a spring has an opening pressure in a valve tester of 600 psig at 60°F. The valve is to be placed in a well at 6000 ft where the temperature opposite the valve, T_v, is 160°F.
(a) If the ratio A_v/A_b is 0.112, what is the bellows pressure at 60°F? *Ans.* 533 psig.

(b) What is the bellows pressure at 160°F? *Ans.* 642 psig.
(c) What is the *surface* closing pressure of the valve? *Ans.* 562 psig.

6.11. An operator needs the *surface* closing pressure of a pressure-operated flow valve without spring to be 600 psig when the valve is installed in a well at 6000 ft at a temperature of 160°F. The ratio A_v/A_b is 0.112.
(a) Determine the gas pressure in the casing opposite the valve at 6000 ft for a surface casing pressure of 600 psig. *Ans.* 685 psig.
(b) What is the pressure inside the bellows at 60°F? *Ans.* 570 psig at 60°F.
(c) Calculate the valve opening pressure (p_{vo}) in a tester at 60°F. *Ans.* 642 psig at 60°F.

6.12. A pressure-operated flow valve has a 75 psi spring pressure effect. The valve opening pressure at 60°F in a tester is 600 psig. The valve is to be installed in a well at 6000 ft where the temperature in the well opposite the valve is 160°F.
(a) Calculate the pressure inside the bellows, p_b, at 60°F if A_v/A_b is 0.112. *Ans.* 466 psig.
(b) What is the bellows pressure at 160°F? *Ans.* 562 psig.
(c) Calculate the closing pressure of the valve, p_{vc}, at 6000 ft. *Ans.* 629 psig.
(d) What is the surface closing pressure of the valve? *Ans.* 550 psig.

6.13. An operator needs the surface closing pressure of a pressure-operated flow valve with a 75 psi spring effect to be 600 psig when the valve is installed in a well at 6000 ft where the temperature is 160°F.
(a) Using again the value 0.112 for A_v/A_b, calculate the pressure opposite the valve at 6000 ft for a surface casing pressure of 600 psig. This is the closing pressure of the valve at the operating temperature of 160°F. *Ans.* 685 psig.
(b) What is the pressure inside the bellows at 160°F? *Ans.* 618 psig.
(c) What is the pressure inside the bellows at 60°F? *Ans.* 515 psig.
(d) Calculate the opening pressure (p_{vo}) of the valve at 60°F which would be specified in ordering from the manufacturer. *Ans.* 655 psig.

6.14. (a) Gas is being injected into the annulus between 1-in. and 2-in. tubing under the following conditions:

Surface gas injection pressure = 700 psig

Injection rate = 750 MSCF/day

Average temperature = 100°F

Average gas gravity = 0.60

Average gas compressibility = 0.97

Annulus area = 1.768 sq in.

Depth to valve = 5000 ft

What is the casing pressure at the valve? *Ans.* 402 psia.
(b) What is the average gradient? *Ans.* 0.0596 psi/ft.

6.15. (a) Define the spread of a gas-lift valve.

(b) What is the spread of a bellows valve having 0.75 sq in. of effective bellows area, 0.10 sq in. of seat area, 800 psig pressure in the dome, and zero back pressure? *Ans.* 123 psi.

(c) What is the spread with a back pressure of 700 psig? *Ans.* 15 psi.

6.16. It is desired to design an intermittent gas-lift installation from the following well data:

Depth to top of perforations = 4750 ft

Depth to packer = 4700 ft

Tubing size = 2 in.

Casing size = $5\frac{1}{2}$ in., 20 lb/ft

Kick-off gas pressure at wellhead = 650 psig

Working gas pressure = 600 psig

Wellhead back pressure = 50 psig

Well full of water

Unloading gradient = 0.5 psi/ft

Valve opening pressures at 60°F = 650, 625, 600, 575, and 550 psig

(a) What are the setting depths of each valve based on valve opening pressure at 60°F? *Ans.* 1200, 2254, 3174, 3970, and 4652 ft.

(b) If the static-fluid level is *known* to be 2500 ft from the surface with tubing in the hole, what is the setting depth of the first valve? *Ans.* 2745 ft.

(c) A closing pressure analysis of the valve string calculated above should be made to be sure that the valves as spaced will have closing pressures decreasing with depth. Complete the closing pressure analysis given in the table.

CLOSING PRESSURE ANALYSIS[a]

p_{vo} at 60°F, psig	Depth of valve, ft	T_v at depth, °F	p_b at 60°F, psig	p_{bT} at T_v, psig	p_{vc} at T_v, psig	p_{vc} at surface, psig
650	1200	89	511	541	608	592
625	2254	106	488	535	602	570
600	3174	121				
575	3970	133				
550	4652	145				

[a] $A_v/A_b = 0.112$; $S_t = 75$ psi.

6.17. (a) The following data are given on a well to be placed on intermittent flow:

Kick-off pressure = 750 psig

Working pressure = 700 psig

Fluid production desired = 50 STB/day

Separator pressure = 50 psig

Depth to operating valve = 4615 ft

Gradient of well loaded with water = 0.5 psi/ft

Injection gas gravity and temperatures as Fig. 6.13.

Spring tension = 100 psig

$A_v/A_b = 0.10$

Desired closing pressures of valves = 660, 650, 640, and 600 psig

Calculate the valve spacing on the basis of closing pressure of the valve above. *Ans.* 1400, 2548, 3622, 4616 ft.
(b) Using Eq. (6.17) to calculate the temperatures opposite the valves, make an opening pressure analysis in tabular form as in Table 6.5. *Ans.* The p_{vo} at 60°F are 711, 700, 683, 637 psig.

6.18. The following well data are given:

Oil production = 106 STB/day

Gravity of oil = 36.2° API

Gas-oil ratio = 906 SCF/STB

Gravity of gas = 0.82

Tubing size = $2\frac{1}{2}$ in.

Tubing pressure = 455 psia

Bottom-hole pressure = 1635 psia

Depth to middle of perforations = 6500 ft

Average temperature = 100°F

No water production

(a) What is the Reynolds Number factor $dv\rho$? *Ans.* 2.7.

(b) With B_t values from Fig. 6.22, evaluate $\int_{p_2}^{p_1} V_m\,dp$. *Ans.* 2.647×10^6 ft-lb/STB.
(c) Calculate the energy loss W_f for this flow. *Ans.* 1014 ft.
(d) Calculate the factor f and compare with Fig. 6.23. *Ans.* 1.5 and 2.00.
(e) Using the value of f read from the graph in part (d), calculate the constant $\overline{K}$ for an average flowing density of 42 lb/cu ft. *Ans.* 0.06127.
(f) Calculate the flowing gradient $\Delta p/\Delta h$ and compare with Fig. 6.25. *Ans.* 0.309 psi/ft.

6.19. The following data are given for a flowing well:

Depth to center of perforated interval = 5000 ft

Static bottom-hole pressure = 1600 psia

Specific gravity of oil = 0.802 (45° API)

Specific gravity of water = 1.07

Specific gravity of gas = 0.60

Initial water-oil ratio = 0

Desired production = 100 STB/day

Tubing size = 2 in. (nominal)

Producing gas-oil ratio = 500 SCF/STB

Initial surface tubing pressure = 350 psia

Productivity index = 4.0 bbl/day/psi

Surface flowing temperature = 114°F

Average flowing temperature = 124°F

Bottom-hole temperature = 134°F

(a) What is the flowing bottom-hole pressure when the well is producing 50 per cent water? *Ans.* 1550 psia.
(b) Using the volumetric balance method, calculate the flowing gradients at 300 psi intervals when the well is producing 50 per cent water. Since this is a flowing well, the procedure is similar to gas-lift calculations below the point of injection (Table 6.8).
(c) From the data of part (b) plot pressure versus depth and determine the flowing tubing pressure. *Ans.* 190 psia.

6.20. (a) For the well of Prob. 6.19, the water-oil ratio has increased to 2:1 and the well is off production. The well is to be placed on continuous-flow gas lift with a surface operating injection pressure of 700 psig and a kick-off pressure of 800 psig. Determine the optimum point of gas injection by the *graphical* method, with a differential across the valve of 100 psi and a 200 psia surface tubing pressure. *Ans.* 2930 ft.
(b) Using an injection gas-oil ratio of 500 SCF per STB, calculate the optimum point of gas injection for Prob. 6.19 using the volumetric balance technique. *Ans.* 2885 ft.

6.21. A well is to be placed on continuous-flow gas lift. The well data are as follows:

Well depth = 4900 ft

Tubing size = 2 in.

Productivity index = 2.0

Static bottom-hole pressure = 2000 psig

Production = 500 bbl of fluid per day

Flowing gradient = 0.5 psi/ft

Tubing pressure = 50 psig

Surface kick-off and operating pressure = 650 psig

Unloading gradient = 0.12 (Fig. 6.19)

Temperature at wellhead = 100°F

Temperature at 4900 ft = 149°F

Point of gas injection = 2530 ft

$A_v/A_b = 0.11$

Spring effect = 75 psig

(a) Using valves with surface closing pressures p_{vc} of 600, 580, and 550 psig, calculate the setting depth for each valve, using the valve closing pressure at valve depth. *Ans.* 1200, 2046, and 2530 ft.
(b) What is the temperature at each valve depth? *Ans.* 112, 120, and 125°F.
(c) What are the pressures inside the bellows at valve depths? *Ans.* 550, 540, and 513 psig.
(d) What are the pressures in the bellows at 60°F? *Ans.* 495, 478, and 450 psig.
(e) What are the valve opening pressures at 60°F? *Ans.* 631, 607, and 581 psig.

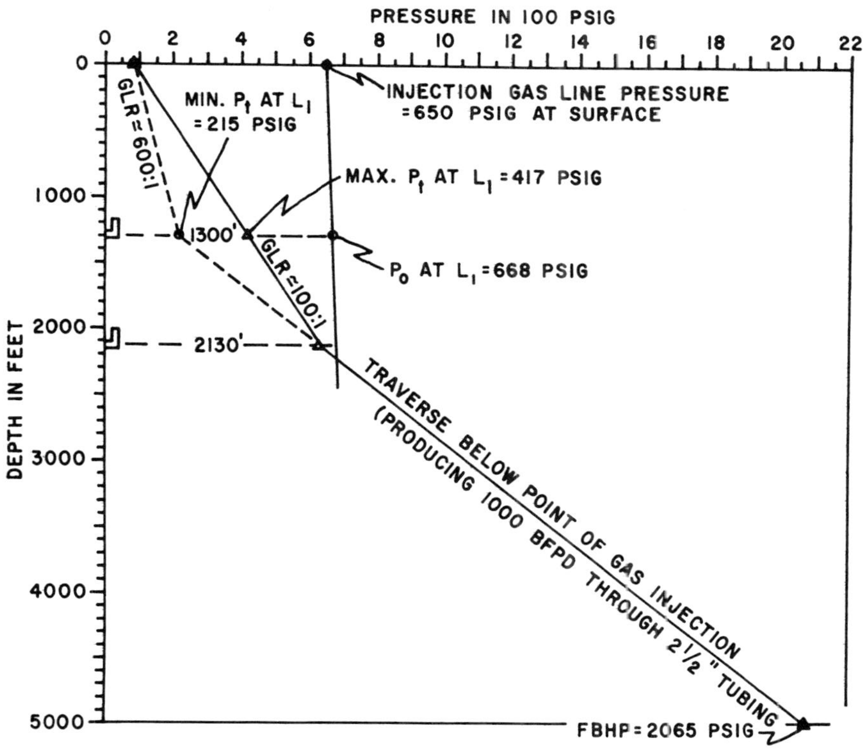

Fig. 6.44 Flowing pressure traverses in continuous-flow gas lift well. (Winkler,[43] *World Oil.*)

6.22. In design calculations for continuous-flow gas lift, the opening pressure of the unloading valves, p_{vo}, at 60°F in a tester should be based on the minimum tubing pressure at valve depth which is required to uncover and allow gas passage through the next lower valve. The bellows pressure at operating temperature for each valve should be calculated for the minimum tubing pressure during unloading. From data[43] shown in Fig. 6.44, the top valve is at a depth of 1300 ft where the pressure is 668 psig and the temperature is 116°F. For injection gas to enter the tubing through the second valve at 2130 ft with a desired differential of 50 psi across the valve, the pressure in the tubing opposite the first valve must be reduced to 215 psig by injecting a sufficient volume of gas through this valve.

(a) Calculate the nitrogen charge pressure at the operating temperature of the top valve p_{bT}, using Eq. (6.3). The spring effect is 75 psig and $A_v/A_b = 0.112$. *Ans.* 551 psig at 116°F.

(b) Calculate the nitrogen charge pressure of the top valve at 60°F. *Ans.* 492 psig.

(c) Calculate the opening pressure at 60°F (p_{vo} at 60°F) of top valve in a tester, using Eq. (6.5). *Ans.* 629 psig at 60°F.

6.23. A well is producing 800 bbl of fluid per day through $2\frac{1}{2}$-in. tubing from a total depth of 6000 ft. The geothermal gradient for the area is 1.2°F/100 ft and the bottom-hole temperature is 142°F. From Fig. 6.27, what is the surface flowing tubing temperature? *Ans.* 92°F.

6.24. (a) A continuous-flow gas-lift analysis is given in Table 6.8. What will be the value of m and $q_o m$ if the injection gas–oil ratio is 2020 SCF per STB? *Ans.* 4440 lb/STB; 177,600 lb/day.

(b) Using pressure points of 100, 300, 500, 700, and 900 psia, calculate the depths to the pressure points. What is the depth to the point of gas injection, and what is the injection pressure? *Ans.* 5000 ft; 1200 psia.

(c) What is the required theoretical adiabatic horsepower for the conditions of part (b)? *Ans.* −10.9 hp.

6.25. A low productivity index, relatively high bottom-hole pressure well is to be placed on intermittent-flow gas lift. The well data are as follows:

Depth to middle of perforations = 9000 ft
Available surface operating pressure = 900 psig
Tubing size = 2 in. (capacity 0.00387 bbl/ft)
Casing size = $5\frac{1}{2}$ in., 20 lb/ft
Gravity of oil = 34° API ($\gamma_o = 0.855$)
Static bottom-hole pressure = 2500 psig
Well productivity index = 0.20
Tubing pressure = 30 psig
Oil production desired = 100 bbl/day
Wellhead temperature = 100°F
Average tubing temperature = 142°F
Minimum cycle time = 35 min

(a) What is the average flowing bottom-hole pressure? *Ans.* 2000 psig.
(b) What is the depth to the static fluid level? *Ans.* 2324 ft.
(c) How many cycles will be required per day? *Ans.* 41 cycles.
(d) What is the depth to the working fluid level? *Ans.* 3675 ft.
(e) Assuming that 40 per cent of the starting slug is produced, what will be the depth to the operating valve? *Ans.* 4463 ft.
(f) What is the pressure in the tubing opposite the valve when it opens? *Ans.* 613 psig.
(g) What is the desired casing pressure opposite the valve? *Ans.* 920 psig.
(h) What is the volume of gas space in the tubing? *Ans.* 11.17 bbl.
(i) What is the minimum injection gas–oil ratio? *Ans.* 1030 SCF per STB.

6.26. (a) What are the two common types of chambers?
(b) When should each be run?
(c) Do you think a chamber installation would entirely deplete a well?
(d) Calculate the chamber length for the following well data:

Injection pressure = 400 psig

Wellhead tubing back pressure = 50 psig

Fluid gradient = 0.4 psi/ft

Insert chamber = 1-in. tubing inside, 4-in. OD tubing

Capacity of 1-in. tubing = 1.07 bbl/1000 ft

Annular capacity between 1-in. dip tube and 4-in. tubing = 13.54 bbl/1000 ft

Ans. 64.1 ft.

6.27. What type of installation would you recommend for the wells specified in the following table?

Well	Depth, ft	Static bottom-hole pressure, psig	Productivity index	Separator pressure, psig	Fluid gradient, psi/ft
(1)	7000	3000	10	50	0.50
(2)	2200	250	0.1	10	0.35
(3)	5000	200	2	30	0.40
(4)	6000	2500	0.02	100	0.35
(5)	12,000	1500	0.2	100	0.45
(6)	1500	65	0.5	10	0.33
(7)	500	10	0.1	0	0.33

REFERENCES

[1]Rev. S. J. M. Eaton, *Petroleum: A History of the Oil Region of Venango County, Pennsylvania* (Philadelphia: S. J. Skelly and Company, 1866), pp. 136–137.

[2]Robert Stirling, "The Air-Lift System of Raising Oil," *J. Inst. Petrol. Technol.*, **6**, 379 (1920).

[3]H. G. Schneider, "Smackover Oil Field, Ouachita and Union Counties, Arkansas," *Trans. AIME*, **70**, 1076 (1924).

[4]S. F. Shaw, "Use of Gas Lifts in the Mid-Continent," *Trans. AIME*, **G25**, 99 (1926).

[5]S. F. Shaw, *Gas Lift Principles and Practice* (Houston: Gulf Publishing Company, 1939), pp. 1–3.

[6]M. E. Nicklin, *Something About Gas Lift* (Waukesha: The Waukesha Motor Company, 1928).

[7]John R. Rayne, B. C. Craft, and Murray F. Hawkins, Jr., "The Reservoir Mechanism of Sulfur Recovery," *Trans. AIME*, **207**, 247 (1956).

[8]S. F. Shaw, "Some Observations on Principles Involved in Flowing Oil Wells," *Trans. AIME*, **86**, 1–47 (1930).

[9]Sylvan Cromer and R. L. Huntington, "Visual Studies of the Flow of Air-Water Mixtures in a Vertical Pipe," *Trans. AIME*, **136**, 79–88 (1940).

[10]S. F. Shaw, "Flow Characteristics of Gas Lift in Oil Production," Bull. No. 113, *Texas Engineering Experiment Station* (1949), pp. 1–9, 26, 47.

[11]C. V. Kirkpatrick, "Advances In Gas-Lift Technology," *Drilling and Production Practice* (API, 1959), p. 56.

[12]F. C. Babson, "The Range of Application of Gas-Lift Methods," *Drilling and Production Practice* (API, 1939), pp. 266–307.

[13]C. V. Kirkpatrick, *The Power of Gas*, 2nd Ed. (Houston: Camco, Inc., 1955) pp. 38, 40, 37, 58.

[14]*Gas Lift Practices and Principles*, Bull. No. 457, Merla Tool Corporation, (Dallas, 1959), pp. 78, 120, 121.

[15]Bruce H. Sage and William N. Lacy, *Thermodynamic Properties of the Lighter Paraffin Hydrocarbons and Nitrogen* (New York: American Petroleum Institute, 1950), pp. 85-97.

[16]B. C. Craft and M. F. Hawkins, *Applied Petroleum Reservoir Engineering* (Englewood Cliffs: Prentice-Hall, Inc., 1959), p. 26.

[17]H. W. Winkler, "Here's How to Improve Your Gas Lift Installations," *World Oil*, **44**, No. 2, 64 (August 1, 1959).

[18]Marshall B. Standing and Donald L. Katz, "Density of Natural Gases," *Trans. AIME*, **146**, 144 (1942).

[19]*Tentative Standard Procedure for the Determination of Superexpansibility and Manometer Factors Used in Measurements of Natural Gas by Orifice Meter at Pressures in Excess of 500 Psig*, Bull. TS-461, California Natural Gasoline Association (1947), pp. 15–31.

[20]Paul Buthod and B. W. Whitley, "Graphic Solutions of Design Problems — Flow of Gases in Vertical Pipes," *The Oil and Gas Journal*, **44**, No. 33, 70 (December 22, 1945).

[21]C. V. Kirkpatrick, "A Study of Wells Producing Through Small Conduits," Paper No. 59-Pet-39 presented before the Petroleum Division, ASME, 1959; pp. 2–29. With the permission of the publisher, The American Society of Mechanical Engineers, 29 W. 39th St., New York, N. Y.

[22]C. V. Kirkpatrick, "Spacing Pressure-Loaded Gas-Lift Valves," *The Oil and Gas Journal*, **53,** No. 40, 110 (February 7, 1955).

[23]*Gas Lift Manual* (Dallas: Otis Engineering Corporation, 1960), Chart No. 4.

[24]Purvis J. Thrash (Otis Engineering Corporation), personal communication.

[25]J. Versluys, "Mathematical Development of the Theory of Flowing Wells," *Trans. AIME*, **86,** 192 (1930).

[26]T. V. Moore and H. D. Wilde, "Experimental Measurement of Slippage in Flow Through Vertical Pipes," *Trans. AIME*, **92,** 279 (1931).

[27]T. V. Moore and R. J. Schilthuis, "Calculation of Pressure Drops in Flowing Wells," *Trans. AIME*, **103,** 170 (1933).

[28]C. J. May, "Efficiency of Flowing Wells," *Trans. AIME*, **114,** 99 (1935).

[29]J. E. Gosline, "Experiments on the Vertical Flow of Gas-Liquid Mixtures in Glass Pipes," *Trans. AIME*, **118,** 56 (1936).

[30]E. C. Babson, "The Range of Application of Gas-Lift Methods," *Drilling and Production Practice* (API, 1939), p. 266.

[31]Fred H. Poettmann and Paul G. Carpenter, "The Multiphase Flow of Gas, Oil, and Water Through Vertical Flow Strings With Application to the Design of Gas-Lift Installations," *Drilling and Production Practice* (API, 1952), pp. 257–270.

[32]W. E. Gilbert, "Flowing and Gas-Lift Well Performance," *Drilling and Production Practice* (API, 1954), p. 126.

[33]M. B. Standing, *Volumetric and Phase Behavior of Oil Field Hydrocarbon Systems* (New York: Reinhold Publishing Corporation, 1952), pocket on back cover.

[34]P. B. Baxendell, "Producing Wells on Casing Flow—An Analysis of Flowing Pressure Gradients," *Trans. AIME*, **213,** 202 (1958).

[35]A. L. Benham and F. H. Poettmann, "Gas-Lifting Through the Annulus of a Well," *The Petroleum Engineer*, **31,** No. 7, B25–B30 (July, 1959).

[36]Harold L. Cook, Jr., with Frank H. Dotterweich, "Report on the Calibration of Positive Flow Beans as Manufactured by Thornhill-Craver Company, Houston, Texas," Department of Engineering, Texas College of Arts and Industries (1946), p. 26.

[37]*Handbook of Gas Lift* (Longview, Texas: Garrett Oil Tools, Division of U. S. Industries, Inc., 1959), p. 837.

[38]H. W. Winkler and George F. Camp, "Down-Hole Chambers Increase Gas-Lift Efficiency," *The Petroleum Engineer*, **29,** No. 6, B87–B94 (June, 1956).

[39]E. D. McMurray, "Use of the Automatic Free Piston in Oil Well Production Problems," *Trans. AIME*, **198,** 165 (1953).

[40]Carrol M. Beeson, Donald G. Knox, and John H. Stoddard, "The Plunger Lift Method of Oil Production," *The Petroleum Engineer*, **30,** No. 6, B96 (June, 1958); No. 7, B68 (July, 1958); No. 9, B58 (August, 1958); No. 10, B76 (September, 1958); No. 11, B106 (October, 1958).

[41]J. M. Lebeaux and L. S. Sudduth, "Theoretical and Practical Aspects of Free Piston Operation," *J. Petrol. Technol. AIME*, **7,** No. 9, 33–37 (September, 1955).

[42]H. W. Winkler, "More Efficient Dual Gas-Lift Installations," *World Oil*, **146,** No. 6, 159 (May, 1958).

[43]H. W. Winkler, "Here's How to Improve Your Gas Lift Installations," *World Oil*, **150,** No. 1, 129 (January, 1960).

CHAPTER SEVEN

Fluid Separators

7.1. Oil and Gas Separators

Closed vessels called *separators* are used to remove liquids from gases. Modern oil-field handling of crude oil includes metering for royalty and tax payments, which requires that all nonsolution gas be removed from the crude oil prior to metering. In addition, sale of the dry gas in many cases is a source of considerable income. Oil-field lease separations range from the removal of slugs of oil from gas to dehydration of the gas; however, in this chapter only the mechanical separation of gases and liquids is considered. Oil and gas separators handle large masses of material with a very high efficiency. For example, with a wellstream of 10 MM SCF per day of gas of specific gravity 0.70 and 1000 bbl per day of 35° API oil, the separator is handling 270 tons of gas and 150 tons of liquid per day. The usual oil-field separator allows a maximum liquid carryover of 0.10 gal of liquid per million standard cubic feet of gas, or a maximum of about 10 parts of liquid carryover per million parts of material handled.

Separator components. For efficiency and stable operation over a wide range of conditions, a gas-liquid separator must have the components to be described*[,1] (see Fig. 7.1).

1. *Primary separation section* for separating the bulk of the liquid from the wellstream. It is desirable to remove quickly liquid slugs and large droplets of liquid from the gas stream in order to minimize turbulence and

*References throughout the text are given at the end of each chapter.

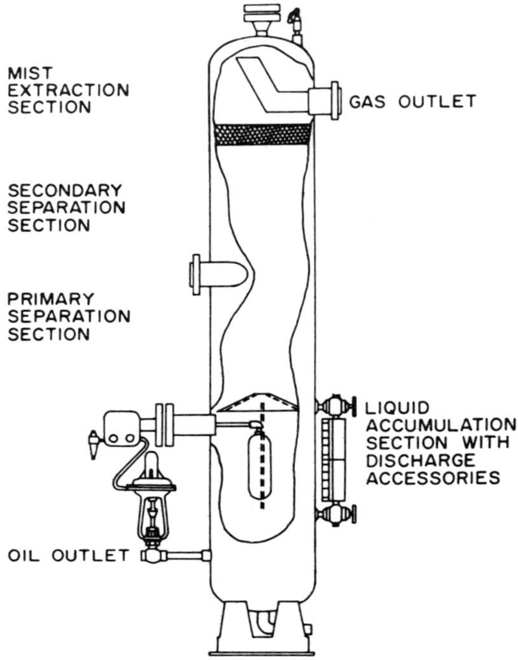

Fig. 7.1 Typical vertical oil and gas separator with mist extractor. (Worley and Laurence, *Journal of Petroleum Technology*, AIME.)

re-entrainment of liquid particles. This is accomplished by the use of a tangential inlet which imparts a circular motion to the fluids.

2. *Liquid accumulation section* for receiving and disposing of the liquids collected. This section should be of sufficient volume to handle fluid surges that may occur in normal operation and should be arranged so that the separated liquid is not disturbed by the flowing gas stream. When large amounts of liquid are handled or severe fluid surges occur, it may be necessary to enlarge the liquid accumulation section.

3. *Secondary separation section* for removing the smaller liquid droplets. The major principle of separation in this section is gravity settling from the gas stream. Since the basic requirement for gravity settling is minimization of turbulence, it is important to decrease the gas velocity right at the entrance to the separator. Properly positioned straightening vanes provide uniform gas flow throughout the separation section.

4. *Mist extraction section* for removing entrained droplets too small to settle by gravity. *Entrained* drops are those carried over from the secondary separation section, when the vapor velocity is greater than the rate of settling of the drops. *Impingement* is the most widely used principle for collection of small particles in liquid-gas separation. In this type of

separation entrained particles strike obstructions which act as collecting surfaces. The usual oil-field separator employs two general types of mist extractors, the difference between the two being the intensity of the centrifugal force. The centrifugal force resulting from a change in direction of flow exerts the same effect on the mist as would an increase in the force of gravity. When a gas stream approaches an obstruction it veers around it, but the liquid droplets, having a greater mass density, offer more

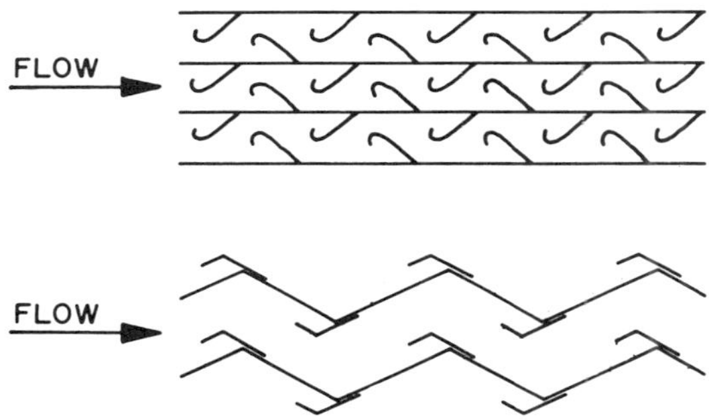

Fig. 7.2 Vane-type mist extractors. (Campbell, *The Oil and Gas Journal.*)

resistance to change in direction and tend to continue in a straight line. Thus the larger particles collide with the confining wall (impingement) and are separated from the gas. The vane-type extractor[2] (Fig. 7.2) consists of a labyrinth formed of parallel metal plates with liquid collection pockets. The coalescence of small particles into droplets large enough to settle by gravity occurs aided by the centrifugal force and the collection surfaces. The liquid that is accumulated within the vanes is returned to the liquid accumulation section through the drain-back pipe.

A second type of separation device is a knitted-wire mesh pad[3] shown in Fig. 7.3. The primary mechanism of separation is impingement; however, centrifugal and gravitational forces are also used in the collection of small liquid particles. A liquid particle striking the metal surface flows downward through capillary space provided by adjacent wires. At these points liquid collects and continues to flow downward. Surface tension holds the drops on the lower face of the pad until

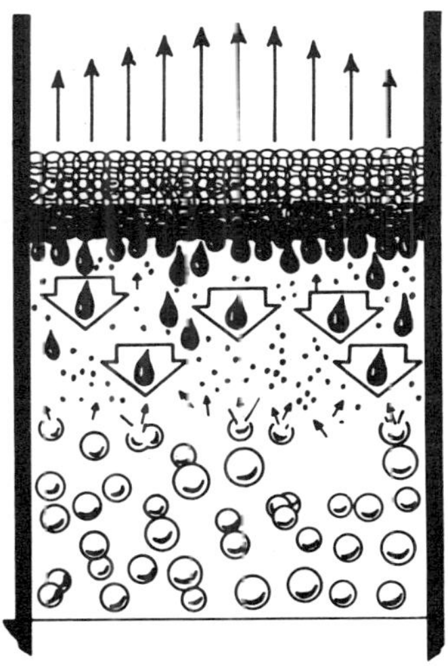

Fig. 7.3 Mist eliminator. (Courtesy of Metal Textile Corporation.)

they are so large that the force of gravity exceeds the forces due to the upward gas velocity and the surface tension. The knitted-wire mesh has high gas and liquid handling capacities, and its removal efficiency increases with gas velocity up to the flooding point.

Types of separators. The separators of various manufacturers differ slightly in detail, but may be classified according to shape as vertical, horizontal, or spherical. In the vertical separator shown in Fig. 7.1, the wellstream enters the separator through the tangential inlet which imparts a circular motion to the fluids. Centrifugal and gravitational forces provide efficient primary separation. A conical baffle separates the liquid accumulation section from the primary separation section to ensure a quiet liquid surface for liquid level float operation and an easy release of solution gas. The separated gas travels upward through the secondary separation section where the heavier entrained liquid particles settle out. The gas then flows through a vane-type or knitted-wire mist extractor where liquid particles 10 microns (1 micron = 0.001 mm or 1/25,400 in.) and larger in diameter accumulate until they are of sufficient weight to fall into the liquid accumulation section. Sediments entering the separator accumulate in the bottom where they are flushed out through the drain connection.

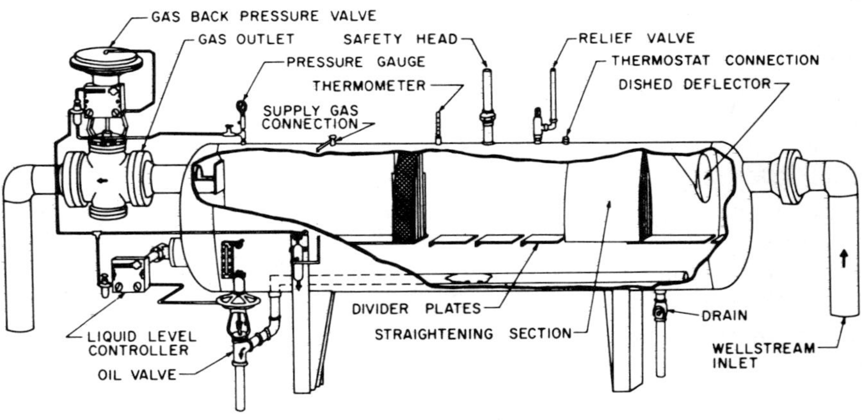

Fig. 7.4 High-pressure horizontal oil and gas separator. (Courtesy of Black, Sivalls, and Bryson, Inc.)

Horizontal separators are of single-tube or double-tube design. In the single-tube horizontal separator (Fig. 7.4), the wellstream enters through the inlet and strikes an angle baffle or dished deflector. The wellstream then strikes the side of the separator, producing maximum initial separation of gas and liquid. The separated liquid drains into the liquid accumulation section. Horizontal divider plates separate the liquid accumulation and gas separation sections to ensure quick removal of solution gas. The separated gas flows through a straightening section which removes turbulence, and the additional liquid particles settle out as the gas stream flows through the secondary separation section. The gas then passes through the mist extractor where liquid particles 10 microns and larger in size are removed.

The horizontal separator[4] shown in Fig. 7.5 is of the double-tube type and consists of an upper separator section and lower liquid chamber.

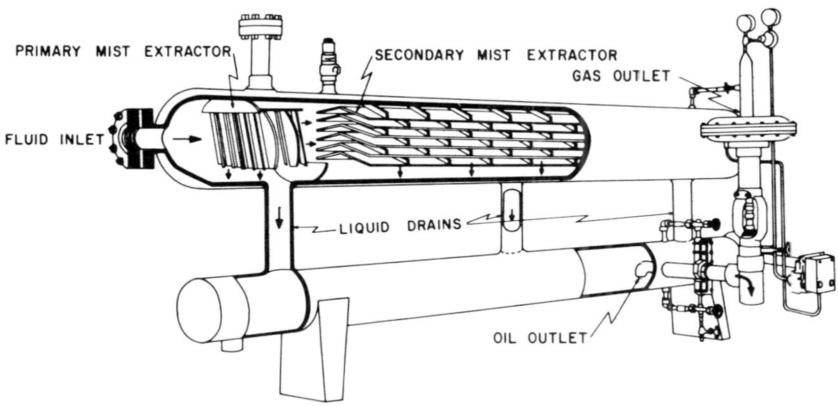

Fig. 7.5 Double-tube horizontal oil and gas separator. (Courtesy of American Petroleum Institute.)

The mixed stream of oil and gas enters the upper tube. Liquids fall through the first connecting pipe into the liquid reservoir and wet gas flows through the upper tube where the entrained liquids separate owing to their difference in density and to the scrubbing action of the mist extractor.

The advantages claimed for single-tube horizontal separators are (1) lower initial cost; (2) they are easier to insulate for cold-weather operation; and (3) the liquid remains warmer, minimizing freezing and paraffin deposition. Proponents of double-tube horizontal separators point to (1) the larger capacity under surging conditions; (2) the better separation of solution gas in the quiescent lower chamber; (3) the better separation of gases and liquids of similar densities; (4) the more stable liquid-level control.

The spherical separator (Fig. 7.6) is designed to make maximum use of all the known means and aids to oil and gas separation, i.e., gravity, low velocity, centrifugal force, and surface contact. The incoming wellstream is split by the inlet flow diverter and directed tangentially against the wall

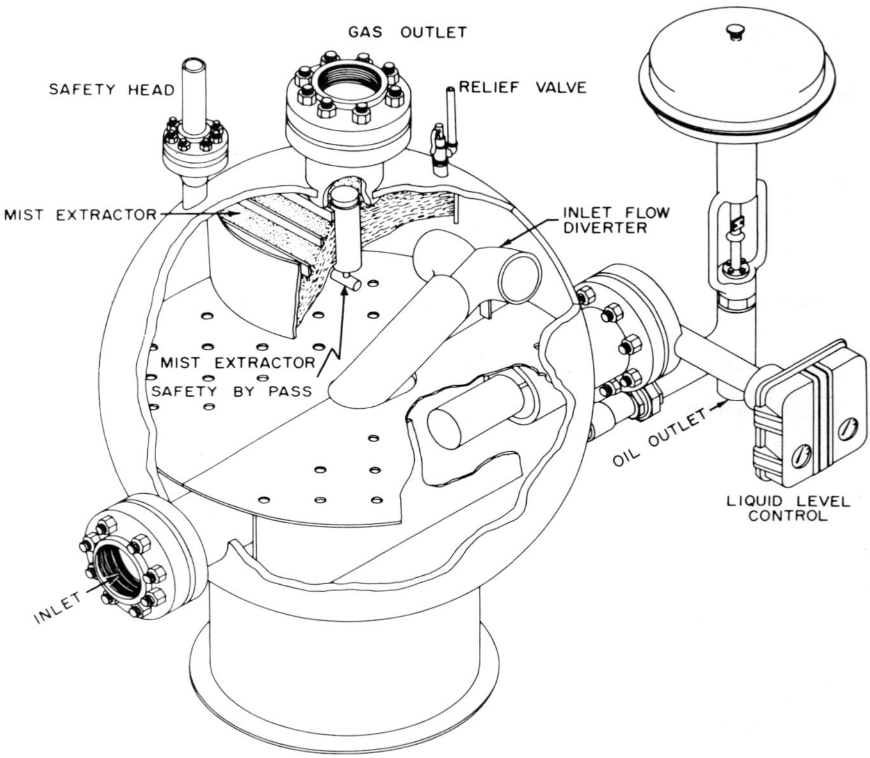

Fig. 7.6 Low-pressure spherical oil and gas separator. (Courtesy of Delta Tank Manufacturing Co., Inc.)

of the separator. The liquid streams come together after flowing 180° around the vessel wall and then fall into the accumulation section to remain there until released through the liquid-level-actuated dump valve. The gas stream, in traveling through the large-diameter central portion of the vessel, loses entrained particles due to its reduced velocity. The final disentrainment takes place as the gas flows through the wire mesh mist extractor. Should plugging of the mist extractor by hydrates or paraffin occur, the safety by-pass protects the wire mesh element by limiting the differential pressure to a safe maximum.

The advantages of each type of separator for the same conditions have been summarized by Campbell[5] as follows:

Vertical Separator

Liquid level control not as critical.
Will handle large quantities of sand.
Easier to clean.
Has greater liquid surge capacity.
Less tendency for revaporization of liquid.

Horizontal Separator

Successfully used in handling foaming crudes.
Cheaper than vertical separator.
Easier to ship on skid assemblies.
More economical and efficient for processing large volumes of gas.
Smaller diameter for a given gas capacity.

Spherical Separator

Cheaper than either horizontal or vertical types.
Better clean-out and bottom drain features than vertical type.
More compact than other types.

Factors influencing separation. The operating pressure of the separator is dependent upon both the flowing tubing pressure and the relative amounts of crude oil and natural gas. A change in pressure effects changes in both the gas and liquid densities, in the allowable velocity, and in the actual flowing volume. The net effect of an increase in pressure is an increased gas capacity of the separator, expressed in standard cubic feet.

Temperature affects gas-liquid separator capacities only as it affects the actual flowing volumes and densities of the gas and liquid. The net effect of an increase in temperature during separation is a decrease in capacity. Temperature control usually involves cooling, as wellstream flowing temperatures are generally above the optimum separation temperature. Cooling is accomplished by one or more of the following means: (1) tubular heat exchangers, (2) cooling tower or evaporative-type cooler, (3) refrigeration, (4) expansion of the wellstream through a choke. Expansion is the cooling system most widely used, because high pressure gas is becoming more common, and little capital outlay is required.

The efficiency of particle collection varies with the gas and liquid densities. A separator operating at constant temperature, pressure, and wellstream composition has a gas capacity proportional to the square root of the difference in the densities of the liquid and gas divided by the density of the gas, $\sqrt{(\rho_l - \rho_g)/\rho_g}$.

Separator capacity ratings assume *gravity* separation of droplets larger than 200 microns in the secondary separation section, smaller droplets being removed in the mist extractor. The effect of gas velocities in excess of that assumed in the separator rating is to increase the size and volume of liquid droplets that reach the mist extractor. Since the volume of droplets varies as the cube of the diameter, a small increase in velocity results in much larger liquid volumes at the mist extractor. When the maximum capacity of the mist extractor is reached, the extractor floods and there is a sudden and large increase in liquid carryover.

The maximum gas velocity for the separation of liquid particles of certain diameters is based on the physical properties of the liquid and the gas. A particle falling under the action of gravity accelerates until the friction on the particle, due to collisions with the gas, equals the particle weight. When these two forces are equal, the particle will fall at a constant rate called the *settling* velocity. When the falling particles are small, they attain their settling velocity quickly, and this velocity is used to determine the time needed for a particle to fall a given distance. Table 7.1 shows the diameter ranges of various types of particles.[1] Particles smaller than 2 microns suspended in air are often considered to be permanent suspensions because they cannot be seen with the naked eye and because of their extremely low settling velocities.

Lapple[6] has presented the sizes of spherical particles suspended in air at atmospheric temperature and pressure (see Fig. 7.7). Also shown are four laws from which settling rates may be calculated depending upon the

Table 7.1. DIAMETER RANGES OF VARIOUS PARTICLES

	Particle diameter, microns[a]
Molecules[b]	Up to 0.004
Smoke	0.0045 to 1
Mist	0.075 to 10
Fog	1 to 100
Dust	1 to 2000
Cocoa	8 to 10
Talcum powder	10
Human hair	50 to 70
Table salt	100
Raindrops	100 to 4000

[a]One micron = 0.0001 cm.
[b]Hydrocarbons up to mol wt $\backsim$ 300.

system and particle diameters. In order of decreasing particle sizes for which they are valid, the laws are: Newton's law, the intermediate law, Stokes' law, and the Stokes-Cunningham law. When the settling particles are large they settle so rapidly that considerable turbulence occurs, and Newton's law is used to calculate the rate of fall. Intermediate-size

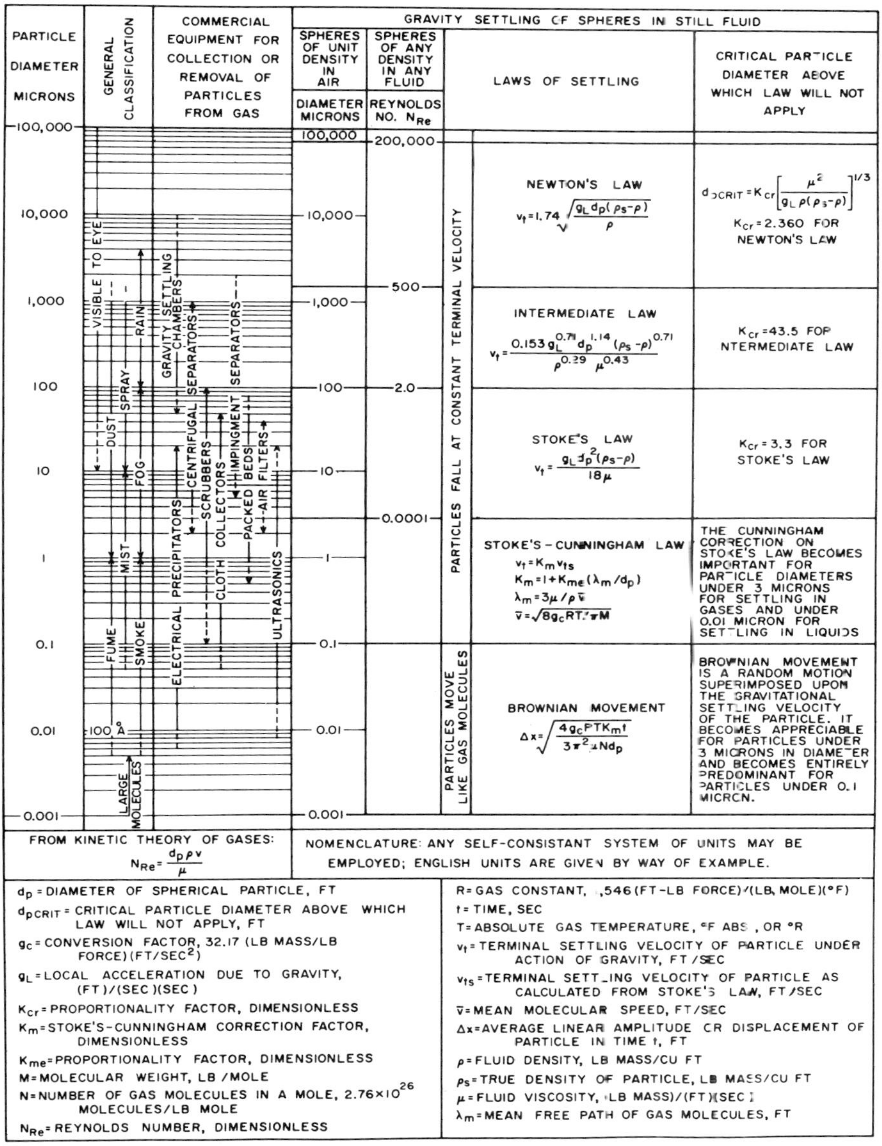

Fig. 7.7 Gravity settling of spheres in still fluid. (After Lapple,[6] *Chemical Engineers' Handbook.*)

particles create some turbulence and the intermediate law is used to calculate settling velocities. Very small particles settle much more slowly and the settling velocities are calculated from Stokes' law. The Cunningham correction to Stokes' law is made where the falling particles are so small that they slip between the gas molecules at a rate faster than that predicted by Stokes' law. Choice of the settling law solely on the basis of particle size range (Fig. 7.7) is valid only for spherical particles having the same density as water and suspended in air. When either (1) the density of the particle, (2) the shape of the particle, (3) the density of the continuous phase fluid, or (4) the viscosity of the continuous phase fluid differs from that assumed in the figure, the particle size range to which the various laws apply also changes. The particular law that best describes the settling conditions is determined by the Reynolds number given in Col. (5) of Fig. 7.7. This dimensionless parameter is obtained by multiplying the particle diameter by the continuous phase fluid density and the velocity of the particle, and dividing by the viscosity of the continuous fluid.

Viscosity affects the various mechanisms of separation in accordance with the settling law that applies. For example, viscosity has no effect on terminal velocities in the range where Newton's law applies; however, it does affect the terminal velocities in the ranges covered by both the intermediate law and Stokes' law, as well as determining the Reynolds number. As the pressure increases or the temperature decreases, the viscosity of the gas increases.

Separator capacity. The capacities of separators are determined under the following conditions:

1. No liquid foaming or undue surging.
2. Adequate line and valve capacities.
3. The operating temperature of 60°F is above the cloud point of the oil and the hydrate point of the gas.
4. The smallest separable liquid particles are spheres 10 microns (0.00039 in.) in diameter.

Under these conditions, the liquid carryover should not exceed 0.10 gal per million standard cubic feet.

The separator's *oil capacity* q is based on the relationship between the normal oil volume V and the retention time t in the separator, which is usually one minute to allow gas and water to separate from the oil:

$$q = \frac{V \text{ cu ft}}{t \text{ min}} \tag{7.1}$$

Since 1 cu ft per min is equal to 257 bbl per day,

$$q = 257 \frac{V}{t} \text{ bbl/day} \tag{7.2}$$

The rated oil capacity is taken to be one half the actual capacity, to allow for normal heading flows. Equation (7.2) becomes

$$q_r = 128\,\frac{V}{t}\,\text{bbl/day} \tag{7.3}$$

The oil volume of a vertical separator is

$$V = 0.785d^2h \tag{7.4}$$

where d is the inside diameter of the separator in feet and h is the height of the oil column in feet above the bottom of the oil outlet; for a shell height of 5 ft, $h = 2.5$ ft; for a 10 ft shell $h = 3.25$ ft; and for 15 ft shell $h = 4.25$ ft. Substituting Eq. (7.4) into Eq. (7.3),

$$q_r = 100.5\,\frac{d^2h}{t} \tag{7.5}$$

for a vertical separator. For a horizontal single-tube separator, the oil volume is

$$V = 0.785d^2\,\frac{L}{2} \tag{7.6}$$

and hence, substituting Eq. (7.6) into Eq. (7.3), the rated oil capacity is

$$q_r = 50.24\,\frac{d^2L}{t} \tag{7.7}$$

For the horizontal, double-tube separator, the oil volume is

$$V = 0.785d^2L \tag{7.8}$$

Substituting Eq. (7.8) into Eq. (7.3), the rated oil capacity is

$$q_r = 100.5\,\frac{d^2L}{t} \tag{7.9}$$

The oil volume of a spherical separator, which is taken as one half the total volume, is given by

$$\begin{aligned} V = \frac{4}{6}\,\pi r^3 \sqrt{\frac{d}{2}} &= \frac{4 \times 3.1416}{6 \times 8}\,d^3 \sqrt{\frac{d}{2}} \\ &= 0.2618d^3 \sqrt{\frac{d}{2}} \end{aligned} \tag{7.10}$$

The volume is increased by the factor $\sqrt{d/2}$ because the spherical separator. owing to its shape, has greater surge capacity. Substituting Eq. (7.10) into Eq. (7.3), the rated oil capacity is

$$q_r = 33.51\,\frac{d^3}{t}\sqrt{\frac{d}{2}} \tag{7.11}$$

The separator's *gas capacity* is related to the entrainment (or suspending) velocity. The upward velocity of a gas stream required for suspension of a particle is determined from the particle's resistance to the moving gas and the force on the particle due to gravity. The resistance of a sphere in a moving gas, neglecting the viscosity of the gas which is small, is given by Souders and Brown[7] as

$$F_a = \frac{k\rho_g \pi d^2 v^2}{4} \tag{7.12}$$

where F_a = total force on the particle

k = constant to be evaluated empirically

d = diameter of particle

ρ_g = density of the gas

v = linear velocity of gas relative to the particle

The force due to gravity on a spherical particle, less buoyancy, is

$$F_g = \frac{\pi d^3}{6} (\rho_o - \rho_g)g \tag{7.13}$$

where ρ_o is the density of the oil particle and g is the acceleration due to gravity. When the gravitational force equals the resistance to the moving gas, the particle remains suspended:

$$F_a = F_g = \frac{k\rho_g \pi d^2 v^2}{4} = \frac{\pi d^3}{6} (\rho_o - \rho_g)g \tag{7.14}$$

and the theoretical suspending velocity is

$$v = \left[\frac{2gd(\rho_o - \rho_g)}{3k\rho_g}\right]^{1/2} \tag{7.15}$$

If d and k are constants in Eq. (7.15),

$$v = C\left[\frac{\rho_o - \rho_g}{\rho_g}\right]^{1/2} \tag{7.16}$$

where v = entrainment velocity, feet per second

C = separation coefficient = $\sqrt{2gd/3k}$

ρ_o = density of the oil under separator conditions, pounds per cubic foot

ρ_g = density of gas under separator conditions, pounds per cubic foot

The volume of gas flowing, q, in cubic feet per second is

$$q = \frac{q_{sc}}{86{,}400} \times \frac{\rho_{gsc}}{\rho_g} \tag{7.17}$$

where q_{sc} is the gas flow rate in standard cubic feet per day, ρ_{gsc} is the gas density at standard conditions in pounds per cubic foot, and ρ_g is the density of the gas at flowing conditions in pounds per cubic foot. The flow area A in square feet is related to the gas flow rate and the allowable gas velocity by

$$A = \frac{q}{v} = \frac{\pi d^2}{4} \tag{7.18}$$

where d is the inside diameter of the separator in feet. Substituting Eqs. (7.16) and (7.17) into Eq. (7.18),

$$\frac{\pi d^2}{4} = \frac{q_{sc}}{86{,}400} \times \frac{\rho_{gsc}}{\rho_g}\left[\frac{\rho_g}{\rho_o - \rho_g}\right]^{1/2}\frac{1}{C} \tag{7.19}$$

The gas density at flowing conditions is given by

$$\rho_g = \rho_{gsc} \times \frac{p}{p_{sc}} \times \frac{T_{sc} + 460}{T + 460} \times \frac{1}{z} \tag{7.20}$$

or

$$\frac{\rho_{gsc}}{\rho_g} = \frac{p_{sc}}{p} \times z \times \frac{T + 460}{T_{sc} + 460} \tag{7.21}$$

where p is the operating separator pressure in psia, p_{sc} is the base pressure in psia, and z is the compressibility factor. Substituting Eq. (7.21) into Eq. (7.19),

$$\frac{\pi d^2}{4} = \frac{q_{sc}z}{86{,}400} \times \frac{p_{sc}}{p} \times \frac{T + 460}{T_{sc} + 460}\left[\frac{\rho_g}{\rho_o - \rho_g}\right]^{1/2}\frac{1}{C}$$

Solving for q_{sc},

$$q_{sc} = \frac{67{,}824Cd^2}{z} \times \frac{p}{p_{sc}} \times \frac{T_{sc} + 460}{T + 460}\left[\frac{\rho_o - \rho_g}{\rho_g}\right]^{1/2} \tag{7.22}$$

Some manufacturers base separator sizes on the outside diameter so that a correction for wall thickness is necessary. From Eq. (7.22), q_{sc} is proportional to d^2 for a given wellstream, or

$$q_1 = q_2(d_1^2/d_2^2)$$

where q_1 is the desired capacity for a separator with inside diameter d_1, and q_2 is the known capacity for a separator with inside diameter d_2.

Equations (7.16) through (7.22) can be rearranged to yield solutions for the diameter of separator or mist extractor required for various operating conditions. The gas density in pounds per cubic foot at operating conditions based on gas gravity γ is given by

$$\rho_g = \frac{\rho_{asc}\,\gamma T_{sc}\,p}{zTp_{sc}} \tag{7.23}$$

where ρ_{asc} is the density of air at standard conditions. For standard conditions of 14.7 psia and 520°R, ρ_{asc} is 0.0764 lb/cu ft, and Eq. (7.23) will reduce to

$$\rho_g = \frac{2.70\gamma p}{zT} \tag{7.24}$$

where γ = gas gravity (for air, equal to 1)

p = operating pressure, psia

z = compressibility factor, dimensionless

T = operating temperature, °R

The allowable velocity is

$$v_a = C\left[\frac{\rho_o - \rho_g}{\rho_g}\right]^{1/2} \text{ ft per sec} \tag{7.25}$$

and the gas volume at operating conditions is

$$q_g = \frac{z \times q_{sc} \times 10.73 \times T}{379.4 \times 86{,}400 \times p} \text{ cu ft per sec} \tag{7.26}$$

Clearing,

$$q_g = \frac{0.3273 z q_{sc} T}{p} \tag{7.27}$$

where q_{sc} is the number of million standard cubic feet of gas per day at 14.7 psia and 60°F. The separator or mist extractor area is

$$A_s = \frac{q_g}{v_a} \text{ sq ft} \tag{7.28}$$

and the diameter is

$$d_s = 13.53\sqrt{A_s} \text{ in.} \tag{7.29}$$

The gas capacity of *spherical* separators is based on the capacity of the mist extractor. The empirical equation for calculating the diameter of the demistor is given by

$$d = 9.93\sqrt{\frac{q_{gsc}}{C}} \times \sqrt[4]{\frac{T\gamma z}{p\rho_o}} \tag{7.30}$$

where d = diameter of mist extractor, inches

q_{gsc} = gas flow rate, MM SCF per day

C = allowable velocity constant (0.35 for knitted-wire mesh extractor)

T = operating temperature, °R

γ = gas gravity

z = compressibility factor

p = operating pressure, psia

ρ_o = liquid density at operating conditions, pounds per cubic foot

In Eq. (7.22), for vertical separators without a mist extractor the allowable velocity constant C is 0.117, and with a mist extractor it is 0.167. In the case of horizontal separators with a mist extractor, C is 0.382. In calculating the gas capacity of single-tube horizontal separators, the cross-sectional area of that portion of the vessel occupied by liquid is subtracted from the total area, to find an equivalent diameter. When this equivalent diameter is used in Eq. (7.22), the gas capacities of horizontal separators are obtained by multiplying the value obtained from this equation by the factor $(L/10)^{0.56}$, where L is the shell length in feet.

In vertical separators the distance from the oil inlet to the bottom of the mist extractor should be at least one diameter, and the distance from the oil inlet to the normal liquid level should be at least 24 in. Since these requirements cannot be satisfied in 5-ft and 7.5-ft separators, a correction factor of 0.80 is applied to the gas capacity of these separators. The specifications covering minimum requirements for the design, fabrication, and testing of oil-field oil and gas separators are given an API Std. 12J [8] Examples 7.1, 7.2, and 7.3 illustrate the sizing of vertical, horizontal, and spherical separators.

Example 7.1. Calculate the capacity of a vertical oil and gas separator, given:

Shell height = 10 ft

Inside diameter of separator = 24 in.

Gas gravity = 0.80

Oil gravity = 45° API at 60°F

Operating pressure = 400 psia

Operating temperature = 60°F

$C = 0.167$

SOLUTION: The density of a 45° API oil is

$$\rho_o = \frac{141.5}{131.5 + 45} \times 62.4 = 50.02 \text{ lb/cu ft}$$

The density of the gas at standard conditions is

$$\rho_{gsc} = \frac{14.7 \times 0.80 \times 28.97}{10.73 \times 520} = 0.0611 \text{ lb/cu ft}$$

Assuming an ideal gas, the density at operating conditions, from Eq. (7.20), is

$$\rho_g = 0.0611 \times \frac{400}{14.7} \times \frac{520}{520} = 1.66 \text{ lb/cu ft}$$

Substituting in Eq. (7.22),

$$q_{sc} = \frac{67{,}824 \times 0.167 \times 4}{1.0} \times \frac{400}{14.7}\left[\frac{50.02 - 1.66}{1.66}\right]^{1/2} = 6.66 \text{ MM SCF/day}$$

As a check, assume that it is desired to calculate the diameter of a vertical separator that will handle 6.66 MM SCF per day under the conditions given above. Then the allowed velocity is

$$v_a = 0.167 \left[\frac{50.02 - 1.66}{1.66}\right]^{1/2} = 0.167 \times 5.4$$

$$= 0.9018 \text{ ft/sec}$$

From the nonideal gas law, the operating rate is

$$q_g = \frac{1 \times 6.66 \times 10^6 \times 10.73 \times 520}{379.4 \times 86{,}400 \times 400} = 2.8344 \text{ cu ft/sec}$$

The area of the gas space is

$$A_g = \frac{q_g}{v_a} = \frac{2.8344}{0.9018} = 3.143 \text{ sq ft}$$

and the separator diameter is

$$d_s = 13.53\sqrt{A} = 13.53\sqrt{3.143} = 24 \text{ in.}$$

With 0.35 taken as the constant in Eq. (7.16), the allowed velocity in the mist extractor is

$$v_a = 0.35 \left[\frac{50.02 - 1.66}{1.66}\right]^{1/2} = 1.890 \text{ ft/sec}$$

Since the operating rate (q_g) is 2.8344 cu ft per sec, the area of the mist extractor is given by

$$A_m = \frac{q_g}{v_a} = \frac{2.8344}{1.890} = 1.5 \text{ sq ft}$$

and the diameter is

$$d_m = 13.53\sqrt{1.5} = 16.6 \text{ in.}$$

From Eq. (7.5), the oil capacity is

$$q_r = \frac{100.5 \times 4 \times 3.25}{1.0} = 1307 \text{ bbl/day}$$

Example 7.2. Calculate the diameter of a horizontal separator, given:

Separator length = 10 ft

Operating pressure = 775 psia

Operating temperature = 90°F

Gas specific gravity = 0.64

Compressibility factor = 0.881

Liquid gravity = 10° API at 60°F

Capacity = 70 MM SCF/day

SOLUTION: From Eq. (8.65),

$$\gamma = 1.00[1 - 0.0004(90 - 60)]$$
$$= 0.988 \text{ at } 90°\text{F}$$

and

$$\rho_o = 0.988 \times 62.4 = 61.65 \text{ lb/cu ft at } 90°\text{F}$$

$$\rho_g = \frac{(0.0764)0.64 \times 520 \times 775}{0.881 \times 550 \times 14.7} = 2\ 77 \text{ lb/cu ft}$$

Using a constant of 0.382 in Eq. (7.16), we find the allowable velocity to be

$$v_a = 0.382 \left[\frac{61.65 - 2.77}{2.77}\right]^{1/2}$$
$$= 0.382 \times 4.61 = 1.761 \text{ ft/sec}$$

The operating rate is calculated from the nonideal gas law,

$$q_g = \frac{0.3273 \times 0.881 \times 70 \times 550}{775} = 14.32 \text{ cu ft/sec}$$

and thus

$$A_g = \frac{q_g}{v_a} = \frac{14.32}{1.761} = 8.132 \text{ sq ft}$$

If the area of the gas space, A_g, is equal to the area of the liquid space, A_l, then

$$\pi d^2/4 = 2 \times 8.132 = 16.26$$

and $d = 4.0$ ft ID.

Example 7.3. Calculate the gas and oil capacities of a spherical separator, given:

Diameter of separator = 30 in.

Operating pressure = 100 psia

Gas gravity = 0.65

Oil gravity = 45° API at 60°F

Operating temperature = 80°F

Assume ideal gas behavior.

SOLUTION: The specific gravity of 45° API oil at 60°F is

$$\gamma = \frac{141.5}{131.5 + 45} = 0.8017 \text{ at } 60°\text{F}$$

From Eq. (8.65), the specific gravity of the oil at 80°F is

$$\gamma = 0.8017[1 - 0.0005(80 - 60)] = 0.7932$$

and

$$\rho_o = 0.7932 \times 62.4 = 49.50 \text{ lb/cu ft}$$

The density of the gas at operating conditions is given by

$$\rho_g = \frac{(0.0764)0.65 \times 520 \times 100}{1 \times 540 \times 14.7}$$
$$= \frac{2.70 \times 0.65 \times 100}{1 \times 540} = 0.325 \text{ lb/cu ft}$$

The constant for a knitted-wire mesh mist extractor is 0.35. The allowable velocity from Eq. (7.16) is

$$v_a = 0.35\left[\frac{49.50 - 0.325}{0.325}\right]^{1/2} = 0.35 \times 12.2$$
$$= 4.27 \text{ ft/sec}$$

Then the operating rate is

$$q_g = \frac{0.3273 \times 540 \times q_{gsc} \times 1}{100} = 1.77 q_{gsc}$$

where q_{gsc} is in MM SCF per day. Assuming q_{gsc} equals 1.0, then q_g is equal to 1.77 cu ft per sec. A spherical separator of 30-in. diameter has a mist extractor of half that diameter. A 15-in. mist extractor has an area A_m of 1.23 sq ft. Then the separator gas capacity is

$$q_{gsc} = A_m\left(\frac{v_a}{q_g}\right) = 1.23\left(\frac{4.27}{1.77}\right) = 2.97 \text{ MM SCF per day}$$

Since the capacity of a spherical separator is based on the capacity of the demistor, the diameter of the separator can also be calculated from Eq. (7.30):

$$d = 9.93\sqrt{\frac{2.97}{0.35}}\sqrt[4]{\frac{540 \times 0.65 \times 1}{100 \times 49.5}}$$
$$= 9.93 \times 2.91 \times 0.518 = 15 \text{ in.}$$

The oil capacity is calculated from Eq. (7.11), or

$$q_r = 33.5 \times (2.5)^3\,(2.5/2)^{1/2}$$
$$= 33.5 \times 15.6 \times 1.12 = 585 \text{ STB/day}$$

The manufacturer's rating will be less than the above capacity, if the separator dumps when the fluid level is below the center.

7.2. Heater Treaters

In an *emulsion*, one liquid is dispersed in another. The particular emulsion we are concerned with is, of course, water in oil. Three conditions are necessary for the formation of an emulsion which will not break down without some form of treating:

1. The liquids must be immiscible.
2. There must be sufficient agitation for one liquid to disperse in the other.
3. An emulsifying agent must be present.

If lubricating oil and water are placed in a bottle and shaken violently, only a small amount of water (1.0 to 2.0 per cent) will form an emulsion with the oil. In order to form stable emulsions with a greater percentage of water in oil, one must add an emulsifying agent, which forms a film around the water droplets.

Almost all methods of treating emulsions are aimed at neutralizing the effect of the emulsifying agent. In the treatment of well fluid emulsions, two steps should be taken: (1) removal of gas, and (2) removal of free water. Both of these materials would cause the emulsion to reform after being broken.

The pressure-type treater[9] is designed to incorporate the functions of three separate units, i.e., the oil-gas separator, the "volume-type" heater with removable internal firebox, and the gun-barrel tank. The gun barrel was the first automatic treating system which provided space for the water to settle out of the oil. With some emulsions a large gun barrel was required to provide sufficient retention time for separation. The volume or indirectly-fired heater was then added to heat the water in the bath section of the gun barrel and thus shorten the settling time and make possible smaller and more economical equipment. In the treater presented in Fig. 7.8, the well-

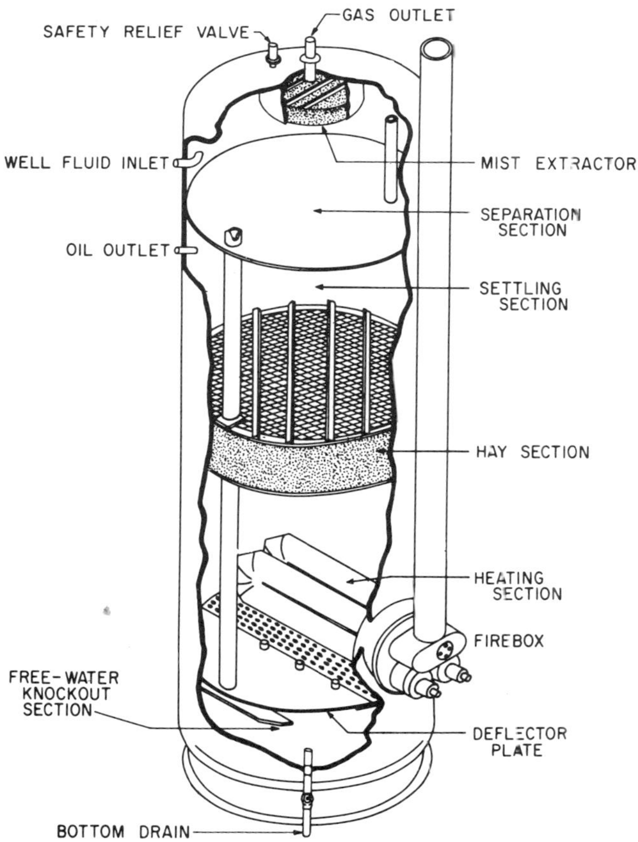

Fig. 7.8 Vertical treater. (Courtesy of Delta Tank Manufacturing Co., Inc.)

stream enters the separation section from whence the gas flows through the mist extractor and the fluid flows through the downcomer to the free-water knockout section. Free water is that water which will settle out of the oil within 5 min. Removal of the free water prevents additional emulsification and decreases the heating load. The oil and emulsion then flow through the deflector plate into the heating (water wash) section. The application of heat has several effects on emulsions: (1) it decreases the viscosity of the oil and the stickiness of the emulsifying agent; (2) it lowers the surface tension of the water; (3) it increases the difference between the specific gravities of oil and water.

The hay section provides for filtering and coalescing. The emulsion flowing through the filter is squeezed between the fibers and the film of the emulsifying agent ruptured. The hay section also acts as a heat barrier between the heating section and the settling section; also it dampens any surges so that the settling section is quiet. After removal of gas and free water and addition of heat, the emulsion flows from the hay section to the settling section. This section should be as large and as free of agitation as possible.

Treater design. The basic steps in determining the size of a treater are as follows:

1. Determine the type of emulsion the well is producing. For a loose emulsion, a treater may have a rated oil capacity of 500 bbl per day; for a tight emulsion, 150 bbl per day will be sufficient.
2. Determine whether the well is producing free water.
3. Determine the amounts and physical characteristics of the oil, water, and gas.
4. Determine the settling time by laboratory analysis. The time required will range from 15 min to more than 12 hr.
5. Determine the end use of the oil, water, and gas to see whether a treater operated at atmospheric pressure can be used.
6. Determine the treating temperature, which will range from 120° to 160°F.
7. Determine the firebox capacity.

The firebox must be large enough to heat the fluid to be treated from the flowing temperature to the treating temperature. There are several factors taken into consideration:

1. The free water will be withdrawn before the fluid is heated.
2. All the oil is to be heated as it flows through the hot-water bath.
3. The emulsified water is also to be heated. If the amount of emulsified water is unknown, assume a value of 20 per cent.

4. The firebox should not be sized for continuous cperation. A firebox that burns intermittently has a much longer life.
5. Sufficient firebox capacity should be provided to take care of heat losses from the treater.

The heating capacity of the firebox is based on a heat balance,

$$Q = wc_p\Delta T \tag{7.31}$$

where Q = heat required, Btu per hour

w = weight of liquid heated, pounds per hour

c_p = specific heat of the liquid, Btu per pound per °F

ΔT = temperature range, °F

The height of the treater is determined both by the size of the settling section and by the head required in order for the oil to flow through the treater to the stock tanks. The following production data are used in sizing a vertical treater:

Total production = 700 STB/day

Oil production = 400 STB/day

Water production = 300 STB/day

Specific gravity of oil = 0.815

Specific gravity of water = 1.02

Emulsion = 20 per cent of water production

Flowing temperature = 90°F

Treating temperature = 150°F

Flowing pressure = 20 psig

Specific heat of oil = 0.50 Btu/lb/°F

Specific heat of water = 1.00 Btu/lb/°F

Type of emulsion = medium loose

Settling time = 2.5 hr

The weight of the oil is

$$w = 8.34 \times 0.815 \times \frac{400}{24} \times 42 = 4758 \text{ lb/hr}$$

The heat load is calculated from Eq. (7.31), or

$$Q = 4758 \times 0.50 \times 60 = 142{,}740 \text{ Btu/hr}$$

The weight of the water is

$$w = 8.34 \times 1.02 \times \frac{60}{24} \times 42 = 893 \text{ lb/hr}$$

and the heat load for the water is

$$Q = 893 \times 1.00 \times 60 = 53{,}600 \text{ Btu/hr}$$

Then the total heat load is

$$142{,}740 + 53{,}600 = 196{,}340 \text{ Btu/hr}$$

When a factor of 25 per cent is introduced to take care of heat losses from the heater, the total heat required is 245,400 Btu per hr. Fireboxes are rated in increments of 1000 Btu per hr,[10] i.e., 200,000, 250,000, 300,000, etc. Therefore, a 250,000 Btu per hr firebox satisfies the minimum requirements in this case. The heat transferred by the firebox to the water bath is taken as 10,000 Btu per hour per square foot of surface area. The heat-transfer area required is 250,000/10,000 or 25 square feet.

The volume of the settling section is determined from the settling time, 2.5 hr. We must select a vessel diameter that will satisfy this settling time without requiring an excessive overall height. Assuming a diameter of 6 ft, the cross section is 28.27 sq ft. The capacity must be such that a volume of 400 bbl of oil and 60 bbl of water per day, or 19.17 bbl of oil plus water per hr, can be handled. Since the settling time is 2.5 hr, the capacity of the section is 48 bbl or 269.52 cu ft, and the length is 269.52/28.27 or 9.5 ft (use 10 ft). To this is added 2 ft for the hay section.

The heating section, which is the space from the bottom of the fire tubes to the hay section, is designed to heat the fluid for 1 hr (i.e., 1 hr retention time). The length of this section is

$$L = \frac{19.17 \times 5.615}{28.27} = 3.8 \text{ or } 4 \text{ ft}$$

The distance from the bottom of the fire tubes to the spreader plate is between 6 and 18 in. In this problem we will use 1 ft.

The free-water section must have sufficient volume to retain all the free water accumulating in 5 min. Since the free water is mixed with both the oil and the emulsion, the total free-water production is retained for 1 min or long enough for the oil and emulsion to reach the spreader plate. This production is 240 bbl per day, and since the retention time is 5 min, the required capacity is 5 × 240/1440 or 0.83 bbl (4.66 cu ft). The oil and emulsion production is 460 bbl per day, and for a 1-min retention time the volume is 5.61 × 1 × 460/1440 or 1.80 cu ft. The total volume of the section is thus 6.46 cu ft and the length is 6.46/28.27 or 0.23 ft. To keep the vessel design in units of feet, the length of this section is taken as 1 ft.

The total production enters the separation section where the entrained gas is separated from the fluid. If the wells are flowing or pumping, the gas volume is small and need not be considered in sizing this section. If, however, the wells are on gas lift or are producing sand which requires a sand jet to keep the section open, allowance for additional space should be made. The retention time for the total liquid production in this section is taken as 1 min. The total flow rate is 700/1440 or 0.49 bbl per min or a volume of 2.75 cu ft. Since some of the wells will flow by heads, this section should have a surge capacity of 100 per cent or more of the total flow. The volume of this section is then 5.50 cu ft and the length is 5.50/28.27 or 0.20 ft. The length of this section will be taken as 1 ft. The vessel length is the sum of the section lengths, 19 ft. Treaters are made in 20-ft, 24-ft, and 28-ft lengths. The extra 1 ft is added to the section requiring additional capacity. The treater required is 6 ft by 20 ft and has a working pressure of 25 psi and a minimum heating capacity of 250,000 Btu per hr. The length is measured from head seam to head seam. This size of heater is usually equipped with a heating capacity of 500 MBtu per hr.

7.3. Oil Skimmers

The oil skimmer shown in Fig. 7.9 is designed to remove oil droplets remaining in the water from an oil treater or a gun barrel prior to disposal into a water injection well. Water containing traces of oil enters the vessel through the flow distribution section containing pea-size coal. This section

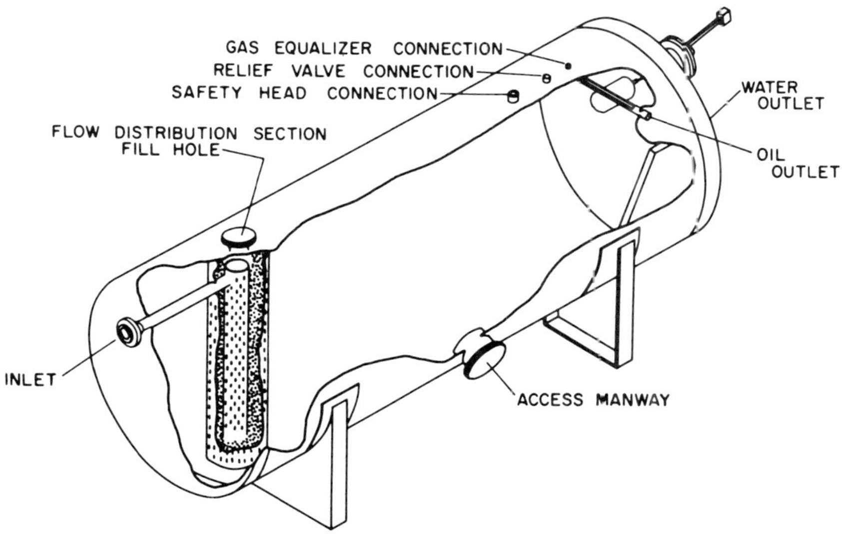

Fig. 7.9 Horizontal oil skimmer. (Courtesy of Delta Tank Manufacturing Co., Inc.)

prevents the inlet stream from causing turbulence in the settling section. In this section of the vessel the water is in nonturbulent flow, allowing the oil particles to rise to the oil-water interface. The oil is skimmed from the surface at the outlet end of the vessel. The oil-water interface is controlled by a loaded float riding the interface and controlling a mechanical water-dump valve. A liquid-operated oil valve is used to dump the oil. By maintaining gas pressure on the skimmer, it is possible to pass the water through a filter or on to the point of disposal. The capacity of the oil skimmer is dependent on several variables, chiefly the densities of the oil and the water. The following production data will be used in calculating the capacity of a skimmer:

Water production = 2500 bbl per day

Oil production = 50 bbl per day

Specific gravity of water = 1.05

Oil gravity = 32.5° API ($\gamma = 0.863$)

Diameter of water particles = 150 microns or

$3.28 \times 10^{-6} \times 150 = 4.92 \times 10^{-4}$ ft

Viscosity of oil = 2.35 cp or 1.578×10^{-3} lb/ft sec

Viscosity of water = 0.80 cp or 5.376×10^{-4} lb/ft sec

The oil capacity is based on the intermediate law of settling given in Fig. 7.7. The terminal settling velocity of a water particle in oil is

$$v_{tw} = \frac{0.153 \times g_l^{0.71} \times d_p^{1.14} (\rho_w - \rho_o)^{0.71}}{\rho_o^{0.29} \mu^{0.43}} \tag{7.32}$$

The density of the water (ρ_w) is 62.4 × 1.05 or 65.5 lb per cu ft and the density of the oil (ρ_o) is 62.4 × 0.863 or 53.9 lb per cu ft. Substituting in Eq. (7.32),

$$v_{tw} = \frac{0.153 \times 32.17^{0.71} (4.92 \times 10^{-4})^{1.14} (65.5 - 53.9)^{0.71}}{(53.9)^{0.29} (1.578 \times 10^{-3})^{0.43}}$$

$$= 8.68 \times 10^{-3} \text{ ft/sec}$$

In the preceding part of this problem, the average diameter of the water particles was taken as 150 microns. In field practice it is found that the oil particles are generally much smaller, i.e., of the order of 100 microns (3.28×10^{-4} ft). The water capacity is based on Stokes' law, or the rate at which the oil particles will rise in water,

$$v_{to} = \frac{g_l d_p^2 (\rho_w - \rho_o)}{18\mu} \tag{7.33}$$

Then the terminal ascending velocity of an oil particle in water, from Eq. (7.33), is

$$v_{to} = \frac{32.17 \times (3.28 \times 10^{-4})^2(65.5 - 53.9)}{18 \times 5.376 \times 10^{-4}}$$
$$= 4.23 \times 10^{-3} \text{ ft/sec}$$

It can be seen from these calculations that the ascending velocity of the oil is less than the decending velocity of the water, and therefore the capacity of the skimmer is primarily a function of the time of retention of the oil.

Assume the skimmer to be 6 ft in diameter and 15 ft long, which gives an effective length of 10 ft downstream from the flow distribution section. A safety factor is provided by neglecting the capacity of the heads. The oil-water interface is assumed to be 4 ft from the bottom center of the vessel and the oil-gas interface is 1 ft above the oil-water interface. The retention time t_r for the oil, in minutes, is given by

$$t_r = \frac{h}{60v_{to}} \tag{7.34}$$

where h is the height of the oil-water interface above the bottom center line in feet. Thus the retention time for the oil in the water section is

$$t_r = \frac{4}{60 \times 0.00423} = 15.8 \text{ min}$$

The water capacity of the vessel q_w, in barrels per day, is given by

$$q_w = \frac{1440A_wL}{5.615t_r} \tag{7.35}$$

where A_w is the area of the water space, L is the effective length of the vessel in feet, and t_r is the retention time in minutes. The area of the water space is given by

$$A_w = \frac{\pi r^2}{2} + y\sqrt{r^2 - y^2} + r^2 \sin^{-1}\frac{y}{r} \tag{7.36}$$

where r is the vessel radius in feet and y is the difference between the height of the oil-water interface, h, and the radius r in feet. Substituting in Eq. (7.36)

$$A_w = \frac{3.14 \times 3^2}{2} + 1\sqrt{3^2 - 1^2} + 3^2 \sin^{-1}\frac{1}{3}$$
$$= 14.13 + 2.83 + 9 \sin^{-1} 0.333$$
$$= 16.96 + 9 \times 0.34 \text{ radians} = 20.02 \text{ sq ft}$$

and from Eq. (7.35) the capacity is

$$q_w = \frac{1440 \times 20.02 \times 10}{5.615 \times 15.8} = 3250 \text{ bbl/day}$$

PROBLEMS

7.1. (a) Each 10 MM SCF of 0.70 specific gravity gas represents about how many tons? *Ans.* 267 tons.
(b) Each 1000 bbl of 35° API oil represents about how many tons? *Ans.* 149 tons.

7.2. (a) List the four essential components for gas-liquid separation equipment.
(b) What are the three separation mechanisms?

7.3. (a) What are the three types of separators?
(b) Under what operating conditions should each be used?

7.4. Discuss the effects of the following factors on separation: pressure, temperature, and gas and liquid densities.

7.5. What are the four laws of settling used to determine maximum liquid velocities?

7.6. A vertical separator has a 24-in. inside diameter, is 10 ft high, and is to operate at a pressure of 400 psia (compressibility factor = 0.909) and 60°F. The oil gravity is 35° API, the gas gravity 0.70, and the separation coefficient 0.167.
(a) What is the oil capacity if the retention time is 1 min? *Ans.* 1307 bbl/day.
(b) What is the gas density at operating conditions? *Ans.* 1.60 lb/cu ft.
(c) What is the gas capacity? *Ans.* 7.69 MM SCF/day.

7.7. A horizontal single-tube separator has an inside diameter of 24 in., is 10 ft long, operates at 400 psia (compressibility factor = 0.909), and is equipped with a knitted-wire mist extractor having a separation coefficient of 0.382. The oil gravity is 35° API, the gas gravity is 0.70, and the operating temperature 60°F.
(a) If the retention time is 1 min, what is the oil capacity of the separator? *Ans.* 2010 STB/day.
(b) If the gas area is half the separator area, what is the equivalent diameter? *Ans.* 1.414 ft.
(c) What is the gas capacity? *Ans.* 8.8 MM SCF/day.
(d) What would be the gas capacity if the separator were 20 ft long? *Ans.* 12.9 MM SCF/day.

7.8. Derive Eq. (7.30) using Eqs. (7.25), (7.27), and (7.29). It is assumed that

$$\left[\frac{1}{1 - \dfrac{2.70\,Gp/Tz}{\rho_o}}\right]^{1/4} \cong 1$$

7.9. A spherical separator is to handle 2 MM SCF/day of gas of specific gravity 0.70. The operating pressure is to be 100 psia, the operating temperature 80°F, and the compressibility factor is 0.980. The oil gravity is 45° API at 60°F. What is the diameter of the mist extractor? *Ans.* 12.4 in.

7.10. It is desired to design a heater treater for a lease with the following production:

Oil production = 240 STB/day

Water production = 160 bbl/day

Oil gravity = 0.802

Water gravity = 1.05

Emulsion = 10 per cent of water production

Flowing temperature = 100°F

Treating temperature = 140°F

Flowing pressure = 15 psig

Specific heat of oil = 0.50 Btu/lb/°F

Specific heat of water = 1.0 Btu/lb/°F

Type of emulsion = medium tight

Settling time = 180 min

(a) Determine the weight of the oil to be heated. *Ans.* 2809 lb/hr.
(b) Calculate the oil heating load on the firebox. *Ans.* 56,180 Btu/hr.
(c) Determine the weight of water to be heated. *Ans.* 245 lb/hr.
(d) What is the water heating load on the firebox? *Ans.* 9800 Btu/hr.
(e) What is the total heating load on the firebox? *Ans.* 65,980 Btu/hr.
(f) Determine the heat load required if the heat loss factor is 20 per cent. *Ans.* 79,180 Btu/hr.
(g) What is the minimum capacity of the firebox for this heater treater? *Ans.* 80,000 Btu/hr.
(h) Determine the heat-transfer area required for the firetubes. *Ans.* 8 sq ft.
(i) Assuming a diameter of 4 ft, what is the length of the settling section? *Ans.* 14.3 ft or 15 ft.
(j) Calculate the length of the free-water section. *Ans.* 0.30 ft or 1 ft.
(k) What is the length of the separation section? *Ans.* 0.124 ft or 1 ft.
(l) On the basis of the answer to part (k), would it be necessary to increase the height of the separation section to take care of surge capacity?
(m) If the distance from the bottom of the firetubes to the spreader plate is 12 in. and the hay section is 2 ft high, what is the seam-to-seam height of the treater? *Ans.* 20 ft. Since this is a standard catalog-size treater except for the calculated firebox capacity, it would be most economical to select the 20-ft treater listed in the catalog, which has a firebox capacity of 350,000 Btu/hr.

7.11. Derive Eq. (7.36).

7.12. A skimmer is to be designed for a West Texas operator. The following information is available:

Water production = 20,000 bbl/day

Water gravity = 1.04

Oil production = negligible

Oil gravity = 45° API ($\gamma = 0.8017$)

Diameter of water spheres = 150 microns

Diameter of oil spheres = 100 microns

Water viscosity = 0.85 cp

Oil viscosity = 2.35 cp

(a) What is the terminal velocity of the oil? *Ans.* 50.06×10^{-4} ft/sec.

(b) Assuming a diameter of 12 ft (Fig. 7.10), the oil-water interface is 11 ft

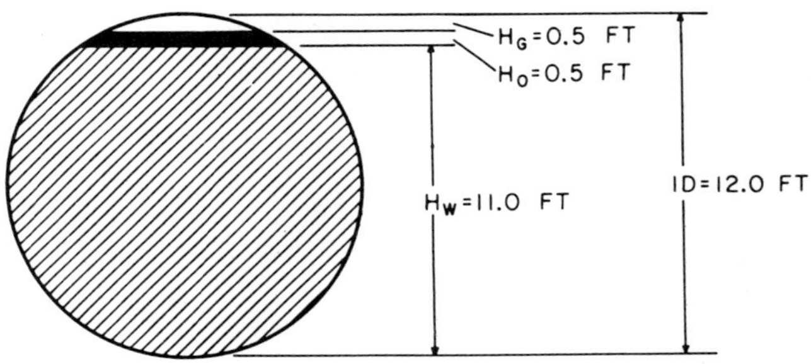

Fig. 7.10

above the bottom center of the vessel and the oil-gas interface is 0.5 ft above the oil-water interface. What is the water area of the vessel? *Ans.* 108.68 sq ft.

(c) What is the retention time for the oil? *Ans.* 36.6 min.

(d) Assuming a 5-ft turbulent zone at the inlet end, what is the length of the vessel? *Ans.* 31.3 ft.

7.13. A three-phase separator has two functions to perform, separating the total liquid from the gas stream and then separating the two liquids. It is desired to calculate the capacity of a three-phase, single-barrel horizontal separator from the following basic data:

Separator length = 15 ft

Separator diameter = 5 ft

Oil-water interface 3 ft above bottom center of vessel

Gas-oil interface 1 ft above oil-water interface

A_w = 12.72 sq ft

A_o = 4.10 sq ft

$A_g = 2.80$ sq ft

Separator pressure = 40 psia

Separator temperature = 80°F

Base pressure = 15.025 psia

Base temperature = 60°F

Density of oil at separator conditions = 53.70 lb/cu ft

Density of gas at separator conditions = 0.1403 lb/cu ft

Density of water at separator conditions = 65.52 lb/cu ft

Diameter of oil sphere in water = 100 microns or

$$(100)(3.28 \times 10^{-6}) = 3.28 \times 10^{-4} \text{ ft}$$

Diameter of water sphere = 150 microns or

$$(150)(3.28 \times 10^{-6}) = 4.92 \times 10^{-4} \text{ ft}$$

Viscosity of oil at separator conditions = 2.35 cp or

$$(2.35)(0.672 \times 10^{-3}) = 1.578 \times 10^{-3} \text{ lb/ft sec}$$

Viscosity of water at separator conditions = 0.80 cp or

$$(0.80)(0.672 \times 10^{-3}) = 5.376 \times 10^{-4} \text{ lb/ft sec}$$

Acceleration due to gravity = 32.2 ft/sec^2

Separation coefficient = 0.382

(a) What is the maximum allowable vapor velocity? *Ans.* 7.46 ft/sec.
(b) What is the gas capacity? *Ans.* 4.54 MM SCF/day.
(c) What is terminal velocity for a water particle in the oil? *Ans.* 0.0089 ft/sec.
(d) The depth of the oil level, from the water-oil interface to the gas-oil interface, h, is 1 ft. Calculate the retention time. *Ans.* 1.87 min.
(e) Calculate the oil capacity. *Ans.* 8436 bbl/day.
(f) Calculate the terminal velocity of an oil particle in water. *Ans.* 0.00423 ft/sec.
(g) The height of the water layer is 3 ft. What is the retention time? *Ans.* 11.82 min.
(h) Calculate the water capacity for the 5 × 15-ft vessel assuming a 3-ft water layer. *Ans.* 4140 bbl/day.

REFERENCES

[1]M. Steve Worley and Lawton L. Laurence, "Oil and Gas Separation Is A Science," *Journal of Petroleum Technology*, **9**, No. 4, 11–16 (April, 1957).

[2]John M. Campbell, "Vane-Type Impingement Mist Extractors," *The Oil and Gas Journal*, **54**, No. 42, 107 (February 20, 1956).

[3]John M. Campbell, "Knitted-Wire Mist Extractors — And How They Work," *The Oil and Gas Journal*, **54**, No. 44, 115 (March 5, 1956).

[4]*Primer of Oil and Gas Production* (New York: American Petroleum Institute, 1954), p. 30.

[5]John M. Campbell, "Know Your Separators," *The Oil and Gas Journal*, **53**, No. 45, 107–111 (March 14, 1955).

[6]C. E. Lapple, *Chemical Engineers' Handbook*, 3rd Ed. (New York: McGraw-Hill Book Company, Inc., 1950), p. 1019.

[7]Mott Souders, Jr., and George Granger Brown, "Design of Fractionating Columns — Entrainment and Capacity," *Industrial and Engineering Chemistry*, **26**, 99 (January, 1934).

[8]*API Specification for Oil and Gas Separators* (Tentative; API Std 12J), 1st Ed. (Dallas: Division of Production, API, April, 1960), p. 4.

[9]R. C. Buchan, "Improvements in the Treatment of Crude-Oil–Water Emulsions," *Drilling and Production Practice* (API, 1942), pp. 154–160.

[10]*API Specification for Indirect Type Oil-Field Heaters* (Tentative; API Std 12K), 1st Ed. (Dallas: Division of Production, API, April, 1960), p. 5.

CHAPTER EIGHT

Hydraulic Fracturing

8.1. Introduction

A fracturing treatment consists essentially in breaking down a producing section hydraulically with a sand-carrying fluid, the sand being used to prop the resulting fracture. The hydraulic fracturing technique of well stimulation is one of the recent major developments in petroleum engineering. The technique was introduced to the petroleum industry by Clark,*[,1] and has become almost a standard practice in many companies for stimulating production in old and new wells. In some areas it is the only technique which will effect commercial production; even some carbonate formations which refuse to respond to multistage acid treatments will, subsequent to hydraulic fracture treatments, yield commercial wells.

The mechanism of this technique is related to three phenomena concerning which an extensive literature has developed: (1) pressure parting in water injection wells, (2) lost circulation during drilling, and (3) the breakdown of formations during squeeze cementing operations. These all appear to involve the formation of open fractures due to pressures applied in a well bore. It has been suggested that in fracturing the pressure parts the formation along a bedding plane and lifts the overburden, disregarding the fact that fracturing pressures, where they have been measured, are significantly less than those due to the total weight of the overburden as determined from its density. The purpose of a fracture treatment is to increase

*References throughout the text are given at the end of each chapter.

the conduction of fluid into the well bore. There are three cases where hydraulic fracturing is beneficial:

1. If the reservoir is composed of a low-permeability, homogeneous rock, fracturing is similar in effect to increasing the size of the hole; i.e., the fluid which formerly flowed through the low-permeability rock is able to move into the high capacity fracture at some distance from the well.
2. Fracturing will eliminate formation damage due to invasion of drilling mud, deposition of mineral matter, or swelling of clays. Since there is damage only in the immediate vicinity of the well bore, only a mild fracture treatment is required.
3. Fractures radiating from the well bore act as gathering lines connecting permeable and porous systems that are otherwise isolated from the well by impermeable barriers.

In order to illustrate the effects of fracturing, assume that a horizontal fracture of relatively high fluid-carrying capacity extends from the well bore. Such a fracture will have a marked influence on the flow pattern and pressure distribution in the formations, the net effect being a tremendous reduction in the resistance to flow. The resulting flow pattern[2] and pressure distribution are illustrated in Fig. 8.1. After production commences, the

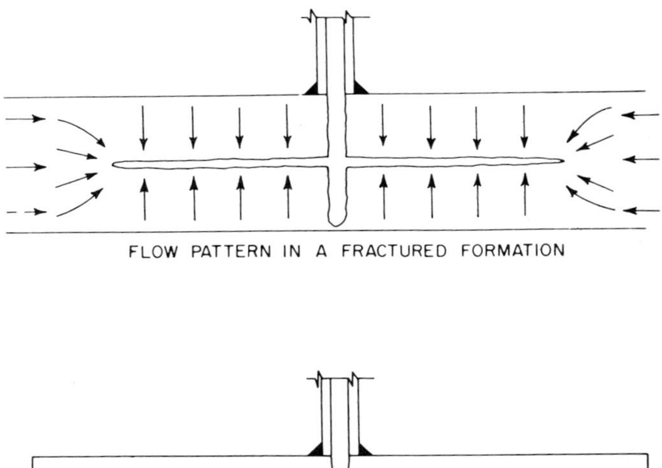

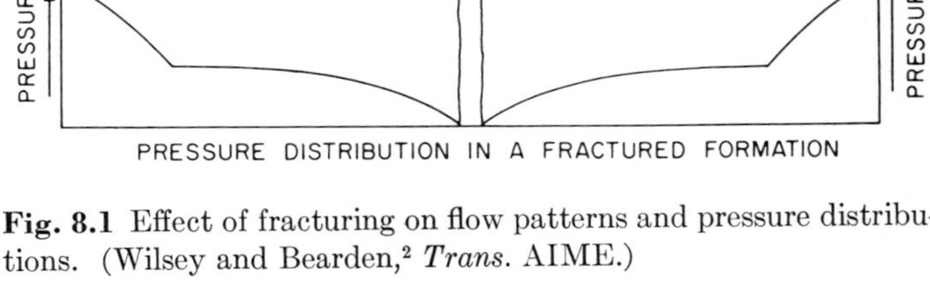

Fig. 8.1 Effect of fracturing on flow patterns and pressure distributions. (Wilsey and Bearden,[2] *Trans.* AIME.)

fluid flows principally vertically from the formation flanking the fracture, into the fracture, and thence to the well. After sufficient oil and gas have been produced to lower the pressure in the formation adjacent to the fracture, there is radial flow into the fracture from the formations beyond the end of the fracture. The high recoveries in very low permeability formations result from changes in the flow pattern in much of the drainage radius, such that most of the fluid flows only a short distance vertically in the tight formation to a high capacity or low resistance avenue to the well.

8.2. Inducing Fractures

Naturally occurring underground stresses resist wellbore fracturing. If we designate as effective stresses those stresses which are normal to planes in which there is no shear, the general stress condition underground can be defined in terms of the effective stresses, σ_z along the vertical Z axis and σ_x and σ_y along the horizontal X and Y axes. In the absence of external forces, the stress at any point is due to the weight of the overburden. Jakosky,[3] in reviewing the work of several investigators, showed that rock densities range from about 2.0 gm per cu cm at shallow depths to approximately 2.6 gm per cu cm at 10,000 ft. Using an average density of 144 lbm per cu ft, the vertical stress at any point is expressed by the equation

$$\sigma_z = 1.0D \text{ psi} \tag{8.1}$$

where D is the depth in feet. Under the influence of this vertical stress, the rock tends to expand laterally but is prevented from doing so by the surrounding rock. This tendency introduces a horizontal stress which under ideal conditions is equal in all horizontal directions. In the elastic zones of the earth's crust, since no horizontal movement has occurred, the horizontal strain ϵ_x equals zero. According to Hooke's law, the horizontal strain is expressed as

$$\epsilon_x = \frac{\sigma_x}{E} - \mu\frac{\sigma_y}{E} - \mu\frac{\sigma_z}{E} \tag{8.2}$$

where E is Young's modulus. For rock in compression, ϵ_x is essentially zero and since the lateral stress σ_x equals the lateral stress σ_y,

$$\sigma_x = \sigma_y = \sigma_h = \frac{\mu}{1-\mu}\sigma_z \tag{8.3}$$

where σ_h is the horizontal stress in general. Since Poisson's ratio μ for consolidated sedimentary rocks[4] ranges from 0.18 to 0.27, the horizontal compressive stress is between 0.22 and 0.37 psi per ft of depth. On the other hand, the soft shales and unconsolidated sands found in the Gulf Coast of Texas and Louisiana can be considered to be in a plastic state of stress and to possess *horizontal* stresses in excess of 0.37 psi per ft of depth.[5] Thus relatively high mud weights are tolerable in Gulf Coast drilling because of the high compressive stresses resisting fracture. In West Texas the relatively low horizontal compressive stresses and existing fractures offer little resistance to fracture induction and extension. In the absence of external forces the horizontal stress is always less than the vertical stress. According to Hubbert and Willis,[6] in geologic regions where normal faulting is taking place, the greatest stress should be approximately vertical and equal to the effective pressure of the overburden, while the least stress should be horizontal, ranging from one half to one third the effective pressure of the overburden. On the other hand, in regions where there is folding or thrust faulting, the least stress should be vertical and equal to the effective overburden pressure, while the greatest stress will be horizontal and between two and three times the overburden pressure. If fluid pressure is applied within rocks and increased until parting of the rocks occurs, that plane along which fracture or parting may first occur is the one perpendicular to the least principal stress (Fig. 8.2).

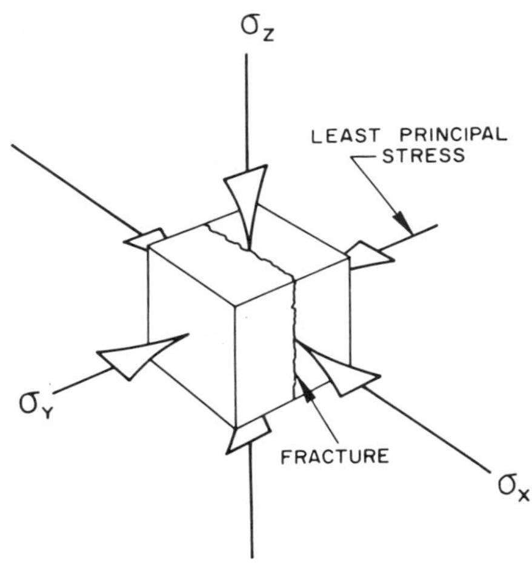

Fig. 8.2 Stress element and preferred plane of fracture. (Hubbert and Willis,[6] *Trans.* AIME.)

When a well is drilled the preexisting stress field in the rock is distorted. An approximate calculation of this distortion has been made[7–11] by assuming the rock to be elastic, the borehole smooth and cylindrical, and the borehole axis vertical. It has been shown[12] that the tangential (circumferential) stress along the periphery of the well bore is equal to twice the horizontal compressive stress at a considerable distance from it. Then the pressure that will fracture the borehole vertically is the sum of that required in order to reduce the compressive stresses on the wall of the hole to zero plus the tensile strength of the rock, or

$$p_f = 2\sigma_h + S_t = \frac{2\mu}{1 - \mu}\,\sigma_z + S_t \tag{8.4}$$

where p_f is the internal pressure in pounds per square inch and S_t is the tensile strength of the rock. Using 0 to 500 psi as the range of tensile strengths for competent sandstones and limestones, we find that the pressure necessary to induce vertical fracturing should lie between

$$p_f = \frac{2 \times 0.18 \times 1.0D}{1 - 0.18} + 0 = 0.44D \text{ psi}$$

and

$$p_f = \frac{2 \times 0.27 \times 1.0D}{1 - 0.27} + 500 = 0.74D + 500 \text{ psi}$$

In addition to the bedding laminations in the rocks across which the tensile strength is a minimum, there usually are intersections by one or more systems of joints with slight normal displacements across which the tensile strength is reduced essentially to zero. In any reasonable-sized section of a well bore, it is probable that many such joints have been crossed. Therefore, the tensile strength of most rocks that are subjected to hydraulic fracturing is effectively zero, and the pressure required to produce a parting in the rocks is simply that required to reduce to zero the compressive stresses across some plane in the walls of the hole.

Once a fracture has been started, the pressure is applied to the walls of the fracture. According to Hubbert and Willis,[6] the minimum down-the-hole injection pressure required to hold open and extend a fracture is slightly in excess of the original stress normal to the plane of the fracture. While no account has been taken of the effect of fluid intrusion on fracturing pressure, it has been shown[8] that loss of fluid slightly decreases the pressure required to produce vertical fractures.

Up to this point, in determining the pressures required to produce vertical fracturing, we have treated the wellbore as a thick-walled cylinder. In the case of horizontal fractures, it is generally agreed that the confining stress holding the fracture planes together is equal to the effective overburden at the depth of the fracture, whereas in the case of vertical fractures, the confining stress holding the planes together is some function of the effective overburden. The lower limit of the pressure necessary to produce horizontal fracturing at the well bore can be determined by assuming the existence of a thin, horizontal, permeable bed or of a naturally occurring horizontal fracture capable of delivering fluid at high pressure to considerable distances from the well bore. Then, in the lower limiting case, horizontal fracturing can take place when

$$p_f = \sigma_z \tag{8.5}$$

The approximate maximum depth at which horizontal fracturing should occur, except in areas still undergoing active compression, can be determined from Eqs. (8.4) and (8.5) by assuming

$$\frac{2\mu}{1-\mu}\sigma_z + S_t > \sigma_z \tag{8.6}$$

Using a vertical stress (overburden) gradient of 1.0 psi per ft, a Poisson's ratio of 0.25, and a tensile strength[13] of 1000 psi, the maximum depth for horizontal fractures is found to be 3000 ft. As indicated in Fig. 8.3, field

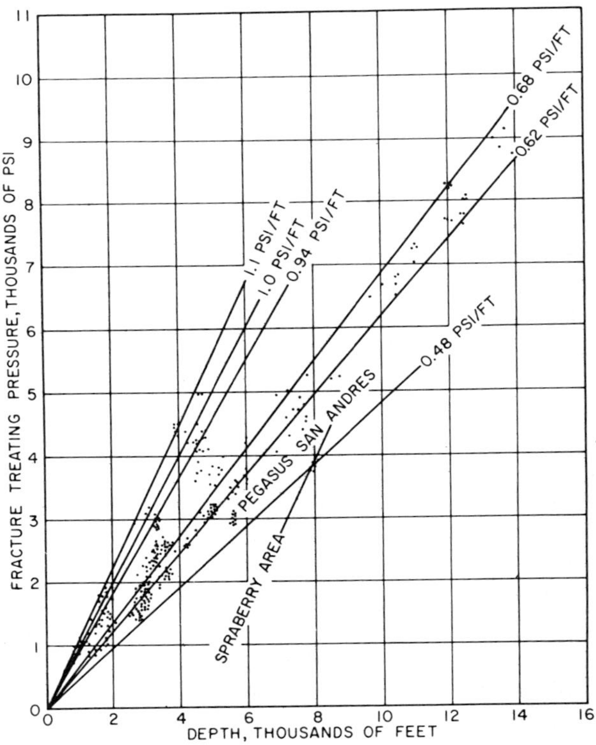

Fig. 8.3 Fracture treating pressure — West Texas Area. (Crittendon,[14] *Journal of Petroleum Technology,* AIME.)

data[14] tend to confirm these limits if it is assumed that the values representing maximum treating gradients for wells less than 5000 ft deep are the result of horizontal fractures. As a general rule, the fracture plane is assumed to be vertical when the fracture gradient is 0.7 psi per ft of depth or less, and horizontal when the fracture gradient is 1.0 psi per ft of depth or greater.

Hubbert and Willis[6] used Mohr diagrams to analyze confining forces within the elastic zone of the earth. For fractures occurring at some angle ϕ from the horizontal, it is possible to obtain an expression for the ap-

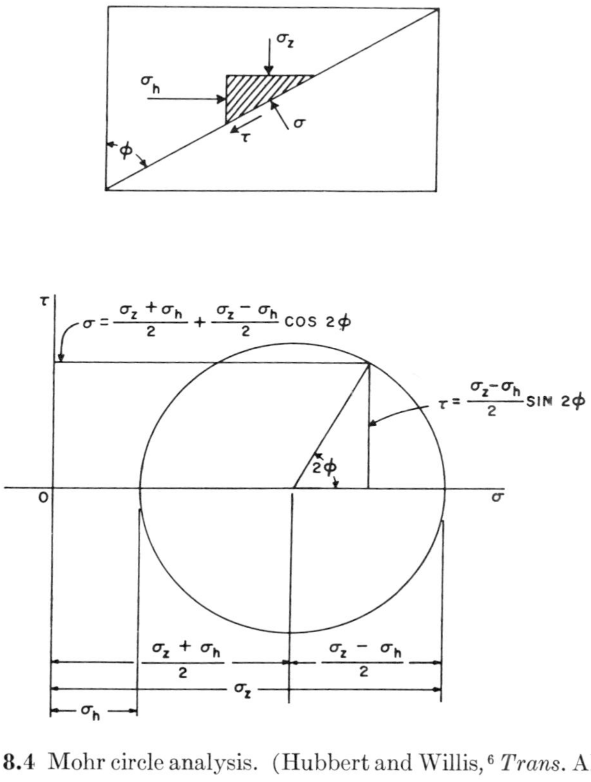

Fig. 8.4 Mohr circle analysis. (Hubbert and Willis,[6] *Trans.* AIME.)

propriate fracture treating pressure. The Mohr circle development (Fig. 8.4) for any angle ϕ yields a normal stress σ of

$$\sigma = \frac{\sigma_z + \sigma_h}{2} + \frac{\sigma_z - \sigma_h}{2} \cos 2\phi \tag{8.7}$$

and a shearing stress τ of

$$\tau = \frac{\sigma_z - \sigma_h}{2} \sin 2\phi \tag{8.8}$$

Utilizing the above analysis and applying stress concentration factors to the horizontal stress component, Crittendon[14] presented a formulation for fracture treating pressure:

$$p_t = \frac{p_{ob} + \frac{2\mu}{1-\mu}\, p_{ob}}{2} + \frac{p_{ob} - \frac{2\mu}{1-\mu}\, p_{ob}}{2} \cos 2\phi$$

or

$$p_t = \frac{p_{ob}}{2}\left[\left(1 + \frac{2\mu}{1-\mu}\right) + \left(1 - \frac{2\mu}{1-\mu}\right)\cos 2\phi\right] \tag{8.9}$$

where p_{ob} is the overburden pressure and ϕ represents the angle measured from the horizontal. Figure 8.5 shows the three types of fracture orientation

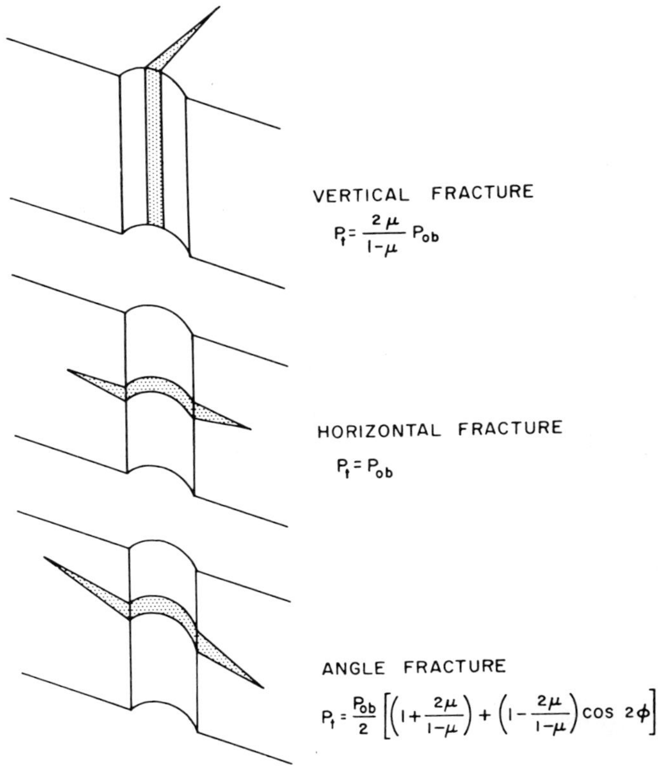

Fig. 8.5 Fracture orientation.

that may be obtained. The type of fracture establishes the directional permeability of the formation to be used in calculating the fluid loss during fracturing as well as the productivity ratio of the fractured well. In addition, the type of fracture determines the advisability of using diverting agents[15].

8.3. Productivity Ratio

The productivity ratio is the ratio of the productivity index of the well after fracturing to that of the well before fracturing, J_f/J. It is impossible to predict exactly the productivity ratio of a fractured well, owing to the fact that every fracture pattern is unique; however, if we assume a uniform radial pattern of fracture, it is possible to estimate the productivity ratios for vertical and horizontal fractures. For the case of a horizontal fracture (fracture gradient ≥ 1.0 psi per ft), an equation for the productivity ratio can be obtained[16] provided it is assumed that there is zero vertical permeability in the fracture zone. It has been shown[17] that

$$\text{PR} = \frac{k_{\text{avg}}}{k} \tag{8.10}$$

where k_{avg} is the average permeability of the fractured formation and k is the permeability of the unfractured formation. It can be seen from

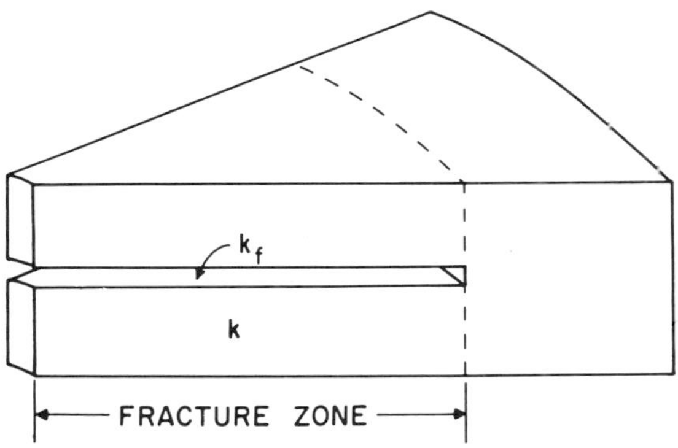

Fig. 8.6

Fig. 8.6 that the average permeability of the fracture *zone* is equal to the average permeability predicted for radial flow in parallel beds,

$$k_{fz} = \frac{k_f W + kh}{h} \tag{8.11}$$

where k_{fz} is the average permeability of the fracture zone, k_f is the permeability of the fracture, W is the thickness of the fracture, k is the formation permeability, and h is the formation thickness.

The average permeability of the fractured formation, k_{avg}, is equal to the average permeability predicted for series beds in radial flow:

$$k_{\text{avg}} = \frac{k k_{fz} \ln (r_e/r_w)}{k_{fz} \ln (r_e/r_f) + k \ln (r_f/r_w)} \tag{8.12}$$

where r_e is the drainage radius of the well, r_w is the well bore radius, and r_f is the radius of the fracture. Substituting Eq. (8.11) into Eq. (8.12),

$$k_{\text{avg}} = \frac{k\left(\dfrac{k_fW + kh}{h}\right) \ln (r_e/r_w)}{\left(\dfrac{k_fW + kh}{h}\right) \ln (r_e/r_f) + k \ln (r_f/r_w)} \tag{8.13}$$

Substituting Eq. (8.13) into Eq. (8.10) and multiplying numerator and denominator by h,

$$\text{PR} = \frac{(k_fW + kh) \ln (r_e/r_w)}{(k_fW + kh) \ln (r_e/r_f) + kh \ln (r_f/r_w)}$$

Factoring,

$$\text{PR} = \left[\frac{k_fW}{kh}\right] \left[\frac{\left(\dfrac{kh}{k_fW} + 1\right) \ln (r_e/r_w)}{\left(\dfrac{k_fW}{kh} + 1\right) \ln (r_e/r_f) + \ln (r_f/r_w)}\right] \tag{8.14}$$

To facilitate rapid calculation of the productivity ratio of horizontal fractures, Fig. 8.7 was constructed with the use of Eq. (8.14). The

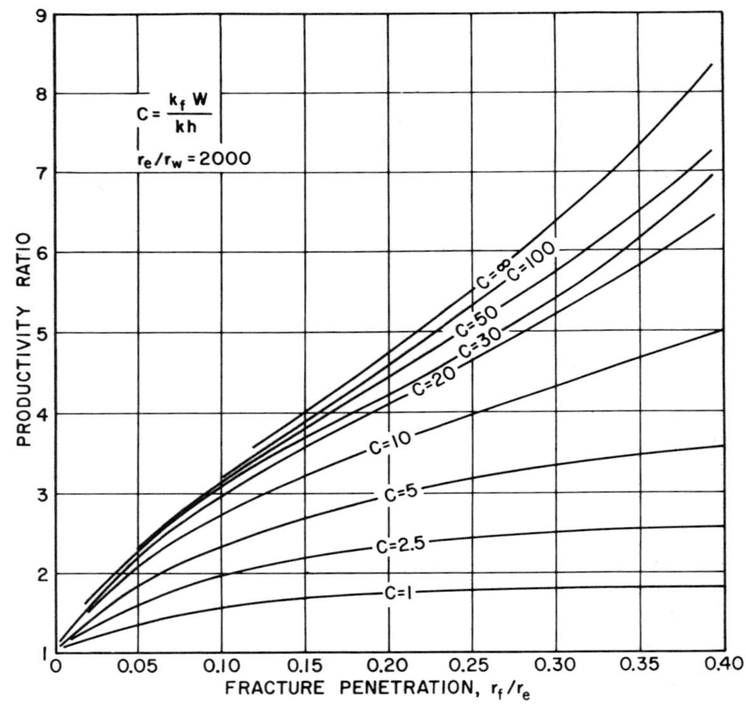

Fig. 8.7 Estimated productivity ratio after fracturing (horizontal fractures). (Courtesy of Mobil Oil Company.)

correlation between Fig. 8.7 and Eq. (8.14) is shown in Example 8.1. Figure 8.8 shows the permeability of commonly used fracture sands.

Example 8.1. Calculate the productivity ratio for a horizontal fracture, given:

Fracture width = 0.1 in.

Net pay thickness = 50 ft

Permeability of propping agent (10-20 mesh) = 32,500 md

Horizontal permeability of formation = 0.54 md

$r_e/r_w = 2000$

Fracture penetration $r_f/r_e = 0.40$

Solution: The value of $k_f W/kh$ in Eq. (8.14) is

$$\frac{32{,}500 \times 0.1 \times 0.083}{0.54 \times 50} = 10.0$$

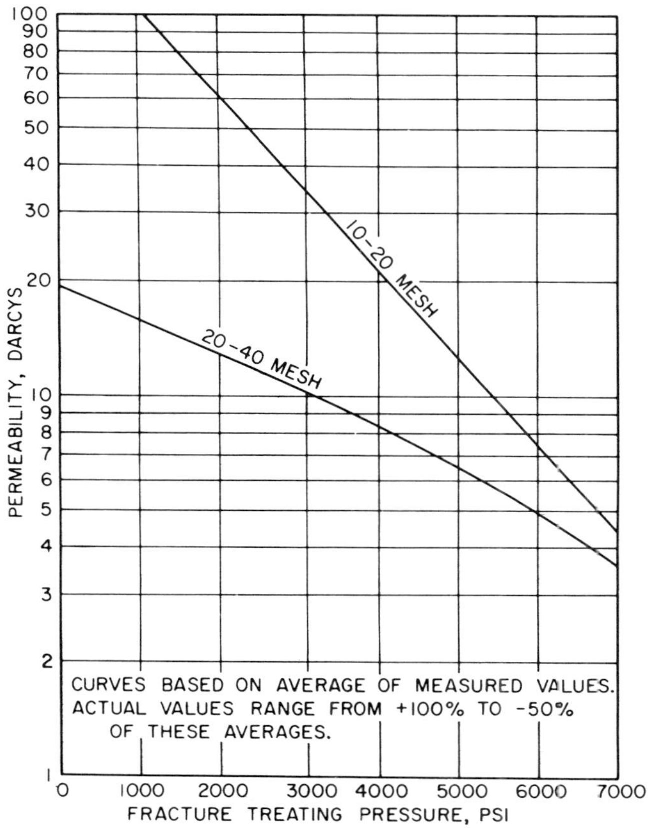

Fig. 8.8 Effect of pressure on frac-sand permeability. (Courtesy of Mobil Oil Company.)

The term $\ln (r_f/r_w)$ in Eq. (8.14) can be expressed as

$$\ln (r_f/r_w) = \ln [(r_e/r_w)(r_f/r_e)]$$

Then substituting in Eq. (8.14),

$$\text{PR} = 10.0 \left[\frac{\left(\frac{1}{10.0} + 1\right) \ln 2000}{(10.0 + 1) \ln (1/0.4) + \ln (2000 \times 0.4)} \right]$$

$$= 4.99$$

In Fig. 8.7, the PR is 5.0.

It is also desirable to estimate the productivity ratio for vertical fractures (fracture gradient ≤ 0.7 psi per ft). On the basis of electric analog data,[18-20] the Mobil Oil Company correlated productivity ratios for various fracture penetrations with the factor $C = k_f W/k$, as shown in Fig. 8.9, where W is the fracture width in feet and k_f and k are the effective fracture and horizontal formation permeabilities in millidarcies, respectively. Prats[41] has also investigated this problem mathematically. His work shows that a high capacity fracture has an effective well radius of $\frac{1}{4}$ the total fracture length. For example, a high capacity fracture of 200 ft

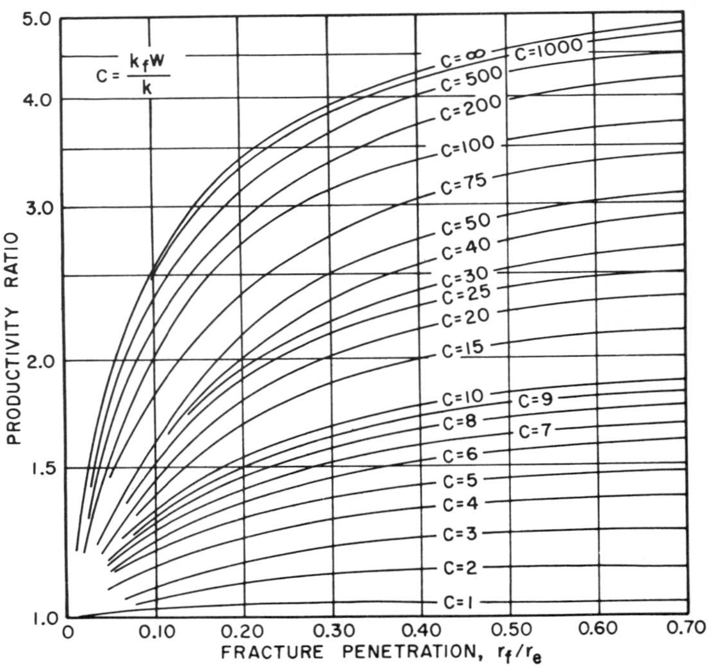

Fig. 8.9 Estimated productivity ratio after fracturing (vertical fracture). (Courtesy of Mobil Oil Company.)

total length in a field of 500 ft radius drained by a well of 0.25 ft radius will have an effective well bore radius of $\frac{1}{4} \times 200$ or 50 ft, and the productivity ratio will be

$$\text{PR} = \ln (500/0.25)/\ln (500/50) = 3.3$$

The same productivity ratio is obtained from Fig. 8.9 for $C = \infty$ and $r_f/r_e = 100/500 = 0.20$.

8.4. Fracture Area

In 1957, Howard and Fast[21] presented a mathematical formula for determining the surface area of a newly-opened fracture on the basis of treating conditions. During the fracturing process, the fracture fluid is injected at the well head at a constant rate q_i. In the fracture this injection rate is split up into two components as shown in Fig. 8.10. Part of the liquid, q_l,

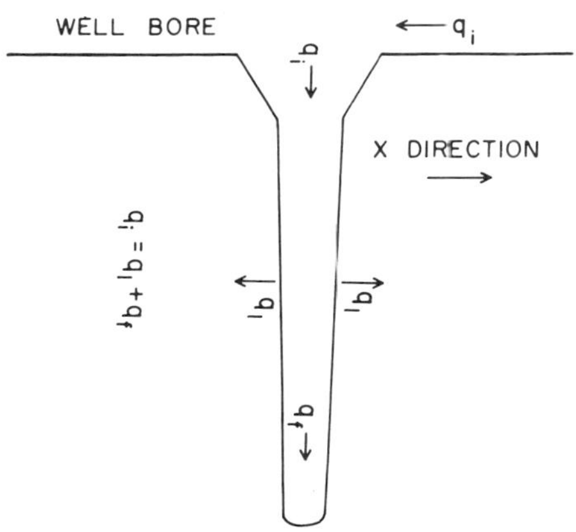

Fig. 8.10

enters the formation as a result of the differential pressure $(p_i - p_e)$ between the fracture and the external boundary, and the remainder, q_f, increases the fracture area, i. e., it increases the volume of the fracture. An expression for the fracture area at any time may be derived by using this basic concept and the following assumptions:

1. The fracture is of uniform width.
2. The flow of fracture fluid into the formation is linear and the direction of flow is perpendicular to the fracture face (Fig. 8.10).
3. The velocity of flow into the formation at any point on the fracture face is a function of the time of exposure of the point to flow.

4. The velocity function $v = f(t)$ is the same for every point in the formation, but the zero time for any point is defined as the instant that the fracturing fluid first reaches it.
5. The pressure in the fracture is equal to the sandface injection pressure, which is constant.

The rate at which the fracture fluid leaks from the two fracture faces into the formation is related to the velocity v, and the fracture area A of one face, by

$$q_l(t) = 2 \int_0^{A(t)} v(t)\, \mathrm{d}A \tag{8.15}$$

where q_l = volume rate of fluid loss to the formation

v = velocity of flow perpendicular to the fracture plane

A = area of the fracture face

Since the extent of fracture increases with time, the fracture area is a function of time. For a given element of area $\mathrm{d}A$ formed at time δ, the velocity of flow into the element is $v(t - \delta)$, correcting for the zero time in the fourth assumption. Since A is also a function of time,

$$\mathrm{d}A = \left(\frac{\mathrm{d}A}{\mathrm{d}\delta}\right) \mathrm{d}\delta \tag{8.16}$$

Equation (8.15) can therefore be written as

$$q_l(t) = 2 \int_0^t v(t - \delta) \left(\frac{\mathrm{d}A}{\mathrm{d}\delta}\right) \mathrm{d}\delta \tag{8.17}$$

The rate of increase of fracture volume is

$$q_f = W \frac{\mathrm{d}A}{\mathrm{d}t} \tag{8.18}$$

where W is the fracture width. It can be seen from Fig. 8.10 that

$$q_i = q_l + q_f \tag{8.19}$$

Substituting Eqs. (8.17) and (8.18) into Eq. (8.19),

$$q_i = \left[2 \int_0^t v(t - \delta) \left(\frac{\mathrm{d}A}{\mathrm{d}\delta}\right) \mathrm{d}\delta\right] + W \frac{\mathrm{d}A}{\mathrm{d}t} \tag{8.20}$$

where q_i = constant injection rate during extension, cubic feet per minute

$v(t - \delta)$ = velocity of fluid flowing into the formation at a given point, feet per minute

t = total pumping time, minutes

δ = time required for the fluid to reach a given point, minutes

$(t - \delta)$ = time interval during which the fluid has leaked from any point, minutes

A = total area of one face of the fracture at any time during injection, square feet

W = fracture width, feet

This equation may be solved for the fracture area at any time $A(t)$ by means of the Laplace transformation,[22] provided q_i is a constant and the equation for $v(t)$ is known. In the next section, the value of v as a function of time will be determined for three specific mechanisms. In all three cases the value of $v(t)$ is

$$v(t) = C/\sqrt{t} \tag{8.21}$$

where C is a constant which is different for each mechanism.

The Laplace transform $L\{F(t)\}$ of any function $F(t)$ is defined as

$$L\{F(t)\} = \int_0^\infty e^{-st}F(t)\,dt \tag{8.22}$$

This integral is readily evaluated and its solution for numerous functions $F(t)$ is found in standard tables of Laplace transforms.[23] One important property of Laplace transforms is the transform of a convolution product. The convolution product of $F(t)$ and $G(t)$, where F and G are some functions of t, is defined as[23]

$$F(t) * G(t) = \int_0^t F(\delta)G(t - \delta)\,d\delta \tag{8.23}$$

It can be shown that the Laplace transform of this product is

$$L\{F(t) * G(t)\} = L\{F(t)\}\,L\{G(t)\} \tag{8.24}$$

The first term on the right-hand side of Eq. (8.20) is the convolution product

$$2\left[\frac{dA}{dt}(t) * v(t)\right]$$

Therefore, the Laplace transform of this term is

$$2\left[L\left\{\frac{dA}{dt}(t)\right\} L\,\{v(t)\}\right]$$

Taking the Laplace transform of both sides of Eq. (8.20),

$$L\{q_i\} = 2\left[L\left\{\frac{dA}{dt}(t)\right\} L\,\{v(t)\}\right] + L\left\{W\frac{dA}{dt}(t)\right\} \tag{8.25}$$

Evaluating the transforms indicated from a standard table of transforms and noting that $v(t)$ is represented by Eq. (8.21), we can write Eq. (8.25) as

$$\frac{q_i}{s} = 2[sL\{A(t)\}][C\sqrt{\pi}\, s^{-1/2}] + WsL\{A(t)\}$$

Solving for $L\{A(t)\}$,

$$L\{A(t)\} = \frac{q_i}{W}\left[\frac{1}{s^{3/2}\left(\dfrac{2C\sqrt{\pi}}{W} + s^{1/2}\right)}\right] \tag{8.26}$$

Letting $b = 2C\sqrt{\pi}/W$, and expanding Eq. (8.26) as a sum of partial fractions,

$$L\{A(t)\} = \frac{q_i}{W}\left[-\frac{1}{b^3(b + s^{1/2})} + \frac{1}{b^3 s^{1/2}} - \frac{1}{b^2 s} + \frac{1}{b s^{3/2}}\right]$$

Grouping the first two terms in the bracket and factoring out $1/b^2$ gives

$$L\{A(t)\} = \frac{q_i}{Wb^2}\left[\frac{1}{s^{1/2}\,(b + s^{1/2})} - \frac{1}{s} + \frac{b}{s^{3/2}}\right] \tag{8.27}$$

Each of the three terms inside the bracket has an inverse transform which can be found in a standard table of transforms.[23] Taking the inverse transform of Eq. (8.27),

$$A(t) = \frac{q_i}{Wb^2}\left[e^{b^2 t}\,\text{erfc}\,(b\sqrt{t}\,) + 2b\sqrt{\frac{t}{\pi}} - 1\right] \tag{8.28}$$

On substitution of the value of $b(= 2C\sqrt{\pi}/W)$, Eq. (8.28) becomes

$$A(t) = \frac{q_i W}{4\pi C^2}\left[e^{(2C\sqrt{\pi t}/W)^2}\,\text{erfc}\left(\frac{2C\sqrt{\pi t}}{W}\right) + \frac{4C\sqrt{t}}{W} - 1\right]$$

or

$$A(t) = \frac{q_i W}{4\pi C^2}\left[e^{x^2}\,\text{erfc}\,(x) + \frac{2x}{\sqrt{\pi}} - 1\right] \tag{8.29}$$

where $x = 2C\sqrt{\pi t}/W$, $A(t)$ is the area, in square feet, of one face of the fracture at time t, q_i is the injection rate in cubic feet per minute, W is the width of the fracture in feet, t is injection time in minutes, C is the fracturing-fluid coefficient in feet per square root of minutes, and erfc (x) designates the complementary error function of (x), defined as[23,24]

$$\begin{aligned}\text{erfc}\,(x) &= \frac{2}{\sqrt{\pi}}\int_x^{\infty} e^{-t^2}\,dt \\ &= 1 - \frac{2x}{\sqrt{\pi}}\left[1 - \frac{x^2}{1\,!\,3} + \frac{x^4}{2\,!\,5} - \frac{x^6}{3\,!\,7} + \ldots\right]\end{aligned} \tag{8.30}$$

The values of the complementary error function[25] have been calculated and are presented in Table 8.1. This table may be compared with logarithm tables which have been worked out also from series expansions such as

$$\ln x = \int_1^x \frac{dx}{x} = 2\left[\frac{x-1}{x+1} + \frac{(x-1)^3}{3(x+1)^3} - \frac{(x-1)^5}{5(x+1)^5} + \ldots\right]$$

Table 8.1. COMPLEMENTARY ERROR FUNCTION OF x^a

x	0	1	2	3	4	5	6	7	8	9
0.0	1.0000	0.9887	0.9774	0.9662	0.9549	0.9436	0.9324	0.9211	0.9099	0.8987
0.1	0.8875	0.8764	0.8652	0.8541	0.8431	0.8320	0.8210	0.8100	0.7991	0.7882
0.2	0.7773	0.7665	0.7557	0.7450	0.7343	0.7237	0.7131	0.7026	0.6921	0.6817
0.3	0.6714	0.6611	0.6509	0.6407	0.6306	0.6206	0.6107	0.6008	0.5910	0.5813
0.4	0.5716	0.5620	0.5525	0.5431	0.5338	0.5245	0.5153	0.5063	0.4973	0.4883
0.5	0.4795	0.4708	0.4621	0.4535	0.4451	0.4367	0.4284	0.4202	0.4121	0.4041
0.6	0.3961	0.3883	0.3806	0.3730	0.3654	0.3580	0.3506	0.3434	0.3362	0.3292
0.7	0.3222	0.3153	0.3086	0.3019	0.2953	0.2888	0.2825	0.2762	0.2700	0.2639
0.8	0.2579	0.2520	0.2462	0.2405	0.2349	0.2293	0.2239	0.2186	0.2133	0.2082
0.9	0.2031	0.1981	0.1932	0.1884	0.1837	0.1791	0.1746	0.1701	0.1658	0.1615
1.0	0.1573	0.1532	0.1492	0.1452	0.1414	0.1376	0.1339	0.1302	0.1267	0.1232
1.1	0.1198	0.1165	0.1132	0.1100	0.1069	0.1039	0.1009	0.0980	0.0952	0.0924
1.2	0.0897	0.0870	0.0845	0.0819	0.0795	0.0771	0.0748	0.0725	0.0703	0.0681
1.3	0.0660	0.0639	0.0619	0.0600	0.0581	0.0562	0.0544	0.0527	0.0510	0.0493
1.4	0.0477	0.0461	0.0446	0.0431	0.0417	0.0403	0.0389	0.0376	0.0363	0.0351

[a]Column headings refer to second decimal place, e.g., erfc (1.25) = 0.0771.

Table 8.1. (CONTINUED)

x	0	1	2	3	4	5	6	7	8	9
1.5	0.0339	0.0327	0.0316	0.0305	0.0294	0.0284	0.0274	0.0264	0.0255	0.0245
1.6	0.0237	0.0228	0.0220	0.0212	0.0204	0.0196	0.0189	0.0182	0.0175	0.0168
1.7	0.0162	0.0156	0.0150	0.0144	0.0139	0.0133	0.0128	0.0123	0.0118	0.0114
1.8	0.0109	0.0105	0.0101	0.0097	0.0093	0.0089	0.0085	0.0082	0.0078	0.0075
1.9	0.0072	0.0069	0.0066	0.0063	0.0061	0.0058	0.0056	0.0053	0.0051	0.0049
2.0	0.00468	0.00448	0.00428	0.00409	0.00391	0.00374	0.00358	0.00342	0.00327	0.00312
2.1	0.00298	0.00285	0.00272	0.00259	0.00247	0.00236	0.00225	0.00215	0.00205	0.00195
2.2	0.00186	0.00178	0.00169	0.00161	0.00154	0.00146	0.00139	0.00133	0.00126	0.00120
2.3	0.00114	0.00109	0.00103	0.00098	0.00094	0.00089	0.00085	0.00080	0.00076	0.00072
2.4	0.00069	0.00065	0.00062	0.00059	0.00056	0.00053	0.00050	0.00048	0.00045	0.00043
2.5	0.00041	0.00039	0.00037	0.00035	0.00033	0.00031	0.00029	0.00028	0.00026	0.00025
2.6	0.00024	0.00022	0.00021	0.00020	0.00019	0.00018	0.00017	0.00016	0.00015	0.00014
2.7	0.00013	0.00013	0.00012	0.00011	0.00011	0.00010	0.00009	0.00009	0.00008	0.00008
2.8	0.000075	0.000071	0.000067	0.000063	0.000059	0.000056	0.000052	0.000049	0.000046	0.000044
2.9	0.000041	0.000039	0.000036	0.000034	0.000032	0.000030	0.000028	0.000027	0.000025	0.000023

8.5. Fracturing Fluid Coefficients

The constant C in Eq. (8.29) is the *fracturing fluid* coefficient, and for any given type of flow system it depends upon the characteristics of the fracturing fluid, the reservoir fluids, and the rock. The fracturing fluid coefficient is the only term in Eq. (8.29) which reflects the properties of the fracturing fluid and is therefore a measure of their relative effectiveness. A low fracturing fluid coefficient means low fluid-loss properties and thus a larger fracture area for a given fluid volume and injection rate. During a fracturing treatment, only the volume of fluid which remains within the walls of the fracture is useful, i.e., the fluid that leaks out into the formation is lost as far as additional fracture extension is concerned. Therefore, the relative effectiveness of different fluids depends upon the properties which control the amount of fluid loss taking place during the fracture treatment. On this basis most fracture fluids fall into three distinct categories: (1) viscosity-controlled fluids, (2) reservoir-controlled fluids, and (3) wall-building fluids.

Viscosity controlled fluids. This category includes viscous or semiviscous fracture fluids, in situations where the viscosity controls the amount of fluid loss taking place during fracturing. Where the viscosity of the fracturing fluid is considerably greater than that of the formation fluid, most of the pressure drop $(p_i - p_e)$ during fracturing will occur in the invaded zone since the fracturing fluid is much less mobile than the formation fluid. Therefore, the viscosity of the fracture fluid controls the amount of fluid lost to the formation.

From Darcy's law the equation for velocity in a linear flow system,[17] in barrels per day per square foot, is

$$v = 1.127 \frac{k}{\mu} \frac{\Delta p}{L} \tag{8.31}$$

where k is the rock permeability in darcies, μ the viscosity of the invading fluid in centipoises, Δp the differential pressure in psi which is active over the variable length (of invaded zone) L in feet. In Eq. (8.31), v is an *apparent* velocity defined in terms of A, the bulk cross-sectional area of flow. In terms of effective flow area $A\phi$, the actual velocity is given by

$$v_{\text{act}} = 5.615 \times 1.127 \frac{k}{\mu\phi} \frac{\Delta p}{L} \text{ ft/day}$$

where ϕ is the porosity of the formation and the conversion 5.615 cu ft/bbl was introduced to obtain v_{act} in ft/day. Since $v_{\text{act}} = dL/dt$, the previous expression can be written as an integral equation,

$$\int_0^L L\,dL = \int_0^t 6.328\,\frac{k\Delta p}{\mu\phi}\,dt \tag{8.32}$$

Integrating Eq. (8.32) and solving for L,

$$L = \left(12.66\,\frac{k\Delta pt}{\mu\phi}\right)^{1/2} \tag{8.33}$$

Substituting Eq. (8.33) into (8.31) gives

$$v = 1.127\,\frac{k}{\mu}\,\frac{\Delta p}{\left(12.66\,\dfrac{k\Delta pt}{\mu\phi}\right)^{1/2}}$$

Simplifying,

$$v = 0.3168\left(\frac{k\Delta p\phi}{\mu t}\right)^{1/2} \tag{8.34}$$

or, if v is expressed in feet per minute and t in minutes,

$$\frac{v \times 1440\text{ min/day}}{5.615\text{ cu ft/bbl}} = 0.3168\left(\frac{k\Delta p\phi}{\mu t/1440\text{ min/day}}\right)^{1/2}$$

and

$$v = 0.0469\left(\frac{k\Delta p\phi}{\mu t}\right)^{1/2} \tag{8.35}$$

Hence

$$v = \frac{C_v}{\sqrt{t}}$$

and the coefficient for this type of fracturing fluid is expressed by

$$C_v = 0.0469\left(\frac{k\Delta p\phi}{\mu}\right)^{1/2}\text{ ft}/\sqrt{\text{min}} \tag{8.36}$$

where k = effective formation permeability to fracturing fluid, in darcies

ϕ = formation porosity, a fraction

μ = fracturing-fluid viscosity at bottom-hole temperature, in centipoises

Δp = differential pressure across the face of the fracture, in psi. This is the product of the fracture gradient and depth, minus the normal reservoir pressure $(G_f \times D) - p_{ws}$

Equation (8.36) simply states that for a viscosity-controlled fluid, the rate of leak-off will depend on the permeability, the porosity, the treating pressure differential, and the fracture fluid viscosity.

The permeability and porosity of the section which may actually fracture should be determined from the most reliable data available. The average effective porosity should be used for calculations. Either an average horizontal permeability or an average vertical permeability should be taken, depending on whether the well can be expected to fracture in predominantly a vertical or a horizontal plane. The object is to determine an average effective permeability through which leak-off will occur.

Example 8.2. Calculate the fracturing fluid coefficient of an oil, given:

Fracture gradient = 0.7 psi per ft

Depth = 4000 ft

Bottom-hole pressure = 1800 psi

Bottom-hole temperature = 100°F

Porosity = 20 per cent

Permeability perpendicular to fracture face = 10 md

Fracturing fluid viscosity at 100°F = 500 cp

SOLUTION: The differential pressure is

$$\Delta p = 0.7 \times 4000 - 1800 = 1000 \text{ psi}$$

Substituting in Eq. (8.36),

$$C_v = 0.0469\left(\frac{0.01 \times 1000 \times 0.20}{500}\right)^{1/2}$$

$$= 3.0 \times 10^{-3} \text{ ft}/\sqrt{\text{min}}$$

Figure 8.11, obtained from Eqs. (8.29) and (8.36), shows the effects of fracturing fluid viscosity on the fracture area for an assumed set of reservoir and treating conditions.[21] The two differential pressure curves show that an increase in absolute viscosity results in an increase in the fracture area — i.e., the more viscous the fracturing fluid, the slower the rate of leak-off — and that a reduction in the differential pressure is equivalent to an increase in the viscosity of the fracturing fluid.

Reservoir-controlled fluids. This group includes those fracturing fluids that have low viscosity and high fluid-loss characteristics, i.e., physical properties identical or nearly identical with those of the reservoir fluid. Fracturing fluids which fall into this classification are lease crude and water, which do not contain additives to reduce fluid loss. The poor properties of these fluids as fracturing fluids are sometimes offset by the favorable properties of the reservoir fluids. Under these conditions, the rate of fluid loss is controlled by the viscosity and compressibility of the reservoir fluid. Any formation which is completely saturated with liquid, such as exists in the vicinity of a water injection well, cannot accommodate additional fluid without the fluid in the reservoir being compressed. The

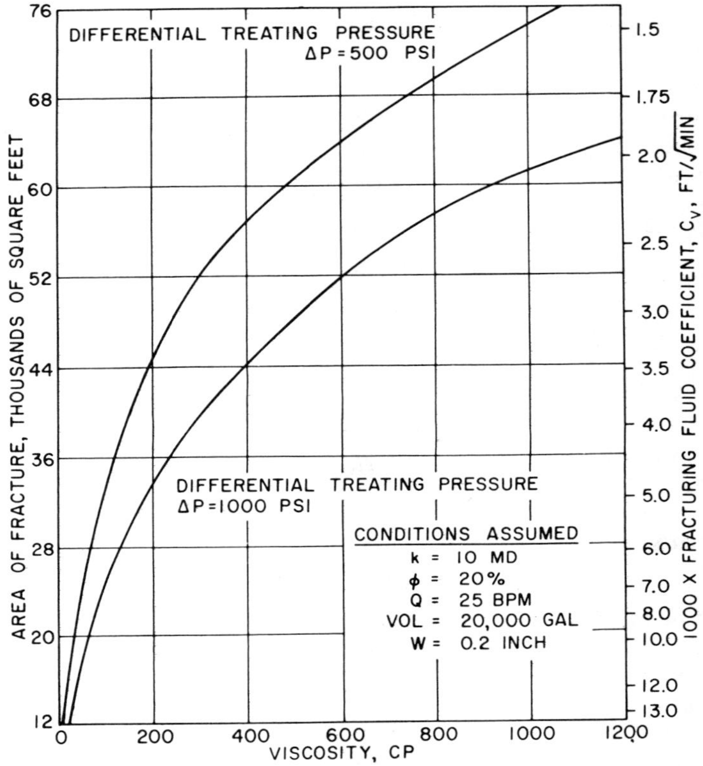

Fig. 8.11 Effect of fracturing fluid viscosity on area of fracture. (Howard and Fast,[21] *Drilling and Production Practice*, API.)

fundamental equation describing the flow of fracturing fluid in this case is the *diffusivity equation* in linear form.

The rate of flow at a distance $(x + dx)$ shown in Fig. 8.12 is greater than that at distance x by an amount dq, owing to both the expansion of the flowing fluid, q, in dropping from pressure $(p + dp)$ to pressure p, and the expansion of fluid in the element $A\, dx$ in which the pressure is dropping at the rate of dp/dt psi per day. The expansion of the flowing fluid may be neglected. The volume of fluid contained in the element is

$$V = \frac{A\phi\, dx}{5.615} \text{ bbl}$$

The change in volume of the element, dV, caused by the pressure drop dp is

$$dV = -c_f V\, dp = -\frac{A\phi c_f\, dx\, dp}{5.615} \text{ bbl}$$

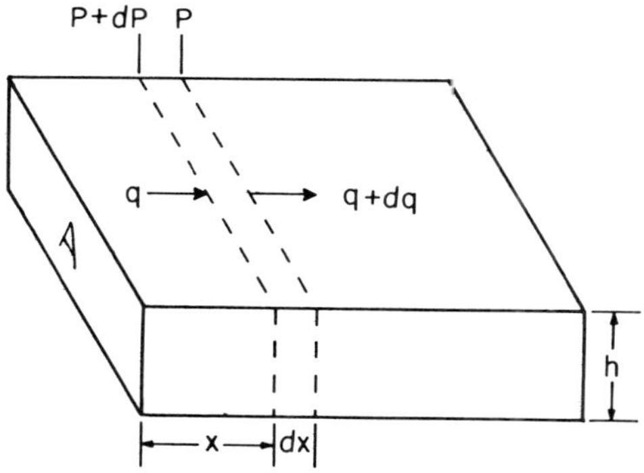

Fig. 8.12

where c_f is the average compressibility of the reservoir fluids. The rate of change of volume with time, $\partial V/\partial t$, at any time is the rate of expansion from the element, or

$$dq = \frac{\partial V}{\partial t} = -\frac{A\phi c_f \, dx}{5.615} \frac{\partial p}{\partial t}$$

whence

$$\frac{\partial q}{\partial x} = -\frac{A\phi c_f}{5.615} \frac{\partial p}{\partial t} \tag{8.37}$$

Darcy's law for linear flow is

$$q = -1.127 \frac{kA}{\mu} \frac{\partial p}{\partial x}$$

Differentiating,

$$\frac{\partial q}{\partial x} = -1.127 \frac{kA}{\mu} \frac{\partial^2 p}{\partial x^2} \tag{8.38}$$

Equating (8.37) and (8.38),

$$1.127 \frac{kA}{\mu} \frac{\partial^2 p}{\partial x^2} = \frac{A\phi c_f}{5.615} \frac{\partial p}{\partial t}$$

Simplifying,

$$\frac{\partial^2 p}{\partial x^2} = \frac{1}{\eta} \frac{\partial p}{\partial t} \tag{8.39}$$

This is the diffusivity equation in linear form where η is the diffusivity constant defined by

$$\eta = \frac{6.328k}{\mu c_f \phi} \tag{8.40}$$

With the Laplace transformation applied to Eq. (8.39), one solution[26] was found to be

$$\frac{p(x, t) - p_e}{\Delta p} = \text{erfc}\left(\frac{x}{2\sqrt{\eta t}}\right) \tag{8.41}$$

where $p(x, t)$ is the pressure at any point x at time t, p_e is the external boundary pressure, and Δp is the pressure differential during fracturing. The velocity of the fracturing fluid at the fracture face for the case of reservoir-controlled fluids is found from Darcy's law to be

$$v_{x=0} = -1.127 \frac{k}{\mu}\left(\frac{\partial p}{\partial x}\right)_{x=0} \tag{8.42}$$

The term $\left(\frac{\partial p}{\partial x}\right)_{x=0}$ is found by differentiating Eq. (8.41),

$$\left(\frac{\partial p}{\partial x}\right)_{x=0} = -\frac{\Delta p}{\sqrt{\pi \eta t}} \tag{8.43}$$

Substituting Eq. (8.43) into Eq. (8.42),

$$v_{x=0} = \frac{1.127 k \Delta p}{\mu \sqrt{\pi \eta t}} \tag{8.44}$$

Substituting Eq. (8.40) into Eq. (8.44) and simplifying,

$$v_{x=0} = 0.2528 \Delta p \left(\frac{k c_f \phi}{\mu t}\right)^{1/2}$$

or, if v is expressed in feet per minute and t in minutes,

$$v_{x=0} = 0.0374 \Delta p \left(\frac{k c_f \phi}{\mu t}\right)^{1/2} \tag{8.45}$$

or

$$v = \frac{C_c}{\sqrt{t}}$$

Hence the equation for the fluid-loss coefficient is

$$C_c = 0.0374 \Delta p \left(\frac{k c_f \phi}{\mu}\right)^{1/2} \text{ ft}/\sqrt{\text{min}} \tag{8.46}$$

where Δp = differential pressure across the face of the fracture, psi
k = effective formation permeability, darcies
ϕ = formation porosity, a fraction
c_f = isothermal compressibility of the reservoir fluid, psi^{-1}
μ = reservoir fluid viscosity, centipoises

It should be noted that the quantities μ and c_f in the above equations are properties of the reservoir fluid and not of the fracturing fluid. The effect of reservoir fluid viscosity on the area of a fracture is shown in Fig. 8.13. Since reservoir fluid compressibility does not vary appreciably, we can assume it to be 10×10^{-6} psi^{-1} for liquid-filled reservoirs. It is seen that Fig. 8.13[21] is similar to Fig. 8.11 in that an increase in viscosity results in an increase in fracture area.

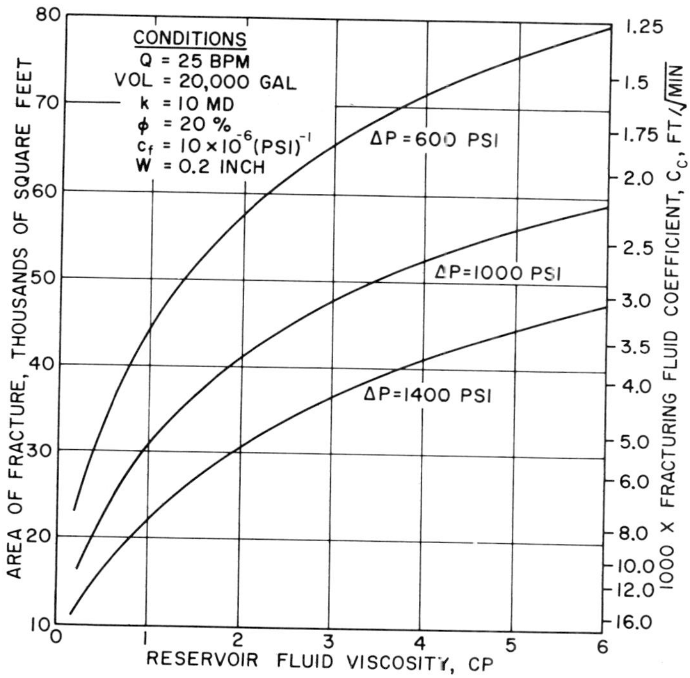

Fig. 8.13 Effect of reservoir fluid on area of fracture. (Howard and Fast,[21] *Drilling and Production Practice*, API.)

Wall-building fluids. The use of modern additives to limit fluid loss[27] (asphaltic-type materials, synthetic gums, and insoluble solids added to oil or water) creates a third class of fracturing fluids. These fluids build a temporary filter cake or wall on the face of the fracture as it is exposed. While a small amount of fluid leaks through to form the wall, once formed, the wall presents quite an effective barrier to further leak-off due to its low permeability. Because of this low permeability, most of the pressure drop $(p_i - p_e)$ will occur across the filter cake, and therefore the rate of flow of the fracturing fluid into the formation is obtained by applying Darcy's law across the filter cake. The volume of fluid which has leaked off through the filter cake at any time is proportional to the volume of the filter cake at that time, or

$$V = cA_fL \tag{8.47}$$

where V is the volume of fluid, A_f the cross-sectional area of the filter cake,

L the filter cake thickness, and c is the proportionality constant. Darcy's law for linear flow is

$$q = \frac{dV}{dt} = 1.127\,\frac{kA_f\Delta p}{\mu L} \tag{8.48}$$

Substituting Eq. (8.47) into Eq. (8.48),

$$\frac{dV}{dt} = 1.127\,\frac{kA_f\Delta p}{\mu V/cA_f}$$

and simplifying,

$$\int_0^V V\,dV = \int_0^t 1.127\,\frac{kA_f^2c\Delta p}{\mu}\,dt$$

Integrating and simplifying,

$$V = \left(2.254\,\frac{kA_f^2c\Delta p}{\mu}\right)^{1/2}\sqrt{t} \tag{8.49}$$

or

$$V = m\sqrt{t} \tag{8.50}$$

where m is $(2.254\,kA_f^2c\Delta p/\mu)^{1/2}$. If a standard fluid-loss test is run on a fracturing fluid and V, the volume of filtrate, is plotted against $\sqrt{t}$, the slope of the curve is m. Differentiating Eq. (8.50),

$$\frac{dV}{dt} = q = \frac{m}{2\sqrt{t}}$$

and the velocity is

$$v = \frac{q}{A_f} = \frac{m}{2A_f\sqrt{t}}$$

or, if v is expressed in feet per minute and t in minutes,

$$v = \frac{0.0164m}{A_f\sqrt{t}} \tag{8.51}$$

where m is the slope of the fluid-loss curve in cu cm/$\sqrt{\text{min}}$ and A_f is the area of the test media in square centimeters, or, finally,

$$v = \frac{C_w}{\sqrt{t}}$$

where

$$C_w = \frac{0.0164m}{A_f}\ \text{ft}/\sqrt{\text{min}} \tag{8.52}$$

In the standard filter cake test, the wall cake is formed against filter paper. Howard and Fast[21] have described a technique in which actual core wafers from producing horizons are substituted for the filter paper in the high temperature, high pressure filter press. Figure 8.14 shows the fluid-loss plot for various producing horizons, from which the fracturing fluid

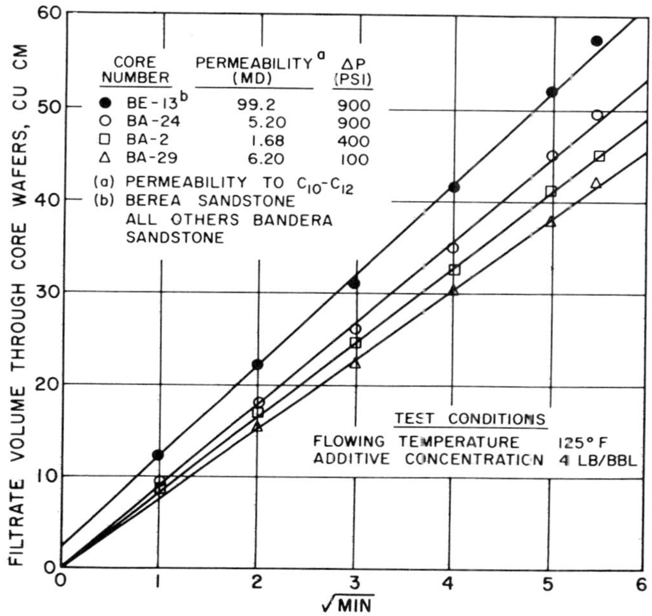

Fig. 8.14 Fluid-loss characteristics of paraffin base crude with fluid-loss additive. (Howard and Fast,[21] *Drilling and Production Practice*, API.)

coefficient C_w is readily calculated.[21] For example, the slope for core BA-2 is 49/6 or 8.17 cu cm/$\sqrt{\text{min}}$. The core cross-sectional area is 18.4 sq cm, and the fracturing fluid coefficient is $0.0164 \times 8.17/18.4$ or 7.28×10^{-3} ft/$\sqrt{\text{min}}$.

If fluid-loss test data are not available at the differential pressure expected for the fracture job, C_w at some other pressure may be corrected to the expected differential. The slope of the fluid-loss curve from Eq. (8.50) is proportional to $(\Delta p)^{1/2}$,

$$m = K\Delta p^{1/2}$$

Therefore,

$$m_{\text{act}} = m\left(\frac{\Delta p_{\text{act}}}{\Delta p}\right)^{1/2}$$

and

$$C_{w\ \text{act}} = C_w\left(\frac{\Delta p_{\text{act}}}{\Delta p}\right)^{1/2} \tag{8.53}$$

In this equation it is assumed that the permeability of the filter cake remains constant with change in differential pressure, which is not exactly true; therefore, Eq. (8.53) should be used only when data for the correct pressure differential cannot be obtained.

Usually, in a fluid-loss test, a certain amount of fluid is forced through the filter paper at the instant the differential pressure is applied, i.e., before any filter cake has begun to form. This loss is called the *spurt loss* (V_{sp}) and is measured in cubic centimeters. For core BE-13 of Fig. 8.14 the spurt loss is 2.5 cu cm, the intercept of the straight line on the y axis at zero time. If a spurt loss is included in the equations above, Eq. (8.50) becomes

$$V = m\sqrt{t} + V_{sp}$$

When this equation is differentiated, V_{sp} will drop out, thus having *no effect* on the fracturing fluid coefficient. However, the spurt loss does affect the calculation of the fracture area, Eq. (8.29). Consider a fracture of area A_f with a spurt loss of V_{sp}. The volume of the fracture is $A_f W$ where W is the *true* fracture width. Since the spurt loss was not considered in Eq. (8.29), the volume of the fracture, as calculated from this equation, is greater than $A_f W$ by $2V_{sp}$. If we define a quantity W' such that the product $A_f W'$ is equal to the volume of the fracture without a spurt loss, then

$$A_f W' = A_f W + 2V_{sp}$$

or

$$W' = W + \frac{2V_{sp}}{A_f}$$

If W and W' are expressed in inches, with A_f in square centimeters and V_{sp} in cubic centimeters, then

$$W' = W + \frac{2V_{sp}}{2.54A_f}$$

or

$$W' = W + 0.787\frac{V_{sp}}{A_f} \tag{8.54}$$

Then if we replace W in Eq. (8.29) by W', the equation will give the correct area of the fracture, A_f, for a fluid with a spurt loss. Example 8.3 shows the calculation of a loss coefficient when a fluid-loss additive is used.

Example 8.3. Calculate the fluid-loss coefficient of a wall-building fluid, C_w, given:

Treating differential pressure = 2700 psi

Slope of fluid-loss curve = 1.2 cu cm/$\sqrt{\text{min}}$

Area of filter = 22.8 sq cm

Spurt loss at 1000 psi = 5 cu cm

SOLUTION: Substituting in Eq. (8.52),

$$C_w = \frac{0.0164 \times 1.2}{22.8} = 0.863 \times 10^{-3}\ \text{ft}/\sqrt{\text{min}}$$

The corrected fluid loss at 2700 psi, from Eq. (8.53), is

$$C_{w\ \text{act}} = 0.000863\sqrt{\frac{2700}{1000}}$$
$$= 1.41 \times 10^{-3}\ \text{ft}/\sqrt{\text{min}}$$

The spurt loss is used in correcting the fracture width by Eq. (8.54). If the fracture width is 0.2 in., the corrected fracture width is

$$W' = 0.2 + \frac{0.787 \times 5}{22.8} = 0.372\ \text{in.}$$

It is seen that the fracturing fluid coefficient may take any of three forms. Each of these forms describes a separate mechanism tending to retard leak-off. In any actual treatment, all of these mechanisms operate to varying extents. Any fracturing fluid is somewhat viscous, and so the C_v mechanism helps retard leak-off. Also, the reservoir contains a compressible fluid, and thus the C_c mechanism will be operative. Finally, many oils without additives will have a wall-building effect, and so the C_w mechanism usually comes into play. The controlling mechanism will always be the most effective one, i.e., that giving the lowest rate of leak-off or the lowest C value.

In partially depleted reservoirs where there is appreciable gas saturation, the calculation of the fracturing fluid coefficient involves special considerations. This difficulty arises from the wide variation in compressibility of the free gas phase. Due to the increase in compressibility, the fracturing fluid coefficient C_v, C_c, and C_w may be all of nearly the same order of magnitude. This means that no one mechanism is in control of the fracturing fluid leak-off, and therefore a combined coefficient must be calculated. The fracture fluid coefficient is analogous to electrical conductance and therefore it has been suggested that a combined coefficient could be calculated similarly to the combined conductance of a series of conductors,

$$\frac{1}{C} = \frac{1}{C_v} + \frac{1}{C_c} + \frac{1}{C_w}$$

This analogy has been tested against field data and found to be reasonably accurate. The following data will be used to illustrate this calculation:

Type of formation = sandstone

Formation permeability at 25 per cent water saturation = 2.6 md

Porosity = 0.06

Formation oil viscosity* = 1.0 cp

*Methods of calculating these properties from known field data can be found in most texts on reservoir engineering.

Connate water saturation, $S_w = 0.25$

Oil compressibility, $c_o = 11 \times 10^{-6}$ psi^{-1}

Gas compressibility, $c_g = 530 \times 10^{-6}$ psi^{-1}

Water compressibility, $c_w = 3 \times 10^{-6}$ psi^{-1}

Gas gravity = 0.75

Gas deviation factor* at 1900 psia = 0.85

Gas viscosity at reservoir conditions* = 0.016 cp

Producing gas-oil ratio = 1450 SCF/STB

Solution gas-oil ratio at 1900 psia = 350 SCF/STB

Formation volume factor* = 1.22 bbl/STB

Reservoir pressure = 1900 psia

Reservoir temperature = 200°F

Fracture treating pressure = 6000 psia

Fracture fluid = lease oil

Fracture fluid gravity = 30° API ($\gamma = 0.876$)

Fracture fluid viscosity at reservoir conditions* = 1.8 cp

Fracture width = 0.1 in.

The viscosity-controlled fracturing fluid coefficient for this system, from Eq. (8.36), is

$$C_v = 0.0469 \left(\frac{0.0026 \times 4100 \times 0.06}{1.8}\right)^{1/2}$$

$$= 27.9 \times 10^{-3} \text{ ft}/\sqrt{\text{min}}$$

To calculate the reservoir-controlled fracturing fluid coefficient, it is first necessary to determine the reservoir gas saturation in order to obtain the compressibility of the combined fluids. The difference between the producing and solution gas-oil ratios is 1450 − 350 or 1100 SCF/STB, and the reservoir volume of 1100 SCF of gas, from the nonideal gas law, is

$$\frac{14.7 \times 1100}{1 \times 520} = \frac{1900 \times V_2}{0.85 \times 660}$$

or

$$V_2 = 9.18 \text{ cu ft/STB}$$

$$= 1.64 \text{ bbl gas/STB}$$

The ratio of gas to oil flow rates in the reservoir is then

$$\frac{q_g}{q_o} = \frac{1.64}{1.22} = 1.34 \text{ bbl gas/bbl oil}$$

It has been shown that this ratio (q_g/q_o) is equal to the mobility ratio M, or

$$\frac{q_g}{q_o} = M = \frac{k_g/\mu_g}{k_o/\mu_o}$$

Solving for k_g/k_o,

$$\frac{k_g}{k_o} = \frac{1.34 \times 0.016}{1.0} = 0.021$$

Arps and Roberts[28] have presented average values of k_g/k_o as functions of total liquid saturations for sandstones and limestones as shown in Fig. 8.15.

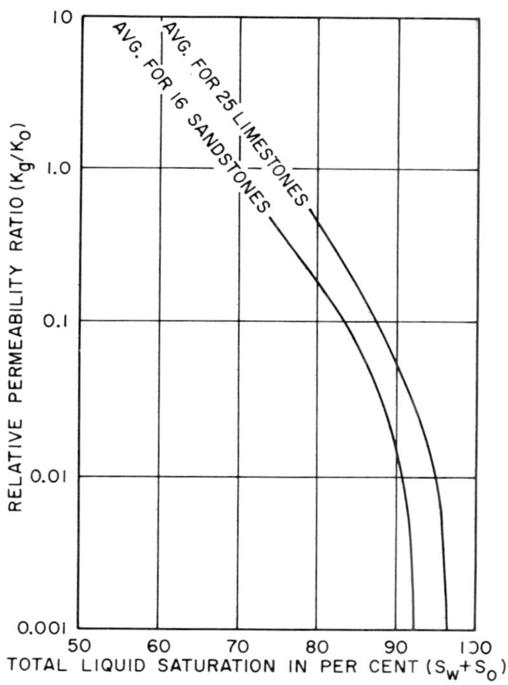

Fig. 8.15 Average k_g/k_o data for sandstones and limestones. (Arps and Roberts,[28] *Trans.* AIME.)

From this figure, the total liquid saturation is 89 per cent and the gas saturation is 11 per cent. Then for a connate water saturation of 25 per cent the oil saturation is 64 per cent. The average compressibility of the formation fluid is

$$c_f = c_g S_g + c_o S_o + c_w S_w$$

or

$$c_f = 530 \times 10^{-6} \times 0.11 + 11 \times 10^{-6} \times 0.64 + 3 \times 10^{-6} \times 0.25$$

$$= 66 \times 10^{-6} \text{ psi}^{-1}$$

The reservoir-controlled fracturing fluid coefficient, from Eq. (8.46), is

$$C_c = 0.0374 \times 4100 \left(\frac{0.0026 \times 66 \times 10^{-6} \times 0.06}{1.0} \right)^{1/2}$$

$$= 15.6 \times 10^{-3} \text{ ft}/\sqrt{\text{min}}$$

Then the combined coefficient is

$$\frac{1}{C} = \frac{1}{27.9 \times 10^{-3}} + \frac{1}{15.6 \times 10^{-3}}$$

or

$$C = 10.0 \times 10^{-3} \text{ ft}/\sqrt{\text{min}}$$

It should be noted that this is not an exact calculation, as the permeability used in the equations for C_c and C_v is the permeability to oil at connate water saturation. Since both oil and gas are flowing, the permeabilities that should be used in these equations would have to be obtained from a complex relative permeability measurement.

8.6. Fracture Efficiency

Having calculated the fracturing fluid coefficient, it is now possible to determine the fracture area from the basic equation

$$A(t) = \frac{q_i W}{4\pi C^2} \left[e^{x^2} \text{ erfc } (x) + \frac{2x}{\sqrt{\pi}} - 1 \right] \tag{8.29}$$

where $x = 2C\sqrt{\pi t}/W$.

Since the solution of Eq. (8.29) is tedious, we shall write it in another form to facilitate the calculation. If we define the efficiency of a fracture job as the volume of the fracture divided by the volume of fluid pumped, then

$$\text{Eff} = \frac{V_f}{V_i} = \frac{WA(t)}{q_i t} \tag{8.55}$$

Substituting Eq. (8.29) into Eq. (8.55),

$$\text{Eff} = \frac{W}{q_i t}\left[\frac{q_i W}{4\pi C^2}\right]\left[e^{x^2}\text{erfc}(x) + \frac{2x}{\sqrt{\pi}} - 1\right]$$

$$= \frac{W^2}{4\pi C^2 t}\left[e^{x^2}\text{erfc}(x) + \frac{2x}{\sqrt{\pi}} - 1\right]$$

or

$$\text{Eff} = \frac{1}{x^2}\left[e^{x^2}\text{erfc}(x) + \frac{2x}{\sqrt{\pi}} - 1\right] \tag{8.56}$$

Since the efficiency is a function of x alone, it is possible to plot efficiency versus x as shown in Fig. 8.16.[29] Example 8.4 shows the method used in constructing the graph.

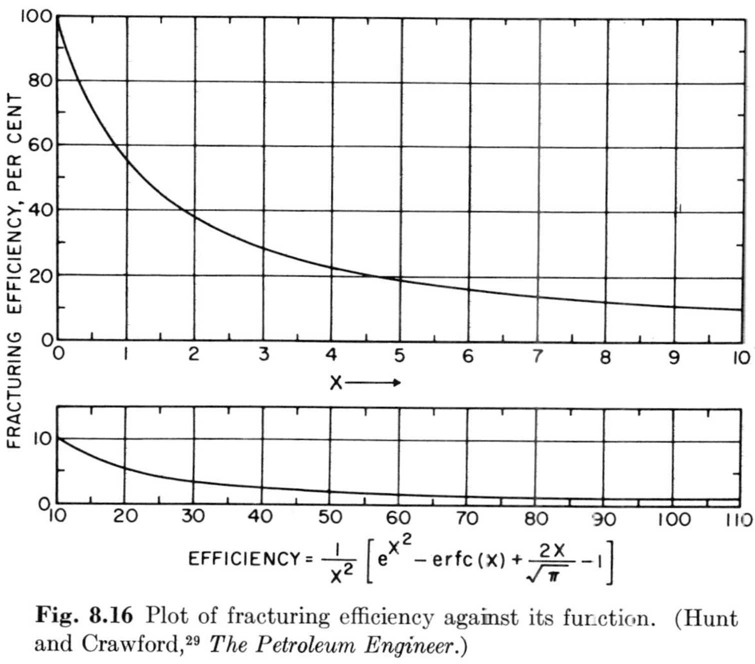

Fig. 8.16 Plot of fracturing efficiency against its function. (Hunt and Crawford,[29] *The Petroleum Engineer.*)

Example 8.4. Calculate fracture efficiency, given:

Injection volume = 40,000 gal

Injection rate = 30 bbl/min

Fracture width = 0.2 in.

Fracturing coefficient = 2.22×10^{-3} ft/$\sqrt{\text{min}}$

SOLUTION: The fracture time is

$$t = \frac{40{,}000}{42 \times 30} = 32 \text{ min}$$

and

$$x = \frac{2 \times 2.22 \times 10^{-3}\sqrt{32\pi}}{0.2/12} = 2.67$$

Substituting in Eq. (8.56),

$$\text{Eff} = \frac{1}{(2.67)^2}\left[e^{(2.67)^2}\,\text{erfc}\,(2.67) + 2\left(\frac{2.67}{\sqrt{\pi}}\right) - 1\right]$$

From Table 8.1, erfc (2.67) = 0.00016, so that the efficiency is

$$\text{Eff} = 0.140[1248 \times 0.00016 + 3.01 - 1] = 31 \text{ per cent}$$

The use of Fig. 8.16 and Eq. (8.55) provide a simplified method of calculating the area of the fracture at any time, $A(t)$. For example, if the injection volume is 20,000 gal of fluid with a coefficient of 2.22×10^{-3} ft/$\sqrt{\text{min}}$, the fracture width is 0.2 in., and the pumping time is 20 min, then

$$x = \frac{2 \times 2.22 \times 10^{-3}\sqrt{20\pi}}{0.2/12} = 2.11$$

From Fig. 8.16, the efficiency is 37 per cent. Then from Eq. (8.55) the fracture area is

$$A = \frac{20{,}000 \times 0.37}{7.48 \times 0.2/12} = 59{,}360 \text{ sq ft}$$

8.7. Fracturing Hydraulics

Fracturing treatments with 100,000 to 200,000 gal injected at rates approaching 100 bbl per min are not only common but are performed with minimum difficulty. The cost of such treatments is, of course, great, and it is imperative that material and equipment be effectively utilized. Therefore, it is desirable to determine the hydraulic horsepower for each fracturing treatment that will keep costs at a minimum. The hydraulic power required to pump q_t barrels of fluid per minute into a well with a surface injection pressure of p_s pounds per square inch is given by the product $p_s q_t$ or, converting to hydraulic horsepower,

$$H_h = \frac{144 \text{ in.}^2/\text{ft}^2 \times 5.615 \text{ cu ft/bbl } p_s q_t}{33{,}000 \text{ ft-lb/min/hp}}$$

$$= 0.0245\, p_s q_t \text{ hp} \tag{8.57}$$

Surface injection pressure. The pump pressure or surface injection pressure, p_s, is equal to the sum of the bottom-hole fracture treating pressure p_t, the frictional pressure drop in the pipe Δp_f, and the pressure drop through the perforations Δp_p, minus the hydrostatic pressure Δp_s, or

$$p_s = p_t + \Delta p_f + \Delta p_p - \Delta p_s \tag{8.58}$$

If, during the treatment, the pumps are stopped, the surface pressure will drop, as there is no flow. In the momentary absence of frictional pressure losses, the bottom-hole fracture treating pressure can be calculated with greater precision from the expression

$$p_t = p_i + \Delta p_s \tag{8.59}$$

where p_i is the instantaneous surface shut-down pressure. In addition, the frictional pressure losses in the system $(\Delta p_f + \Delta p_p)$ can be measured indirectly, by use of the equation

$$\Delta p_f + \Delta p_p = p_s - p_i \tag{8.60}$$

Since the pressure drop across the perforations is usually small in comparison with the other pressure terms, one can assume that it is negligible and obtain the surface injection pressure as

$$p_s = p_t + \Delta p_f - \Delta p_s \tag{8.61}$$

Since Δp_f and Δp_p in Eq. (8.58) are influenced by the flow rate, the surface injection pressure will also be dependent upon rate. Safety should always be considered in selecting the injection rate, since the surface injection pressure is limited by the bursting pressure of the casing. For example, the bursting pressure of $4\frac{1}{2}$-in., 11.6-lb, J-55 casing is 5350 psi. Generally, 70 per cent of the bursting pressure is taken as the maximum injection pressure, and therefore p_s should not exceed 3745 psi. If it does, a lower injection rate should be selected to reduce the frictional losses.

Bottom-hole treating pressure. The bottom-hole treating pressure p_t is determined from the fracture gradient G_f and the depth D to the fracture, since

$$p_t = G_f D \tag{8.62}$$

Therefore, combining Eq. (8.58) with Eq. (8.62),

$$G_f = \frac{p_s + \Delta p_s - \Delta p_f - \Delta p_p}{D} \tag{8.63}$$

As pointed out previously, the fracture gradient G_f expected for horizontal fractures (above 2000 ft) is of the order of 1.0 psi per ft or higher, and for vertical fractures (below 4000 ft), it is 0.7 psi per ft or lower.

Hydrostatic pressure. The hydrostatic pressure Δp_s is obtained from the density of the fluid including the propping agent. The density of the fluid – sand mixture, ρ, in pounds per gallon will be the total number of pounds divided by the total number of gallons; on the basis of one gallon of liquid,

$$\rho = \frac{8.34\gamma + x}{1 + \dfrac{x}{2.63(8.34)}}$$

$$= \frac{8.34\gamma + x}{1 + 0.0456x} \tag{8.64}$$

where γ is the specific gravity of the fracture fluid, x is the concentration of sand in pounds per gallon of fluid, and 2.63 is the specific gravity of sand. If the specific gravity of the fracturing fluid is given at 60°F, the specific gravity at the average well temperature T is calculated from

$$\gamma_T = \gamma_{60}[1 - \beta(T - 60)] \tag{8.65}$$

where β is the thermal coefficient of expansion of the fluid. If the fracture fluid is a crude oil, the value of β is obtained from volume correction tables.

The hydrostatic pressure Δp_s is equal to the hydrostatic gradient times the depth, or

$$\Delta p_s = \frac{0.433}{8.34}\left(\frac{8.34\gamma_T + x}{1 + 0.0456x}\right) D$$

$$= 0.052\rho_T D \tag{8.66}$$

Pressure drop across perforations. The flow velocity and hence the capacity of a perforation is theoretically limited by the *critical velocity* which is the velocity of sound in the medium. However, attainment of this critical velocity through casing perforations is unlikely since the pressure required for development of a critical velocity is in excess of the pipe strength.[30] Therefore, the pressure drop across the perforations is obtained directly from Bernoulli's equation[31]

$$\frac{p_1}{\rho} + \frac{v_1^2}{2g_c} = \frac{p_2}{\rho} + \frac{v_2^2}{2g_c} \tag{8.67}$$

where p_1 = casing pressure, pounds per square foot

p_2 = pressure at perforations, pounds per square foot

ρ = density, pounds per cubic foot

v_1 = velocity in the casing, feet per second

v_2 = velocity in the perforation, feet per second

Rearranging Eq. (8.67),

$$\frac{\Delta p_p}{\rho} = \frac{v_2^2 - v_1^2}{2g_c}$$

But $v_2^2 - v_1^2 \cong v_2^2$, and hence

$$v_2^2 = 2g_c\frac{\Delta p_p}{\rho} \tag{8.68}$$

The ideal flow rate is $q_i = A_2 v_2$ and the actual flow rate is $q = C_d q_i$, where C_d is the coefficient of discharge for the perforation, found to be approximately 0.82 for bullet perforations.[32] With these substitutions, Eq. (8.68) becomes

$$\frac{q^2}{(0.82A_2)^2} = 2g_c \frac{\Delta p_p}{\rho}$$

Solving for Δp_p,

$$\Delta p_p = \frac{\rho q^2}{1.345 g_c A_2^2}$$

or, converting ρ to pounds per gallon, q to gallons per minute, A to square inches, and pressure to pounds per square inch,

$$\Delta p_p = \frac{\rho q^2}{8090 A_2^2} \tag{8.69}$$

Frictional pressure drop. Any satisfactory method of calculating friction drop in a field application must yield fairly accurate data and must utilize, in a simple fashion, available fluid properties. Fluids used in fracturing can be divided into three groups on the basis of their rheological properties, as discussed in Chapter 1. These are:

1. Oils, water, and acids (Newtonian)
2. Drilling muds and emulsions (non-Newtonian)
3. Gels (non-Newtonian)

For each class of fluid, equations for flow in the tubing and annulus have been developed in Chapter 1.

Since lease crude is used extensively in fracturing operations and in many cases the API gravity of the crude is the only fluid property readily available, the Socony Mobil Oil Company Research Division has developed a correlation for viscosity of crude at various temperatures, as shown in Fig. 8.17.

8.8. Fracture Design Calculation

The objective of any type of fracture design calculation is to plan the most economical treatment that will result in the desired increase in productivity. This type of calculation involves consideration of the following variables:

1. Fracturing fluid coefficient C.
2. Injection rate q.
3. Total injection volume V.
4. Area of the fracture A.
5. Weight of propping agent S.
6. Surface injection pressure p_s.
7. Horsepower required.
8. Productivity ratio of the well.

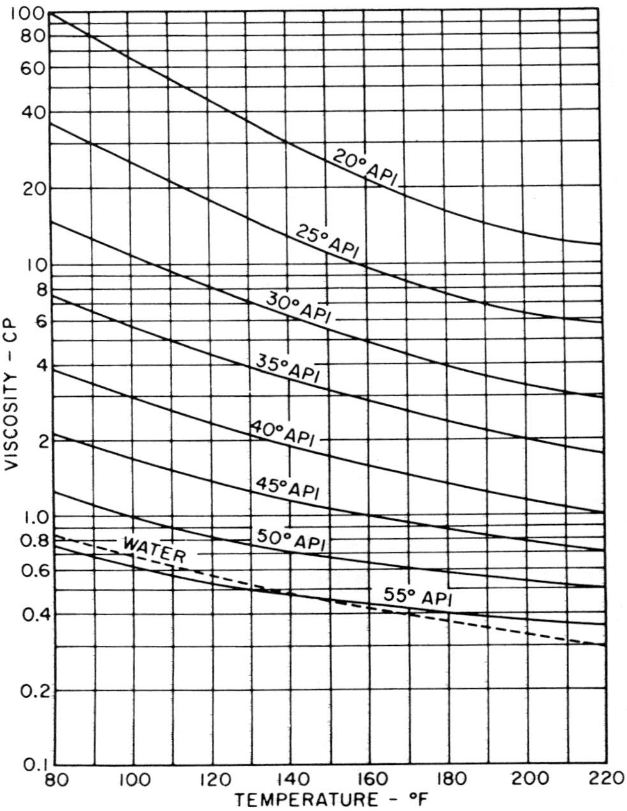

Fig. 8.17 Viscosity of gas-free crude oils. (Courtesy of Mobil Oil Company.)

From the previous discussion of these variables, it can easily be seen that values for some of these parameters must be known or assumed before it is possible to determine the others. This suggests two possible methods of solution. The first approach is to assume an injection rate, a size of treatment (total volume of fracture fluid), and a fracture gradient based on previous fracture jobs in the area. From these assumptions it is possible to obtain the surface pressure, horsepower requirements, maximum quantity of propping agent necessary, and productivity ratio. In the second method all of these parameters can be determined from the desired productivity ratio and an *assumed* fracture gradient. In the first approach, due to the assumptions made, we will have a design which will be economically feasible; the second approach, however, will yield a design which is practical only if the desired productivity ratio is economically practical.

Therefore, in the second method, calculations are made for several assumed productivity ratios to provide a cost comparison.

These two approaches are illustrated in Examples 8.5 and 8.6.

Example 8.5. Design a fracture treatment assuming an injection rate, size of treatment, and fracture gradient, given:

Well depth = 2000 ft

Fracture fluid = lease oil

Oil gravity = 35° API at 60°F ($\gamma = 0.85$)

Propping agent = 10–20 mesh sand

Fracture width = 0.1 in.

Average porosity = 0.135

Average formation permeability = 0.9 md

Formation thickness = 50 ft

Viscosity of reservoir fluid (and lease oil) = 4.0 cp at reservoir conditions

Average compressibility of reservoir fluids = 10×10^{-6} psi^{-1}

Static bottom-hole pressure = 300 psig

Average well temperature = 80°F

Casing size = $5\frac{1}{2}$ in., J-55, ID 4.892 in.

Tubing size = 2-in. nominal, OD 2.375 in.

Fracturing through the annulus

Well spacing = 40 acres (r_e = 660 ft)

SOLUTION: A review of past treatments in the area shows that a treatment of 40,000 gal of lease oil at an injection rate of 30 bbl per min has given effective results in this reservoir; also, it is noted that the fracture gradients have been approximately 1.0 psi per ft. With this fracture gradient, the bottom-hole treating pressure, from Eq. (8.62), is

$$p_t = 1.0 \times 2000 = 2000 \text{ psig}$$

Then the differential pressure across the fracture face is

$$\Delta p = 2000 - 300 = 1700 \text{ psi}$$

Since lease oil is to be used as the fracturing fluid, the fracturing fluid coefficient is C_c. Substituting in Eq. (8.46),

$$C_c = 0.0374 \times 1700 \left(\frac{0.0009 \times 10 \times 10^{-6} \times 0.135}{4}\right)^{1/2}$$

$$= 1.11 \times 10^{-3} \text{ ft}/\sqrt{\text{min}}$$

The pumping time for 40,000 gal of fluid at a rate of 30 bbl per min is

$$t = \frac{40{,}000}{30 \times 42} = 31.7 \text{ min}$$

Solving for x in Eq. (8.29),

$$x = \frac{2 \times 1.11 \times 10^{-3}\sqrt{31.7\pi}}{(0.1)/12} = 2.66$$

and the efficiency from Fig. 8.16 is 31 per cent. Then from Eq. (8.55) the fracture area is

$$A = \frac{40{,}000 \times 0.31}{7.48 \times (0.1)/12} = 198{,}900 \text{ sq ft}$$

From the fracture area and width, it is possible to calculate the maximum amount of sand required, i.e., the amount of sand needed to completely pack the fracture. The volume per unit area of the fracture is

$$V = 1.0 \times (0.1)/12 = 0.00833 \text{ cu ft/sq ft}$$

The porosity of most fracturing sand is of the order of 35 per cent. The specific gravity of sand is 2.63 and the density is 2.63 × 62.4 or 164 lb per cu ft. Then the weight of sand, S, necessary to pack one square foot of fracture is

$$S = 0.00833(1 - 0.35) \times 164 = 0.888 \text{ lb/sq ft}$$

The maximum amount of sand needed is

$$0.888 \times 198{,}900 = 176{,}600 \text{ lb}$$

and the sand concentration in pounds per gallon of oil is 176,600/40,000 or 4.42 lb per gal. The actual flow rate (including sand) is the total volume (oil plus sand) divided by the total fracture time, or

$$q = \frac{40{,}000 + \dfrac{176{,}600}{2.63 \times 8.34}}{31.7} = 1516 \text{ gal/min}$$

$$= \frac{1516}{42} = 36.1 \text{ bbl/min}$$

The coefficient of thermal expansion for oil is approximately 0.0005 per degree. The specific gravity of the oil at the average well temperature, from Eq. (8.65), is

$$\gamma_{80} = 0.85[1 - 0.0005(80 - 60)] = 0.842$$

and the density, from Eq. (8.64), is

$$\rho_{80} = \frac{8.34 \times 0.842 + 4.42}{1 + 0.0456 \times 4.42} = 9.53 \text{ lb/gal}$$

The hydrostatic pressure developed at 2000 ft, from Eq. (8.66), is

$$\Delta p_s = 0.052(9.53)2000 = 991 \text{ psi}$$

The annular geometry parameter α is 2.375/4.892 or 0.485. From the Crittendon correlation given in Fig. 1.6,

$$d_e = 0.74 \times 4.892 = 3.62 \text{ in.}$$

where d_e is the diameter of the circular pipe whose flow rate – pressure loss characteristics will duplicate those of the annulus. The average flow velocity in the equivalent pipe, from Eq. (1.21a), is

$$v = \frac{17.16 \times 36.1}{3.62^2} = 47.3 \text{ ft/sec}$$

The Reynolds number of the flow in the equivalent pipe is found from Eq. (1.15a),

$$N_R = 928 \frac{3.62 \times 47.3 \times 9.53}{4} = 3.79 \times 10^5$$

From Fig. 1.4 the Fanning friction factor is 0.0040, and the frictional pressure drop from Eq. (1.18a) is

$$\Delta p_f = \frac{0.0040 \times 2000 \times 9.53 \times (47.3)^2}{25.80 \times 3.62} = 1826 \text{ psi}$$

When the line efficiency correction for the Crittendon correlation is made,

$$\Delta p_f = \frac{1826}{(0.90)^2} = 2254 \text{ psi}$$

Then the surface injection pressure, neglecting the pressure drop across the perforations, from Eq. (8.61) is

$$p_s = 2000 + 2254 - 991 = 3263 \text{ psig}$$

Since for $5\frac{1}{2}$-in., J-55 casing 70 per cent of the bursting pressure (5320 psi) is 3724 psi, the injection rate used is within the safe limit.

The hydraulic horsepower is calculated from Eq. (8.57), or

$$H_h = 0.0245 \times 3263 \times 36.1 = 2886 \text{ hp}$$

On the basis of 600 hydraulic horsepower per pump truck, five pumpers are required for the job.

To calculate the productivity ratio, several assumptions must be made. Since the gradient is 1.0 psi per ft and the well depth is 2000 ft, it is assumed that the fracture plane is horizontal. Also, if the fracture extends uniformly in a circular pattern, the fracture radius can be found from the fracture area by $\pi r_f^2 = 198{,}900$ ft² or $r_f = 252$ ft. The ratio of fracture radius to external radius is $r_f/r_e = 252/660 = 0.381$. From Fig. 8.8 the permeability of 10-20 mesh sand for a fracture-treating pressure of 2000 psig is 60,000 md, the value of $k_f W/kh$ in Eq. (8.14) is

$$\frac{60{,}000 \times (0.1)/12}{0.9 \times 50} = 11.1$$

From Fig. 8.7 the PR is 5.0, so that, if the initial production was 10 bbl per day, the production after fracturing is 50 bbl per day.

Example 8.6. Design a fracture treatment assuming productivity ratio, given:

Well depth = 7000 ft

Fracture fluid = crude oil with 0.1 lb/gal of additive to reduce fluid loss

Oil gravity = 30° API at 60°F ($\gamma = 0.876$)

Slope of fluid-loss curve $= 1.2$ cu cm/$\sqrt{\text{min}}$

Fluid loss measured at 1000 psi and 125°F

Area of filter paper $= 22.8$ cm^2

Viscosity of the crude $= 7$ cp at well conditions

Propping agent = 10-20 mesh sand

Fracture width $= 0.1$ in.

Average permeability $= 1.0$ md

Formation thickness $= 50$ ft

Static bottom-hole pressure $= 2500$ psig

Average well temperature $= 125$°F

Casing size $= 5\frac{1}{2}$ in., J-55, ID 4.892 in.

Well spacing $= 40$ acres ($r_e = 660$ ft)

Fracturing through casing

Fracture gradient $= 0.70$ psi/ft

SOLUTION: The well before fracturing is capable of producing 30 bbl of oil per day; it is desired to increase the production rate to the depth bracket allowable of 90 bbl per day, i.e., the desired productivity ratio is 3.00. Since the well is 7000 ft deep, a vertical fracture is expected and the fracture treating pressure is 0.7×7000 or 4900 psig. From Fig. 8.8 the permeability of 10-20 sand at a pressure of 4900 psig is 13,000 md, and the correlation factor C, in Fig. 8.9 is

$$C = \frac{13{,}000 \times (0.1)/12}{1.0} = 108$$

From Fig. 8.9, for a productivity ratio of 3.00 the fracture penetration required is

$$r_f/r_e = 0.25$$

or

$$r_f = 0.25 \times 660 = 165 \text{ ft}$$

Assuming that the vertical extent of the fracture is 50 ft (formation thickness) and that the fracture extends an equal distance on both sides of the well bore, the area of one fracture face is

$$A = 2 \times 165 \times 50 = 16{,}500 \text{ ft}^2$$

Since the assumed fracture width is the same as that in Example 8.5, the weight of sand that will completely pack one square foot of fracture is the same as that calculated in Example 8.5 ($S = 0.888$ lb/ft^2). The maximum amount of sand needed is then

$$0.888 \times 16{,}500 = 14{,}650 \text{ lb}$$

It is now desirable to obtain the flow rate and total volume of fracture fluid needed to yield the required area. There are an infinite number of combinations of flow rate and volume that will yield the desired area; the combination selected for the job will be the one that gives the most economical treatment. The method used to determine these parameters consists of assuming a flow rate and determining the total volume of fluid required. We assume a flow rate of 30 bbl per min, and calculate the volume of fracture fluid required by trial and error, using Fig. 8.16. The fracturing fluid coefficient from Eq. (8.52) is

$$C_w = \frac{0.0164 \times 1.2}{22.8} = 0.863 \times 10^{-3} \text{ ft}/\sqrt{\text{min}}$$

The pressure differential across the fracture face is

$$\Delta p = 4900 - 2500 = 2400 \text{ psi}$$

Substituting in Eq. (8.53),

$$C_{w\,\text{act}} = 0.863 \times 10^{-3}\left(\frac{2400}{1000}\right)^{1/2}$$

$$= 1.34 \times 10^{-3} \text{ ft}/\sqrt{\text{min}}$$

The value of x in Eq. (8.29) is

$$x = \frac{2 \times 1.34 \times 10^{-3}\sqrt{\pi t}}{(0.1)/12} = 0.570\sqrt{t}$$

From Eq. (8.55) the fracturing efficiency is

$$\text{Eff} = \frac{0.1/12 \times 16{,}500}{30 \times 5.615t} = \frac{8.16}{t}$$

Using the two equations above and Fig. 8.16, the time t for the job is found by trial and error to be 30 min. Then the volume of fracture fluid required is

$$V = qt = 30 \times 42 \times 30 = 37{,}800 \text{ gal}$$

and the sand concentration in pounds per gallon of fluid

$$x = \frac{14{,}650}{37{,}800} = 0.39 \text{ lb/gal}$$

The total flow rate q_t (including sand) for a sand of specific gravity 2.63 is given by

$$q_t = 30 + \frac{30 \times 0.39}{2.63 \times 8.34} = 30.5 \text{ bbl/min}$$

The specific gravity of the crude at the average well temperature, from Eq. (8.65), is

$$\gamma_{125} = 0.876[1 - 0.0005(125 - 60)] = 0.848$$

From Eq. (8.64) the density is found to be

$$\rho_{125} = \frac{8.34 \times 0.848 + 0.39}{1 + 0.0456 \times 0.39} = 7.33 \text{ lb/gal}$$

and the hydrostatic pressure

$$\Delta p_s = 0.052 \times 7.33 \times 7000 = 2668 \text{ psi}$$

The velocity in the casing, by Eq. (1.21a), is

$$v = \frac{17.16 \times 30.5}{(4.892)^2} = 21.9 \text{ ft/sec}$$

and the Reynolds number for this flow, from Eq. (1.15a), is

$$N_R = \frac{928 \times 4.892 \times 7.33 \times 21.9}{7} = 104{,}100$$

From Fig. 1.4 the friction factor is 0.0046, and the frictional pressure drop, from Eq. (1.18a), is

$$\Delta p_f = \frac{0.0046(7000)7.33(21.9)^2}{25.80(4.892)} = 897 \text{ psi}$$

The surface injection pressure, from Eq. (8.61), is

$$p_s = 4900 + 897 - 2668 = 3129 \text{ psig}$$

and the hydraulic horsepower is calculated from Eq. (8.57),

$$H_h = 0.0245 \times 3129 \times 30.5 = 2338 \text{ hp}$$

These calculations should now be repeated for different assumed flow rates for the selection of the most economical treatment.

Propping agents. One of the predominant factors affecting the outcome of a hydraulic fracturing job is the propping of the fracture. Several authors have pointed out the effectiveness of various sand propping-agent concentrations in increasing fracture capacity. Dehlinger *et al*[33] concluded that sand concentration optima exist but do not seem to be very sharp. Huitt and McGlothlin,[34] from point-loading tests, observed that in horizontal fractures sand used as a propping agent tends to be crushed or embedded in the fracture faces, and recommended the use of rounded-crushed walnut shells. In another study, Darin and Huitt[35] found that fractures of higher flow capacity are obtained by using a partial monolayer of a large-sized propping agent, rather than packing the fracture with a smaller-sized propping agent, to obtain greater fracture width. Of course, the actual achievement of a partial monolayer in the fracture is open to question regardless of the extent of the mechanics of treatment. In present designs sand concentrations range from 85 to 137.5 lb per 100 sq ft of fracture area using either 20-40 or 10-20 mesh sand; however, where greater fracture capacity is required, the designs are for 70 per cent 10-20 mesh sand followed by 30 per cent of 8-12 mesh angular sand.

Hydraulic notching, described by Huitt,[36] is a technique whereby a fluid with abrasive solids is pumped through an orifice and directed against the casing or formation, so that a notch is eroded from which the fracture is initiated. By positioning these notches in the correct plane, it is possible to initiate either a horizontal or a vertical fracture. According to Brown and Loper,[37,38] hydraulic perforation, with water containing 1 to 2 lb per gal of fracturing sand, is superior to conventional perforation. When the well is to be fractured subsequently, hydraulic perforation is less expensive and has, in many cases, resulted in increased production, less cement shattering, and better fracture control in stimulation of problem wells.

Kern, Perkins, and Wyant[39] investigated sand deposition in a vertical fracture and concluded that below a critical velocity the sand, upon entering the fracture, settled to the bottom. Figure 8.18 shows the growth of a settled sand bed. It can be seen from this figure that the sand settles to the low point of the fracture and the upper part of the fracture "heals." Refracturing in the same section should considerably increase the oil production.

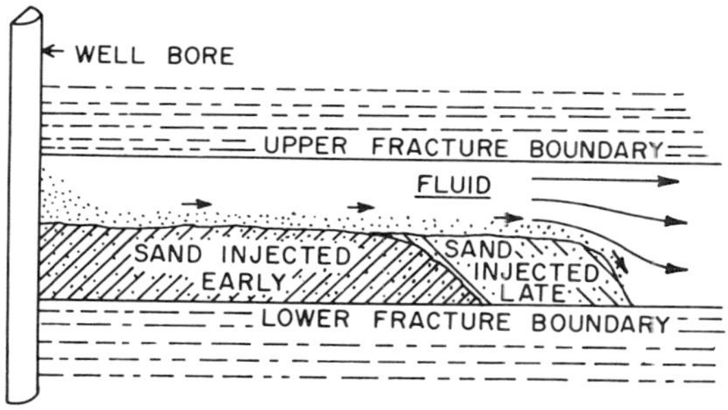

Fig. 8.18 Lineal growth of settled sand bed in a vertical fracture. (Kern, Perkins, and Wyant,[39] *Jour. Pet. Tech.*, AIME.)

Economics of fracturing treatments. After calculating the fluid volume, injection rate, horsepower required, and after considering various fracturing fluids, one can compare these factors on a cost basis to select the most economical treatment. Table 8.2 shows the costs of various treatments designed to provide 103,500 sq ft of fracture area.[40] The costs range from $1655 to $6725.

PROBLEMS

8.1. What is the range of bottom-hole fracturing pressures in a vertical plane if Poisson's ratio is taken to be 0.25 for a formation at 5000 ft in which the tensile strength is known to be between 800 psi and 1600 psi? *Ans.* 4130–4930 psi.

8.2. What is the maximum depth at which horizontal fractures are expected for a formation where Poisson's ratio is 0.25 and the tensile strength is 1400 psi? *Ans.* 4200 ft.

8.3. A well on 40-acre spacing was completed with a diameter of 8 in. in a 40-ft sand with a permeability of 2 md. The well was fractured to 40 per cent of the drainage radius and the fracture was completely packed with 10-20 mesh sand. The fracture-treating pressure was 2000 psig. Calculate the PR of the well from Eq. (8.14),

Table 8.2. COSTS OF DIFFERENT FRACTURING TREATMENTS

(After Reference 40)

Fluid	Fracturing fluid coefficient	Injection rate, bbl/min	Fluid required, gal	Surface pressure, psi	Hydraulic horsepower required	Fluid cost	Pumping cost	Total cost
Lease oil	$C_c = 5.0$	10	375,000	2400	590	\$3750	\$ 725	\$4475
		20	200,000	2750	1350	2000	1570	3570
		30	140,000	3150	2310	1400	2355	3755
Blend 2 refined oil	$C_v = 3.5$	10	200,000	2500	610	6000	725	6725
		20	100,000	3050	1490	3000	1570	4570
		30		3850[b]				
Lease oil with additive	$C_w = 1.1$[a]	10	31,000	2400	590	930	725	1655
		20	23,000	2750	1350	690	1570	2260
		30	20,000	3150	2310	600	2355	2955

[a] Spurt loss = 4 cc.

[b] Exceeds maximum pressure limitation.

(a) Assuming a fracture width of 0.1 in. *Ans.* 4.05.
(b) Assuming fracture width of 0.2 in. *Ans.* 5.23.
(c) Obtain comparable results from Fig. 8.7. *Ans.* 4.00, 5.30.
(d) Explain the difference between the first two results and those of part (c).

8.4. A fracture treatment similar to that of Problem 8.3 is planned for a well where the fracture treating pressure is 4000 psig.
(a) Assuming a vertical fracture, calculate the productivity ratio if the fracture width is 0.2 in. *Ans.* 3.6.
(b) Show that if the sand had a permeability of 100 md or higher, no appreciable increase in production would result from the fracturing treatment.

8.5. Show that Eq. (8.43) is valid, by differentiating Eq. (8.41). [Hint: Use the series expansion for erfc (x).]

8.6. Show that the constant 0.0374 in Eq. (8.45) is valid for v expressed in ft/min and t in min.

8.7. The following information was obtained for a well in the Tubb Field in Ector County, Texas:

Average porosity = 0.10

Average permeability = 1 md

Bottom-hole pressure = 1800 psig

Bottom-hole temperature = 100°F

Pressure differential expected across the fracture face = 2000 psi

Fracture fluid viscosity at bottom-hole temperature = 94 cp

(a) Calculate the fracturing fluid coefficient. *Ans.* 2.16×10^{-3} ft/$\sqrt{\text{min}}$.
(b) If the well is fractured with 40,000 gal of fluid in 30 min, calculate the fracture area from Eq. (8.29) assuming a fracture width of 0.2 in. *Ans.* 103,300 ft^2.

8.8. A well has just been completed in a reservoir which has the following properties:

Average porosity = 0.20

Average effective permeability to oil = 2 md

Viscosity of reservoir fluid = 4 cp (at bottom-hole conditions)

Estimated compressibility of reservoir fluid = 10×10^{-6} psi^{-1}

The fracturing fluid is lease oil, and a pressure differential of 1200 psi across the fracture face is expected. Calculate the fracturing fluid coefficient. *Ans.* 1.42×10^{-3} ft/$\sqrt{\text{min}}$.

8.9. A fracturing oil containing an additive to reduce fluid loss was tested in a filter press with an area of 22.8 sq cm at 1000 psig, and the following data were obtained:

Time, min	*Filtrate volume, cc*
1	4
4	6
16	10
25	12
30	13

The fracturing oil is to be used in a 6000-ft well where the bottom-hole pressure is 2200 psig. From previous fracture jobs in the area, the fracture gradient has been found to be 0.65 psi/ft. The well has a drainage radius of 660 ft and a permeability of 1 md. The propping agent is 10-20 mesh sand.
(a) Calculate the fracturing fluid coefficient for a differential pressure across the fracture face of 1000 psi. *Ans.* 1.44×10^{-3} ft/$\sqrt{\text{min}}$.
(b) Recalculate part (a) for the field differential expected. *Ans.* 1.88×10^{-3} ft/$\sqrt{\text{min}}$ for $\Delta p = 1700$ psi.
(c) Assuming a fracture width of 0.2 in., calculate the fracture width to be used for area calculations. *Ans.* 0.269 in.
(d) Determine the fracturing efficiency for a fluid volume of 30,000 gal and an injection rate of 30 bbl/min. *Ans.* 42 per cent.
(e) Calculate the fracture area. *Ans.* 75,140 ft^2.
(f) Determine the productivity ratio assuming the fracture to extend out from the well bore equally in both directions and to have a vertical extent of 100 ft (formation thickness). *Ans.* 4.3.

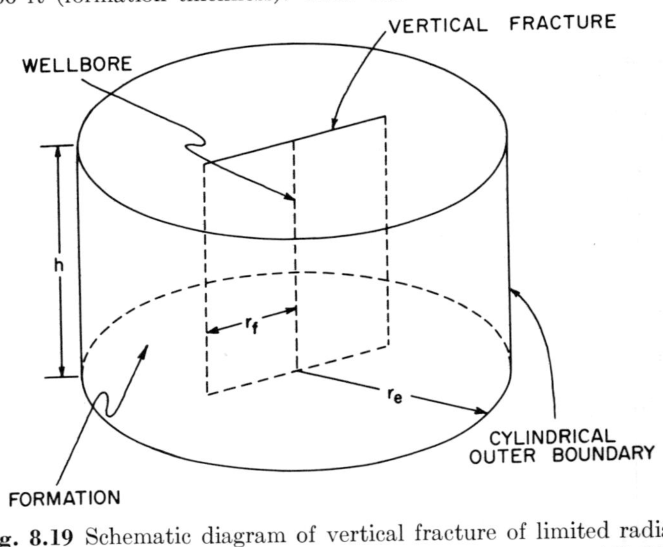

Fig. 8.19 Schematic diagram of vertical fracture of limited radial extent. (After Prats,[41] *Society of Petroleum Engineers,* AIME.)

8.10. The term q_t is the total flow rate including liquid and sand. Derive an expression for q_t in terms of q, the liquid flow rate, and x, the sand concentration in lb/gal. The specific gravity of sand is 2.63. *Ans.* $q_t = q[1 + 0.0456x]$.

8.11. It is sometimes desired to induce more than one fracture; to accomplish this, ball sealers are employed to selectively plug the perforations that are accepting fluid so that a new fracture can be initiated. On a fracture job in which this technique was used the following information is available:

Casing size = $5\frac{1}{2}$ in., ID 4.892 in.

Newtonian fracturing fluid

Viscosity of fracturing fluid = 4 cp (at well temperature)

Specific gravity of fracturing fluid = 0.82 (at well temperature)

Depth to formation = 2000 ft

Surface treating pressure = 2050 psig

Instantaneous shut-down pressure = 1050 psig

Injection rate = 45 bbl/min

Sand concentration = 2 lb/gal

Perforation diameter = 3/8 in.

(a) Calculate q_t, the total injection rate including sand. *Ans.* 49.1 bbl/min.
(b) Calculate ρ, the fluid density in lb/gal. *Ans.* 8.1 lb/gal.
(c) Determine the friction factor. *Ans.* 0.0039.
(d) Calculate the frictional pressure drop. *Ans.* 620 psi.
(e) Calculate the pressure drop across the perforations. *Ans.* 380 psi.
(f) Calculate the flow rate through each perforation. *Ans.* 1.62 bbl/min.
(g) How many ball sealers are required to plug all perforations accepting fluid? *Ans.* 30.
(h) Discuss the accuracy of calculations required in this problem.

8.12. The following well data are given:

Well depth = 5000 ft

Fracture fluid = oil-base gel

Fracturing-fluid coefficient = 1.2×10^{-3} ft/$\sqrt{\text{min}}$

Specific gravity of gel = γ = 0.84 at 110°F

Flow behavior index n' = 0.39

Consistency index K'_p = 0.084 lbf-sec^n/ft^2

Propping agent = 20-40 mesh round sand

Fracture width = 0.1 in.

Average permeability = 1 md

Formation thickness = 200 ft

Static bottom-hole pressure = 2000 psig

Average well temperature = 110°F

Casing size = $5\frac{1}{2}$ in., N-80, 17 lb/ft, ID 4.892 in.

(bursting pressure = 7740 psig)

Well spacing = 40 acres (r_e = 660 ft)

Fracture gradient in the area = 0.60 psi/ft

For an injection rate of 30 bbl/min and a total volume of 40,000 gal, calculate:
(a) The fracture efficiency. *Ans.* 30 per cent.
(b) The fracture area. *Ans.* 193,000 ft^2.
(c) The maximum weight of sand required. *Ans.* 171,000 lb.
(d) The flow rate including the sand. *Ans.* 35.9 bbl/min.
(e) The total fluid density. *Ans.* 9.46 lb/gal.
(f) The hydrostatic pressure. *Ans.* 2460 psi.
(g) The frictional pressure drop. *Ans.* 971 psi.
(h) The surface injection pressure. *Ans.* 1511 psig.
(i) The hydraulic horsepower required. *Ans.* 1329 hp.
(j) The productivity ratio, assuming the fracture to have a vertical extent of 250 ft. *Ans.* 3.40.

8.13. It is desired to plan the most economical treatment for a well. To obtain the desired production, a fracture area of 100,000 sq ft is required. The following well data are available:

Well depth = 7000 ft

Fracture fluid = lease oil

Fracture fluid coefficient = 5×10^{-3} ft/$\sqrt{\text{min}}$

Gravity of oil = 36° API at 60°F (γ = 0.845)

Average well temperature = 120°F

Fracture gradient = 0.7 psi/ft

Casing size = $5\frac{1}{2}$ in., J-55, 17 lb/ft, ID 4.892 in.

Maximum allowable pressure = 3700 psig

Static bottom-hole pressure = 1500 psig

Fracture width = 0.1 in.

Sand required = 88,000 lb

(a) Determine the viscosity of the crude. *Ans.* 4.41 cp.
(b) Calculate the volume of lease oil required for injection rates of 10 bbl/min and 30 bbl/min. *Ans.* 378,000 and 126,000 gal.
(c) Calculate q_t, the total injection rate, for the two rates given in part (b). *Ans.* 10.10 and 30.96 bbl/min.
(d) Calculate the density of fluid for the two rates in part (b). *Ans.* 7.21 and 7.51 lb/gal.

(e) Calculate the hydrostatic pressure for the two rates of part (b). *Ans.* 2620 and 2730 psi.

(f) Calculate the frictional pressure drop for both cases. *Ans.* 112 and 888 psi.

(g) Calculate surface injection pressure at each rate. *Ans.* 2392 and 3058 psi.

(h) Calculate the hydraulic horsepower required for each rate. *Ans.* 592 and 2320 hp.

8.14. Recalculate Problem 8.13 assuming that a material to reduce fluid loss is added to the lease oil, giving a fracturing fluid coefficient of 1.1×10^{-3} ft/$\sqrt{\text{min}}$ with no spurt loss. *Ans.* (a) 4.41 cp; (b) 25,200 and 15,100 gal; (c) 11.6 and 38.0 bbl/min; (d) 9.09 and 10.2 lb/gal; (e) 3300 and 3710 psi; (f) 172 and 1685 psi; (g) 1772 and 2875 psi; (h) 504 and 2677 hp.

8.15. The production cost of lease oil for the well in Problem 8.13 is 1 cent per gallon. The cost of treating the lease oil with a fluid-loss additive is 4 cents per gallon. The service company charge for pump trucks rated at 600 hydraulic horsepower is $600 per truck.

(a) Calculate the costs for the job at each flow rate under the conditions of Problems 8.13 and 8.14, respectively. *Ans.* at 10 bbl/min, $4380 and $1860; at 30 bbl/min, $3660 and $3755.

(b) Arrange your results in a form similar to Table 8.2.

(c) What treatment would you recommend?

REFERENCES

[1]J. B. Clark, "A Hydraulic Process for Increasing the Productivity of Oil Wells," *Trans.* AIME, **186,** 1 (1949).

[2]L. E. Wilsey and W. G. Bearden, "A Method of Oil Recovery from Extremely Low Permeability Formations," *Trans.* AIME, **201**, 169–174 (1954).

[3]J. J. Jakosky, *Exploration Geophysics* (Los Angeles: Trija Publishing Company, 1950), pp. 263–266.

[4]Francis Birch, J. F. Schairer, and H. Cecil Spicer, *Handbook of Physical Constants*, Geological Society of America Special Paper No. 36 (January, 1942), p. 76.

[5]Eugene Harrison, W. F. Kieschnick, Jr., and W. J. McGuire, "The Mechanics of Fracture Induction and Extension," *Trans.* AIME, **201,** 254–255 (1954).

[6]M. King Hubbert and David G. Willis, "Mechanics of Hydraulic Fracturing," *Trans.* AIME, **210,** 154 (1957).

[7]A. J. Miles and A. D. Topping, "Stresses Around a Deep Well," *Trans.* AIME, **179,** 186–191 (1949).

[8]P. P. Scott, William G. Bearden, and G. C. Howard, "Rock Rupture as Affected by Fluid Properties," *Trans.* AIME, **198,** 111 (1953).

[9]James M. Cleary, "Hydraulic Fracture Theory Part I — *Mechanics of Materials*," Illinois State Geological Survey Circular 251 (1958).

[10]James M. Cleary, "Hydraulic Fracture Theory Part II — *Fracture Orientation and Possibility of Fracture Control*," Illinois State Geological Survey Circular 252 (1958).

[11]James M. Cleary, "Hydraulic Fracture Theory Part III — *Elastic Properties of Sandstone*," Illinois State Geological Survey Circular 281 (1959).

[12]S. Timoshenko, *Strength of Materials* (Part II: Advanced Theory and Problems) 3rd Ed. (New York: D. Van Nostrand Company, Inc., 1956), pp. 304–305.

[13]B. C. Craft, "Tensile and Compressive Strength of Formations," *The Oil and Gas Journal*, **55**, No. 26, 138 (July 1, 1957).

[14]Bob C. Crittendon, "The Mechanics of Design and Interpretation of Hydraulic Fracture Treatments," *Journal of Petroleum Technology*, AIME, **11**, No. 10, 21–29 (October, 1959).

[15]Robert Wade Brown and Raymond G. Loper, "Stimulation Treatment Selectivity Through Perforation Ball Sealer Technology," *The Petroleum Engineer*, **31**, No. 6, B82 (June, 1959).

[16]H. K. van Poollen, John M. Tinsley, and Calvin D. Saunders, "Hydraulic Fracturing — Fracture Flow Capacity vs. Well Productivity," *Trans.* AIME, **213**, 91–95 (1958).

[17]B. C. Craft and M. F. Hawkins, *Applied Petroleum Reservoir Engineering* (Englewood Cliffs: Prentice-Hall, Inc., 1959), pp. 304, 276.

[18]A. B. Dyes, C. E. Kemp, and B. H. Caudle, "Effect of Fractures on Sweep-out Pattern," *Trans.* AIME, **213**, 245–249 (1958).

[19]H. K. van Poollen, "Productivity vs. Permeability Damage in Hydraulically Produced Fractures," *Drilling and Production Practice* (API, 1957), p. 103.

[20]Paul B. Crawford and Bobby L. Landrum, "Effect of Unsymmetrical Vertical Fractures on Production Capacity," *Trans.* AIME, **204**, 251–254 (1955).

[21]George C. Howard and C. R. Fast, "Optimum Fluid Characteristics for Fracture Extension," *Drilling and Production Practice* (API, 1957), pp. 261–270.

[22]R. D. Carter, "Derivation of the General Equation for Estimating the Extent of the Fractured Area," *Drilling and Production Practice* (API, 1957), pp. 267–268.

[23]Ruel V. Churchill, *Operational Mathematics*, 2nd Ed. (New York: McGraw-Hill Book Company, Inc., 1958), pp. 35–37, 67, 323–331.

[24]Herbert Bristol Dwight, *Tables of Integrals and Other Mathematical Data*, 3rd Ed. (New York: The MacMillan Company, 1947), p. 129.

[25]Eugene Jahnke and Fritz Emde, *Tables of Functions with Formulae and Curves*, 4th Ed. (New York: Dover Publications, Inc., 1945), p. 24.

[26]Donald L. Katz, David Cornell, Ribi Kobayashi, Fred H. Poettmann, John A. Vary, Jack R. Elenbaas, and Charles F. Weinaug, *Handbook of Natural Gas Engineering* (New York: McGraw-Hill Book Company, Inc., 1959), p. 410.

[27]J. B. Stewart and A. W. Coulter, "Increased Fracturing Efficiency by Fluid-Loss Control," *The Petroleum Engineer*, **31**, No. 6, B43 (June, 1959).

[28]J. J. Arps and T. G. Roberts, "The Effect of the Relative Permeability Ratio, the Oil Gravity, and the Solution Gas-Oil Ratio on the Primary Recovery from a Depletion Type Reservoir," *Trans.* AIME, **204**, 120 (1955).

[29]Davis D. Hunt and H. R. Crawford, "Engineered Hydraulic Fracture Treatments," *The Petroleum Engineer*, **31**, No. 4, B78 (April, 1959); *ibid.*, **31**, No. 5, B25 (May, 1959).

[30]Robert Wade Brown and Bruce Gilbert, "Pressure Drop Across Perforations Can Be Computed," *The Petroleum Engineer*, **29**, No. 10, B82 (September, 1957).

[31]Carl Gatlin, *Petroleum Engineering Drilling and Well Completions* (Englewood Cliffs: Prentice-Hall, Inc., 1960), p. 97.

[32]Jack W. Kraemer, "Fluid Flow Rate Through Perforations," *The Petroleum Engineer*, **31**, No. 1, B44 (January, 1959).

[33]Peter Dehlinger, W. H. Browne, C. O. Bundrant, "Optimum Sand Concentrations in Well Treatments," *Trans.* AIME, **201**, 273–278 (1954).

[34]J. L. Huitt and B. B. McGlothlin, Jr., "The Propping of Fractures in Formations Susceptible to Propping-Sand Embedment," *Drilling and Production Practice*, (API, 1958), p. 115.

[35]S. R. Darin and J. L. Huitt, "Effect of a Partial Monolayer of Propping Agent on Fracture Flow Capacity," *Journal of Petroleum Technology*, AIME, **12**, No. 3, 31–37 (March, 1960).

[36]J. L. Huitt, "Hydraulic Fracturing with the Single-Point Entry Technique," *Journal of Petroleum Technology*, AIME, **12**, No. 3, 11–13 (March, 1960).

[37]Robert Wade Brown and Jack L. Loper, "A New Completion Technique — Sand Erosion," Paper No. 906-5-N presented before the Division of Production at the Spring Meeting of the API, March 2, 1960, Preprint Vol. p. 2.

[38]Robert Wade Brown and Jack L. Loper, "Theory of Formation Cutting Using the Sand Erosion Process," Paper No. 1572-G presented at the Fall Meeting of the Society of Petroleum Engineers, AIME, September 1960, Preprint Vol. p. 9.

[39]L. R. Kern, T. K. Perkins, and R. E. Wyant, "The Mechanics of Sand Movement in Fracturing," *Journal of Petroleum Technology*, AIME, **11**, No. 7, 55 (July, 1959).

[40]*Frac Guide — An Engineered Approach to Oil and Gas Well Fracturing* (Tulsa: Dowell Division of the Dow Chemical Company, 1959), p. 15.

[41]M. Prats, "Effect of Vertical Fractures on Reservoir Behavior — Incompressible-Fluid Case," Paper No. 1575-G presented at the Fall Meeting of the Society of Petroleum Engineers, AIME, September, 1960, Preprint Vol. p. 6.

CHAPTER NINE

Acidizing

9.1. Introduction

The practice of introducing acid into oil wells to increase oil and gas production has been in use since 1894, when it was first used in the Pennsylvania fields.*[1] The earliest acid treatment consisted of dumping raw hydrochloric acid down the casing without subsequent application of pressure. It was not until 1930 that better treating materials and improved techniques in directing the acid into the desired pay sections proved successful in increasing production. Oil-production rocks which are treated with hydrochloric acid are limestones (composed of calcium carbonate $CaCO_3$), dolomites (mixtures of calcium and magnesium carbonates, $CaMg(CO_3)_2$), and dolomitic limestones.

The chemical equation for the reaction between hydrochloric acid and limestone is as follows:

$$\underset{73}{2HCl} + \underset{100}{CaCO_3} \rightarrow \underset{111}{CaCl_2} + \underset{18}{H_2O} + \underset{44}{CO_2} \tag{9.1}$$

Except in the case of the two molecules of HCl, the figures shown below each of the compounds are their molecular weights. They represent the relative weights of the individual molecules and therefore the relative weights of substances which react or are formed as products of the reaction. Equation (9.1) may be interpreted as saying that 73 lb of hydrochloric acid react with 100 lb of calcium carbonate to form 111 lb of calcium

*References throughout the text are given at the end of each chapter.

chloride, 18 lb of water, and 44 lb of carbon dioxide. The usual acid treating solution contains 15 per cent hydrochloric acid by weight and has a specific gravity of 1.075. In 1000 gal of this acid solution there are $1000 \times 8.34 \times 1.075 \times 0.15 = 1344.8$ lb of hydrochloric acid. From the relative weights of the substances involved in reaction (9.1), it can be calculated by proportion that 1344.8 lb of hydrochloric acid reacts with 1842.2 lb of calcium carbonate to form 2044.8 lb of calcium chloride, 331 6 lb of water, and 810.6 lb of carbon dioxide. As the average density of limestone is 170 lb per cu ft, the 1842.2 lb of calcium carbonate represents a volume of 10.84 cu ft.

When dolomite reacts with hydrochloric acid, the products are the same as in the limestone reaction, except that magnesium chloride is also formed, as shown by the following equation:

$$\underset{146}{4HCl} + \underset{184.3}{CaMg(CO_3)_2} \rightarrow \underset{111}{CaCl_2} + \underset{95.3}{MgCl_2} + \underset{36}{2H_2O} + \underset{88}{2CO_2} \tag{9.2}$$

Magnesium chloride is quite soluble in water, and so, as in the case of pure limestone, all the products of the reaction are either soluble in spent acid, or are gaseous. From the relative weights of the molecules shown below Eq. (9.2), it can be calculated that 1000 gal of 15 per cent hydrochloric acid solution (1344.8 lb of hydrochloric acid) will react with 1697.6 lb of pure dolomite to form 1022.4 lb of calcium chloride, 877.7 lb of magnesium chloride, 331.5 lb of water, and 810.4 lb of carbon dioxide. Limestone and dolomitic formations frequently contain small quantities of sand, anhydrite, and gypsum, together with smaller quantities of other impurities which are mostly insoluble. It is now considered good practice to withdraw the spent acid within one hour after it is introduced as there is danger of insoluble residues precipitating from the suspension and plugging the pores.

9.2. Types of Treatments

The concepts and derivations which follow are based on the analysis of acidizing techniques, in correlation with reservoir data and a backlog of past treatments, presented by Hendrickson, Hurst, and Wieland.[2] Acidizing treatments are classified by three categories.

Matrix acidizing. In this category are treatments in which acid is injected into the pores and flow channels of carbonate rocks at a bottom-hole pressure considerably less than the fracturing pressure, the purpose being to increase uniformly the permeability of the formation. Under these conditions, it is assumed that the acid enters only the natural pores and flow channels, and reacts with the walls of the pores, dissolving the rock and enlarging the pores. This reaction slows down as the acid is spent, until finally additional radial penetration produces no additional benefit. The maximum radial penetration of unspent acid is a function of the acid velocity in the pore and its spending time (which, for a given acid, depends on the acid volume and the rock porosity). Since the spending time of acid does not change appreciably during a specific treatment, maximum penetration is attained when the first increment of injected acid is completely spent. Assuming a constant injection rate, and that the spending time remains constant for all increments of acid, later increments of acid entering the pore will penetrate no deeper before spending than the first increment, i.e., instead of lengthening the treated flow channel additional acid will only enlarge the cross-sectional area.

In evaluating this type of acidizing, the following basic assumptions are made:

1. The formation is homogeneous.
2. The pores are of uniform size.
3. The acid penetrates uniformly and radially.
4. The reaction rate declines uniformly with decreasing acid concentration.
5. The weight of limestone dissolved per increment of distance declines uniformly until the acid is completely spent.

On the basis of the above assumptions, the radial distance the acid will penetrate before being totally spent depends on the equation

$$\text{Volume injected, ft}^3 = \text{Pore volume invaded, ft}^3$$

or

$$q_i t = \pi \phi h (r_a^2 - r_w^2)$$

Solving for r_a,

$$r_a = \sqrt{\frac{q_i t}{\pi \phi h} + r_w^2}$$

If q_i is expressed in barrels per minute and t in seconds,

$$r_a = \sqrt{\frac{5.615 q_i t}{60 \pi \phi h} + r_w^2}$$

or

$$r_a = \sqrt{\frac{0.0936 q_i t}{\pi \phi h} + r_w^2} \tag{9.3}$$

where r_a = radial distance acid will penetrate until it is spent, feet

ϕ = fractional porosity

q_i = acid injection rate, barrels per minute

t = spending time, seconds

r_w = wellbore radius, feet

h = formation thickness, feet

The only unknown factor in Eq. (9.3) is the spending time t, which must be measured for the particular acid in the laboratory. The spending time for an acid depends upon the ratio of the area of the rock exposed to the acid to the volume of acid, here denoted as specific surface area s_ϕ (square centimeters per cubic centimeter). An acid will have the same spending time in rocks of the same composition with the same specific surface area (at the same temperature and pressure). For matrix acidizing, the specific surface area can be obtained from the Kozeny equation as modified by Pirson,[2,3]

$$k = \frac{10^8}{2Fs_\phi^2} \tag{9.4}$$

or

$$s_\phi = 10^4\sqrt{\frac{1}{2Fk}} \tag{9.5}$$

where k = permeability, darcies

s_ϕ = specific surface area, square centimeters per cubic centimeter

F = formation resistivity factor, dimensionless

The formation resistivity factor[4] is related to the porosity by

$$F = \phi^{-m}$$

where the cementation factor m varies from 1.3 for unconsolidated sands and oolitic limestones to 2.2 for dense limestone. From laboratory investigations, Hendrickson *et al.*[2] found that in matrix acidizing s_ϕ is so great that the spending time for most acids is less than 15 sec. To obtain greater penetration during matrix acidizing, it is necessary to either decrease the reaction rate or increase the rate of injection of the acid into the formation. Table 9.1 shows injection rate and differential pressure required to obtain various penetrations for matrix acidizing.[2] The injection rates and differential pressures were calculated from Eq. (9.3) and the Darcy equation for steady-state radial flow. From this table it is readily seen that

Table 9.1. Matrix acidizing

	Necessary minimum injection rate and pressure differential at indicated formation permeabilities	
Desired penetration, ft	5 md	100 md
0.5	0.56 bbl/min/ft 26,800 psi	0.26 bbl/min/ft 624 psi
1	1.7 bbl/min/ft 117,000 psi	0.77 bbl/min/ft 2710 psi
5	—	14.2 bbl/min/ft 94,000 psi

in order to obtain an appreciable penetration into the formation, a pressure much greater than the fracturing pressure must be applied. The spending time of an acid is also affected by pressure, temperature, velocity of acid in the rock, acid concentration, physical and chemical formation structure, and the use of various retarding additives. Since the spending time is affected by so many variables, it can be found only by laboratory measurements on each rock. Acid treatments in tight formations at less than fracturing pressures will increase the permeability only in the vicinity of the well bore, and are most applicable for overcoming formation damage.

Acidizing through preexisting fractures. Treatments may be conducted in formations containing natural fractures. The purpose of this type of treatment is to remove secondary deposition or loose particles in the fracture and to dissolve rock from the fracture faces. The injection rate is controlled during such treatments so as to not exceed the formation fracturing pressure. The maximum penetration of the acid into the fracture is dependent on the spending time of the acid under reservoir temperature and pressure, and on the injection rate. The evaluation of these treatments requires the following basic assumptions:

1. The fractures are horizontal and of uniform width, and extend radially from the well bore.
2. The acid leak-off into the formation is considered negligible.
3. The rate of reaction of the acid is proportional to its concentration, and the quantity of rock dissolved from the fracture face decreases with increased acid penetration until the acid is spent. At constant injection rate, additional increments of acid passing through the fracture will not extend the acidized area but merely increase the fracture width.

At an injection rate q_i, the radial distance the acid will penetrate a horizontal fracture until it is spent at time t is found from the equation

$$\text{Volume of fractures} = \text{Volume of acid injected}$$

or

$$n\pi W(r_a^2 - r_w^2) = q_i t$$

Solving for r_a,

$$r_a = \sqrt{\frac{q_i t}{\pi n W} + r_w^2}$$

If q_i is expressed in barrels per minute, t in seconds, and W in inches,

$$r_a = \sqrt{\frac{5.615 \times 12\, q_i t}{60\, \pi n W} + r_w^2}$$

or

$$r_a = \sqrt{\frac{1.123\, q_i t}{\pi n W} + r_w^2} \tag{9.6}$$

where q_i = injection rate, barrels per minute
t = spending time, seconds
W = fracture width, inches
n = number of fractures
r_a = distance acid will penetrate before being spent, feet
r_w = wellbore radius, feet

As in matrix acidizing, the spending time for an acid in a fracture is dependent on the specific surface area. In order to obtain s_ϕ, consider a fracture of width W ft with an area of one square foot. The area exposed to acid is 2 sq ft and the volume of acid is W cubic feet, so that

$$s_\phi = \frac{2}{W}\ \text{ft}^2/\text{ft}^3$$

For natural fractures, it is safe to assume a fracture width of 0.1 mm or less. In this case, Hendrickson *et al.*[2] found that the spending time for most acids is less than one minute. Since most natural fractures are very narrow, have low permeability, and may not extend appreciable distances from the well bore, excessive bottom-hole differential pressures will be required to achieve appreciable formation penetration before the acid is spent. Maximum penetration cannot be calculated accurately because it is impossible to determine the number of fractures exposed to the well bore, their inclination, and their average width.

High pressure acidizing through fractures. This type of stimulation consists of injecting acid into fractures or zones of weakness at injection pressures exceeding the fracturing pressure, in order to connect high permeability channels to the well bore. In evaluating this type of treatment the following assumptions are made:

1. A single fracture is created which is either vertical or horizontal.
2. The major portion of the acid enters the fracture, and therefore acid entering the matrix *from the well bore* is negligible.
3. The acid solutions do not contain propping agents.
4. The volume and area of the fracture depend on the volume of acid, the injection rate, the fracture width during treatment, the viscosity and fluid loss of the acid, and the physical characteristics of the rock and reservoir fluid.

The depth of penetration of the acid prior to complete spending will depend upon the reaction rate, the velocity through the fracture, and the specific surface area of the fracture. Since widths of hydraulically formed fractures are much greater than those of natural fractures ($\sim$ 0.1 in. as compared to $\sim$ 0.1 mm), the specific surface area of hydraulic fractures is much less than that of a natural fracture ($s_\phi = 2/W$); thus, the spending time will be greater in the hydraulic fracture. The acidizing benefits are defined by the maximum penetration of the acid before it is spent. Although additional fracture extension may be obtained from the hydraulic action of the spent acid, no permanent benefit is obtained because the fracture can be expected to "heal" in the absence of a propping agent. However, in the portion of the fracture reached before the acid is spent, formation irregularities due to differences in composition, solubility, and reaction rate result in nonuniform dissolving of the fracture face. These irregularities prevent the fracture from closing completely on release of the fracturing pressure even in the absence of propping agents. The acid which leaks off into the matrix of the formation during fracturing will also provide some additional benefit; however, very little penetration will occur because of the short spending time (matrix acidizing). In high pressure acidizing through fractures, much greater penetration is obtained because (a) the spending time is increased due to the lower specific surface area and (b) the injection rate is greater due to the higher permeability (greater width).

Since the depth of penetration in the first two types of acid treatments discussed is extremely small, such treatments are generally applicable for overcoming formation damage, i.e., increasing the formation permeability in the immediate vicinity of the well bore. The high pressure acid-fracturing technique is used to increase the productivity of reservoirs of extremely low permeability.

9.3. *Acid-Fracturing Design*

An acid fracture treatment is designed in the same manner as the fracture treatment in Chapter 8, with the additional condition for the spending time of the acid. Also, as in fracturing, the design is controlled by economic factors. Comparative treatment costs must be balanced against payout times in order to obtain the most economic treatment for any specific well.

The main difference between fracturing and acidizing design is the absence of propping agents in the latter case. For this, Hendrickson *et al.*[2] have suggested that an acidizing productivity ratio be calculated considering the capacity of the fracture to be infinite. Example 9.1 shows an acid-fracture design calculation.

Example 9.1. Design an acid fracture, given:

Well depth = 2000 ft

Acid = 15 per cent hydrochloric acid with a fluid-loss additive

Fracturing-fluid coefficient = 1.20×10^{-3} ft/$\sqrt{\text{min}}$

Spurt loss = 5 cu cm

Area of filter press = 22.8 sq cm

Acid density = 1.08 gm/cu cm

Acid viscosity = 86 cp

Spending time with retarder = 45 min

Average well temperature = 90°F

Fracture gradient = 1.0 psi/ft

Tubing size = 2 in., ID 1.995 in.

Shut-in bottom-hole pressure = 500 psi

Fracture width = 0.1 in.

Well spacing = 40 acres (r_e = 660 ft)

SOLUTION: The well before acidizing is producing 20 bbl of oil per day; it is desired to increase the production to 86 bbl per day, the depth-bracket allowable for the well. The desired acidization productivity ratio is thus 4.3. From Fig. 8.10, for a horizontal fracture of infinite capacity and a productivity ratio of 4.3, the ratio r_f/r_e is 0.167 and the fracture radius r_f is 0.167 × 660 or 110 ft. Then the fracture area required is

$$A = \pi(110)^2 = 38{,}000 \text{ sq ft}$$

The fracture width to be used in area calculations, from Eq. (8.54), is

$$W' = 0.1 + 0.787\,\frac{5}{22.8} = 0.117 \text{ in.}$$

The term x in Eq. (8.29) is

$$x = \frac{2 \times 1.20 \times 10^{-3}\sqrt{45\pi}}{0.117/12} = 2.92$$

Then from Fig. 8.18, the fracturing efficiency is 29 per cent, and the total volume of acid required, from Eq. (8.55), is

$$V = \frac{7.48 \times 38{,}000 \times 0.117/12}{0.29} = 9560 \text{ gal}$$

and the injection rate,

$$q = \frac{9560}{42 \times 45} = 5.06 \text{ bbl/min}$$

The acid density is

$$\rho = 8.34 \times 1.08 = 9.01 \text{ lb/gal}$$

and the hydrostatic pressure,

$$\Delta p_s = 0.433 \times 1.08 \times 2000 = 935 \text{ psig}$$

The bottom-hole treating pressure from Eq. (8.62) is

$$p_t = 1.0 \times 2000 = 2000 \text{ psig}$$

and the velocity in the tubing, from Eq. (1.21a), is

$$v = \frac{17.16 \times 5.06}{(1.995)^2} = 21.8 \text{ ft/sec}$$

The value of the Reynolds number, from Eq. (1.15a), is

$$N_R = \frac{928 \times 1.995 \times 21.8 \times 9.01}{86} = 4228$$

The friction factor from Fig. 1.4 is $f = 0.0099$ and the frictional pressure drop is

$$\Delta p_f = \frac{0.0099 \times 2000 \times 9.01 \times (21.8)^2}{25.80 \times 1.995} = 1657 \text{ psi}$$

The surface injection pressure, from Eq. (8.61), is

$$p_s = 2000 - 935 + 1657 = 2722 \text{ psi}$$

and the hydraulic horsepower required, from Eq. (8.57), is

$$H_h = 0.0245 \times 2722 \times 5.06 = 337 \text{ hp}$$

It should be realized that this design technique is merely a guide to acid fracturing and should not be used as an exact method of evaluation. It has been assumed that formation irregularities will prevent the complete healing of the fracture, thereby giving infinite capacity. In the absence of formation irregularities complete healing can occur, in which case essentially no increase in productivity is realized.

PROBLEMS

9.1. A 50-ft producing section of pure limestone with a porosity of 0.15 has been acidized to increase the matrix permeability. The treatment consisted of 500 gal of 15 per cent hydrochloric acid (spending time 15 sec and specific gravity of 1.075) pumped at a rate of 10 bbl per min. The wellbore radius is 4 in.
(a) Calculate the depth of penetration of the unspent acid. *Ans.* 0.508 ft.
(b) Calculate the weight of limestone dissolved. *Ans.* 921 lb.
(c) Assume that the acid had been only placed opposite the producing section without being pumped into the matrix, and calculate the new wellbore radius if the density of limestone is 170 lb per cu ft. *Ans.* 0.382 ft.

9.2. Recalculate Problem 9.1 assuming the producing section to contain 100 preexisting fractures at the well bore. The average width of the fractures is 0.004 in. and the acid spending time is 45 sec. *Ans.* (a) 20.1 ft, (b) 921 lb, (c) 0.382 ft.

9.3. Calculate the specific surface area for limestone having a porosity of 0.17 and a permeability of 5 md,
(a) for matrix acidizing (assume a cementation factor of 2). *Ans.* 5.88×10^5 sq cm/cu cm.
(b) for acidizing in preexisting fractures (assume 20 fractures at the well bore with an average width of 0.004 in.). *Ans.* 196 sq cm/cu cm.
(c) for acid fracturing (assuming a fracture width of 0.1 in.). *Ans.* 7.87 sq cm/cu cm.

9.4. It is desired to design an acidizing treatment to provide a productivity ratio of 3.0. The following information is available:

Well depth = 5000 ft

Formation thickness = 50 ft

Acid = 15 per cent hydrochloric with fluid-loss additive and retarder

Fracture fluid coefficient = 2.5×10^{-3} ft/$\sqrt{\text{min}}$

Spurt loss = 7 cu cm

Area of filter press = 22.8 sq cm

Acid density = 1.10 gm/cu cm (at avg. well temperature)

Acid viscosity = 70 cp

Spending time = 40 min

Average temperature = 120°F

Fracture gradient = 0.65 psi/ft

Tubing size = 2 in., ID 1.995 in.

Shut-in bottom-hole pressure = 1000 psi

Fracture width = 0.1 in.

Estimated vertical extent = 50 ft

Well spacing = 40 acres (r_e = 660 ft)

(a) Calculate the required fracture area. *Ans.* 9900 sq ft.
(b) Calculate the total volume of acid required. *Ans.* 5670 gal.
(c) Calculate the injection rate. *Ans.* 3.38 bbl/min.
(d) Calculate the hydrostatic pressure. *Ans.* 2380 psig.
(e) Calculate the frictional pressure drop. *Ans.* 1970 psi.
(f) Calculate the surface injection pressure. *Ans.* 2840 psig.
(g) Calculate the hydraulic horsepower required. *Ans.* 235 hp.

9.5. Recalculate Problem 9.4 for an acid with the same properties but with the spending time reduced to 15 min. *Ans.* 9900 sq ft; 4196 gal; 6.66 bbl/min; 2380 psig; 6420 psi; 7290 psig; 1190 hp.

9.6. If the retarded acid in Problem 9.4 costs \$0.30 per gal and that in Problem 9.5 costs \$0.19 per gal, what is the total cost of the job in each case? The cost of a pump truck capable of delivering 275 hydraulic horsepower is \$300.00. *Ans.* \$2001; \$2297.

REFERENCES

[1]F. B. Plummer and R. B. Newcombe, Jr., "Laboratory Investigations on Acid Treatment of Oil Sands," *Trans.* AIME, **118**, 100 (1936).

[2]A. R. Hendrickson, R. E. Hurst, and D. R. Wieland, "Engineered Guide for Planning Acidizing Treatments Based on Specific Reservoir Characteristics," *Journal of Petroleum Technology,* AIME, **12,** No. 2, 16–23 (February, 1960).

[3]S. J. Pirson, *Oil Reservoir Engineering,* 2nd Ed. (New York: McGraw-Hill Book Company, Inc., 1958), pp. 101, 108.

[4]G. E. Archie, "The Electrical Resistivity Log as an Aid in Determining Some Reservoir Characteristics," *Trans.* AIME, **146,** 56 (1942).

APPENDIX A

Cementing Tables

Table A.1. CAPACITY OF CASING
(Courtesy Halliburton Oil Well Cementing Co.)

OD, in.	Weight per foot with couplings, lb	ID, in.	Capacity			Sacks of cement/ lin ft[b]
			gal/lin ft	bbl/lin ft	cu ft/lin ft	
$4\frac{1}{2}$	9.50	4.090	0.6825	0.0162	0.0912	0.0829
$4\frac{1}{2}$	11.60	4.000	0.6528	0.0155	0.0873	0.0793
$4\frac{1}{2}$	13.50	3.920	0.6270	0.0149	0.0838	0.0761
$4\frac{3}{4}$[a]	16.00	4.082	0.6798	0.0162	0.0909	0.0826
5	11.50	4.560	0.8484	0.0202	0.1134	0.1031
5	13.00	4.494	0.8240	0.0196	0.1102	0.1001
5	15.00	4.408	0.7928	0.0189	0.1060	0.0963
5	18.00	4.276	0.7460	0.0178	0.0997	0.0907
5[a]	21.00	4.154	0.7040	0.0168	0.0941	0.0856
$5\frac{1}{2}$	13.00	5.044	1.0380	0.0247	0.1388	0.1261
$5\frac{1}{2}$	14.00	5.012	1.0249	0.0244	0.1370	0.1246
$5\frac{1}{2}$[a]	15.00	4.974	1.0094	0.0240	0.1349	0.1227
$5\frac{1}{2}$	15.50	4.950	0.9997	0.0238	0.1336	0.1215
$5\frac{1}{2}$	17.00	4.892	0.9764	0.0232	0.1305	0.1187

[a]Not API standard. Shown for information only.
[b]Based on 1.1 cu ft of slurry per sack of cement.

Table A.1 (CONTINUED)
(Courtsey Halliburton Oil Well Cementing Co.)

OD, in.	Weight per foot with couplings, lb	ID, in.	Capacity gal/lin ft	Capacity bbl/lin ft	Capacity cu ft/lin ft	Sacks of cement/ lin ft[b]
$5\frac{1}{2}$	20.00	4.778	0.9314	0.0222	0.1245	0.1132
$5\frac{1}{2}$	23.00	4.670	0.8898	0.0211	0.1189	0.1081
$5\frac{3}{4}$[a]	14.00	5.290	1.1418	0.0272	0.1526	0.1388
$5\frac{3}{4}$[a]	17.00	5.190	1.0990	0.0262	0.1469	0.1336
$5\frac{3}{4}$[a]	19.50	5.090	1.0570	0.0252	0.1413	0.1285
$5\frac{3}{4}$[a]	22.50	4.990	1.0159	0.0242	0.1358	0.1235
6	15.00	5.524	1.2450	0.0296	0.1664	0.1513
6[a]	16.00	5.500	1.2342	0.0293	0.1650	0.1500
6[a]	17.00	5.450	1.2119	0.0288	0.1620	0.1473
6	18.00	5.424	1.2003	0.0285	0.1605	0.1459
6	20.00	5.352	1.1687	0.0278	0.1562	0.1420
6	23.00	5.240	1.1203	0.0266	0.1498	0.1361
$6\frac{5}{8}$	17.00	6.135	1.5356	0.0365	0.2053	0.1866
$6\frac{5}{8}$	20.00	6.049	1.4929	0.0355	0.1996	0.1814
$6\frac{5}{8}$[a]	22.00	5.989	1.4634	0.0348	0.1956	0.1778
$6\frac{5}{8}$	24.00	5.921	1.4304	0.0341	0.1912	0.1738
$6\frac{5}{8}$[a]	26.00	5.855	1.3987	0.0333	0.1870	0.1700
$6\frac{5}{8}$	28.00	5.791	1.3683	0.0326	0.1829	0.1663
$6\frac{5}{8}$[a]	29.00	5.761	1.3541	0.0322	0.1810	0.1646
$6\frac{5}{8}$	32.00	5.675	1.3140	0.0312	0.1757	0.1597
7	17.00	6.538	1.7440	0.0415	0.2331	0.2119
7	20.00	6.456	1.7005	0.0405	0.2273	0.2067
7[a]	22.00	6.398	1.6701	0.0398	0.2235	0.2031
7	23.00	6.366	1.6534	0.0393	0.2210	0.2009
7[a]	24.00	6.336	1.6379	0.0390	0.2190	0.1991
7	26.00	6.276	1.6070	0.0383	0.2148	0.1953
7[a]	28.00	6.214	1.5754	0.0375	0.2106	0.1915
7	29.00	6.184	1.5602	0.0371	0.2085	0.1896
7[a]	30.00	6.154	1.5452	0.0368	0.2066	0.1878
7	32.00	6.094	1.5151	0.0360	0.2025	0.1841
7[a]	34.00	6.040	1.4884	0.0354	0.1989	0.1808
7	35.00	6.004	1.4707	0.0350	0.1966	0.1787
7	38.00	5.920	1.4298	0.0340	0.1911	0.1738
7[a]	40.00	5.836	1.3896	0.0330	0.1858	0.1689
$7\frac{5}{8}$	20.00	7.125	2.0716	0.0493	0.2769	0.2517
$7\frac{5}{8}$	24.00	7.025	2.0135	0.0479	0.2692	0.2247
$7\frac{5}{8}$	26.40	6.969	1.9815	0.0472	0.2649	0.2408
$7\frac{5}{8}$	29.70	6.875	1.9284	0.0459	0.2578	0.2344
$7\frac{5}{8}$	33.70	6.765	1.8672	0.0445	0.2496	0.2269
$7\frac{5}{8}$	39.00	6.625	1.7907	0.0426	0.2394	0.2176

Table A.1 (CONTINUED)

(Courtsey Halliburton Oil Well Cementing Co.)

OD, in.	Weight per foot with couplings, lb	ID, in.	Capacity gal/lin ft	Capacity bbl/lin ft	Capacity cu ft/lin ft	Sacks of cement/lin ft[b]
8$\frac{5}{8}$	24.00	8.097	2.6759	0.0367	0.3576	0.3251
8$\frac{5}{8}$	28.00	8.017	2.6223	0.0624	0.3506	0.3187
8$\frac{5}{8}$	32.00	7.921	2.5599	0.0610	0.3422	0.3111
8$\frac{5}{8}$	36.00	7.825	2.4982	0.0595	0.3340	0.3036
8$\frac{5}{8}$[a]	38.00	7.775	2.4664	0.0587	0.3297	0.2997
8$\frac{5}{8}$	40.00	7.725	2.4348	0.0579	0.3254	0.2959
8$\frac{5}{8}$[a]	43.00	7.651	2.3883	0.0569	0.3193	0.2902
8$\frac{5}{8}$	44.00	7.625	2.3721	0.0565	0.3171	0.2883
8$\frac{5}{8}$	49.00	7.511	2.3017	0.0548	0.3077	0.2797
9$\frac{5}{8}$	29.30	9.063	3.3512	0.0797	0.4480	0.4073
9$\frac{5}{8}$	32.30	9.001	3.3055	0.0787	0.4419	0.4017
9$\frac{5}{8}$	36.00	8.921	3.2470	0.0773	0.4341	0.3946
9$\frac{5}{8}$[a]	38.00	8.877	3.2151	0.0765	0.4298	0.3907
9$\frac{5}{8}$	40.00	8.835	3.1847	0.0758	0.4256	0.3869
9$\frac{5}{8}$	43.50	8.755	3.1273	0.0745	0.4183	0.3803
9$\frac{5}{8}$	47.00	8.681	3.0747	0.0732	0.4110	0.3737
9$\frac{5}{8}$	53.50	8.535	2.9721	0.0708	0.3973	0.3612
10$\frac{3}{4}$	32.75	10.192	4.2382	0.1009	0.5666	0.5151
10$\frac{3}{4}$[a]	35.75	10.136	4.1917	0.0998	0.5604	0.5094
10$\frac{3}{4}$	40.50	10.050	4.1209	0.0981	0.5509	0.5008
10$\frac{3}{4}$	45.50	9.950	4.0393	0.0962	0.5400	0.4909
10$\frac{3}{4}$	51.00	9.850	3.9585	0.0943	0.5292	0.4811
10$\frac{3}{4}$[a]	54.00	9.784	3.9056	0.0930	0.5221	0.4746
10$\frac{3}{4}$	55.50	9.760	3.8865	0.0925	0.5196	0.4723
13$\frac{3}{8}$	48.00	12.715	6.5962	0.1571	0.8818	0.8016
13$\frac{3}{8}$	54.50	12.615	6.4928	0.1546	0.8680	0.7891
13$\frac{3}{8}$	61.00	12.515	6.3903	0.1522	0.8543	0.7766
13$\frac{3}{8}$	68.00	12.415	6.2886	0.1497	0.8407	0.7642
13$\frac{3}{8}$	72.00	12.347	6.2199	0.1481	0.8318	0.7559
13$\frac{3}{8}$[a]	83.00	12.175	6.0478	0.1440	0.8085	0.7350
14[a]	50.00	13.344	7.2649	0.1730	0.9712	0.8829

Table A.2. ANNULAR CAPACITIES FOR VARIOUS TUBING OR PIPE AND HOLE DIAMETERS[a]

(Courtesy Halliburton Oil Well Cementing Co.)

Tubing or drill pipe OD, in.	Diameter of hole, in.	Annular Capacities			
		gal/lin ft	bbl/lin ft	cu ft/lin ft	Sacks of cement/lin ft[b]
$2\frac{3}{8}$	4	0.4226	0.0101	0.0565	0.0514
	$4\frac{1}{4}$	0.5068	0.0121	0.0676	0.0615
	$4\frac{1}{2}$	0.5961	0.0142	0.0797	0.0725
	$4\frac{3}{4}$	0.6904	0.0164	0.0923	0.0839
	5	0.7899	0.0188	0.1056	0.0960
	$5\frac{1}{4}$	0.8944	0.0213	0.1196	0.1087
	$5\frac{1}{2}$	1.0041	0.0239	0.1342	0.1220
	$5\frac{3}{4}$	1.1188	0.0266	0.1496	0.1360
	6	1.2386	0.0295	0.1656	0.1505
	$6\frac{1}{4}$	1.3636	0.0325	0.1823	0.1657
	$6\frac{1}{2}$	1.4937	0.0356	0.1997	0.1815
	$6\frac{3}{4}$	1.6288	0.0388	0.2177	0.1979
	7	1.7691	0.0421	0.2365	0.2150
	$7\frac{1}{4}$	1.9144	0.0456	0.2559	0.2326
	$7\frac{1}{2}$	2.0649	0.0492	0.2760	0.2509
	$7\frac{3}{4}$	2.2204	0.0529	0.2968	0.2698
	8	2.3810	0.0567	0.3183	0.2894
	$8\frac{1}{4}$	2.5468	0.0606	0.3405	0.3095
	$8\frac{1}{2}$	2.7177	0.0647	0.3633	0.3303
	$8\frac{3}{4}$	2.8936	0.0689	0.3868	0.3516
$2\frac{7}{8}$	4	0.3155	0.0075	0.0422	0.0383
	$4\frac{1}{4}$	0.3997	0.0095	0.0534	0.0486
	$4\frac{1}{2}$	0.4889	0.0116	0.0654	0.0594
	$4\frac{3}{4}$	0.5835	0.0139	0.0780	0.0709
	5	0.6828	0.0163	0.0913	0.0830
	$5\frac{1}{4}$	0.7873	0.0187	0.1053	0.0957
	$5\frac{1}{2}$	0.8969	0.0214	0.1199	0.1090
	$5\frac{3}{4}$	1.0117	0.0241	0.1352	0.1229
	6	1.1315	0.0269	0.1513	0.1375
	$6\frac{1}{4}$	1.2565	0.0299	0.1680	0.1527
	$6\frac{1}{2}$	1.3865	0.0330	0.1854	0.1685
	$6\frac{3}{4}$	1.5217	0.0362	0.2034	0.1849
	7	1.6620	0.0396	0.2222	0.2020
	$7\frac{1}{4}$	1.8073	0.0430	0.2416	0.2196
	$7\frac{1}{2}$	1.9578	0.0466	0.2617	0.2379
	$7\frac{3}{4}$	2.1133	0.0503	0.2825	0.2568

[a]No allowance made for couplings.
[b]Based on 1.1 cu ft of slurry per sack of cement.

Table A.2 (CONTINUED)

(Courtesy Halliburton Oil Well Cementing Co.)

Tubing or drill pipe OD, in.	Diameter of hole, in.	Annular Capacities			
		gal/lin ft	bbl/lin ft	cu ft/lin ft	Sacks of cement/lin ft[b]
$2\frac{7}{8}$	8	2.2739	0.0541	0.3040	0.2763
	$8\frac{1}{4}$	2.4397	0.0581	0.3261	0.2965
	$8\frac{1}{2}$	2.6105	0.0622	0.3490	0.3173
	$8\frac{3}{4}$	2.7865	0.0663	0.3725	0.3386
$3\frac{1}{2}$	5	0.5202	0.0124	0.0695	0.0632
	$5\frac{1}{4}$	0.6248	0.0149	0.0835	0.0759
	$5\frac{1}{2}$	0.7344	0.0175	0.0982	0.0892
	$5\frac{3}{4}$	0.8491	0.0202	0.1135	0.1032
	6	0.9690	0.0231	0.1295	0.1178
	$6\frac{1}{4}$	1.0940	0.0260	0.1462	0.1329
	$6\frac{1}{2}$	1.2240	0.0291	0.1636	0.1487
	$6\frac{3}{4}$	1.3592	0.0324	0.1817	0.1652
	7	1.4994	0.0357	0.2004	0.1822
	$7\frac{1}{4}$	1.6447	0.0392	0.2199	0.1999
	$7\frac{1}{2}$	1.7952	0.0427	0.2400	0.2182
	$7\frac{3}{4}$	1.9508	0.0464	0.2608	0.2371
	8	2.1114	0.0503	0.2823	0.2566
	$8\frac{1}{4}$	2.2771	0.0542	0.3044	0.2767
	$8\frac{1}{2}$	2.4480	0.0583	0.3273	0.2975
	$8\frac{3}{4}$	2.6239	0.0625	0.3508	0.3189
	5	0.3672	0.0087	0.0491	0.0446
	$5\frac{1}{4}$	0.4718	0.0112	0.0631	0.0573
	$5\frac{1}{2}$	0.5814	0.0138	0.0777	0.0707
	$5\frac{3}{4}$	0.6962	0.0166	0.0931	0.0846
	6	0.8160	0.0194	0.1091	0.0992
	$6\frac{1}{4}$	0.9410	0.0224	0.1258	0.1144
	$6\frac{1}{2}$	1.0710	0.0255	0.1432	0.1302
	$6\frac{3}{4}$	1.2062	0.0287	0.1612	0.1466
	7	1.3464	0.0321	0.1800	0.1636
	$7\frac{1}{4}$	1.4917	0.0355	0.1994	0.1813
	$7\frac{1}{2}$	1.6422	0.0391	0.2195	0.1996
	$7\frac{3}{4}$	1.7978	0.0428	0.2403	0.2185
	8	1.9584	0.0466	0.2618	0.2380
	$8\frac{1}{4}$	2.1242	0.0506	0.2840	0.2581
	$8\frac{1}{2}$	2.2950	0.0546	0.3068	0.2789
	$8\frac{3}{4}$	2.4710	0.0588	0.3303	0.3003

Table A.2 (CONTINUED)

(Courtesy Halliburton Oil Well Cementing Co.)

Tubing or drill pipe OD, in.	Diameter of hole, in.	Annular Capacities			
		gal/lin ft	bbl/lin ft	cu ft/lin ft	Sacks of cement/lin ft[b]
$4\frac{1}{2}$	5	0.1938	0.0046	0.0259	0.0236
	$5\frac{1}{4}$	0.2984	0.0071	0.0399	0.0363
	$5\frac{1}{2}$	0.4080	0.0097	0.0545	0.0496
	$5\frac{3}{4}$	0.5228	0.0124	0.0699	0.0635
	6	0.6426	0.0153	0.0859	0.0781
	$6\frac{1}{4}$	0.7676	0.0183	0.1026	0.0933
	$6\frac{1}{2}$	0.8976	0.0214	0.1200	0.1091
	$6\frac{3}{4}$	1.0328	0.0246	0.1381	0.1255
	7	1.1730	0.0279	0.1568	0.1426
	$7\frac{1}{4}$	1.3183	0.0314	0.1762	0.1602
	$7\frac{1}{2}$	1.4688	0.0350	0.1964	0.1785
	$7\frac{3}{4}$	1.6244	0.0387	0.2171	0.1974
	8	1.7850	0.0425	0.2386	0.2169
	$8\frac{1}{4}$	1.9508	0.0464	0.2608	0.2371
	$8\frac{1}{2}$	2.1216	0.0505	0.2836	0.2578
	$8\frac{3}{4}$	2.2976	0.0547	0.3071	0.2792
	9	2.4786	0.0590	0.3313	0.3012
	$9\frac{1}{4}$	2.6648	0.0634	0.3562	0.3238
	$9\frac{1}{2}$	2.8560	0.0680	0.3818	0.3471
	$9\frac{3}{4}$	3.0524	0.0727	0.4080	0.3709
	10	3.2538	0.0775	0.4350	0.3954
	$10\frac{1}{4}$	3.4604	0.0824	0.4626	0.4205
	$10\frac{1}{2}$	3.6720	0.0874	0.4909	0.4462
	$10\frac{3}{4}$	3.8888	0.0926	0.5199	0.4726
	11	4.1106	0.0979	0.5495	0.4996
5	6	0.4488	0.0107	0.0600	0.0545
	$6\frac{1}{4}$	0.5738	0.0137	0.0767	0.0697
	$6\frac{1}{2}$	0.7038	0.0168	0.0941	0.0855
	$6\frac{3}{4}$	0.8390	0.0200	0.1122	0.1020
	7	0.9792	0.0233	0.1309	0.1190
	$7\frac{1}{4}$	1.1245	0.0268	0.1503	0.1367
	$7\frac{1}{2}$	1.2750	0.0304	0.1704	0.1549
	$7\frac{3}{4}$	1.4305	0.0341	0.1912	0.1739
	8	1.5912	0.0379	0.2127	0.1934
	$8\frac{1}{4}$	1.7569	0.0418	0.2349	0.2135
	$8\frac{1}{2}$	1.9278	0.0459	0.2577	0.2343
	$8\frac{3}{4}$	2.1037	0.0501	0.2812	0.2557

Table A.2 (CONTINUED)

(Courtesy Halliburton Oil Well Cementing Co.)

Tubing or drill pipe OD, in.	Diameter of hole, in.	Annular Capacities			
		gal/lin ft	bbl/lin ft	cu ft/lin ft	Sacks of cement/lin ft[b]
5	9	2.2848	0.0544	0.3054	0.2777
	9¼	2.4710	0.0588	0.3303	0.3003
	9½	2.6622	0.0634	0.3559	0.3235
	9¾	2.8585	0.0681	0.3821	0.3474
5½	6	0.2346	0.0056	0.0314	0.0285
	6¼	0.3596	0.0086	0.0481	0.0437
	6½	0.4896	0.0117	0.0655	0.0595
	6¾	0.6248	0.0149	0.0835	0.0759
	7	0.7650	0.0182	0.1023	0.0930
	7¼	0.9103	0.0217	0.1217	0.1106
	7½	1.0608	0.0253	0.1418	0.1289
	7¾	1.2164	0.0290	0.1626	0.1478
	8	1.3770	0.0328	0.1841	0.1673
	8¼	1.5428	0.0367	0.2062	0.1875
	8½	1.7136	0.0408	0.2291	0.2083
	8¾	1.8896	0.0450	0.2526	0.2296
	9	2.0706	0.0493	0.2768	0.2516
	9¼	2.2568	0.0537	0.3017	0.2743
	9½	2.4480	0.0583	0.3273	0.2975
	9¾	2.6444	0.0630	0.3535	0.3214
7	8	0.6119	0.0146	0.0818	0.0744
	8¼	0.7777	0.0185	0.1040	0.0945
	8½	0.9486	0.0226	0.1268	0.1153
	8¾	1.1245	0.0268	0.1503	0.1367
	9	1.3056	0.0311	0.1745	0.1587
	9¼	1.4917	0.0355	0.1994	0.1813
	9½	1.6830	0.0401	0.2250	0.2045
	9¾	1.8793	0.0447	0.2512	0.2284
	10	2.0808	0.0495	0.2782	0.2529
	10¼	2.2873	0.0545	0.3058	0.2780
	10½	2.4990	0.0595	0.3341	0.3037
	10¾	2.7157	0.0647	0.3630	0.3300
	11	2.9376	0.0699	0.3927	0.3570
	11¼	3.1645	0.0753	0.4230	0.3846
	11½	3.3966	0.0809	0.4541	0.4128
	11¾	3.6337	0.0865	0.4858	0.4416

Table A.2 (CONTINUED)

(Courtesy Halliburton Oil Well Cementing Co.)

Tubing or drill pipe OD, in.	Diameter of hole, in.	Annular Capacities			
		gal/lin ft	bbl/lin ft	cu ft/lin ft	Sacks of cement/lin ft[b]
7	12	3.8760	0.0923	0.5181	0.4710
	$12\frac{1}{4}$	4.1233	0.0982	0.5512	0.5011
	$12\frac{1}{2}$	4.3758	0.1042	0.5850	0.5318
	$12\frac{3}{4}$	4.6334	0.1103	0.6194	0.5631
$9\frac{5}{8}$	11	1.1570	0.0275	0.1547	0.1406
	$11\frac{1}{4}$	1.3840	0.0330	0.1850	0.1682
	$11\frac{1}{2}$	1.6161	0.0385	0.2160	0.1964
	$11\frac{3}{4}$	1.8532	0.0441	0.2477	0.2252
	12	2.0955	0.0499	0.2801	0.2547
	$12\frac{1}{4}$	2.3428	0.0558	0.3132	0.2847
	$12\frac{1}{2}$	2.5953	0.0618	0.3469	0.3154
	$12\frac{3}{4}$	2.8528	0.0679	0.3814	0.3470
	13	3.1155	0.0742	0.4165	0.3786
	$13\frac{1}{4}$	3.3832	0.0806	0.4523	0.4112
	$13\frac{1}{2}$	3.6561	0.0871	0.4887	0.4443
	$13\frac{3}{4}$	3.9340	0.0937	0.5259	0.4781
	14	4.2170	0.1004	0.5637	0.5125
	$14\frac{1}{4}$	4.5052	0.1073	0.6023	0.5475
	$14\frac{1}{2}$	4.7984	0.1142	0.6415	0.5831
	$14\frac{3}{4}$	5.0968	0.1214	0.6813	0.6194
	15	5.4003	0.1286	0.7219	0.6563
	$15\frac{1}{4}$	5.7088	0.1359	0.7632	0.6938
	$15\frac{1}{2}$	6.0224	0.1434	0.8051	0.7319
	$15\frac{3}{4}$	6.3412	0.1510	0.8477	0.7706
$10\frac{3}{4}$	$11\frac{1}{2}$	0.6808	0.0162	0.0910	0.0827
	$11\frac{3}{4}$	0.9180	0.0219	0.1227	0.1116
	12	1.1603	0.0276	0.1551	0.1410
	$12\frac{1}{4}$	1.4076	0.0335	0.1882	0.1711
	$12\frac{1}{2}$	1.6601	0.0395	0.2219	0.2017
	$12\frac{3}{4}$	1.9176	0.0457	0.2563	0.2330
	13	2.1803	0.0519	0.2915	0.2650
	$13\frac{1}{4}$	2.4480	0.0583	0.3273	0.2975
	$13\frac{1}{2}$	2.7208	0.0648	0.3637	0.3307
	$13\frac{3}{4}$	2.9988	0.0714	0.4009	0.3644

Table A.2. (CONTINUED)

(*Courtesy Halliburton Oil Well Cementing Co.*)

Tubing or drill pipe OD, in.	Diameter of hole, in.	Annular Capacities			
		gal/lin ft	bbl/lin ft	cu ft/lin ft	Sacks of cement/lin ft[b]
$10\frac{3}{4}$	14	3.2818	0.0781	0.4387	0.3988
	$14\frac{1}{4}$	3.5700	0.0850	0.4772	0.4339
	$14\frac{1}{2}$	3.8632	0.0920	0.5164	0.4695
	$14\frac{3}{4}$	4.1616	0.0991	0.5563	0.5058
	15	4.4650	0.1063	0.5969	0.5426
	$15\frac{1}{4}$	4.7736	0.1137	0.6381	0.5801
	$15\frac{1}{2}$	5.0872	0.1211	0.6801	0.6182
	$15\frac{3}{4}$	5.4060	0.1287	0.7227	0.6570
	16	5.7298	0.1364	0.7660	0.6963
	$16\frac{1}{4}$	6.0588	0.1443	0.8099	0.7363
	$16\frac{1}{2}$	6.3928	0.1522	0.8546	0.7769
	$16\frac{3}{4}$	6.7320	0.1603	0.8999	0.8181
	17	7.0763	0.1685	0.9460	0.8600
	$17\frac{1}{2}$	7.7801	0.1852	1.0400	0.9455
	18	8.5043	0.2025	1.1369	1.0335
$13\frac{3}{8}$	14	0.6980	0.0166	0.0933	0.0848
	$14\frac{1}{4}$	0.9862	0.0235	0.1318	0.1198
	$14\frac{1}{2}$	1.2794	0.0305	0.1710	0.1555
	$14\frac{3}{4}$	1.5778	0.0376	0.2109	0.1917
	15	1.8812	0.0448	0.2515	0.2286
	$15\frac{1}{4}$	2.1898	0.0521	0.2927	0.2661
	$15\frac{1}{2}$	2.5034	0.0596	0.3347	0.3042
	$15\frac{3}{4}$	2.8222	0.0672	0.3773	0.3430
	16	3.1460	0.0749	0.4206	0.3823
	$16\frac{1}{2}$	3.8090	0.0907	0.5092	0.4629
	17	4.4925	0.1070	0.6006	0.5460
	$17\frac{1}{2}$	5.1963	0.1237	0.6946	0.6315
	18	5.9205	0.1410	0.7915	0.7195
	$18\frac{1}{2}$	6.6650	0.1587	0.8910	0.8100
	19	7.4300	0.1769	0.9933	0.9030
	$19\frac{1}{2}$	8.2154	0.1956	1.0982	0.9984
	20	9.0212	0.2148	1.2060	1.0963
	$20\frac{1}{2}$	9.8474	0.2345	1.3164	1.1967

Table A.3

(Courtesy Halliburton Oil Well Cementing Co.)

Diameter of hole, in.	Capacities of Open Hole			
	gal/lin ft	bbl/lin ft	cu ft/lin ft	Sacks of cement/lin ft[a]
4	0.6528	0.0155	0.0870	0.0791
$4\frac{1}{8}$	0.6942	0.0165	0.0926	0.0842
$4\frac{1}{4}$	0.7369	0.0175	0.0983	0.0893
$4\frac{3}{8}$	0.7809	0.0186	0.1044	0.0949
$4\frac{1}{2}$	0.8262	0.0197	0.1106	0.1006
$4\frac{5}{8}$	0.8727	0.0208	0.1168	0.1062
$4\frac{3}{4}$	0.9206	0.0219	0.1230	0.1118
$4\frac{7}{8}$	0.9696	0.0231	0.1297	0.1179
5	1.0200	0.0243	0.1364	0.1240
$5\frac{1}{8}$	1.0716	0.0255	0.1432	0.1302
$5\frac{1}{4}$	1.1246	0.0268	0.1504	0.1367
$5\frac{3}{8}$	1.1788	0.0281	0.1576	0.1433
$5\frac{1}{2}$	1.2342	0.0294	0.1650	0.1500
$5\frac{5}{8}$	1.2909	0.0308	0.1726	0.1569
$5\frac{3}{4}$	1.3489	0.0321	0.1803	0.1639
$5\frac{7}{8}$	1.4083	0.0335	0.1883	0.1711
6	1.4688	0.0350	0.1963	0.1785
$6\frac{1}{8}$	1.5306	0.0364	0.2046	0.1860
$6\frac{1}{4}$	1.5938	0.0380	0.2132	0.1937
$6\frac{3}{8}$	1.6581	0.0395	0.2217	0.2015
$6\frac{1}{2}$	1.7238	0.0410	0.2304	0.2095
$6\frac{5}{8}$	1.7908	0.0426	0.2394	0.2176
$6\frac{3}{4}$	1.8590	0.0443	0.2485	0.2259
$6\frac{7}{8}$	1.9284	0.0459	0.2578	0.2343
7	1.9992	0.0476	0.2673	0.2430
$7\frac{1}{8}$	2.0712	0.0493	0.2769	0.2517
$7\frac{1}{4}$	2.1445	0.0511	0.2867	0.2606
$7\frac{3}{8}$	2.2191	0.0528	0.2967	0.2697
$7\frac{1}{2}$	2.2950	0.0546	0.3068	0.2789
$7\frac{5}{8}$	2.3722	0.0565	0.3171	0.2883
$7\frac{3}{4}$	2.4505	0.0584	0.3276	0.2978
$7\frac{7}{8}$	2.5302	0.0602	0.3382	0.3075
8	2.6112	0.0622	0.3491	0.3173
$8\frac{1}{8}$	2.6935	0.0641	0.3601	0.3273
$8\frac{1}{4}$	2.7769	0.0661	0.3712	0.3375
$8\frac{3}{8}$	2.8617	0.0681	0.3826	0.3478
$8\frac{1}{2}$	2.9478	0.0702	0.3941	0.3583
$8\frac{5}{8}$	3.0351	0.0723	0.4057	0.3688
$8\frac{3}{4}$	3.1237	0.0744	0.4176	0.3796
$8\frac{7}{8}$	3.2136	0.0765	0.4296	0.3905

[a]Based on 1.1 cu ft of slurry per sack of cement.

Table A.3 (CONTINUED)

(Courtesy Halliburton Oil Well Cementing Co.)

Diameter of hole, in.	Capacities of Open Hole			
	gal/lin ft	bbl/lin ft	cu ft/lin ft	Sacks of cement/lin ft[a]
9	3.3048	0.0787	0.4418	0.4016
9⅛	3.3972	0.0809	0.4542	0.4129
9¼	3.4910	0.0831	0.4667	0.4243
9⅜	3.5859	0.0854	0.4794	0.4358
9½	3.6822	0.0877	0.4922	0.4475
9⅝	3.7797	0.0900	0.5053	0.4593
9¾	3.8785	0.0924	0.5185	0.4714
9⅞	3.9786	0.0947	0.5319	0.4835
10	4.0800	0.0971	0.5454	0.4958
10⅛	4.1826	0.0996	0.5592	0.5083
10¼	4.2865	0.1021	0.5730	0.5209
10⅜	4.3917	0.1046	0.5871	0.5337
10½	4.4982	0.1071	0.6013	0.5467
10⅝	4.6059	0.1097	0.6157	0.5597
10¾	4.7150	0.1123	0.6303	0.5730
10⅞	4.8252	0.1149	0.6451	0.5864
11	4.9368	0.1175	0.6599	0.5999
11⅛	5.0497	0.1202	0.6750	0.6137
11¼	5.1637	0.1230	0.6903	0.6276
11⅜	5.2791	0.1257	0.7057	0.6415
11½	5.3958	0.1285	0.7213	0.6557
11⅝	5.5137	0.1313	0.7371	0.6701
11¾	5.6329	0.1341	0.7530	0.6846
11⅞	5.7535	0.1370	0.7691	0.6992
12	5.8752	0.1399	0.7854	0.7140
12⅛	5.9982	0.1428	0.8018	0.7289
12¼	6.1225	0.1458	0.8184	0.7440
12⅜	6.2482	0.1488	0.8353	0.7593
12½	6.3750	0.1518	0.8522	0.7748
12⅝	6.5031	0.1548	0.8694	0.7903
12¾	6.6326	0.1579	0.8867	0.8061
12⅞	6.7632	0.1610	0.9041	0.8219
13	6.8952	0.1642	0.9217	0.8380
13⅛	7.0284	0.1673	0.9395	0.8541
13¼	7.1630	0.1706	0.9576	0.8705
13⅜	7.2988	0.1738	0.9757	0.8870
13½	7.4358	0.1770	0.9940	0.9036
13⅝	7.5741	0.1803	1.0125	0.9205
13¾	7.7138	0.1837	1.0312	0.9374
13⅞	7.8546	0.1870	1.0500	0.9545

Table A.4. TUBING CHARACTERISTICS

(Courtesy Halliburton Oil Well Cementing Co.)

Nominal size, in.	OD, in.	ID, in.	Weight per foot with couplings, lb	Capacities			
				gal/lin ft	bbl/lin ft	cu ft/lin ft	Sacks of cement/ lin ft[b]
$\frac{3}{4}$	1.050	0.824	1.20	0.02769	0.00066	0.00370	0.00336
1	1.315	1.049	1.80	0.0449	0.00107	0.00600	0.00545
$1\frac{1}{4}$	1.660	1.380	2.40	0.0777	0.00185	0.01039	0.00944
$1\frac{1}{2}$	1.900	1.610	2.90	0.1057	0.00252	0.01414	0.01285
2[a]	2.375	2.041	4.00	0.1699	0.00405	0.02272	0.02065
2	2.375	1.995	4.70	0.1623	0.00387	0.02171	0.01973
$2\frac{1}{2}$	2.875	2.441	6.50	0.2431	0.00579	0.03250	0.02954
3[a]	3.500	3.068	7.70	0.3840	0.00914	0.05133	0.04666
3	3.500	2.992	9.30	0.3652	0.00870	0.04883	0.04439
3[a]	3.500	2.922	10.20	0.3483	0.00829	0.04657	0.04235
$3\frac{1}{2}$[a]	4.000	3.548	9.50	0.5136	0.01223	0.06866	0.06242
$3\frac{1}{2}$	4.000	3.476	11.00	0.4929	0.01774	0.06590	0.05991
4	4.500	3.985	12.75	0.6391	0.01522	0.08544	0.07768

[a]Nonupset tubing.

[b]Based on 1.1 cu ft of slurry per sack of cement.

Table A.5. Characteristics of Internal Upset Drill Pipe[a]

(*Courtesy Halliburton Oil Well Cementing Co.*)

Size OD, in.	ID, in.	Weight per foot with couplings, lb	Capacities gal/lin ft	bbl/lin ft	cu ft/lin ft	Sacks of cement/lin ft[c]
$2\frac{3}{8}$[b]	2.000	4.80	0.1632	0.00389	0.0218	0.0199
$2\frac{3}{8}$	1.815	6.65	0.1344	0.00320	0.0180	0.0163
$2\frac{7}{8}$[b]	2.469	6.45	0.2487	0.00592	0.0332	0.0302
$2\frac{7}{8}$[b]	2.323	8.35	0.2202	0.00524	0.0294	0.0267
$2\frac{7}{8}$	2.151	10.40	0.1888	0.00450	0.0253	0.0230
$3\frac{1}{2}$[b]	3.063	8.50	0.3828	0.00911	0.0511	0.0465
$3\frac{1}{2}$[b]	2.900	11.20	0.3431	0.00817	0.0459	0.0417
$3\frac{1}{2}$	2.764	13.30	0.3117	0.00742	0.0417	0.0379
$3\frac{1}{2}$	2.602	15.50	0.2769	0.00659	0.0369	0.0335
4	3.340	14.00	0.4551	0.01083	0.0608	0.0553
4[b]	3.240	15.70	0.4286	0.01020	0.0573	0.0521
$4\frac{1}{2}$[b]	4.000	12.75	0.6528	0.01554	0.0873	0.0793
$4\frac{1}{2}$[b]	3.958	13.75	0.6392	0.01522	0.0855	0.0777
$4\frac{1}{2}$	3.826	16.60	0.5972	0.01422	0.0798	0.0726
$4\frac{1}{2}$	3.640	20.00	0.5408	0.01288	0.0723	0.0657
5	4.276	19.50	0.7460	0.0178	0.0997	0.0907
$5\frac{1}{2}$	4.778	21.90	0.9314	0.0222	0.1245	0.1132
$5\frac{1}{2}$	4.670	24.70	0.8898	0.0211	0.1189	0.1081
$5\frac{9}{16}$[b]	4.975	19.00	1.0098	0.02404	0.1350	0.1227
$5\frac{9}{16}$[b]	4.859	22.00	0.9633	0.02294	0.1288	0.1171
$5\frac{9}{16}$[b]	4.733	25.25	0.9140	0.02176	0.1222	0.1111

[a]No allowance made for internal restrictions of upsets and tool joints.

[b]Not API standard. Shown for information only.

[c]Based on 1.1 cu ft of slurry per sack of cement.

APPENDIX B

Constants and Conversion Factors

Basic Constants

Absolute temperature corresponding to 0°F .... 459.688°R
Maximum density of water (39.15°F) .... 0.999973 gm per cm^3
Density of water at 60°F .... 0.999014 gm per cm^3
Average molecular weight of dry air 28.97

Derived Constants

Gas constant .... $10.732 \frac{(\text{psi})(\text{cu ft})}{(\text{lb mole})(°\text{R})}$
Volume of 1 lb mole of gas at 14.7 psia and 60°F .... $379.41z$ cu ft[a]
Density of water at 60°F .... 62.366 lbm per cu ft
Pressure due to 1 ft of water at 60°F .... 0.43310 psi
Weight of water at 60°F .... 8.33727 lbm per gal
Density of air at 60°F .... 0.0764 lbm per cu ft

Conversions

Mass

1 oz = 28.34953 gm
1 lbm = 453.59243 gm
1 lbm = 16 oz

Area

1 acre = 43,560 sq ft
1 acre = 208.71 ft sq
1 sq mi = 640 acre

Length

1 cm = 0.3937 in.
1 ft = 30.4801 cm
1 m = 39.370 in.
1 m = 3.2808 ft
1 mi = 5,280 ft
1 in. = 2.54001 cm

Volume

1 bbl = 42 U.S. gal
1 bbl = 5.61458 cu ft
1 cu ft = 1728 cu in.
1 cu ft = 7.4805 gal
1 gal = 231 cu in.
1 gal = 3785.43 cu cm
1 qt = 946.35 cu cm

Density

1 gm per cu cm = 62.428 lbm per cu ft
1 gm per cu cm = 8.3455 lbm per gal
1 gm per cu cm = 350.51 lbm per bbl
1 lbm per cu ft = 0.0160184 gm per cu cm

Pressure

1 atm = 760 mm Hg (0°C)
1 atm = 29.921 in. Hg (0°C)
1 atm = 14.696006 psi

Viscosity

1 poise = 100 cp
1 cp = 6.7197×10^{-4} lbm per ft sec
1 cp = 2.0886×10^{-5} lbf sec per sq ft

Power

1 hp = 550 ft-lbf per sec
1 hp = 33,000 ft-lbf per min
1 hp = 745.7 watts

Temperature

Degrees Fahrenheit (°F) = 1.8 (°C) + 32
Degrees Centigrade (°C) = (1/1.8)(°F − 32)
Degrees Kelvin (°K) = °C + 273.16
Degrees Rankine (°R) = °F + 459.7

Data given in "Phase Relations of Gas Condensate Fluids," by C. Kenneth Eilerts and others, *U.S. Bureau of Mines Monographs*, **1**, 427–434 (New York: American Gas Association, 1957).

[a] z is the gas deviation factor.

Author index

AGNEW, B. G., 347, 348
Alves, G. E., 79, 80
Anderson, F. A., 178, 244
Angel, R. R., 47
Archie, G. E., 539
Arps, J. J., 513
Aspdin, J., 158

BABBIT, H. E., 40, 48
Babson, E. C., 372, 403
Baker, O., 80
Bartok, A. W., 111
Baxendell, P. B., 417
Bearden, W. G., 183, 484, 486, 487
Beck, R. W., 40, 48
Beesom, C. M., 439, 440
Benham, A. L., 417
Berdine, D., 208, 213, 217
Bergman, W. E., 196, 202
Bertuzzi, A. F., 80, 81, 82, 83
Bielstein, W. J., 52
Bingham, E. C., 32
Binkley, G. W., 221
Birch, F., 485
Bird, R. B., 41, 70
Blanks, R. F., 159, 161, 194
Blenkarn, K. A., 262, 328, 329, 331, 332
Blum, H., 143, 144, 145
Bobo, R. A., 47
Bogue, R. H., 159, 160
Boudreaux, G. S., 47
Bouman, C. A., 262
Brown, G. G., 464
Brown, R. W., 490, 518, 526
Browne, W. H., 526
Buchan, R. C., 471
Bundrant, C. O., 526
Burba, J. L., 219
Burdyn, R. F., 244
Buthod, P., 387

CALDWELL, D. H., 40, 48
Camp, G. F., 437
Campbell, J. M., 455, 459
Cannon, G. E., 178, 215, 219
Cardwell, W. T., 219
Carpenter, P. G., 403, 405, 411, 413, 414
Carter, L. G., 160, 176, 186, 196
Carter, R. D., 497
Caudle, B. H., 494

Chavez, J. A., 80
Churchill, R. V., 497, 498
Clark, E. H., 219, 220, 221
Clark, J. B., 208, 215, 217, 218, 243, 483
Clark, R. C., 177, 178
Cleary, J. M., 486
Clinedinst, W. O., 125
Coberly, C. J., 288, 361
Collins, R. E., 221
Cook, H. L., 427
Corley, C. B., 217
Cornell, D., 505
Coulter, A. W., 507
Cox, W. R., 262, 263, 272, 273
Craft, B. C., 175, 198, 368, 382, 488, 491, 501
Crawford, H. R., 515
Crawford, P. B., 494
Crittendon, B. C., 30, 488, 490, 522, 523
Cromer, S., 368
Crowe, R. H., 25
Cullender, M. H., 22

Dale, O. O., 184, 189
Darin, S. R., 526
Davidson, R. D., 201
Davis, E. L., 163
Davis, S. H., 187, 188, 190, 229, 237, 238
Dawson, D. D., 219
Dehlinger, P., 526
Dodge, D. W., 67, 68, 74
Doherty, W. T., 241, 242, 305
Dotterweich, F. H., 427
Dublin, J. R., 229
Dumbauld, G. K., 186, 221, 229
Dunlap, I. R., 186, 202
Dunlop, C. A., 101, 127, 131
Dunn, T. H., 40, 48
Dwight, H. B., 498
Dyes, A. B., 494

Eaton, S. J. M., Rev., 368
Eckel, J. R., 52
Edwards, S. H., 125
Eilerts, C. K., 561
Elenbaas, J. R., 505
Emde, F., 498
Ernsberger, F. M., 162
Eubanks, J. M., 349, 350, 351, 352, 353, 354, 355, 356, 357

Fagg, L. W., 335, 343, 344
Farris, R. F., 166, 176, 177, 198
Fast, C. R., 229, 230, 495, 503, 504, 506, 507, 508, 509
Faulk, J. H., 187, 188, 190, 229
Fisher, H. B., 196, 202
Folmar, L. W., 223
Foster, W. R., 41
Frame, W. M., 125
France, W. G., 162
Franks, B. L., 349, 350, 351, 352, 353, 354, 355, 356, 357
Frederickson, A. G., 41, 70

Gatlin, C., 518
Gilbert, B., 518
Gilbert, W. E., 403
Godbey, J. K., 178
Goins, W. C., 219, 241
Gosline, J. E., 403

Harrison, E., 230, 486
Hassebrock, W. E., 215
Havenarr, I., 41, 52
Hawkins, M. F., 198, 368, 382, 491, 501
Hayward, J. T., 259
Hedstrom, B. O. A., 41, 47
Helmick, W. E., 212, 213
Hendrickson, A. R., 537, 539, 541, 542
Hills, J. O., 125, 141
Hilton, A. G., 213, 214
Hoch, R. S., 47
Hodges, H. D., 178
Hodges, J. W., 231, 235
Holden, W. R., 244
Holmquist J. L., 125, 136
Howard, G. C., 192, 202, 203, 208, 215, 217, 218, 229, 230, 243, 486, 487, 495, 503, 504, 506, 507, 508, 509
Hower, W. F., 201
Hubbert, M. K., 486, 487, 489
Huber, T. A., 229
Huitt, J. L., 526
Hunt, D. D., 515
Huntington, R. L., 368
Hurley, J. R., 196, 202
Hurst, R. E., 537, 539, 541, 542

Jahnke, E., 498
Jakosky, J. J., 485
Jasper, T. M., 125
Johnson, D. O., 344

Jones, P. H., 208, 213, 217

Katz, D. L., 383, 505
Kelley, H. L., 300, 305, 306, 313, 314, 315, 316
Kemler, E. N., 101, 343
Kemp, C. E., 494
Kennedy, H. B., 159, 161, 194
Kern, L. R., 527
Kieschnick, W. F., 230, 486
Kirkpatrick, C. V., 370, 371, 373, 374, 375, 376, 377, 378, 379, 386, 388, 390, 391, 392, 393, 409, 410, 417, 418, 420, 421, 430, 433, 434, 435, 436, 438, 439
Klinkenberg, A., 262
Knox, D. G., 439, 440
Kobayashi, R., 505
Kraemer, J. W., 519
Krieger, I. M., 74

Lacy, W. N., 381
Laird, W. M., 41
Lamb, H., Sir, 26
Lamberger, E. H., 310, 361
Landrum, B. L., 494
Lange, N. A., 200
Langer, B. F., 310, 361
Lapple, C. E., 460, 461
Laurence, L. L., 453, 454, 460
Lawrence, D. K., 349, 350, 351, 352, 353, 354, 355, 356, 357
Leach, G. L., 203
Lebeaux, J. M., 439, 441
Leidner, E. J., 244
Lévy, M., 125
Lockhart, R. W., 79, 80
Longley, A. J., 212, 213
Loper, J. L., 526
Loper, R. G., 490
Lubinski, A., 261, 262, 328, 329, 331, 332
Ludwig, N. C., 158, 162, 201

McGhee, E., 202
McGlothlin, B. B., 526
McGuire, W. J., 230, 486
McMurray, E. D., 439
Main, W. C., 125
Mangold, G. B., 194
Maron, S. H., 74
Marsh, H. N., 288, 352, 358, 361
Martinelli, R. C., 79, 80, 83
Masters, R. W., 262
Maxwell, T. E., 349, 350, 351, 352, 353, 354, 355, 356, 357
May, C. J., 403
Melrose, J. C., 41
Melton, L. L., 60
Merryman, C. J., 349, 350, 351, 352, 353, 354, 355, 356, 357
Metzner, A. B., 59, 61, 63, 65, 67, 68, 74
Miles, A. J., 486
Miller, C. P., 125
Mills, K. N., 343, 361
Montgomery, P. C., 201
Moody, L. F., 22
Mooney, M., 58, 65
Moore, T. V., 403
Morgan, B. E., 186, 192, 229

Nadai, A., 125, 136
Newcombe, R. B., 536
Nicklin, M. E., 368
Nuss, W. F., 40, 48
Nussbaumer, F. W., 191, 192

Oberg, C. H., 262
Ormsby, G. S., 219
Owsley, W. D., 185, 210

Parish, E. R., 41
Patchen, F. D., 186, 202
Peret, J. W., 124, 260, 261
Perkins, T. K., 527
Pirson, S. J., 539
Plummer, F. B., 536
Poettmann, F. H., 80, 81, 82, 83, 403, 405, 411, 413, 414, 417, 505
Prats, M., 494, 530

Rabinowitsch, B., 58, 70
Rayne, J. R., 368
Reed, J. C., 59, 61, 65
Richardson, T. W. G., 140
Rieniets, R. W., 309, 361
Rike, J. L., 217
Roberts, T. G., 513
Robinson, W. W., 161, 178
Rollins, J. T., 201
Roper, W. F., 56
Rordam, S., 182, 183
Rule, R. B., 163
Russell, J. H., 337, 339, 342, 343

SAGE, B. H., 381
Saunders, C. D., 60, 191, 192, 491
Savins, J. G., 41, 56, 70, 76
Saye, J. E., 140
Schairer, J. F., 485
Schilthuis, R. J., 403
Schmoe, W. P., 340, 341, 342
Schneider, H. G., 368
Scott, J. O., 175
Scott, P. P., 192, 202, 203, 486, 487
Shaw, S. F., 368, 369
Shell, F. J., 196, 197, 202
Silcox, D. E., 163
Singer, F. L., 257, 258
Slagle, K. A., 196
Slonneger, J. C., 322, 337, 338
Smith, D. K., 160, 176, 186
Smith R. V., 22
Souders, M., 464
Spicer, H. C., 485
Standing, M. B., 383, 408
Stearns, G. M., 229
Sterne, W. P., 194
Stewart, J. B., 507
Stewart, R. T., 126, 127
Stirling, R., 368
Stoddard, J. H., 439, 440
Stout, C. M., 202
Sudduth, L. S., 439, 441

TAUSCH, G. H., 229, 230, 235
Tek, M. R., 80, 81, 82, 83
Teplitz, A. J., 215, 219
Thomas, P. D., 111
Thompson, M. L., 340
Thrash, P. J., 400
Timoshenko, S., 486
Tinsley, J. M., 491
Topping, A. D., 486
Torrey, P. D., 229

UREN, L. C., 102, 341

VAN POOLLEN, H. K., 491, 494
Van Wingen, N., 79
Vary, J. A., 505
Versluys, J., 403

WAHL, W. W., 202
Walker, A. W., 229
Watts, E. V., 352, 358
Weichent, J. P., 219
Weinaug, C. F., 505
Wescott, B. B., 101, 127, 131
Whitley, B. W., 387
Wieland, D. R., 537, 539, 541, 542
Wilde, H. D., 403
Willis, D. G., 486, 487, 489
Willis, R. M., 300, 305, 306, 313, 314, 315, 316
Wilsey, L. E., 484
Winkler, H. W., 383, 384, 437, 442, 447, 448
Worley, M. S., 453, 454, 460
Wyant, R. E., 527

YOCUM, B. T., 80

ZABA, J., 285, 305, 321

Subject index

ACIDIZING, 536–44
acid fracturing design, 542–44
chemistry of, 536
treatment, types of, 537–42
high pressure fracturing, 541
matrix, 537–40
through preexisting fractures, 540
Annular flow:
Bingham plastics, 41–48
effective viscosity, 46
laminar flow, 41, 43, 46
Reynolds criterion, 46
turbulent flow, 46
Newtonian liquids, 26–31
average velocity, 26
equivalent diameter, 30
laminar flow, 26
Reynolds criterion, 30
turbulent flow, 29
power-law fluids, 73–74
Laminar flow, 74
Reynolds criterion, generalized, 73
turbulent flow, 74

BERNOULLI's equation, 13, 518

Bingham plastic fluids:
defined, 32
limitations as a drilling fluid model, 57
viscometry, 56
(*see also* Annular flow, Drilling hydraulics, Pipe flow, Slot flow)

CALIPER SURVEY, 223
Casing:
API collapse equations, 127
axial loading, effects, 133–39
centralizers, 212
classification, 101
collapse resistance (*table*), 129, 130
combination string, design procedure, 146–55
design factors, 139–43
differential pressure sticking, 213
dimensions, 103–105
API casing list (*table*), 103–105
cross-sectional (*table*), 134–35
length ranges (*table*), 103
elastic failure limits at zero axial stress (*table*), 140
external pressure, effects of, 124–30

Casing (*cont.*)
forces acting on, 102
functions of, 101
grades, 102–103
internal pressure, effects, 130–33
internal yield pressures (*table*), 132–33
joint strengths (*tables*), 111, 114, 115
liners, 124
make-up loss, 109 (*table*), 110
selection of sizes, 143–45
string, defined, 101
threads and couplings, 106–15
dimensions (*table*), 107–108
weights, 115–24
average, 116–21
nominal (*table*), 120–21
plain end, 116
Casing cementing, 215–29
caliper survey, 223
displacement velocities, 218
pressure surges while running casing, 219–23
temperature survey, 224, (*tables*), 226, 229
Casing landing:
buckling criteria, 261–70
design considerations, 272–76
general practices, 256–61
wellhead loads, 270–72
Cement:
additives, 183–206
bentonite, 185
calcium chloride, 198
diatomaceous earth, 196
fluid loss, 202
gilsonite, 196
Latex, 200
lost circulation, 202
perlite, 191
physical properties (*table*), 185
pozzolanic materials, 194
sodium chloride, 198
weighting materials, 203
composition, 159–63
Bogue constituents, 159
compound analysis (*table*), 160
fineness, 178
manufacture, 158-59
modified, 186
permeability (*table*), 186
rheology, 244
slurry weight, 180, (*tables*), 204–205
Cement (*cont.*)
standardization of, 163–75
API classification, 168
API test schedules, basis for (*table*), 168
strength, 175-83
bonding strength (*table*), 176–77
compressive, 187, (*tables*), 195, 197, 199, 201
sulfate resistance, 182
thickening time:
API Class A, 187
calcium chloride (*table*), 198
diatomaceous earth (*table*), 197
pozzolan cement (*table*), 196
sodium chloride (*table*), 201
weighting materials (*table*), 204
thickening-time testers, 163–66
water-cement ratios, 178–82
API water, 180
maximum water, 179
minimum water, 179
WOC time, 177 (*see also* Casing cementing)
Cementing equipment, 206–15
centralizers, casing, 212
float collars, 208
guide and float shoes, 206
cementing heads, 206
scratchers, 213
stage collars, 209
squeeze packers, 231
wiper plugs, 210
Cementing hydraulics, 243
Cementing tables:
capacity, annular, 550–55
capacity, casing, 547–49
capacity, drill pipe, 559
capacity, open hole, 556–57
capacity, tubing, 558
Choke flow, 427
Consistency index:
defined, 63
equivalent:
annular flow, 73
rotational viscometer, 75
slot flow, 71, 72
measurement, 62, 63
Constants and conversion factors (*table*), 560

DILATANT fluids, 33

Dimensions, 2–9 (*see also* Units)
Dimensions, casing (*see* Casing dimensions)
Drilling hydraulics, 48-55
  critical velocities, 49
  downhole frictional losses, 48, 51
  mud pumps, 54
  surface connection losses, 50, (*table*), 51
  (*see also* Jet bits)
Drill pipe, dimensions and capacity (*table*), 559
Dynamometer:
  installation, 333
  interpretation of cards, 336–49
  pumping efficiencies, determination, 352–59
  pumping well testing, 349, 352

EMULSIONS, oil and water, 470
Error function, complementary (*table*), 499–500

FANNING equation, 21, 406
Flow-behavior index:
  annular flow equivalent, 73
  defined, 62
  measurement, 60–63
  rotational viscometer equivalent, 75
  slot flow equivalent, 71, 72
Fluid flow, basic laws, 9–16
  equations (*table*), 45
  head loss, 16
  lost work, 13
  material balance, 9
  mechanical energy balance, 12, 405
  pressure balance, 14
  total energy balance, 10
Formation volume factor, two phase, correlation, 410
Fracturing, hydraulic:
  area calculations, 495–501
  defined, 483
  design calculations, 519–26
  economic considerations, 527 (*table*), 528
  efficiency, 514–16
  fluid coefficients:
    combined coefficients, 511
    fluid-loss characteristics, 509
    reservoir-controlled fluids, 503
    viscosity controlled fluids, 501
    wall-building fluids, 507
Fracturing hydraulic (*cont.*)
  fracture induction, 485–90
    orientation, 487, 490
    treating pressures, typical, 488
  pressure:
    frictional loss, 519
    hydrostatic, 517
    perforation loss, 518
    surface injection, 516
    treating, bottom-hole, 517
  productivity ratio:
    horizontal fractures, 491
    vertical fractures, 494
  propping agents, 526
Friction factors:
  defined, 21
  fittings losses, 25
  Newtonian fluids, 22, 23, 25
  non-Newtonian, 67, 68
  pipeline efficiency, 25
  two-phase vertical flow, 411

GAS compressibility factors, 383, 384
Gas lift:
  accumulation chambers, 434–39
  applications, 372–77
  choke size, 427
  continuous flow:
    defined, 372
    power requirements, 420, 426
    system design, graphical, 400–403
    system design, volumetric method, 403–27
    terminology, 375
  free piston system, 439
  historical review, 368–72
  intermittent flow:
    defined, 372
    injection cycle, selection (*table*), 376
    pressure gradients, flowing, 429, 431
    system design, 431–34
  pressure gradients, flowing:
    2-in. tubing, 413
    $2\frac{1}{2}$-in. tubing, 414
    master curves, 428
  valve mechanics, 377–91
    closing pressure, 380
    opening pressure, 378
    valve spread, 381
    unloading, 389
  valve spacing, 391–400
    closing pressure analysis (*table*), 398

Gas lift, valve spacing (*cont.*)
opening pressure analysis (*tables*), 399, 405
Gas-oil ratio, correlation, 409

HAGEN-Poiseuille law, 20
Heater treaters:
components, 471
design, 472–75

JET bits:
discharge coefficients, 52, 53
jet velocity, 53
nozzle sizes and areas (*table*), 53
pressure drop across, 51

KARMAN number, 23, 24

LAMB's equation, 26
Laminar flow, defined, 17 (*see also* Annular flow, Pipe flow, Slot flow)
Liner cementing, 236–39
Liners, 124 (*see also* Casing)

MAKE-UP loss, 109 (*see also* Casing)

NEWTONIAN fluids, defined 17, 31 (*see also* Annular flow, Pipe flow, Slot flow)
Non-Newtonian fluids:
apparent viscosity, 33
consistency index, 62, 63
defined, 17, 32
flow-behavior index, 60–62
generalized equations of flow, 55–79
laminar, pipe flow, 35–37, 65–67
laminar, slot flow, 42, 70
time-dependent behavior, 33
viscometry, 58–64, 75
(*see also* Annular flow, Pipe flow)

PERMEABILITY, relative, 513
Pipe flow:
Bingham plastics, 34–41
effective viscosity, 39
laminar flow, 35, 38, 39
Reynolds criterion, 40
turbulent flow, 40
Newtonian liquids, 18–26
friction factors, 22, 23, 25
laminar flow, 20
noncircular pipe, 29
pipeline efficiency, 25

Pipe flow, Newtonian liquids (*cont.*)
Reynolds criterion, 18, 21
turbulent flow, 21
velocity profiles, 18
non-Newtonian fluids, 64–70
friction factors, 68
laminar flow, 65–67
Reynolds criterion, generalized, 65, 67
turbulent flow, 67–70
Pipeline efficiency, 25
Plugback cementing, open-hole, 239–43
design calculation, 242
Power-law fluids:
defined, 34
types used as drilling and completion fluids, 58
(*see also* Annular flow, Pipe flow, Slot flow)
Pressure, gas column:
flowing, 387
static, 382
Pseudoplastic fluids, 33
Pump:
displacement volumes (*table*), 172, 173
power requirements, 54, 305
Pumping equipment, subsurface:
design equations, summary, 312–20
general considerations, 281
API classification, 282
plunger stroke, effective, 288–93
pump displacement, 299
pump plunger data (*table*), 291
pump production rate, 299
(*see also* Sucker rods)
Pumping equipment, surface:
counterbalance design, 300
design equations, summary, 312, 320
gear box:
speed reduction, 304
torque, 302–304
general considerations, 284–86
polished rod:
horsepower, 305
loads, 293–95
prime mover:
general considerations, 286
power requirements, 304
Pumping installation, design, 312–20
design data for various API units and stroke lengths, (*tables*), 314–16
example calculations, 318–20
general considerations, 312, 316

Pumping installation, design (*cont.*)
  summary of calculations, 317

REYNOLDS number:
  Bingham plastic, 40
  generalized non-Newtonian, 65, 72, 73
  Newtonian fluid, 19, 30
Rheological classification of fluids, 31–34
  Bingham plastic, 32
  dilatant, 33
  model-fluid limitations, 55
  Newtonian, 31
  power-law, 34
  power-model, 34
  pseudoplastic, 33
  rheology, defined, 32
Rheopectic fluids, 33

SEPARATORS, oil and gas:
  basic considerations, 459–62
  capacity, 462–67
  components, 453–56
  design problems, 467–70
  laws of gravity settling, 460
  types, 456–59
Skimmers, oil:
  capacity, 476
  design, 477
  function of, 475
Slot flow:
  Bingham plastic, 43
  Newtonian fluid, 28, 43
  non-Newtonian fluid, 42, 70
  generalized Reynolds number, 72
Sodium chloride solutions, densities (*table*), 199
Squeeze cementing, 229–36
  amount of cement, 235
  design calculations, 236
  squeeze packers, 231–35
  squeeze pressures, 235
Sucker rods:
  analysis of rod motion, 286
  corrosion, 321
  design equations, summary, 312–20
  design of tapered string, 295–99

Sucker rods (*cont.*)
  elongation under load, 288, 291
  endurance limits (*table*), 322
  failure, 320–33
  general considerations, 283
  physical data (*table*), 290
  rod guides, 332
  synchronous pumping speeds, 322-26

TEMPERATURE gradients, 382, 418
Temperature survey, 166, 224, (*tables*), 226, 229
Thixotropic fluids, 33
Transmission factor, 23, 24
Tubing:
  anchors, 327, 330
  annulus area ratios and annular volumes (*table*), 394, 395
  capacity (*table*), 558
  dimensions (*table*), 291
  failure in pumping wells, 320, 326
Turbulent flow, defined, 17 (*see also* Annular flow, Pipe flow, Slot flow)
Two-phase flow, 79–87
  frictional pressure losses, 80, 82
  types, 79

UNITS, 2–9
  consistent use in equations, 7
  defined, 2–5
  symbols and practical engineering units (*table*), 44
  sytems of, 5, (*table*), 6

VISCOMETERS:
  absolute, capillary, 58–61
  Fann V-G Meter, 55–57, 77, 244
  rotational, 74
Viscosity:
  absolute, 16, 31
  apparent, 33
  effective, 39, 46
  gas-free crude oils correlation, 520
  plastic, 33 (*see also* Bingham plastic fluids)